R. A. Fisher

The Life of a Scientist

Fisher at his desk calculator at Whittingehame Lodge, 1952.

R. A. Fisher
The Life of a Scientist

JOAN FISHER BOX

JOHN WILEY & SONS, New York • Chichester • Brisbane • Toronto

Library of Congress Cataloging in Publication Data

Box, Joan Fisher, 1926–
   R. A. Fisher, the life of a scientist.

   (Wiley series in probability and mathematical
statistics)
   "Bibliography of R. A. Fisher": p.
   Includes bibliographical references and index.
   1. Fisher, Ronald Aylmer, Sir, 1890–1962.
2. Statisticians—Great Britain—Biography.   I. Title.

QA29.F57B68   519.5'092'4 [B]   78-1668
ISBN 0-471-09300-9

Printed in the United States of America

10 9 8 7 6 5 4 3

Preface

Sir Ronald Fisher was my father. This might seem to disqualify me as his biographer, and I certainly had no notion of becoming one when, 5 years after his death, I began to ask his friends and colleagues for their recollections of him. I thought only to preserve the personal record before it was irretrievable. Then, just as Fisher's discourse used to fascinate his hearers in his lifetime, so the flow of his ideas began to fascinate me. His life took form as one continuous discourse from Fisher's lips. It had the vitality of his immense pleasure in the process of thinking, the play of ideas, the solution of puzzles. I recognized that, having known the man, I knew Fisher even as an undergraduate, when he stated the thesis of his life. From this starting point, I began to follow what, being his, was a logical and coherent argument, full of complexities, subtleties, and intriguing subplots. His conversation was not always easy to follow, but it was always fresh and stimulating. It is my hope that the reader of this biography also may take pleasure in the unfolding of a brilliant mind.

Fisher was an idealist, committed to the establishment of truth and the advancement of mankind. Born in 1890 in East Finchley, London, his mathematical ability was quickly apparent. His biological interests became important during his schooldays at Harrow. He accepted Charles Darwin's theory of evolution by natural selection, and he shared Francis Galton's concern that selective effects in society should tend to improve the biological inheritance of Man. However, the scientific facts were not well established: the nature of human inheritance was a matter of debate between geneticists and biometricians; Darwin's evolutionary theory had largely given way to theories involving genetic mutation, and most biologists assumed the effects of selection to be negligible. Above all, the means of establishing the truth of such matters on a firm, quantitative basis were practically nonexistent. So, while still an undergraduate, Fisher began his lifework on the quantification of the variable phenomena of biology; he began his services to the eugenic movement and his involvement with the elucidation of genetical and evolutionary questions.

Rejected for national service in World War I because of poor eyesight, he taught in public schools for 5 years. During this period he made two scientific advances, notable both for the method and the result: using an argument from n-dimensional geometry, he found the distribution of the correlation coefficient; and, using variance components, he demonstrated that human inheritance was entirely consistent with Mendelian principles. In 1919 he was hired as statistician at Rothamsted Experimental Station, where the work he had already done was immediately applicable. He continued investigations into distributions of statistics which, like the correlation coefficient, were in common use; with small experimental samples, it was essential to know the theoretical distribution in assessing the actual results. At the same time, he responded to the stimulus of actual problems in the field by inventing statistical methods that used all the information in the data to draw correct inferences, and he conceived the whole idea of the design of experiments, by which to gain fuller and more precise information for a given amount of experimental effort. He laid the foundations, coined the language, and developed the methodology of modern biometry.

During the 1930s the influence of his statistical work spread throughout the English-speaking world, and often beyond, so that, after World War II, with the formation of the International Biometric Society in 1947, a new field of scientific research was recognized. As a result of the introduction of quantitative methods and the refinements of inductive reasoning for which Fisher had been primarily responsible, a transformation had taken place in the very way a biologist conceived of his research.

Meanwhile, in 1917, Fisher was married to Ruth E. Guinness, younger daughter of Dr. Henry Grattan Guinness, a preacher and evangelist who, at the time of his death in 1915, had been head of the missionary training college and the mission his father had started. The Fishers read together in the history of civilizations, noting evidences that the decay of past civilizations had resulted from the pattern of social selection arising in a moneyed economy. Fisher advocated practical measures by which to reduce this deleterious selection in contemporary society and thus save Western Civilization from the fate of its predecessors. Eugenic practice accorded with theory and, for the Fishers, started with the investment of their own physical and economic resources in the next generation. Eventually they raised a family of two sons and six daughters.

In addition, at home Fisher pursued work in population genetics that led to his formulation of *The Genetical Theory of Natural Selection* (1930) and to his election in 1933 to the Chair of Eugenics at University College, London, where he initiated genetical researches, in particular, in human blood group serology. In 1943 he moved to Cambridge University as Balfour Professor of Genetics. Combining genetical work in the laboratory with evolutionary field work, in collaboration with E. B. Ford, he demonstrated the reality of the effects of natural selection and developed statistical methods by which they could be measured. He also enjoyed the more theoretical aspects of his work, the theory of inbreeding arising from his work with mice, and the combinatorial aspects of his work with polyploid plants. Logical difficulties in the theory of inductive inference continued to engage him after his retirement in 1957. He died in Adelaide in 1962.

Given its subject, this biography is necessarily scientific. However, I have not assumed that the reader is familiar with any of the sciences involved, and I have avoided mathematical elaboration. The symbols of mathematics are used mainly to clarify logical connections necessary to the argument. I hope that any reader prepared to follow through what is sometimes fairly stiff argument may understand its substance without difficulty. At the same time I have tried to make the account technically correct.

My ignorance of mathematics and statistics would have prevented me from attempting the present work without the help of my husband, George E. P. Box. Patiently, he has expounded and discussed statistical matters with me and interpreted the mathematics. He has read and corrected several drafts of the manuscript. The errors are mine; much of the clarity of the statistical discussion I owe to him.

I have greatly benefited from the publication of the *Collected Papers of R. A. Fisher* (Volumes 1–5), J. H. Bennett, Ed., University of Adelaide, 1971–1973. These comprise 294 of Fisher's contributions to the scientific literature between 1912 and 1962, with amendments or prefaces Fisher had made in his lifetime, the biographical memoir written by K. Mather and F. Yates, and a very complete bibliography (included in this volume). Nearly half the collected papers are mainly mathematical or statistical, and most of the others concern topics in genetics, evolution, or eugenics. The broad scope of many of the papers, however, makes any grouping according to subject matter unsatisfactory, and the papers have been arranged chronologically according to year of publication, with the more statistical ones placed ahead of the more genetical. In his Preface, the editor writes, "It is hoped that with the material presented in this way, readers will be better able to appreciate the extraordinary diversity of Fisher's contributions to science and to trace their overall development and interrelationships." The *Collected Papers* constitute a resource no student of Fisher's work can afford to be without. In this volume, therefore, references in the text are given numerically as they appear in the *Collected Papers*.

I am deeply indebted to a number of Fisher's friends for their help. Prof. J. Henry Bennett gave me access to the Fisher papers at Adelaide University and put every facility at my disposal during my 3 months with the Genetics Department there in 1970. He has read and commented on the manuscript, has checked quotations from the papers in Adelaide, and brought source material to my attention. Professor E. B. Ford recorded for me a priceless account of his long and close association with Fisher and has read and commented on relevant portions of the manuscript. Prof. G. A. Barnard arranged a valuable series of seminars on Fisher's statistical papers during our stay at Essex University in the academic year 1970–1971. He has commented on my discussion of scientific inference. Prof. D. J. Finney has read and commented on a draft of the early chapters, and Prof. J. F. Crow on the chapter on the evolution of dominance.

Many have contributed their own recollections, written records, photographs, and even personal correspondence with Fisher. For such generous assistance I wish to express my gratitude to the following: A. E. Brandt, Sir Edward Bullard, L. L. Cavalli Sforza, the late E. A. Cornish, Gertrude M. Cox, A. W. F. Edwards, D. J. Finney, T. N. Hoblyn, S. B. Holt, E. Irving, Sir Bernard Keen, R. T. Leslie, the late P. C. Mahalanobis and Mrs. Mahalanobis, Besse B. Mauss, J. R. Morton, E. S. Pearson, R. R. Race, Stuart A. Rice, P. R. Rider, W. A. Roach, S. K. Runcorn, the late H. Fairfield Smith, Sir Gerard Thornton, Helen N. Turner and F. Yates. I thank Rothamsted Experimental Station, the Agricultural Research Council, the Genetical Society and the Royal Society for supplying information from their records at my request and K. Mather for permission to quote from his contribution to the Royal Society's Biographical memoir of R. A. Fisher.

Special thanks are due to my aunt, Geraldine Heath (Gudruna), for an account she alone could have given of Fisher during the years 1910–1920. The clarity of her perceptions and memories illuminates the pages of the long letters, often hand-written, in which in her mid-eighties she recalled Fisher and commented on my account of him in the first two chapters of this volume. Above all, I thank my mother for opening the gates of sometimes painful memories on my behalf and for her full and enthusiastic discussion of my questions.

JOAN FISHER BOX

Madison, Wisconsin
February 1978

Acknowledgments

I am indebted to the University of Adelaide, which holds the copyright on most of the publications of the late Sir Ronald Fisher, for permission to quote various passages, as indicated in the text, from

Statistical Methods for Research Workers (14th ed., Hafner Press, New York)
The Design of Experiments (8th ed., Hafner Press, New York)
Statistical Methods and Scientific Inference (3rd ed., Hafner Press, New York)

and also from *Collected Papers of R. A. Fisher,* Volumes 1–5 (J. H. Bennett, ed.), University of Adelaide, Adelaide 1971–1973. I am also indebted to the University of Adelaide for permission to consult and to quote various passages, as indicated in the text, from *R. A. Fisher Correspondence* held at the University of Adelaide.

I am indebted to the various societies and publishers shown below who have kindly given permission for the reproduction of material published elsewhere. The exact source is given by numerical reference in square brackets to references listed at the end of this volume or to the *Collected Papers* [CP xx] listed in the Bibliography of R. A. Fisher.

American Academy of Arts and Sciences, *Science*
American Naturalist
American Scientist
Trustees of *Biometrika*

A. S. Eddington Memorial Lecture (Cambridge University Press)

Eugenics Review

Great Britain, Ministry of Agriculture, *Journal* (by permission of the Controller of H.M.S.O.)

Journal of Agricultural Science

Linnean Society of London, *Proceedings*

Philosophy of Science

Royal Society of London, *Philosophical Transactions*

Royal Society of London, *Proceedings*

Royal Society of London, *Biographical Memoirs of Fellows of the Royal Society*

Royal Statistical Society, *Journal*

Genetical Theory of Natural Selection (Dover Publications)

                                                     J.F.B.

Contents

R. A. Fisher

The Life of a Scientist

Prologue

At eight thirty on the evening of November 10, 1911 a group of students gathered for the second undergraduate meeting of the Cambridge University Eugenics Society to hear a short address by their Chairman, Ronald Aylmer Fisher. He was 21, a little below middle height, with a wiry figure and good shoulders carried squarely on a straight back; and he was shabbily dressed. What was impressive in his appearance was his head: abundant rufous hair crowned the high forehead, blue eyes glinted through very thick eyeglasses, and the strong line of the jaw thrust forward eagerly as he made his point.

He spoke slowly and deliberately, explaining technical terms where necessary with examples that developed through a logical progression from elemental ideas of Mendelism and biometry to the complexities of human evolution. But ideas crowded the canvas; it was too much to take in at a single sitting. The composition had, in effect, the richness and depth of Botticelli's "Adoration of the Magi"; each well-defined idea contributed its own form and color and was worth consideration alone, but in their relationship with each other transcended individuality and made submission to a single grand design in which each tended to the contemplation of the central truth.

The more his hearers knew of either of the subjects included under his title, "Mendelism and Biometry," [1] the more stunned they must have been. In a 15-minute talk he not only described the essential elements of these two new and, as was then thought, incompatible sciences but combined them as complementary ways of studying the phenomenon of inheritance. In the light of the established facts of both, he defined the primary conditions required in so-

1

ciety for the realization of immeasurably grand aspirations for the future of the human race. From Mendelian (genetical) considerations, he concluded:

Suppose we knew, for instance, twenty pairs of mental characters. These would combine in over a million pure [homozygous] mental types. In practice each of these would naturally occur rather less frequently than once in a billion, or in a country like England, about once in 20,000 generations. It will give some idea of the excellence of the best of these types when we consider that the Englishmen from Shakespeare to Darwin (or choose whom you will) have occurred within ten generations; the thought of a race of men combining the illustrious qualities of these giants, and breeding true to them, is almost too overwhelming, but such a race will inevitably arise in whatever country first sees the inheritance of mental characters elucidated.

Of course, he was not proposing experimental or controlled breeding in man and, in fact, went on to discuss biometrical considerations, from which it became clear that the desired effects could be obtained in man without the experimental breeding associated with Mendelism.

The paragraph quoted above is typical of the man in his idealism and in the range of his scientific vision. It is a weather vane too, indicating the prevailing direction of his scientific interest towards genetics as the mechanism of human inheritance and towards statistics as the appropriate way of thinking about genetical and other population problems.

His theme was *eugenic* and, in this paragraph, wholly so: he was considering the possibility of improvements in the biological inheritance of mankind. Implicit in his thought is an evolutionary panorama extending from the humble beginnings of man to the possibilities of the indefinite future. The unchanging laws of nature, revealed in the Mendelian process of inheritance and in progressive evolution by means of selection, will determine the future of mankind; under these laws the race may progressively realize its genetic potential and improve it; and it is in the power of man, by a wise understanding of natural law, to ensure that evolution shall, in fact, continuously improve the quality of succeeding generations. Optimism and idealism color the vast canvas. It glows with his appreciation of mankind, his confidence that men will not only elucidate the genetics of mental characters but will use the knowledge with wisdom.

Fisher spoke of human evolution in terms of Mendelian *genetics*. The work of Abbé Gregor Mendel had established the fact of particulate biparental inheritance of a number of factors (genes) in the garden pea: the plants were tall or short, the seeds smooth or wrinkled; and these genes assorted independently of each other. As a plant and animal breeder, Professor Bateson had taken up the subject with enthusiasm because Mendelism explained the *discontinuities of inheritance* that he had observed. Being unconvinced by

Darwinian theory, he had welcomed it also as supporting the quite distinct notion of *discontinuous evolution*. In contrast, Fisher accepted both Mendelian genetics and the Darwinian theory of continuous evolution.

Moreover, a bitter controversy raged between Mendelians and biometricians, led, respectively, by their colorful champions, Professor Bateson and Professor Pearson, each shooting off volleys of heavy sarcasm in the direction of the other. Most of the known human characters do not exhibit discrete types like Mendel's peas but show continuous variation, and Pearson's work was *biometrical*, dealing with the distributions of such continuous variables. Pearson rejected Mendelism, and in 1903 [2] he had shown to his satisfaction that the theoretical expectations of Mendelism were not realized in his analysis of biometrical data. In 1906, however, G. U. Yule [3] made the same sort of analysis, with somewhat different assumptions. He did not find the data incompatible with Mendelian theory, and Fisher (as implied by his example in which twenty Mendelian factors together determine intelligence) did not hesitate to accept both Mendelian theory and biometrical method as being relevant to the elucidation of continuous variation.

Thus he disagreed with Bateson about evolutionary theory and with Pearson about genetical theory. Yet he had followed a logical train of thought that had led him to a synthesis of the essential ideas of both. It is remarkable, indeed, that this 21-year-old undergraduate could present his own ideas as almost self-evident truths, in the face of expert opinion that on both sides fiercely opposed him.

In considering genetical factors he employed the idea of probabilities. It is a *statistical* concept to calculate the rate of occurrence of certain combinations of genetic factors, as he does the rate of occurrence of individuals who will breed true in the twenty supposed mental characters. It is a statistical concept to compare the rate of occurrence of a hypothetical type with the actual occurrence of intellectual giants in ten generations of Englishmen—to use the comparison to illustrate the extreme rarity of the most desirable combinations and hence to suggest the heights to which humanity could rise if all its best qualities were to be combined in a single individual. To get a measure of the meaning of the scientific facts, he applied a statistical yardstick to them.

All this inheres in one short paragraph of straight-forward English prose, and there are many such paragraphs in this talk. Only later did Fisher realize that an audience could not absorb so much so fast, and then he tried to limit himself in each talk, according to Mr. Gladstone's advice, to the consideration of no more than two main points: but it was a hard discipline. He was later also to realize the paralysis that seizes a nonmathematical mind when faced with mathematical concepts, and he could explain to his biological colleagues the principles they needed to grasp, without raising the specter of incomprehensibles. Again, it was a conscious effort, because he himself was equally at home in the two worlds of thought and hardly perceived the boundary where his colleagues stopped short.

Vividly his numbers tell how rarely will the genetic factors combine even to give us a single giant. Such a giant was the speaker himself on that occasion, for whom a life's work was foreshadowed in his little talk to his fellow students in Cambridge in 1911.

1
Nature and Nurture

The occurrence of genius is quite unpredictable, so rare as to be unexpected even in the most illustrious families. It was unexpected in the Fisher family, who were tradesmen and later professionals who had risen to prosperity, though hardly to distinction, through their enterprise and general ability during the nineteenth century.

The family emerged from obscurity in the first half of the eighteenth century, when George Fisher (previously perhaps a farm laborer in Lincolnshire) set up a poulterer's shop in the parish of St. James's, Piccadilly. John, his only child, inherited the business and, in turn, left it to his only child, George. By this time the business was prosperous. George Fisher became a churchwarden of the parish church and eventually retired to live a few miles away in the celebrated village of Hampstead. His eldest son, George, died in young manhood and, when he summoned his second son, John, from his medical studies to take over the family business, he could afford to buy a Life Governorship of Christ's Hospital in compensation for the sacrificed career. It was a handsome present for a man just come of age.

John, R. A. Fisher's grandfather, is the only one of his ancestors known to have been inclined to a scientific career. At 23, John married the "girl from next door," Emma Mortimer, daughter of Thomas Jackson Mortimer, gunsmith of St. James's Street, Piccadilly. The young couple continued to live above the poulterer's shop in Duke Street and to keep shop until the death of John's father in 1855 (after slipping on a piece of orange peel). Then John invited his younger brother to take over the business and himself retired in

1857, at the age of 40, to the life of a city gentleman, a retirement that lasted 50 years. In public life he was a member (and for 6 years chairman) of the Chelsea Board of Guardians and a member of the Metropolitan Asylums Board and other public bodies. In family life he was a patriarchal figure. He supported his widowed sisters and packed off successive nephews to farm in New Zealand when they became burdensome at home. After the death of his wife, he summoned one or another of his nieces to keep house for him, and when he eventually left London at the age of 74 to join his daughter Mary Paulley and her family in Norfolk, he had a large house built to accommodate them all. Every morning half a dozen children lined up in order of their age and filed into his bedroom to greet their grandfather Fisher and to receive a little Tom Thumb sweet from the jar by his bedside.

John Fisher had ten children, all but the last born above the poulterer's shop. Only five of them survived the infectious diseases of infancy and childhood, and this meant that the eldest son, George, who was to father R. A. Fisher, came to be separated by 9 years from the oldest of his brothers.

George Fisher broke new ground by taking up fine arts as a business. His father enjoyed fine possessions, as he enjoyed his fine tenor voice, as an amateur. (He was once called on to sing with Jenny Lind when her professional partner had cancelled a London engagement.) There were family portraits and some good pieces in the house. George Fisher, however, took up the arts with zest as partner in the firm of Robinson and Fisher, and the business prospered. The firm enjoyed a reputation at the end of the century comparable with that of Sotheby's or Christie's and owed it in no small measure to George Fisher's initiative and expertise.

In 1875, George Fisher, being 32 years old and well established in business, married Katie Heath, 20-year-old daughter of Samuel Heath, the fourth of that name to follow the law. Sam Heath was a popular solicitor among the City Livery Companies, lively and socially inclined. Of his four children his only son, Alfred, became a lawyer, emigrated to America, and became sheriff of Rawlings, Wyoming, in the days of the wild west. He died there in his forties as Judge Heath. Some years after George had married Katie, his brother Arthur, a Wrangler at Cambridge and longtime Rector at Skelton, Yorkshire, married Katie's sister, Dora. This left Jessica, John Fisher's favorite among the three Heath girls, to be snapped up by a Mr. Armstrong.

In 1896 George Fisher moved from his home in Finchley to Heath House, a mansion he had built near the top of Hampstead hill, set in five acres of parkland and gardens. There were ponies for the children and a goat-chaise which they drove round the grounds and a carriage and pair for the parents. There were a succession of country houses where, for fear of infection, they kept a cow to provide milk exclusively for the family, and there were horses, tennis courts, and boats.

In Yorkshire and in London, a large double cousinship grew up to display very considerable ability and diversity of interest, and Ronald, George's youngest son, and the subject of this story, in his career as a research scientist, displayed recurrent family interests in antiquity and history, medicine and farming, travel and exploration. Like many of his cousins Ronald was not satisfied with a single profession: the pattern seems to have been to travel and to farm, to be doctor and naval officer, cleric and antiquarian. Ronald was a scientist and, although his scientific interest was professionally channeled into statistics and genetics, it is significant that these particular subjects are diffusive. One may today specialize as a medical, or ecological, or population geneticist, a plant or animal breeder, bacteriologist or biometrician; Ronald involved himself in all these aspects of the subject. One may be a theoretical mathematical statistician or deal with special applications in genetics or agriculture or astronomy, in economics or industry. In fact, the choice of these two subjects did not exclude any sort of scientific problem from his range.

As one of Ronald's cousins commented, "Some Fishers were brilliant, some were dull, some very sane and responsible, some were brilliant but went off the rails, some just went off the rails." George Fisher was one of the brilliant ones, and his brilliant son came to resemble him markedly in appearance as well as in the wideranging curiosity he had about life and the enthusiasm with which he took up particular projects and carried them through. Ronald believed that intellectual aptitudes are closely related. He assumed that any man with a good mind could apply it successfully to any subject that interested him, including mathematics. This impression was doubtless fostered by the mental affinity that existed between himself and his father, even though their talents found expression in very different fields.

□ □ □

For all his business acumen, George Fisher was a romantic. He loved Katie with the devotion of a mediaeval knight, humbly, protectively, possessively. His letters carried his love to her whenever she was away in the country while he remained in town. He delighted in giving her gifts whose taste and extravagance showed the quality of his love for her and in seeing her appreciation of himself in her enjoyment of all he gave. And Katie rewarded his love. Her dress and carriage played up to the feminity of her small, plump figure, crowned with a magnificent head of curling red-gold hair. She was an accomplished pianist and displayed her slender wrists and hands to advantage at the keyboard. The large household and numerous servants emphasized the native authority in the tones of her light, assured voice. In fact, prosperity suited her, and she accepted the increasing grandeur of their style of life and her social role as naturally as she accepted and occupied her domestic role as wife and mother.

George and Katie were of one mind with their times in desiring a large family, preferably of sons. George was the eldest son and had an example of patriarchal family feeling in his father, and family for him meant boys bearing his name.* In 1876 the first hoped for son, Geoffrey, was born, followed by a daughter, Evelyn, in 1877, and another son, Alan, in 1878. Alan died in his third year, however, and Katie felt superstitiously that there should have been a "y" in his name. The younger members of the family were, therefore, named Sibyl, Phyllis, and Alwyn. On February 17, 1890, Katie was again brought to bed, and George Fisher awaited the birth of his latest child, hoping that it might be another boy.

The nurse entered but he could read no verdict on her face. Then it came: "A beautiful baby boy," she said, "he was stillborn." She excused herself, leaving him with the comfort at least of knowing that his wife was doing well. Fifteen minutes later the nurse entered again, flustered and radiant: "There's a second baby, a tiny boy, alive!" Twins: there had been no suspicion that a multiple birth was to be expected. Upstairs the cry of the puny baby cracked the quiet of night as George Fisher took the body of his nameless son and buried it in the garden. The survivor they called Ronald Aylmer.

Thus, Ronnie, as he was called, at first, was special from the beginning, not only as the desired son but because he came unexpectedly like Isaac from the sacrifice, a living son where a son had been given up for dead. He was the youngest child, his mother's baby and special darling. How could he fail to be spoiled? His three sisters, 8 to 13 years older than he was, all in their different ways acknowledged the domestic fact that boys were of the first importance, but, seeing Geoffrey spoiled by over-indulgence, they formed a sort of Greek chorus, observing, inspiring, and commending the performance of their little brothers. Also, Ronnie had healthy competition from Alwyn, 3 years his senior, whose advantage in years was accentuated by the contrast in physical appearance. Alwyn was dark, but Ronnie had the fair complexion common with his auburn hair; Alwyn was tall and strong, whereas Ronnie, for all his physical toughness, was slight and small.

Ron was born into the special emotional environment provided by his mother. His father adored Katie. His sisters admired her greatly and with affection. Ron idolized her. Though he mentioned her rarely, he scattered her memorials through his life: her devotion to the Church of England was reflected in his unquestioning adherence to that religious tradition; her ideal was reflected in his often gallant behavior toward women and in his expectations of them—he had a soft spot for plump little women, "bunjy, like the

---

*Katie seems fully to have shared his bias in this direction, and they named their sons in this spirit: George Geoffrey, Alan Heath, and George Alwyn.

women in our family"; and he propagated her image by naming his eldest daughters Katie and Margaret Katie and causing them all to grow their hair long to be, like hers, their "crowning glory." Yet, outside her immediate family, Katie was positively disliked by her in-laws. John Fisher had been disappointed that his favorite son had chosen her rather than her warmhearted sister, Jessica. Fred Fisher and Mary Paulley, with their respective families, thought her poisonous. In their occasional meetings they found her selfish, indolent, and arrogant, with ideas above her station; and her daughters they thought as bad as herself. The myth grew: the George Fisher women were deemed extravagant and callous; they had, it was later said, bled George Fisher to bankruptcy and brought him to ultimate suicide. This, of course, was sheer nonsense. After the death of his wife and his subsequent business failure, two of his daughters cared for him devotedly, and he them, through life. Within a few months of his death from natural causes at the age of 77, he was full of the enjoyment of life—taking a hand at the cross-cut saw, discoursing widely with his son with undiminished animation, and dandling his grandson and namesake delightedly on his knee.

Why, then, the scurrilous tradition? Fred Fisher and Nat Paulley were country doctors, less wealthy than the George Fishers and accustomed to country style of life and dress. There was some envy. Certainly the country cousins were astonished that Aunt Katie should wear on a weekday the black silk which was obviously to them "Sunday best." Probably, also, Katie did not hide her personal dislike of Mary. Still, these things alone seem inadequate to account for the unanimous and vitriolic dislike of Katie in these families.

Katie's reputation was shared by her daughters and, in considering their personalities, one perceives a possible origin for the myth. Sibyl, in her sixties and seventies was a round, furry, quiet-spoken, fascinating puss. She was feline in her poise, exquisite, self-contained. She had the authority of assured self-control in whose presence one felt oneself to be all puppy-fat and bumbling enthusiasm. She was superior and she knew it; it is easy to imagine that as a beautiful, passionate, and clever girl she had wrought havoc on weaker mortals, her wit playing cat and mouse with its victims and sometimes striking cruelly. But, most devastatingly, she offered no lifelines at an emotional level which could draw one up over the sheer face of her individuality into her fellowship. One might learn what she thought (and it was often wise), know what she did (and it was largely good), but never a clue what she felt. She walked on her wild lone. If Katie was like her in her selfassurance and in her inability to communicate her sympathies, it is no wonder that she inspired both devotion and bitter resentment.

Ron was his mother's darling but it may be his mother's love was a matter of faith rather than of feeling. Katie is always described in passive terms, a figure seated apart from the romping children, receiving their enthusiasm but not embracing them with her own. Her presence in the nursery was an occasion, though a daily one, and she presided in an Olympian world beyond theirs, which the children saw only, for example, when they were admitted to the dining room for dessert and she, at the table's end, dispensed a home-grown peach or bunch of grapes before they retired from the company. Again, her husband's devotion made her something of a goddess at home. To outsiders she seemed cold and insensitive, and it is quite possible that even a beloved child became aware of her love not through her maternal warmth but indirectly.

If this interpretation is correct, Katie passed on to her son her own inadequate emotional vocabulary. There was no easy flow of unspoken sympathy and understanding with either of them. Their communication at an emotional level, that is, by indirect intimations, was stilted and unconvincing, and they were curiously oblivious of the feelings of others. In Ron the relative isolation at this level fostered selfreliance and a strong individuality, but the need to transcend the inadequate communicative process was undeniable. His conventional and ritual actions became loaded with the inexpressible emotional messages. His emotional insecurity was evidenced by his sometimes almost paranoic reaction to the frustration of his will and to those whom he associated with such frustration.

Although emotional communication with his mother was constricted, intellectual channels were wide open, and intimacy could be established at a purely intellectual level. He relied on this line. For him, straight thinking came to have all the virtue of sincere feeling, and shared enthusiasm about ideas became a form of love. Frustration in the attempt to communicate through the intellect was a really painful experience, but success in the meeting of minds and in their mutual exploration brought with it a pleasure like physical intimacy in its intensity. Katie adored her son and he adored her: their love made an *intellectual* hothouse to which his nature was capable of phenomenal response. When he was five she opened the heavens to his comtemplation as she lay on the sofa reading Sir Robert Ball's astronomy to him and he sat enthralled beside her on a velvet cushion on the floor.

Thus Ron grew up without developing a sensitivity to the ordinary humanity of his fellows. He was unaware of the effects of his own behavior, and he often expressed his love ineptly. To give an example from his middle age, he showed what is justly described as "rather touching solicitude" for an old friend when he arrived in Cambridge on a summer's day:

He said to me, "You look tired. I shall take you on the river. You shall rest in the canoe while I paddle." Dear Fisher was half blind, the river crowded and the canoe unstable. He would charge boat after boat, calling out, "Look where you are going, sir!" I cannot think how it was we were not upset. Since he and I were both in our normal clothes that would have been unwelcome. It was a harrowing experience.

All the while Fisher, in the kindness of his heart, was unaware of anything but the shared pleasure in an afternoon of relaxation on the river! And in this he was transparently honest; what he missed was framed in a language he had never learned to hear, which his mother perhaps had never used or taught him to listen for. Nor could he speak it; he assumed that his own good will and his desires were obvious, and when they were misunderstood, he felt that malice or stupidity must be to blame.

Being frustrated, he erupted. There was no forewarning of his passionate outbursts which struck, like the thunderbolts of Jove, with sudden fierce terror on some unsuspecting head, who was not always the one who had failed to understand or to do the thing that was close to his heart. But his anger passed, the thunder died away, and he was himself again. At one moment in fury against a laboratory assistant, he crushed to death the mouse in his hand, then, realizing, muttered, "See what you made me do," cast the mouse from the open window, and left the room; but a moment later, reopening the door as if to enter and seeing the girl facing him in an attitude of rude gesticulation, he grinned, friendly again, and slipped away.

These incidents illustrate something of the insecurity underlying Fisher's personal relationships throughout his life. Early in life he was recognized as being different from others both because of his intelligence and in consequence of his poor eyesight. The effect of being set apart was to strengthen his love of companionship, which he ranked with the highest forms of human happiness. He needed friends imperiously, companions with whom he could share his thoughts, colleagues who could accept his goodwill and his bad temper and sympathize with his aims if not always with his conclusions.

He gave much to friendship, and he demanded much. His manner was often charming and his discourse fascinating; he was kind and helpful and generous in most of his dealings. At the same time, his friends felt compelled to abandon their own wishes in his presence, allowing themselves to be fascinated by his conversation and to accept kindnesses as ill-conceived as the harrowing canoe trip, for they recognized that Fisher would not take any modification of his plans easily and, if they insisted, he would be hurt and angered as if he personally had been rejected and abandoned in his need.

His loyalty was absolute: practically nothing could persuade him to reverse his opinion of any man, or to let down a friend. Having formed a largely intuitive judgement of any man, usually but not always sound, he was fully committed. A similar loyalty bound him to his country, his church and his profession. He was a patriot, a political Conservative, a member of the Church of England, always a scientist, and, though he was often revolutionary in his approach, his most subversive views expressed his individual loyalty to ideals he perceived in the various establishments.

The peculiarity of his blindness to emotional tones was to set him apart as, in some sense, a difficult person to know, to some natures baffling, to some intolerable, to some "beyond good and evil." Yet the deficiency was offset by such patent honesty, such passionate integrity, that in some sense also he made himself transparent. He was at once exceedingly self-centered and utterly self-forgetful, charming and impossible. And his friends learned to accept his inconsiderate and irritable behavior because they perceived the greatness of his character behind the screen, ruled by the great simple principles of truth and compassion.

If he was less sensible than most men to emotional overtones, the fact made him a better scientist, for, as he pointed out, emotional considerations are irrelevant to scientific work (although they are deeply important to individual scientists). Science is concerned with the coherent formulation of the evidence of nature. He thought through the logical implications of the evidence as if the clouds of opinion did not exist and, having thought it through, he withstood the storms of disapprobation that raged against him, intellectually unmoved unless new facts changed the logical situation. He did not often have to change his position. We suggest that it was not only his greater vision but also his greater blindness that enabled him to pass through the fog without stumbling, not only that he knew and trusted the intuitions of thought but that he did not recognize or trust the infections of feeling.

□ □ □

Ron differed from others in two ways which greatly influenced his schooling. The first was evident almost from the beginning: he was a precocious child. By the time he went to school, the second became noticeable: his eyesight needed special care.

The Fisher family were staying at their country home at St. Peter's Court* on the Isle of Thanet in the summer of 1893. A visiting friend asked Katie how old her little boy was. She replied "three," but Ronnie said he was not. The friend asked him, then, how old he was and received the emphatic answer, "I am three years, four months, and five days."

At about this age when he had been set up in his high chair for breakfast, he asked: "What is a half of a half?" His nurse answered that it was a quarter. After a pause, he asked, "And what's a half of a quarter?" She told him that it was an eighth. There was a longer pause before he asked again, "What's a half of an eighth, Nurse?" When she had given her reply there was a long silence. Finally, Ronnie looked up, a plump pink and white baby face framed

---

*Later made St. Peter's Court school, attended by the young princes and future kings of England, Edward VIII and George VI.

with waving red-gold hair, and said slowly, "Then, I suppose that a half of a sixteenth must be a thirty-toof."

The whole incident is so characteristic of the man he was to become that one suspects at first the story must be apocryphal. But two of his sisters who were present at the time volunteered the story on very different occasions, Phyllis telling the young Mrs. Fisher in the 1920s very much what Sibyl told in the 1960s for the record, and vouched for. It certainly has Fisher's hallmarks: an interesting problem set by himself in mathematical terms, and not an unsophisticated problem either, for it involves the inversion of a geometrical progression; his inquiries which gave him the data he needed to establish the operative principles; the manner in which he took time successively to grasp each step in the argument; finally, a calculation worked out in his head to produce not only the numerically correct but the logically inevitable conclusion, which he expressed by a characteristic conjunction, "I suppose . . . must be."

The boy was indeed precocious in his mathematical development. Two or three years later, on hearing Sir Robert Ball's astronomy lectures read aloud at home, he seemed to take in his stride the descriptions of astronomical dimensions and movements, of eclipse and nutation and parallax. In fact, this was the beginning of his lifelong interest in astronomy.

There was also a certain logical order in the way he expressed his sentiments, which no doubt helped him hold his own with his bigger brother. Thus on one occasion when they were at table, Alwyn repeatedly put his knees up until their nurse exclaimed in exasperation, "Oh, Alwyn, whatever shall I do with you!" With a glance at the offending objects, Ron unhesitatingly answered the rhetorical question: "I should smack them if I were you, Miss White: they are big enough, and dirty enough and ugly enough!"

Because of his evident brilliance, Ron was entered at Mr. Greville's day school in Hampstead Village with his brother Alwyn, at a younger age than his school-fellows. He was probably something of a school pet. A photograph shows him nestled between the headmaster and his wife, surrounded by the bigger boys. On entrance to preparatory school at Stanmore Park, when he was ten, he was at first placed in the lowest form but after one term was moved to one of the highest forms, among boys 2 or 3 years his senior, as he had been earlier at Mr. Greville's school. Now he had to exert himself to maintain first place. The term reports consistently read, for his position among a class of eleven boys: Mathematics, 1 st; Latin 1 st; Science 1 st; and so on to French 11 th. He never did have much patience with irrelevancies.

As a student in mathematics he had a disconcerting habit of producing the correct answer without showing how he had arrived at it. Despite repeated demands that he "show his working," it seemed impossible to elicit more. Apparently, he had not formulated the problem to himself in the conventional manner, for there was none of the usual process of deduction to show. This is the first instance of a continuing difficulty for his teachers, colleagues, and students. As E. B. Ford comments:

Fisher was of course a fine mathematician but difficult to follow. He would leap over intermediate stages in a calculation, leaving his colleagues floundering. I have several times heard a distinguished mathematician say, "He has evidently solved the problem correctly, but I don't see how he has done it."

Ron had an excellent and sympathetic mathematics master in W. N. Roe. Roe was a county cricketer and extended his interest in his boys to the playing field. With him Ron acquired the nickname, "Piggie," which stuck through his years at school and college and sometimes beyond. In 1928 Fisher wrote to his old master, still at Stanmore Park, to inquire about the possibility of sending his own son there and received a reply addressed to him under this title, promising to do all he could for the son of his star pupil and recalling the old names and the old places: the path where Ron had won the 200-yard handicap, the fall of the great trees they had called Castor and Pollux.

At preparatory school the second influence on his education came into play, the protection of his weak eyes. His mother was quite nearsighted, and he doubtless inherited the defect from her, although a severe attack of measles at an early age may have further impaired his sight. He began to wear spectacles of increasing power, until the lenses were so thick that they resembled transparent pebbles. Without them he was practically blind. His eyes were an anxiety to his parents and teachers throughout his school days.

At home it became customary to read aloud for his sake. His sister Sibyl records:

I formed his childish taste in literature during his boyhood. I read aloud to him an hour a day the best of the classics and some Conan Doyle, not the Sherlock Holmes but the historical novels. I remember meeting him at the station on the day he sat for the Harrow entrance exam. He took the highest scholarship in maths. into Harrow and was amused at the name on the paper—Bumble. I was reading Oliver Twist at the time. Ron passed so high into Harrow that his housemaster suggested his taking a term's rest of his eyes, which he took at our country house on the River Dart in the Spring and Summer of 1904.

The first time I became aware of Ron as a personality, not as a clever child, was in Dartmouth that year. He was 14. We boated and managed our rowing boat on the Dart and swam in the Dart and the sea and took long walks over the hills and surrounding country. Losing our mother in June only drew us closer together—a great blow to us both. Ron was the youngest and adored son. He felt it very much, at 14. My mother was only 49 and ill only two days."

The death of his mother of acute peritonitis, together with his transition to Harrow, closed childhood firmly behind him.

At Harrow the same considerations for his brilliance and for his poor eyesight prevailed. He threw himself into the work and took the Neeld Medal in 1906. Since this competitive essay in mathematics was open to all the school irrespective of age, he was very proud of his early trophy. He did not enter the competition again.

To avoid eyestrain, the doctor advised that he should not work by lamplight. Consequently, he received tuition in mathematics in the evenings without the use of pencil, paper, or any other visual aid. Tradition has it that, being given the choice, he selected spherical trigonometry as the subject for these tutorials. This meant that the natural aptitude of a young mathematician to form mental images of his problems, instead of being channeled, as it is with most schoolboys, into algebraic expressions and restricted to representations on a flat sheet of paper, was unrestricted. Indeed, the power of mental representation was constantly strengthened by exercise. He developed an exceptional ability to solve mathematical problems in his head and also a strong geometrical sense. He gained a facility in thinking in more than two dimensions, which was to influence his manner of attacking mathematical problems throughout his life; and it was later to make possible, in particular, the geometrical derivation of a number of important distributions which had until then defied algebraic elucidation. Other mathematical statisticians without his geometrical insight, therefore, found his work difficult to follow and often criticized him for inadequate proofs and the use of "intuition."

At Harrow, again, he was fortunate in having excellent mathematics masters in W. M. Roseveare and his tutor, G. H. P. Mayo; he had a happy relationship with both of them. Roseveare, later acknowledging a gift of *Statistical Methods for Research Workers*, recalled their doings at school. Writing from Pietermaritzburg, Natal in May 1926, he congratulated his former pupil on his current work and, before telling of himself how elliptical functions had been carrying him away on aetherial wings, reverted to a joy of the same sort at school:

I assure you I have not forgotten you: your dear little $(a + 2)^4$ is in my mind and often on my lips when ears worthy of it have minds behind them. I have had a very pleasant time out here since 1910 . . . . If only you had accompanied me to the Cape in 1906* I am sure you would be longing to be in this sunny country.

His reference to $(a + 2)^4$ recalled the occasion when Fisher had brought his mathematics master the following exposition. A straight line is a line and two

*A trip cancelled probably because of the financial collapse of George Fisher.

points. Let us represent it by $a + 2$. A square has an area, four lines and four points, and can be represented by $a^2 + 4a + 4 = (a + 2)^2$. A cube has a surface, six faces, twelve edges, and eight points and so can be represented by $a^3 + 6a^2 + 12a + 8 = (a + 2)^3$. Accordingly, a cube in four dimensions can be specified completely by $(a + 2)^4$ and, in general, a cube in n dimensions by $(a + 2)^n$.

Ron also sent a copy of his first book to Mayo. Later, in congratulating Ron on his election to the Royal Society, Mayo also referred to his schooldays:

If I were to try and recall the past there would come to me very vividly full memories of our work together, sometimes in that old form-room in the old school, sometimes in my own dining room. I recall too the dark December evening when you brought the telegram from Caius to say you had won an £80 Scholarship.

By that time the scholarship was not only an honor but a necessity. Within 18 months of Katie's death, George Fisher had lost his fortune. Several big transactions had proved costly mistakes. Possibly, as he believed, he had been cheated; more likely, his judgement had failed in critical cases. The shock of the sudden death of his beloved wife may have left him more shaken in mind than he recognized. The partnership with Robinson was dissolved, and he was forced to agree never in future to trade within 25 miles of London. The leasehold of Heath House* and most of its contents were sold, and George Fisher retired to Streatham with Evelyn and Phyllis. He had little left when he had paid his creditors 20 shillings to the pound, and he could not recoup his losses by the practice of his trade. He had no funds to send Ron to college. Ron's brothers and sisters still depended on their father: even Geoffrey, at the age of 30 still the feckless spoiled son, relied on his father to bail him out. Alwyn took his college degree, then set out to seek his fortune ranching in Argentina, and Ron took his place as a man beside his father. His first reaction was to offer his father the money that had been put by for him from the Harrow scholarship. He was determined hereafter to earn his own way. The scholarship to Cambridge made the first steps to independence possible.

It had been a toss-up whether he should take the scholarship in mathematics or natural sciences. His teacher in biology, Arthur Vassal, had formed the highest opinion of his ability. Just how highly he thought of Ron at school one learns from E. B. Ford's story:

I was once talking to Arthur Vassal, for many years a schoolmaster at Harrow; he was no more than an acquaintance of mine and certainly not aware of my friendship with Fisher. I asked him if it would have been possible to name the ten or twelve cleverest boys who had passed through his hands. He said it would be difficult to do so but

---

*Sold to Lord Leverhulme who lived in it for many years. It is now a hospital. Heath Gardens are maintained as a public park of the Village of Hampstead.

added that on sheer brilliance he could divide all those whom he had taught into two groups: one contained a single outstanding boy, R. A. Fisher, the other all the rest.

Vassal was keen that this boy should continue the biological studies in which he excelled.

Ron said later that it was the sight of the skull of a codfish he had happened to see exhibited in a museum that had dissuaded him. The great number of skull bones displayed and labeled, whose names were no doubt to be committed to memory by the biological student, threatened him with a boring and bootless task. What particularly appealed to him were the wide-ranging lectures in which Vassal abandoned for the moment the required syllabus for young medical and biological students and discussed the curiosities of nature, its unexplained phenomena, new problems, and new work. Some of these problems interested Ron enormously, like the description of the numerical oddity of the neck vertebrae of the sloth and a species of manatee. These cropped up in his later work so that, sending Vassal *The Genetical Theory of Natural Selection,* he recalled Vassal's lecture and the train of thought leading from it, until the sloth took its place in the theory and entered the exposition as an example.

The choice of mathematics never detracted from Ron's biological pursuits. As he wrote to Vassal in March 1929,

The fact is that nearly all my statistical work is based on biological material and much of it has been undertaken merely to clear up difficulties in experimental technique.

Naturally I have had most of the fun myself as there are very few mathematicians with any serious interest in biology or contact with working biologists.

It would have worked out much the same, I fancy, if I had taken your suggestion and taken biology for scholarship purposes at school. I still think the scholarships would have been more chancy and I suppose, without being sure, that a mathematical technique with biological interests is a rather firmer ground than a biological technique with mathematical interests, like d'Arcy Thompson.

What I want to know is why can you not cut out all the stocks and shares and quadratic equations from the compulsory mathematics for biologists and doctors and put in the elements and the ideas of statistics.

When the time came to select two prizes at Harrow in his last year, Ron did not choose mathematical or astronomical texts; instead he chose a collection of Greek plays in translation and the collected works of Charles Darwin. There was some debate as to whether the latter were suitable works to award the schoolboy but they were given, and he went up to Cambridge in possession of volumes he was to read and reread with loving care throughout his life.

Ron went up to Cambridge as a scholar at Gonville and Caius College in October 1909. The scholarship money was sufficient for his needs, and he was not inclined to let the lack of luxuries trouble him: money was not important. He relegated financial success to an inferior status among the good things in life, and thereafter he did not change his mind but sought rewards and left a legacy of a different kind. His attitude throughout his life was expressed [*C P* 3] before he left Cambridge:

Like all healthy philosophies, eugenics urges us to simplify our lives, and to simplify our needs; the only luxury worth having is that of a worthy human environment. We must be ready to sacrifice social success, at the call of nobler instincts. And, even as regards happiness, has any better way of life been found than to combine high endeavor with good fellowship?

At college he combined high endeavor with good fellowship, and he was very happy.

He seemed unembarrassed by his clothes or any other material possession, but there were other students who were more obviously embarrassed and on whose behalf it was worth taking a stand: and he had the courage to do it. All students were compelled to pay the college an amalgamated subscription, chiefly, if not solely, for the cost of athletics. The great consumers of this money were the rowing men, especially when, as was usual, the college ran a boat at Henley in addition to one on the Cam. Squabbles often broke out over this. Ron felt that the charge was an unfair imposition on many students. He took the antirowing line and made a forthright speech about it. The rowing men were not pleased, and a friend recalls, "They wrecked his room for him and (I regret to say) poured his own water over his own bed. He did not care."

At Cambridge, Ron and a few like-minded young men formed a little coterie sometimes referred to as The We Frees. They were a lively and talented group, with high ideals, varied interests, and very little money. They used to gather under the lime trees in Tree Court or in each other's rooms to read and talk together. They evolved their own idiom, talking in the phraseology of Nietzsche's *Thus Spake Zarathustra*. They read the Icelandic Sagas, and the fashion of their talk incorporated phrases and names from the sagas. They made expeditions together on bicycle or boat, picnicked and walked, and drank their beer in the spirit of Hilaire Belloc, whose songs they sang together in a rollicking chorus along the way.

It was not long before the medical student among them suggested an innovation: to bring into their fellowship a friend of his, Ian Mackenzie, a student from Guy's Hospital, together with his young wife. They had trouble persuading Ron, who was convinced that a woman would spoil all their fun.

However, he was persuaded or lured along to a meeting at which she was present and found, to his surprise, that she actually added to the occasion. She was a new sort of woman in his experience: in contrast to the small women of his family she was tall (5'10"), vivacious, and quite unconventional. She was also imaginative and clever. They called her Gudruna, from their admiration of the splendid Norsewomen, and for the next 4 years they were all to visit her home on many occasions. She recalls:

One of the group was a man who would have been a great poet*, I think, but he died at Neuve Chapelle; there was too a fresh-faced motorcycle enthusiast nicknamed The Censor who contributed chiefly to our out-door activities (punting, picnics etc.); the medical student I remember chiefly as a precocious philanderer; my husband contributed to the group from his very slightly more wordly experience: the mixing of rum punches, celebration of cup tie night in the metropolis, and his experiments in the new technique of hypnotism. My own contribution was chiefly domestic: our Cambridge excursions had to be planned around my baby girl; and later, when the friends visited our home at Sydenham Hill, their voices would wake her and she emerged to be dandled by one of the young men. The poet sketched her in innumerable moods. He and Ron were the first outsiders to see my second baby, twelve hours after her birth. I think they all formed their ideas of married life from what they saw of us.

The earliest evening sessions at Cambridge were chiefly literary. The poet was passing through an aesthetic stage and wore bow ties. He and the medical student were very well read. Ron was very much at home in the group, although his scholastic interests were mathematical rather than literary. Those were the days when no evening was long enough for all we had to say, and we found each other excitingly stimulating.

We all took it for granted that Ron was an intellectual giant, curiously uncouth, and, even in that innocent youthful society, marked as "unspotted by the world." He had a top-heavy appearance due to his great head; his clothes always managed to appear disreputable and he was known invariably as "Piggy." He was very good company and was extremely popular, getting in consequence more than his share of teasing. (I gathered that his repertory of the cruder kind of smoke-room stories was formidable!) The poet made for me a caricature of Ron as a Norse savage, short, phenomenally broad, with a noble head, a naked pink skin and a mace. He had captured the characteristic pugnacity of his stance.

This was a happy time, and Ron delighted in all that Cambridge showed him. He loved the old buildings, especially his own college, with its warm

---

*The poet was Tom Renton. The Censor, H. V. Neilson, was a lifelong friend. A curate in Winnipeg when Fisher visited in 1913, on the outbreak of war he volunteered for military service and was rejected as a consumptive. By persistent application, he managed to get into the army, was wounded in France and invalided out with a damaged leg. A clergyman until his death in 1948, he was godfather to Fisher's daughter, Margaret.

Cotswold stone and its gates symbolizing his college career, for one entered
by the Gate of Humility, passed from one court to the next under the Gate of
Virtue and emerged into the world beyond through the Gate of Honour, the
oldest part of the existent college buildings. He loved the narrow streets, the
"perlieus" between the walls of neighboring colleges, where once the open
sewers ran from the colleges into the River Cam. He loved the river itself,
"our little ditch," and the fens and river meadows along the water's edge, the
great old trees breaking the flat horizon, and the trees and gardens and grassy
courts within the college walls.

   The companionship of the little coterie was meat and drink to him. It was
good to be young and carefree together. They were idealists, full of confidence and patriotism, earnest even in their wit and whimsy, and they laughed
together at human pretensions and at their own. By association with Caius
College, the We Frees became "Waius Fraius." Their freedom took the form,
for instance of adding verses to Hilaire Belloc's song "His hide is covered with
hair." Belloc [4] himself had given them their cue, "For all the beasts of the
field, and creeping things, and furred creatures of the sea come into this
song," he had written, "and towards the end of it the Hairy Ainu himself.
There are hundreds upon hundreds of verses." Several of the students'
verses were witty in a literary way, but Ron's wit was peculiarly biological. His
verses, quoted here with the closing verse of G. P. Thompson of Trinity, recall
the sloth and manatee whose vertebral peculiarities had impressed him in
Vassal's lecture; they pick on some interesting adaptations to the arboreal life
of our ancestors; and they cut near the bone by choosing the *galeopithecus* (a
flying lemur), the primate called the slow loris, the mandrill ape, and the
original mermaid (the dugong or the manatee):

> The flying fox and the vampire bat,
>   Their hide is covered with hair;
> The loris is slow and exceedingly fat,
>   But his hide is covered with hair.

> The gay dugong and the manatee,
> They wallow about in the tropical sea,
> But the galeopithecus lives in a tree
>   And his hide is covered with hair.   R.A.F.

Chorus:

> Oh, I thank my God for this at the least,
> I was born in the West and not in the East,
> And He made me a human instead of a beast,
>   Whose hide is covered with hair!   H. Belloc

The Indian bullock is proud of his hump,
　　But his hide is covered with hair.
The mandrill ape has a purple rump,
　　But *his* hide is covered with hair.

The sloth at leisure among the trees,
Gracefully swaying in every breese,
He passes his time in proverbial ease,
　　But his hide is covered with hair.　　　R.A.F.

Chorus:　Oh, I thank . . . etc.

(Any additional verses to be added here as the Ainu comes last)

The duck-billed platypus lives in the mud,
　　His hide is covered with hair.
The last of the mastodons died with the flood,
　　But his hide was covered with hair.

The Ainu inhabits as best he can
An insular spot in the north of Japan,
And he looks like a monkey instead of a man
　　For his hide is covered with hair.　　　G.P.T.

Chorus:　Oh, I thank . . . etc.

□ □ □

Fisher is said to have worked for his degree by spurts of concentrated thought, excusing himself from the company on some evenings, "to do in two hours the work of two months." Certainly, he managed to take interest in matters unconnected with his mathematical syllabus, enjoying not only the literary extravagances of his society but keeping abreast of the biological innovations and discussions.

Cambridge was the center in England where the new subject of Mendelism had its focus, for William Bateson of Cambridge had made himself its protagonist. From the first moment he had seen Mendel's work described in 1900, Bateson had been enthusiastic in publicizing the new ideas and initiating experimental breeding along Mendelian lines. At a meeting of the Horticultural and Plant Breeding Association at Cambridge in 1906 he had given the subject its modern name "genetics."

The work of Abbé Gregor Mendel, first published in 1865, had been rediscovered in 1900 simultaneously by three eminent European scientists, Tschermak in Austria, Correns in Germany, and de Vries in Holland. The time was ripe for the discovery. Recent discoveries about cell structure and cell division offered new information about physical inheritance. A physical basis was now known to exist, in animals and plants alike, that paralleled the pattern of inheritance Mendel had discovered in garden peas. In 1902 Sutton had pointed out the connection between the facts of cytology and Mendel's hypothesis of biparental particulate inheritance. The heritable factors of Mendel could be physical particles carried on the dark-staining bodies, called chromosomes, visible in the cell nucleus during cell division. These factors were first named genes by Johannsen in 1909. It is convenient to anticipate the use of the word and to say Mendel had proposed that peas to contain a double set of genes, one set derived from each of its parents and that they passed on either one of each pair of genes (alleles) to their progeny. The new cytological work showed that pairs of homologous chromosomes come together in the zygote at fertilization and separate again in the formation of the gametes giving rise to the next generation. Thus the genes might be imagined as strung along the chromosomes like beads on a necklace.

Mendel had found that genes controlling different characters assorted independently of each other, and each pair of chromosomes was seen similarly to assort independently of other pairs. Genes located on the same chromosome would not be expected to exhibit independent assortment. The chromosome theory was, therefore, strengthened when first W. Bateson and R. C. Punnett in Cambridge and then T. H. Morgan in New York discovered some genes that were linked in their inheritance. The association of genes with chromosomes, predicted by Sutton in 1902, was, in fact, established in 1911 by Morgan, who introduced the term "linkage."

During Fisher's undergraduate years the subject of genetics was accorded academic recognition in Cambridge University. A professorship in biology was created for William Bateson in 1908, and the chair was permanently endowed as the Arthur Balfour Chair of Genetics in 1912. By that time, however, Bateson had accepted directorship of the newly created John Innes Horticultural Institute at Merton (1910) and unhesitatingly refused the tempting offer to return to Cambridge. Consequently, R. C. Punnett, formerly Bateson's chief assistant in Cambridge succeeded to the chair in Cambridge. Fisher probably had no personal contact with Bateson during the year in which their residence in Cambridge overlapped, but Punnett was a Fellow of Caius, and his presence there throughout Fisher's years in Cambridge doubtless stimulated some of the genetical discussion in which Fisher participated.

Thus, when Fisher was in college (1909—1913) the new subject was becoming established and fundamental discoveries were being made. From 1900 on, the relationship between Mendelism and evolutionary theory gave prominence to the new science, for it aroused loud and bitter controversy. Darwin's theory was reflected in the title of his book, *On the Origin of Species by Means of Natural Selection*. Natural selection resulted from the facts (i) that characters are inherited (ii) that variation occurs within species, and (iii) that many more young are born than survive to reproduce themselves. Genetics thus provides the mechanism by which characters are inherited and by which variation occurs in these inherited characters which, under natural selection, must result in progressive evolution. At the time, however, the logical connections were not clearly appreciated; genetics did not necessarily seem to imply evolution, nor did Darwinism seem necessarily to require genetics.

Darwin's theory had been convincing to scientists because it was an evolutionary theory that relied only on the operation of natural causes. Once the fact of evolution was accepted, however, the mechanism of natural selection was generally abandoned. Older theories of a teleological nature were revived. In particular, Lamarck had proposed early in the nineteenth century that the will or desire of an organism, or its needs in a particular environment, would stimulate physical changes in the organism and that these would be passed on in the inheritance of its progeny. According to the neoLamarckians of the twentieth century, the environment directly affected the genetic make-up so that evolution took place directly, without the indirect influence of natural selection.

Fisher later [*C P* 217, 1947] recalled the attitude of the times toward evolution by natural selection:

I first came to Cambridge in 1909, the year in which the centenary of Darwin's birth and the jubilee of the publication of *The Origin of Species* were being celebrated. The new school of genetics using Mendel's laws of inheritance was full of activity and confidence, and the shops were full of books good and bad from which one could see how completely many writers of this movement believed that Darwin's position had been discredited.

There were three main contestants for alternative positions: William Bateson, a Mendelian who had doubts about progressive evolution; Hugo de Vries who, having discovered genetic mutation (as he thought), attributed it to environmental influences and opposed natural selection; and Karl Pearson, a traditional Darwinian in accepting evolution by natural selection, who rejected Mendelism and the theories of de Vries. Each had his own problems which seemed explicable on his own theory, and the claims of the three men became mutually exclusive. The biological world rocked with their controversy. It was to be many years before it was possible to suggest, with any hope of carrying conviction, that Mendelism, mutation, and natural selection were not incompatable but complementary contributors to organic evolution.

It is of interest historically that it was the fragment of the truth each man held that led him to an erroneous view of the universe. William Bateson knew inheritance to be discontinuous and so thought that evolution must be; de Vries recognized that mutation of the genes was necessary for progressive evolution and so concluded that natural selection was not important; Pearson knew that natural selection acted on small differences that arise from continuous variables and, he assumed, not from Mendelian pairs. Only when all the pieces were joined in a coherent pattern did the relevance of each truth emerge.

Fisher came to the new ideas already convinced that the theory of evolution by natural selection was incontrovertible. He saw that the facts of genetics supplied what was lacking to Darwin's theory and found no difficulty in recognizing that discontinuous inheritance was not inconsistent either with continuous variation or continuous evolution but would lead to just these things. His view is evident from his discussion of Mendelism and biometry in 1911, with which this biography began.

□ □ □

Fisher was obviously interested in the scientific discoveries associated with genetics in these years and aware of the potential contribution of biometrics in elucidating the genetics of continuous variables. Nevertheless, both genetics and statistics appear, at that time, as tributaries to his interest in a subject that he saw as more vitally important to mankind, the subject of eugenics, eugenics being the name given by Francis Galton [5] to the improvement of the biological inheritance of man.

Francis Galton, a half-cousin of Charles Darwin, was a man of versatile genius whose interest had turned, after the publication of *The Origin of Species* in 1859, to the study of human inheritance. He had grasped clearly not only that mankind was an evolving and therefore mutable species but that changes for better or worse would take place in any race by means of its hereditary composition. He had begun the study of human heredity with a view to gaining the sort of understanding of the subject that would enable the framing of deliberate policies tending toward the improvement of human inheritance.

Galton's interest had resulted in his own studies of human inheritance and his beginnings in the elucidation of statistical concepts and techniques appropriate to the analysis of the continuous variables found in man. He had interested younger men, including Karl Pearson, in the statistical problems of the subject. This had led to his sponsorship in 1901 of the statistical journal *Biometrika,* edited by Pearson, and in 1904 to his endowment of a research fellowship in national eugenics at University College, London, the nucleus of the eugenics department (under Karl Pearson) created by his endowment at his death in 1911 and named in his honor the Galton Laboratory.

Galton's interest had further led to his becoming president when the Eugenics Education Society of London was formed in 1907. The laboratory and the society were planned to complement the work of each other, for although the scientific research must be fundamental to eugenic understanding, the knowledge gained through research would be of little practical worth without applications that must lie in the realm of public policy. Galton was conscious of the lack of an informed public on matters affecting the race, and the society was charged with the task of publicizing the scientific facts among those responsible for the making of policy and for alerting those concerned to the racial effects of any policies under consideration.

Eugenics is concerned with the improvement of inherited qualities in human society; when the Eugenics Society was formed in 1907, the improvement of society, by all means, was stirring the imagination of many Englishmen. Social reforms were in the air. The public conscience was aroused to the reality of poverty and degradation in the city slums, to the problems presented by such varied social ills as crime, alcoholism, feebleness of mind and insanity, syphilis, and pauperism itself. Private charities, religious organizations, and legislative bodies were actively engaged in the struggle to improve the condition of the neglected classes, and their actions both revealed the magnitude of the task and drew attention to it. The Labour Party, then for the first time represented in Parliament, and the Suffragette Movement further publicized the unsatisfactory conditions affecting workers and women. The Liberal government then in power was to introduce radically new legislation on behalf of the needy: for example, to set up a Royal Commission on the Care and Protection of the Feebleminded (1904), to provide an organization and a fund to assist a coordinated improvement in the depressed agricultural industry (1909), and to introduce health and unemployment insurance and provision for disability and retirement of some industrial workers (1912). Thus the Eugenics Education Society appealed to a society concerned for the improvement of human life, and it attracted to its membership both noble and intellectual reformers, doctors, politicians, and some few enthusiasts for special causes.

Naturally, in view of the current ignorance of scientific facts and the interest in social amelioration, the emphasis tended to be placed on environmental rather than genetic improvement. An improved environment was widely desired, and something could be done about it quickly with immediate, if short-term, results. Many felt, too, that an improved environment would benefit hereditary factors, for it was widely believed that acquired characters were transmitted from one generation to the next. Improved heredity, on the other hand, was a long-term undertaking; the very idea was vague; and the simple-minded enthusiasm of some socially conscious Mendelians fostered the unwelcome notion that eugenic policy would entail the control by law of individual marriage and procreation. There were extremists on behalf of both environmental and genetic improvement within the Eugenics Education Society.

Fisher followed Galton in awareness of the new opportunities evolutionary theory provided in respect to man. To man, the most important product of evolution must be himself, his species; the evolutionary destiny of mankind had now, for the first time, been brought within the range of rational consideration, through the elucidation of the fundamental principles of its process. Man was in a new way the master of his fate.

Fisher's interest thus centered on the operation of biological laws: selection was perceptible in the nation if any class left more decendants relative to other classes. To ensure progressive improvement required only that the more admirable qualities of men should always be represented in a slightly greater proportion of the children than they had been in the parents, and this would happen wherever people of higher attainment had on the average slightly more children than those who were less successful. He knew that in England the reverse was true; the middle and upper classes were not so prolific as the lower classes; selection was favoring the wrong end of the scale. If the situation continued, it would lead progressively to the weakening of the nation in all the qualities that were most admired: strength and skill, imagination and intelligence, foresight, courage, and honor were all in the balance: the writing was on the wall. Given only that these qualities were to any extent heritable, and the work of Galton and Pearson had shown that they were, continued selection in this direction would do the rest. In man, however, selection operates through the medium of society, and that man can control—if he will.

Like Galton, Fisher felt the importance of presenting the ideas of eugenics to his fellows, so that they might be given the consideration they deserved. In his second year as an undergraduate, therefore, supported by his friends C. S. Stock of Clare and G. P. Balzarotti of Trinity, he began consulting senior members of the university about the possibility of forming a Cambridge University Eugenics Society. They were fortunate enough to obtain the assistance of Mr. W. C. Dampier Whetham (later Sir William Dampier), and a small meeting was called as a result of which the society was formed. Fisher was on the council along with Dampier Whetham and such notable men as

R. C. Punnett, John Maynard Keynes, and Horace Darwin. The first meeting was held in May 1911 at Emmanuel College. Prof. W. R. Inge presented an address on "The Menace of Shirked Parenthood," and Fisher proposed the vote of thanks, seconded by the Master of Emmanuel. Prof. A. C. Seward, in the chair, recounted how the society had been formed at the instigation of a few undergraduates whose desire was to interest and educate themselves and others in the subject of eugenics. Their second aim was to carry out certain pieces of research, although, as Prof. Seward remarked, they could not do a great deal of this work at first: some of their members, particularly the junior ones, had a great deal of other work to do.

That autumn the undergraduate committee of the Cambridge University Eugenics Society began meeting monthly in the rooms of undergraduate members at various colleges in turn. Fisher was chairman, Stock secretary of this committee. The first meeting, held in Fisher's rooms, was addressed by Stock. Fisher spoke at the second meeting, in November 1911, and again at the sixth undergraduate meeting in March 1912.

Membership of the society grew rapidly, including both undergraduates and College Fellows and, before Fisher left Cambridge, there were some 150 members. As early as February 1912 Major Leonard Darwin, who the previous year had become president of the Eugenics Education Society of London, came to Cambridge to address the society on "First Steps towards Eugenic Reform." This link with the London Society was facilitated, perhaps, by Horace Darwin who was a member of the council of the Cambridge University Eugenics Society. The link was strengthened that summer when the First International Eugenics Congress took place in London under Major Darwin's presidency, when Fisher, with other Cambridge members, served as stewards at the meetings. Fisher's leadership and his growth in stature within the eugenic movement are further demonstrated in that he was the principal speaker at the Second Annual Meeting of the Cambridge Society in November 1912, and the following year he was asked to present the same address to the Eugenics Society in London.

His success with the Cambridge Eugenics Society owed something to his personality, the enthusiasm and conviction he brought to the enterprise, and something also to the intellectual atmosphere of the times, which was receptive to eugenic ideas. His personal eminence owed more to the flow of his developing ideas, beginning with his firm grasp of the scientific principles and extending to his consideration of the special and complex evolution of the higher powers of man and the selective effects operating in society.

As we have seen, when he spoke at the second undergraduate committee meeting in 1911, he immediately introduced the scientific foundations of eugenics: expounding the principles of genetics and its relevance to eugenics; explaining the main concepts and methods of biometry as a technique complementary to Mendelian research, by which the continuous variables in man could be studied and by the understanding of which eugenic improvement could be effected in the population. He concluded [1]:

Biometrics then can effect a slow but sure improvement in the mental and physical status of the population; it can ensure a constant supply to meet the growing demand for men of high ability. The work will be slower and less complete than the almost miraculous effects of Mendelian synthesis but on the other hand, it can dispense with experimental breeding, and only requires that the mental powers should be closely examined in a uniform environment, for instance of the elementary schools, and that special facilities should be given to children of marked ability. Much has been done of late years to enable able children to rise in their social position; still we may as well remember that such work is worse than useless while the birth rate is lower in the classes to which they rise than in the classes from which they spring.

In this talk Fisher had brought his evolutionary interest into focus on the subject of eugenics, his mathematical interest into focus on statistics and his biological interest into focus on genetics. Eugenics, statistics, genetics: all three of them, like himself, had just started in life. They hardly predated the century; and although their potentialities were as yet unrealized, they were recognizably powerful young sciences, excitingly full of promise, puzzles, and confusions. It was difficult to keep a clear head among the tumultuous surge of hypothesis and conjecture, the special pleading on behalf of one aspect or another of each subject within itself, and the mutual hostility of workers in the different subjects. Individually incoherent, mutually incompatible, the most notable theories and theorists repelled each other. As a eugenist, as a geneticist, as a statistician, Fisher's task was to clarify the basic principles, to reveal the logical structure that controlled the empirical facts of each science, and to exhibit their mutual coherence. The synthesis between the three subjects was already made in his own mind when the Cambridge University Eugenics Society was formed.

□ □ □

In March 1912 he spoke again, this time on "Evolution and Society." He considered the opposite trends in human society—toward individual versatility and toward communal organization—and the ultimate choice selection makes between them, solely on the basis of greater efficiency in the struggle for existence. Practically nothing was known about how selection operates in human society; indeed, few people took selection seriously enough to consider it. Thus Fisher's perception of the scientific processes at work on man, as evidenced in philosophy, history, art, psychology, and sociology, was highly original. In this talk he formulated ideas regarding the evolution of conscience and the conquest of empires, which he was later to develop. A few paragraphs are quoted here, because they are his first formulation of these ideas; because he was just 22 and about to take his Tripos in Mathematics and, one might think, had no business with connections between the history of female infanticide, the most pleasurable experience of tigers, the sources of poetry, and the problem of free will; and because of a characteristic care about terminology—because, in effect, they give some measure of the man.

We may say that every form of symbiosis in the animal kingdom is paralleled usually with greater complexity and more perfect development in human society. The terminology too in the latter case is so much more varied and complete that it is at first difficult to see that all the modern social problems, for instance, of centralisation or decentralisation, of personal freedom or regimentation, of differentiation of the sexes and specialisation of the classes have been faced under other conditions in the animal kingdom and solved in nature's provisional, tentative way by the simple pragmatic method of trial and error. And it is worth noting that the solution which commends itself to nature and which is of interest to us as that which will be adopted in the future, is characterised not by the greatest happiness or by the most magnificent realization of human ideals of this age or of any other, or by any other practical consideration, but solely by fertility and power of survival.

An instinct from the external point of view is the tendency to perform some act or series of actions under the stimulus of a suitable train of circumstances. The term is rightly restricted to those acts which have some purpose by which the animal benefits directly or indirectly, in furthering some symbiotic alliance. From the psychological point of view it is a motive or desire depending upon the idea that man's estate is more desirable, more pleasant, more happy, if the instinct is obeyed than if it is not. Pleasure is nature's bribe to persuade a conscientious man to obey its instinct. The terms "pleasure", "happiness", "contentment" refer to states which differ in their duration and differ in their activity. It is as well to emphasize the similarity of their origin as due to the need of persuading a free will to conform to the courses that selection has found to be best.

Now, if the object were the greatest human happiness, would we succeed in producing a race whose instincts exactly coincided with their economic needs? It will help us to answer this question if we observe that the more complicated an instinct is and the more difficult to perform, the greater is the pleasure derived from it. Indeed it is necessary that an animal's interest should be centred on those objects which are hardest to obtain. The greater effort requires the greater reward. Among carnivorous animals the greatest problem is to obtain food and their highest pleasure seems to be hunting and eating. Among men selection seems to have acted more ruthlessly by failure to obtain a woman, especially over the immense periods during which female infanticide, often combined with polygamy, seems to have prevailed, and the result is that half our poets devote their labours to the pleasures of love. This consideration alone suggests that pleasure will be of a tepid nature if ever our instincts become easy to obey. But we have another sidelight on the problem. The very existence, real or apparent of free will implies a multitude of possible courses, a conflict of instincts. Whenever the course is inevitably followed, we shall have no choice, no need for motives, rewards and penalties, nothing but an automatic reflex action.

Thus, for Fisher, will and conscience were products in man's consciousness of his biological inheritance and susceptible to selective modification. As he said later that year [C P 3],

Not only the organisation and structure of the body, and the cruder physical impulses, but the whole constitution of our ethical and aesthetic nature, all the refinements of beauty, all the delicacy of our sense of beauty, our moral instincts of obedience and compassion, pity or indignation, our moments of religious awe, or mystical penetration—all have their biological significance, all (from a biological point of view) exist in virtue of their biological significance.

In considering whether ultimate human socialization would prove more efficient than the collaboration of individuals, he said, in part,

We may admit that efficiency in the petty duties prescribed for him by the state is a factor which may determine the usefulness of the ordinary man in times of peace, and it is possible when armies become more elaborately organised no higher qualities will be required of him in time of war; although, here, history is against our argument in showing several instances of enormous, wealthy, organised nations having broken themselves in trying to subdue some small, poor, high-spirited race to whom such social organisation would smack too much of servility and who valued their personal liberty more than wealth.

He dismissed the idea of the socialized community very briefly in his closing words, "Men bred and specialised for this purpose might be contented but they would not be men." But the success of "some small, poor, high-spirited race" in breaking the power of established civilization continued to haunt him, and it found a central place in his last talk to the Cambridge University Eugenics Society, at the Second Annual Meeting in November 1912.

Before this talk was repeated a year later before the Eugenics Education Society and published [C P 3, 1914], an article by J. A. Cobb appeared in the *Eugenics Review* in January 1913, to which Fisher gave prominence in the later talk; for, as he said in introduction, "the original conclusions to which I had arrived, primarily by inductive methods, appear to be strengthened by finding themselves incorporated in a much wider deductive scheme." This scheme he presented, with his original contribution as corollary, as follows:

What appears to be the underlying principle in the decadence of civilized races has been revealed in an article by Mr. J. A. Cobb, which, if my faith in it is justified, must be regarded as containing the greatest addition to our eugenic knowledge since the work of Galton.

Mr. Cobb points out that in any society which is so organised that members of small families enjoy a social advantage over members of large ones, the qualities of all kinds, physical, mental, and moral which go to make up what may be called "resultant sterility" tend, other things being equal, to rise steadily in the social scale; so that in such a society, the highest social strata, containing the finest representatives of ability, beauty, and taste which the nation can provide, will have, apart from individual inducements, the smallest proportion of descendants; and this dysgenic effect of social selection will extend throughout every class in which any degree of resultant sterility provides a social advantage.

It is this principle, vital in its importance and almost universal in its application, which explains to us why civilisations in the past, with one notable exception (Footnote: The Nordic civilisation of the 10th century, in which the ruling classes had very large families, and as we should expect from Cobb's principle, it seems to have been a very material social advantage to have many near relations.) and especially urban civilisations, in which the value of wealth is greatly accentuated, have ultimately collapsed owing to the decay of the ruling classes to which they owed their greatness and brilliance. And it is this principle which must underlie the reconstruction of our own civilisation if it is not to share the fate of those which have preceded it.

After the war he was to emphasize the immediacy of the eugenic danger undermining our own civilization. In 1913 he sounded this warning note lightly and not insistently. The talk as a whole was, rather, a clarion call to the higher good. He thought too well of men to doubt they would choose the eternal good of mankind rather than their own temporary and personal comfort. Seriously, goodhumoredly, by attributing to potential opponents the wisest and most honorable motives, he discovered them also to be ranged on his side.

The eugenic reconstruction of Western civilization required a simplification of lives and needs, but he had no quarrel with the enjoyment of wealth:

People indulge in luxuries not because they feel a need for them but in order to maintain themselves socially. And what is more, they make a very good investment; it is a cheap price to pay for the company of pleasant people; the pity is that so universal a tax should serve no useful purpose. . . . The only luxury worth having is that of a worthy human environment. . . . We require a new pride of birth . . . and a new confidence in our instinctive [as opposed to economic] judgements of human worth.

Although his thesis was based on genetical inheritance, he refused to quarrel with environmentalists:

The supposed conflict between environment and heredity is quite superficial; the two are connected by double ties: first that the surest and probably the quickest way to improve environment is to secure sound stock; and secondly that, for the eugenist, the best environment is that which effects the most rapid racial improvement. The or-' dinary social reformer sets out with a belief that no environment can be too good for humanity; it is without contradicting him that the eugenist may add that man can never be too good for his environment.

Though the practice of eugenics was a social undertaking, he did not admit the necessity of specialization, which implies the limitation of a man's function, with each segregated within his range. Man is born versatile:

We are as far as ever from any ultimate solution of the specialisation problem; but I would like to suggest that, for the moment, the problem is not too acute to be met by a greatly increased versatility—if possible actual adult versatility. So that not only in youth, but throughout life, we may retain full sympathy with our neighbours.

Above all, although the eugenic reform would be effected by free individual choice, he was confident of the support of all right-thinking men. The cause of the fall of civilizations having been uncovered apparently in the dysgenic effects of social selection, these effects would be prevented in the future by sincere eugenists:

We do not dub ourselves knights of a new order. But necessarily, inevitably, it might be unconsciously, we are the agents of a new phase of evolution. Eugenists will on the whole marry better than other people, higher ability, richer health, greater beauty. They will, on the whole, have more children than other people. Their biological type, characterised by their solicitude for human betterment, their scientific insight, above all their intense appreciation of human excellence, has a strong tendency to improve and to survive. Many will fail; many will forget; that is how we shall become more steadfast and more successful. And those that remain, an ever increasing number, absorbing more and more the best qualities of our race, will become fitted to spread abroad, not by precept only, but by example, the doctrine of a new, natural nobility of worth and birth.

These were stirring words, based on a convincing argument. They must have moved many of his young audience to intellectual assent. Yet how unworldly was his idealism, how unreasonably optimistic his hopes of humanity! What an innocent he seems, to place trust in his hearers' loyalty throughout life to a distant, impersonal goal. It would have been wonderful if any single person present on that occasion had lived up to his eugenic ideal and actually "sacrificed social success at the call of nobler instincts." But one did, Fisher himself, being, as his fellow eugenist, Stock, wrote 50 years later with emphasis, *"the only man I knew* to practice eugenics." His innocence was in supposing that other men were like himself, committed to follow the dictates of his reason.

□ □ □

The year 1912 was a busy one, with the continuing series of eugenics society meetings, the publication in April of Fisher's first mathematical paper, and his Tripos examination in June, from which he emerged a Wrangler with distinction in the optical paper in Schedule B. During the summer he was in touch with "Student" and sent him a mathematical proof of the formula for "Student's" distribution. He also led his band of Stewards to the International Eugenics Congress in London. Because he had been awarded a studentship in physics, he was able to return to Cambridge in October for a year of graduate study of statistical mechanics and quantum theory under James Jeans and the theory of errors under F. J. M. Stratton. At this time he was preparing a second mathematical paper [*C P* 2, 1913] for publication the following year.

In 1913 his studies and meetings continued until the end of the academic year, but he was again experiencing trouble with his eyes. In the winter he had fallen while skating and had had a slight concussion, with subsequent severe headaches, which were brought on by close work. He was advised to occupy himself out of doors and away from books, and thus he decided to use his remaining scholarship funds to spend the summer on a Canadian farm. It proved a rewarding experience for the young scholar. He shipped among immigrants from Liverpool and lived with them in close proximity during the slow, uncomfortable journey to Montreal. He worked on a farm near Winnipeg; he grew thinner, tougher; his fair skin burned and peeled and burned again. He experienced thirst, almost as a panic sensation, after sweating profusely under the harvest sun. The fine, dry dust seeped into his lungs, and he returned to England thin, sunburned, and coughing still. But he was young and fit and found the life good. He had unusual resources: when he collapsed on his bed after a day's work and closed his eyes, he loosed his mind to contemplate the fascinating complexity of conic sections. His training in mental geometry gave him a means of enjoying intellectual nourishment even in the absence of books.

He had a more practical problem to consider: how he should support himself now that he had left Cambridge. His real interests were in the eugenical and statistical work which were combined only at the Galton Laboratory, but he had yet to win recognition of Professor Pearson there. He had allied himself with the Eugenics Education Society, but the society did not offer full-time employment and he needed to earn a living at once. Farming was a good life, but, to set up as a farmer, he needed to make a large capital investment which he could not afford, and, with his inexperience of country life, it would be a big gamble. He had been offered a job at an aircraft factory when one of his college friends had taken him around the family concern.* It was a spontaneous offer made on the spot when they saw his ready interest and grasp of their problems; he had even written a little article (unpublished) "On the stability of veering flight," a highly mathematical treatment of a practical flying problem, but he did not feel confident that aircraft design would satisfy him permanently, and he had turned down the offer. There was no opening in the sort of area where he felt he might settle.

Temporarily, he could take one of the statistical jobs in business which were available to a man with a first-class degree in mathematics, and he may have hoped to discover something of statistical or eugenical interest in insurance work; but, again, he foresaw a different future for himself if the international situation deteriorated, as seemed all too likely. He was marking time until war should be declared; then he would join the regular army and, with his Officers' Training Corps and Territorial Army training, he would be ready to be sent at once into battle. Meanwhile, he took what he could get, a statistical job with the Mercantile and General Investment Company in London.

*Short's at Rochester.

2

In the Wilderness

From the time Fisher left Cambridge, 6 years were to pass before he found himself in a post where he could employ his abilities. From the age of 23 until he was nearly 30, he was isolated in an alien environment. After a triumphant school and university career he found himself unwanted and unappreciated in the world he longed to serve and, as he described it later, "an egregious failure in two occupations." After the companionship of college he found himself alone, excluded from participating in the lives of his friends because he was excluded from active service in the war. It was a testing time, for the old circle was broken, and he was barred from the circles he wished to enter. He needed all the self-sufficiency of his nature to make his isolation tolerable. It was the time, too, when two of the peripheral friendships of his college days sustained him and extended their scope: when the faith of a young woman and an old man in him personally, their full acceptance, lightened the darkness of the years; when the girl whom he called Gudruna stood by him in personal loneliness and Major Leonard Darwin stood by him professionally.

He was penniless when he left college. He took the job in the City of London and shared rooms with his college friend, Stock, in Talgarth Road, West Kensington. But he was not conformable to city ways: his employers were not so ready as he to accept his poverty, and one day he was called into a director's room and addressed from the sofa where Mr. Boulter was accustomed to relax after a heavy lunch. "Young man," the director said, "you are a statistician working for the Mercantile and General Investment Company. You must dress the part. Baggy grays and dirty sportscoat won't do here." Fisher returned to his rooms quite depressed; where was the money to come from? Stock proposed a way: "If you cannot get credit from Harrow Tailors, where will you get it?" So Fisher went to them and returned with top hat, tramline trousers, pale coat, and all. ("Even then," Stock recalls, I don't think they trusted him with a brolly.")

Next day they breakfasted and Stock bade him goodbye. Then, as he made for the door, Stock stopped him, "Don't you think Boulter will be annoyed if you turn up with *no tie?*" It was not the last time that friends were to save him from going out imperfectly equipped for the occasion. Sometimes the worst happened and he appeared, a white-haired professor at an evening function in bedroom slippers, or a Territorial Army Officer in 1914 setting off on maneuvers without his rifle.

For him the City job was merely a temporary way of earning his living; soldiering was important. All his Cambridge friends belonged to the Officers' Training Corps and realized the danger that world events were sweeping the country into war but Fisher was the only one who enjoyed army life for its own sake. As August 4 approached, he was more and more occupied with a future in the army. Gudruna remembers seeing him off to a training camp earlier that summer: "As he set off with his army boots dangling behind him, he was in triumphant spirits, like a boy loosed from school."

When war came, he presented himself as a volunteer, trained and ready to go. His medical test showed him A.1 on all points except his eyesight which rated C.5. He was rejected. He tried again and again to win admission; he offered to take three spare pairs of spectacles so as never to be at a loss for them. As late as 1918 he was still applying, but in vain. He had to stand by while his friends donned uniform and disappeared across the English Channel, while his brother arrived from Argentina and departed for France, while his sister with Voluntary Aid Detachment training was accepted in an emergency hospital on the South Coast and served as nursing sister, for a while in charge of a ward full of casualties from the front. All he could do for his country was to stand in for a man deemed more worthy than him of the honor of active service. In this spirit he became a school teacher; it was his war work, and until the end of the war he continued to do his duty. He taught physics and mathematics for a year each at Rugby, at Haileybury, and on H.M. Training Ship "Worcester" and for 2 years at Bradfield College in Kent.

Fisher was a poor teacher, a poor disciplinarian, for he did not have that resonance with his students that a teacher needs to gain a sympathetic hearing, and he failed to arouse curiosity about his subject. It was not enough to feel in himself the intellectual fascination of the ideas he taught: he needed to convey it to his students, and he could not do this by purely intellectual means. So he battered his head against a brick wall of boyish boredom and mischief or against passive incomprehension. And he hated it. There were moments when some boy responded with an unexpected flash of intelligence or interest, there were a few boys it was a pleasure to teach, but on the whole the classroom was purgatory for him.

He found he was hardly better off among his fellow teachers. Each had a narrow interest, bounded by the syllabus he taught and by the teaching craft, and was incurious beyond the set formula of ideas acceptable for examination purposes. It was an attitude of mind that barred intercourse with Fisher's broad synthetic approach and his speculative and challenging attitude to "the established facts." Again, there were exceptions, but he formed a low opinion of teachers as a professional body: to be incurious, uncritical, and at the same time didactic was to be oppressively dull and often misleading.

The depth of Fisher's feeling about schools, developing through the misery of these years, may be gauged from the fact that during the whole period of the education of his own eight children, he was never once persuaded to visit their schools and, with a single inescapable exception, never saw any of the staff at his home. School-teaching was a brackish side water to which his bad eyesight condemned him.

Early in the war, Gudruna writes,

Ron was missing his old friends, all now in the army and soon at the Front. He made no new ones. Wartime occupations made our meetings rare. He was teaching at Rugby; my husband was working for his finals; and I was running a Refugee Hostel in London. It was a relief for him to escape from teaching schoolboys and talk to some one about farming. But it was not easy to arrange circumstances which would give him a chance to be himself. In strange company he was awkward and tongue-tied—a young intellectual who was unused to society and found his ignorance excruciating. Whenever *savoir faire* was demanded, he lost his head and had recourse to rudeness. At petty annoyances he was liable to go berserk.

There were times too when his youthful high spirits were tried by the serious pretensions of adults, like the occasion when he invited me to attend the annual dinner of the Dalton Association. He looked embarrassed in unaccustomed evening clothes and masked his awkwardness with an arrogant air although the youngest present. I remember we sat opposite Dean Inge at the narrow dining table; and that the subsequent lecture so bored Ron that he led the way out past all the knees in the row, and escaped from the hall to race like a madman down the Fleet to the fresh air of the embankment. The poet received my description of this incident in the trenches, and it gave him a delighted fit of laughing. To him that was characteristic of "Piggy." Fellow Dalton Associates, however, must have limited themselves to deprecatory remarks about a young man who was completely oblivious to other peoples' feelings.

The shock of being refused admission to the regular army was a serious factor in his life. He could not get over it. In time he reconciled himself to the disappointment by a determination to farm as soon as he could afford it and by developing theories that farming provided the only normal life. It is difficult to imagine what a sergeant major would have made of the young scholar as a military recruit: Fisher's logical mind attached naturally to the basic principles and ultimate goals of action, and he was absent-minded and negligent of routine details, such as dress, except as they contributed to the primary objective; he was impatient of irrelevancies and of petty authority. It is no easier to imagine him turned farmer, a racehorse pulling a hay-wain. Yet, throughout the war, soldiering and farming were his two ambitions, and as one possibility faded, the other filled his vision.

The transition from one ambition to the other was consistent with his view of life. He had said that a man cannot be too good for his environment, and he meant what he said. He had experienced farming as a good life, a man's life, and he was anxious to serve as a man. If the soldier embodied the physical fortitude, the adventurous mind, and the courageous spirit of the primitive hunter or the Nordic explorer, the farmer embodied no less strength and endurance, alertness of observation, and depth of understanding of Nature's ways. He contributed no less to the excellent qualities of the race; indeed, he was no less essential to its survival, as soon became apparent when wartime difficulties in importing food laid the burden of feeding the people of England on the shoulders of the English farmers. Soldiering was an occasional, farming a perennial necessity, but they were similar in proving a man in action: his weakness could not be glossed over in the field by his possession of wealth or Blarney; there, the prentensions of social grace or useful learning were stripped away to reveal the real worth of a man.

He had said we need a new faith in our instinctive judgements of human worth, and he was anxious to justify his own self-estimate and the good opinion of his friends, being tried in action by nature's criterion of fitness to survive. Farming had preceded and would outlive the urban criterion of success. A farmer's reputation did not depend on a show of wealth. His wealth was buried with his seed: his acres wore glad rags for him, his animals won his prizes in life; and his children, brought up to robust health, to practical skills, and to the pertinacity of good workers in whatever field of endeavor they might choose, gave to the nation men and women of sound physical and moral traditions, aware of the roots from which their country drew its strength. Thus the idea of farming satisfied both his eugenic ideal and his desire to serve.

On a visit to Gudruna during his year at Rugby, Ron enlarged on the importance of subsistence farming to eugenics. Prior to the industrial revolution it had been the norm of rural life, practiced on farms and small holdings of all sizes. As distinct from commercial farming, it was a way of life not a way of earning money. Its policy was to grow all the crops needed for supporting the family and its livestock and to supply the necessary cash by selling a surplus of one product. It was the only profession in which a large family was a social advantage. Enterprising urban families, he said, could trust to their versatility in facing this life without the specialized knowledge of country people. Their children would profit. A subsistence farm offered them responsible jobs, obviously vital to the economy, and graded to their increasing powers, so offsetting the evils of an educational system with exclusively academic aims.

Frustrated by the teaching experience, Ron had come to think that practice of eugenics demanded from him an assay in subsistence farming. He spoke in terms of a farm of 100 acres or more; he made no reference to the capital necessary, and Gudruna knew him too well to trespass on the male enclave of finance. Finally, he asked if she would join him in the venture, taking over responsibility for dairy and pig breeding while he managed the arable land. At the time, Gudruna's marriage was breaking up. The life proposed was what she would like for the early education of her children. Ron's plea for the eugenic cause appealed to her, and she agreed to prepare for her part in the plan. She learned dairy work at the Dairy School of Reading University, and pig lore through a job at a Midland farm.

This decision was a remarkable instance of the dynamic power of Ron's arguments. Gudruna was a young intellectual, the oldest of a family of seven boys and two girls, gently reared in the Missionary Training College at Harley House in the East End, which her grandfather, Henry Grattan Guinness, had founded and her father of the same name continued. When war broke out Gudruna was 26, widely traveled, the author of a book on Peru, a freethinking Christian, the mother of two children, a very capable, charming, and intelligent young woman. But she had never laid hand to any kind of manual work. Ron's claims for versatility were certainly to be tested in the case of his first disciple.

□ □ □

The "eugenic life" would be incomplete until Fisher married, of course, and Gudruna saw more clearly than he that Ron was not at ease with women nor likely to become so in the sort of life he led. With the thought of helping him break his awkwardness with girls, therefore, she arranged that her younger sister, Ruth Eileen Guinness, should be with her if possible when he visited. Later, before either Ron or Eileen realized it, she foresaw that it might come to a match between them and tactfully encouraged this desirable outcome.

Ron had met Eileen first when she was hardly more than a child. In the year before war broke out, Mrs. Guinness paid a last visit to her aged mother in Tasmania and, during her absence, Gudruna returned with her family to keep an eye on her brothers and sisters at her parents' home, then at Sydenham Hill. This was a convenient plan for her husband who (like her brother Gerald at that time) was attending medical studies at Guy's Hospital. Eileen was 13, a tomboy playing Red Indians and scouting games with her brothers in the woods attached to Crescentwood Road and competing with them in cricket, croquet, stilt walking, and high jumping on the great lawn. She was also something of a dreamer who, in her nest at the top of one of the Cedars of Lebanon, looked over the wide world below and read poetry or sang her own compositions aloud among the birds.

The grown-up visitors of her sister were outsiders, subjected to tracking but usually above noticing it. Ron was different; he turned and chased his pursuers. One lad, Howard, fled up a cedar. Ron followed. Eileen grabbed at his foot as he climbed but, meeting his glance, and suddenly aware of her sex, she let go. They climbed up and up. Then Howard did the unrepeatable: climbing out to the far end of a branch, he dropped to the tip of the branch below. Ron, behind him, grabbed his coat and it tore in his hand, while the lad, below him now and out of reach, escaped to safety.

It was through her sister's hospitality that Ron and Eileen came to know each other on the rare occasions when they both visited Gudruna's home during the war. There Eileen felt afresh how different Ron was from her brothers: he had brains and what he did he did with zest, working hard, caring intensely, exploring eagerly—it was an adventure to be with him. An occasional correspondence sprang up between them. Once Eileen criticized her sister. (She had been shocked by the marital quarrelling and inwardly swore, if she married, never to answer her husband in such terms as Gudruna used.) Ron's reply startled her: he said she must never speak of her sister, his dear friend and hers, in blame. His letter, exhibiting his loyalty to the sister she loved, called forth the impetuous response that she loved him for backing his friend. His next letter flabbergasted her: did she, he then asked, love him enough to marry him? She answered, "No, she did not . . . yet," and he seemed content with that reply.

He did not court her when they met. There were no soft words, no urgent appeals. He walked with her, as he had with college friends, and sang the same songs, broadening her evangelical background by a hearty rendering of Belloc's [4] "Song of the Pelagian Heresy," which tells of the violence with which Germanus, Bishop of Auxerre, marshalled his flock into orthodoxy, and ends with a noisy reiteration of the last phrase of the verse:

> And thank the Lord
> For the temporal sword,
> And howling heretics too;
> And whatever good things
> Our Christendom brings,
> But especially barley brew!

Arriving at an inn for lunch, he introduced her to "barley brew," and they "put it away to infallible truth" to such effect that she fell asleep, overwhelmed after their walk in the cold by a Christmas dinner and half a pint of beer.

Later, they collected celandine flowers to classify the plants by petal numbers. Ron was already considering variation in relation to natural selection and, in particular, its measurement in terms of the intraclass and interclass correlation coefficients. In 1916 he could at least get a notion of the extent of these correlations in certain cases from his own observations and share his interest with the girl he intended to marry.

They made a memorable visit to Kew Gardens in all the beauty of springtime, and she treasured the volume on British trees he then gave her. They saw a performance of "The Magic Flute" in London and were both delighted when the Birdcatcher, Papageno, sang with his betrothed of their future joys, of adding "another little Papageno . . . another little Papagena" to their family. Even when Ron excused himself from her company during one of their brief meetings, the sympathy between them strengthened, for Eileen felt, as he did, that the correlation paper on which he wished to work that night took precedence over the ephemeral pleasure of an evening together.

□ □ □

In view of Ron's idealistic nature and his eugenic convictions, his choice of a wife is significant. It was wholly a love match, and yet it would be impossible to say that love or judgement most guided him, for they emerged together in his consciousness. Writing on sexual selection, in 1915 [C P 6], he gives a clue to his feelings.

In the infinite variety of human experience, which the past history of the race has afforded, in complete savagery, in nomadic barbarism, and in settled civilisation, both rural and urban, in warfare and love, in hunting, agriculture, lawsuits and commerce, the general grounds of our judgements of human excellence have built themselves up, and entwined themselves with our sexual nature. It is only under the influence, and during the growth of sexual attraction, that we can hope that these judgements will fully reveal themselves. Only the strongest passion could possibly free us from the bias of aesthetic and moral ideals, which accident and inappropriate teaching and surroundings may have ingrained in the character. It is probable that only the intense personal interest of growing love can arouse that acuteness of perception, that freedom and certainty of interpretation, by which alone the finer, rarer and more elusive traits of human excellence may be apprehended.

He was experiencing the intense personal interest of growing love, and he felt the heightened perceptions of the lover not as overwhelming his judgement but as releasing a judgement more basic to his nature than anything he had himself learned, more sure and instinctive, more sensitive and idealized.

Later in the same paper, he wrote:

In the deepest minds the idea of beauty links itself with one of altogether higher significance; in fact, with nothing less than the mystical appreciation of human personality. Here is the highest plane, and the source from which all our valuations in the lower categories take their value.

He felt his whole personality to be involved with the whole personality of Eileen. He perceived her beauty and her character not separately but as manifested together in her unique being. The reality of the synthesis in his mind is illustrated by the fact that he called her after the heroine of the old French Romance, *Aucussin and Nicolette*. It epitomized his feelings for her. The romance expressed, of course, the chivalric devotion he felt to the womanly ideal and exulted, as he did, in the physical beauty and sexuality of the young lovers. At the same time, Nicolette had courage, and it was he who pointed out that it was she and not the hero who took the initiatives which brought the lovers together. For Ron, the beauty and character were so fused that he expressed his pleasure in the poetic description of the fictional Nicolette's feet, whiter than the daisies she trod upon, as a compliment to the feet of his own Nicolette, which were actually broad and brown. Something in his Nicolette's whole personality made her feet worthy of poetic celebration in the same terms as those of the fictional heroine.

Just as he referred to Gudruna and addressed her by the name he had given her, so throughout their years together he knew Eileen (whose given names he disliked) by the name of Nicolette.

As Eileen grew into a tall adolescent she was unconscious of any particular beauty or ability in herself—her brothers left her few vain illusions about herself—and her lack of affectation or vanity was, in itself, charming. Left alone with Ron, and ashamed to be blushing, she raised her hands to her face and when Ron begged her "not to hide her lovely cheeks"; she was filled with confusion, unable to believe he meant he found her lovely, as she was. She had her father's vitality, his joy in living and in physical activity (all the family were keen athletes); his love of words, his imagination, and appreciation of beauty. She had her mother's ability for clear thinking and puzzle solving, and her shyness socially spontaneously yielded her husband first place. She had the emotional stability of a secure childhood during which she had never known her parents to quarrel. And she captivated Ron.

Ron himself seemed to have skipped the usual emotional developments of adolescence. The discovery of love took him by surprise and astonished him by its relevance to his inner life. He accepted it precipitantly, unquestioningly, with wonder and zest. He was like an enterprising city boy plunging into a country river, exploring the ecstasy of ordinary life. He found it crucial to the development of his ego. As he said in a moment of rare self-revelation, it was for him emergence from the dust of library shelves and an arming for the conflict which was before him.

He was fortunate in finding a suitable partner to his eugenic commitment, one who experienced with him the theoretical evolution of his ideas and was prepared to enjoy—and to endure—the results of eugenic practice. His large family, in particular, reared in conditions of great financial stringency, was a personal expression of his genetic and evolutionary convictions. Eileen took after her mother in her love of babies, of family—that was a part of their shared idealism. She also felt the need to give her life to some worthy cause, to serve some ideal, as she had seen her parents serve God. As a girl she could find no proof of the existence of God and could not center her life on Him. Then Ron presented himself, an idealist with very real and worthwhile ambitions, and she found her life's work in him, in his person, and in the pursuit with him of great aims for the increase of human excellence and the advance of scientific truth. His suit was won. Eileen left school to study poultry keeping and to attend agricultural college. She did love him. She would marry him.

Dr. Guinness had died in 1915, and Eileen knew her mother would not approve of her marriage, so young, to such a man as Ron. It was agreed, therefore, that they should be married from her sister's home without her mother's knowledge. They waited until her seventeenth birthday, and a few days later, on April 26, 1917, at the little church at Greenhithe, Kent, near to Rons's work on H.M. Training Ship "Worcester," Gudruna gave away her sister to a very happy groom. That day they went to Streatham and were received by George Fisher and Ron's sisters, Evie and Phyllis, who had prepared a wartime feast to celebrate the wedding. A few months later they moved to their new home, a former gamekeeper's cottage, Great House Cottage, Bradfield.

□ □ □

In the autumn of 1917 Ron began teaching at Bradfield College. He was still hoping to transfer from teaching to farming, and so he leased what had been the gamekeeper's cottage on the Bradfield estate, together with adjoining land, so that the family might gain experience of subsistence farming. Gudruna, with one little daughter, joined them in this enterprise. Eileen managed the poultry and garden, Gudruna the pig breeding and dairy work. Finances were tight, and they lived simply and frugally. No bought food, with half a dozen basic exceptions, was allowed. Throughout the war they lived without using their food coupons, and "feasted regally," in proportion as their cooking improved, on homemade bread and butter, cheese and jam, goat's milk, home-cured bacon and ham, eggs, and occasional pork and veal. Perhaps twice in a year they indulged in the extravagance of butcher's meat.

Ron was fully occupied with school duties, reviewing books for the Eugenics Society, publishing articles, and preparing material for the book he was later to publish in 1930. He could not hope to take part in the routine of the holding. But he outlined its economic plan, incited every innovation made, and refrained from criticism of amateurish efforts by Eileen and Gudruna.

The farming was a hobby providing pleasure and exercise for his weekends and school holidays. He began by building a pigsty, using the foot-square timbers from the old coach house. Then he cut down a piece of woodland to make space for a pightle, or pigyard, which he fenced with the wood he had cut. The wood supplied material with which he built another log shelter for pigs and a small henhouse; and, as the trees were felled by the dozen, spare wood was cut up and stacked to dry for burning in the kitchen grate. Sometimes a schoolboy would be summoned to take his punishment for school offenses in the form of struggling at one end of the crosscut saw, with Ron at the other end. George Fisher visited often and gladly took his punishment at the same task.

Ron and his father were a pair: physically they matched, in their medium height, broad shoulders, and wiry frame; in his zest for work, George Fisher was not outdone; and when they paused, the conversation between them, on all sorts of subjects, took up the flow; they were equal in ranging curiosity and companionship. Before the wood cutting was completed, Ron also began to dig the garden, double-trenching the whole area within the year.

With the henhouse built, they bought half a dozen pullets, which Ron characteristically named after other females forsaken by their menfolk: Clytemnestra, Ariadne, Dido, Persephone, Penelope. Sophie, the sow, foraged around in her pightle; Marsyas, the goat, and his spouse were tethered out to graze, later joined by a calf which Ron had been pleased to acquire illegally (because in this case the law deserved to be broken) from a farmer who was permitted neither to kill nor to sell the calf for which he could not legally obtain feed. Naturally, Ron broke the law again in procuring feed for the calf.

Sometimes the animals escaped. The calf pulled up its stake and careered through the vegetable garden, terrified by the stake crashing behind it and, getting the rope entangled at last with the runner bean stakes, stood waiting to be rescued by Eileen who saw, between laughter and pity, the clumsy innocent standing in the garden he had wasted, terror in his great soft eyes and ruin in his wake. At a later date, Sophie the sow was found nosing into the galvanized tub where their new-butchered meat lay temporarily in the yard, and Eileen, by that time near term and heavy with her first child, took up a stave and struggled with the huge beast, finally beat her off, and stood a trembling guard until Gudruna returned from an expedition (illegally, of course) to sell other joints in the village. No expectant mother ever enjoyed her chitterlings more than Eileen did those she had rescued by her epic struggle with the sow.

Soon Ron began to make innovations. He took great pleasure in Sophie, enjoying the simplicity of her animal appetites and the directness of their gratification. He sympathized with her contentment, grunted out as he scratched her back with a rake; with her hunger for food and her immediate pleasure in eating; with her fecundity and evident satisfaction in maternity, and the artistry with which she circled round, gradually letting herself down into a comfortable position to give suck, while avoiding the danger of crushing the piglets clamoring around her. She was a magnificent mother, and he admired how, even as she farrowed, the newborn piglets attacked her teats and afterward how quickly they grew. She was a triumph of evolved instinct and physiology.

Ron began daily weighing of the piglets to estimate the sow's milk yield. Since sucking pigs depend on their mother's milking capacity, it was a quality for which pigs should perhaps be selected and bred. So each piglet in turn was caught, squealing and kicking, placed in a netting bag, and suspended from a spring balance. The weight indicated by the leaping needle was set down, and the thrashing hooves of one piglet were disentagled from the net to make room for the next. Sophie's milk yield was prodigious. Ever afterward Ron showed respectful affection for a sow.

Again, in dairy work, Ron had ideas for more economical use of limited resources. There was a wartime shortage of sugar, and they attempted to boil whey to a syrup to supply this need, without success. Butter was also short, and Ron proposed, therefore, to skim off the butterfat to be made into butter, then to replace the fat content of the skim milk by added vegetable fat and to use the supplemented skim milk for cheese making. Emulsification of the milk with the added fat was a problem, but so keen was he that Eileen and Gudruna raised £100 from their slender capital and bought a homogenizer. In the outcome Ron's enthusiasm led to useless expenditure of capital. He could not possibly spare the time to experiment with the huge, complicated machine. When it arrived and was set up in a room of its own, he was persuaded to glance at it, but he never again entered the room or referred to the homogenizer. The experiment in cheese making was never tried, and the machine was eventually donated to the Dairy Research Institute at Reading. The farming was in every sense experimental!

Meanwhile, the routine work of the small holding devolved on the women. Judged by suburban standards the home would have appeared more than a little squalid and the life one of unrelieved drudgery, but they were proud of the experiment and competed eagerly to make it a success. Their only care was to accomplish the daily round without impinging on the evening hours, which were reserved for reading and scientific discussion.

The possibility of initial expenditure on adapting Great House Cottage to dairy work was never mooted. Yet, in the unoccupied rooms of two cottages they were proposing to carry on the crafts traditional in an eighteenth century farm house with dairy, cellars, curing room with slate shelves for dry salting, tubs for brine, and an available smokehouse. In a kitchen without any adequate oven, with no running hot water, no refrigerator or washing machine, they had to learn to compress domestic chores into the fewest possible hours daily to leave time for their farming duties. They had to be able to handle the heavy cooking incidental to the killing of a pig or calf and to produce dishes of which a surplus would be sold.

The main difficulty in learning rural crafts from books, they found, was the unreliability of traditional recipes. By following 'old wives' tales' one could never be sure how the final product would turn out. Even with good facilities, a standard quality could not be achieved. At the Guinness brewery of their cousins in Dublin unsuccessful brews had to be drained into the Liffey, just as in Huntingdonshire cartloads of spoiled Stilton cheese had to be tipped into the river. (Reading University had just offered Gudruna facilities if she would undertake research on Stilton.) No one knew the nature of the changes that took place in cheese or wine or bacon hung to cure; thus no one could control the processes of manufacture. Important variables like the temperature of brewing and storing were not fully under control and often there were no instruments available for measuring chemical and bacteriological workings.

The facilities at Great House Cottage were minimal. Eileen still remembers a side of bacon that was long tasted with suspicion and eventually discarded, an optimistically "Stilton" cheese that was thrown to the pigs, and a brew of mead that burst its bottles and cascaded down the staircase. On the other hand, mouths still water at the memory of the best Great House Stilton and Port de Salut, its home cured hams, homemade sausages, Bath chaps (pig's cheeks), raised pork pies, and brawns (head cheese). Even at his cottage table Ron was made aware of the practical effects of variability in biological material.

To the young people the cottage with its garden and buildings was idyllic. It lay a mile from the nearest building, among woods thick with white violets and periwinkle and populous with birds. When occasion demanded, they rode a bicycle over the railway sleepers (or ties) which paved the long driveway to the road leading to the village (a practice that put an end to Eileen's first pregnancy), but for the most part the cottage was a world in itself for the women, and it absorbed all their powers. They had no outings, no neighbors, no radio; their only paper was *The Farmer and Stockbreeder;* yet their social life was satisfying.

There were occasional visits from the world outside. Old Mr. Fisher paid several visits. He rejoiced in his son and especially in his daughter-in-law: when a tree was ready to be felled, she was the "strong man" on whom he called to bring it down; when she put on her pretty frock, she was "the Queen of Sheba" shimmering in beauty. His flattery was full of love and pride. When his grandson, George, was born, he was touchingly proud and delighted. Ron's laconic telegram, stating no more than the sex, weight, and date of arrival, brought Mr. Fisher down to Bradfield bearing the few pieces of silver he had reserved until his seventy-sixth year for the grandson who had, at last, arrived.

There was an evening, too, when they entertained the senior classics master at Bradfield and his wife. Since it was St. John's Eve, they re-enacted for them the dance of the feast of the lovers of St. John, Ron leaping over the bonfire in fine style in this curious ritual, whose symbolic significance and primitive associations doubtless added piquancy to Ron's pleasure in his chosen role in its celebration.

□ □ □

In the evenings they retired to the sitting room and there read aloud. From the beginning the ladies took the lion's share of reading to relieve Ron's eyes. The twelve volumes of Frazer's *Golden Bough* were followed by Burton's *Arabia* (unexpurgated) and other readings from the same author, Doughty's *Arabia Deserta,* Gibbon's *Decline and Fall of the Roman Empire* (complete), Prescott's *History of the Conquest of Mexico,* and other volumes borrowed from the London Library. They read various Norse sagas: *Gisli the Outlaw, Burnt Njal,* and the collected stories translated by William Morris. Once, after a visit to London, Ron brought back three exquisitely produced books: *Aucassin and Nicolette,* in compliment to Eileen, and *Ancient Arabian Poetry* and *Albarelli** in acknowledgement of the research Gudruna was doing for him on changes in the distribution of a high birthrate in Arabian society. *Albarelli* had notable plates, and the pages of *Aucassin and Nicolette* were imitations of mediaeval manuscript, with jewel-like illustrations and borders decorated with gold.

In the main, the focus of their reading was on the rise and fall of civilizations. The Arabian civilization held a particular fascination for Ron: it was relatively well documented, with initially brilliant military and intellectual achievements, followed within a few short centuries by a stagnation which had

---

*Albarelli are exquisite majolica jars designed by Saracenic potters to hold preserves. Introduced into Europe through Sicily and Spain, they inspired Italian majolica, which became very elaborate and valuable but was never as choice as the original whose technical secrets were never all discovered.

allowed the Arabian peoples in all the civilized arts to fall progressively behind the European nations to whom they had once supplied the very elements of their civilization.

With such topics to occupy them, they had none of the small talk—of weather, domestic environment, health, finance, and daily news—which makes up so much of ordinary human converse. Nor did they share with each other their earlier experiences and relationships. Ron never mentioned his mother, his childhood, his visit to Winnipeg, his job in the City, or the conditions of his teaching work; nor did he show any interest in his wife's family or her past. This was always to be his way. His brother Alwyn was killed in France in 1915, his sister Evie died in 1917, his father in 1920. Though he had loved them in their lives, he did not refer to them afterward. His reticence, his apparent indifference not only to the past but to so much of current life, made him appear self-contained.

It was obviously a relief and pleasure to him to discuss his current thoughts. At Great House Cottage he shared the development of his eugenic ideas so that the ladies knew them intimately from their inception. The intensity of his feelings, his thought, and his application was irrepressible. He needed an audience on whom to try out his ideas and to share the joy of their exploration, yet, ultimately, in this, too, he was alone, self-contained in proposing and in criticizing his formulation of the theory.

Moreover, he expected others to be as self-contained as himself, as is evident from his strong opinions on childrearing. Surprisingly in one who trusted instinct as much as he did, he underestimated the wisdom of maternal instinct. He believed that parents should ignore the physical and psychological limitations of children: the children should discover their own limitations as they discovered the physical world around them, how to manage themselves in it and how to adapt to the environment of an adult society. Danger to life and limb were natural, children would learn judgement by experience and were not to be frightened into timidity by expressions of parental affection, anxieties, and fears (he did not add common sense). Even a baby a few months old could learn to cry if it were rewarded by instant loving care. Eileen, therefore, was prevented, forcibly if necessary, from such self-indulgent protectiveness and was continually guarded against yielding to the more soft-hearted maternal feelings.

The child was to be treated as an autonomous individual from the beginning. Consistent with this view, Ron was often seen to give his full attention to childish opinions and answer them seriously and simply, according the child a respect which fonder parents often fail to give. On occasions, outside his own family, he would support a young rebel's right to his own opinion or action when it opposed parental authority; in his view the child had as much right to his choice as his elders had to theirs.

This mode of upbringing, rigidly enforced, could be traumatic for mother and child, as Gudruna and her daughter Kestrel found. For instance, when Kestrel was 3 years old, Ron insisted that she should be trusted to deliver a Port de Salut at a house in the village, a mile away. For three hours Gudruna controlled her maternal anxiety and then set out in search of the child. Kestrel was wandering homeward from the direction of a wood where she had been inveigled by a band of older village children. She could not describe the interrogation to which she had been subjected but was shaken by the sudden introduction to life among other children.

Ron's view of the educational value of life on a subsistence farm was justified in Kestrel's case. At the age of 5 she was in charge of the goats and could, unaided, devise the arrangement of the shed for a parturition and assist efficiently at the birth. Yet Gudruna realized that Kestrel was beginning to suffer from Ron's immature theories, as she was herself from his interventions in matters of educational principle.

At this time Gudruna was harrassed by a series of visits from inspectors of education, who would not admit that she was qualified to teach her 5 year old. She realized that while she lived in a workman's cottage, she would be forced to submit Kestrel to the national educational system. She decided, therefore, without further delay to found her own progressive boarding school, where the child would have companionship and enlightened teachers. She left the Fisher household in 1920 and successfully carried out her plan.

□ □ □

What Gudruna and Eileen gave to Ron in support and faith in a personally lonely period of his life, Major Leonard Darwin gave him in his professional isolation. Major Darwin, a younger son of Charles Darwin, had retired early from an army career and had interested himself as an amateur in scientific problems raised by the work of his father and Francis Galton. In 1911 he accepted the office of honorary president of the Eugenics Education Society, from which Galton had retired in 1909. He was to be President until 1928, his eighty-ninth year, when he resigned his difficult task.

In 1912, Fisher had met the president personally and had become a member of the Eugenics Education Society. In the autumn of 1913, when Fisher and Stock had left college and were sharing rooms in London, they both attended the meetings and interested themselves in serving the society, and in 1914 the first of many book reviews bearing Fisher's name appeared in the *Eugenics Review*. Perhaps the acquaintance with Major Darwin was no more, at this time, than one of mutual respect between associates in a common cause, but before the World War I was over Major Darwin had conceived a very great faith in Fisher's character and ability, and Fisher honoured, loved, and revered him.

Fisher was predisposed to revere Darwin for his name and to respect him in his office, but it was Darwin's character which won his unwavering adherence. Major Darwin's personal humility made Fisher his champion, his sincerity made him his companion, and his appreciation of others—of Fisher himself—made him his devoted supporter. It is significant that it was Darwin, whose mind boggled over mathematics, who believed in the genius of the young mathematician; Darwin, the amateur biologist, who appreciated the value of his genetical contributions, for all their mathematics, at a time when Fisher was an obscure schoolteacher and his contributions largely rejected by the professionals. Major Darwin, by his faith and his personal generosity, stood godfather to Fisher's scientific career, and in the 30 years of deepening friendship until Darwin's death in 1942, Fisher honored him as a father.

With Darwin, he reflected some of Darwin's character. He responded to Darwin's tender consideration of others by himself showing rare consideration for all that concerned Darwin's feelings. Because Darwin shrank from controversy, Fisher curbed his impatience with difficult personalities in the Eugenics Society to a remarkable degree to save Darwin from strife. Though he himself was involved in professional controversy, he modified the acerbity of his public statements to meet Darwin's standard of personal courtesy. Both men were of a generous nature, and both were sincerely concerned for eugenics. One can hardly say that it was on Darwin's account that Fisher served the Eugenics Education Society loyally through 20 years of growing frustration and disillusionment, but in so doing he certainly expressed his commitment not only to eugenics but to Darwin personally.

The collaboration between the old man and the youth began with the book reviews, and over 200 such reviews appeared between 1914 and 1934 while Fisher was a regular reviewer for the *Eugenics Review*. It was an excellent way of keeping a finger on the pulse of new developments in the subjects that interested him; not only current views of eugenical topics but news of archeology, anthropology, ethnology, genetics, statistics, sociology, and medicine came under his eye. A gradual increase is perceptible in the proportion of genetical and statistical work reviewed, and this may have been a result of Fisher's reaction against the wide currency of biological opinion and eugenical propaganda that lacked a sound scientific basis. It was also a side effect of the financial straits of the Eugenics Society during the 1920s, on which account Fisher volunteered to undertake for the journal all of the reviewing of the statistical and genetical literature.

As an assistant master at Rugby in 1914–1915, Fisher could not afford to come up to London to work with the society until Major Darwin appointed him to part-time position and paid him a salary. When Major Darwin renewed the offer of part-time work toward the close of 1915, his tact camouflaged the fact that his motives included helping a young man without private resources:

(undated)

My Dear Fisher,

I am sorry not to have been able to write to you before concerning the Eugenics Educ. Soc. The Society could and would pay you a salary at the rate of £100 a year for the year 1916 if you were to continue to work for it. You will naturally ask me what you would be expected to do. I do not think it any use trying to lay this down very precisely. To give you an indication what is in our minds, I suggest that you should aim at giving us about one quarter of your working time, this time to be spent on the Society's business or, in default of that, on eugenic investigation. There is no hurry in coming to a decision on this matter.

I have not yet heard whether you will be able to come up one day a week this term. I hope so.

Whether it is worth your while staying on next year you must judge. I should be very glad if you were sticking to eugenics.

Yours sincerely,

Leonard Darwin

Fisher was certainly sticking to eugenics and to his president. The first year's work had already given them a new respect for each other, and it had pointed up the need for an investigation into the genetical relevance of the correlation coefficient. Fisher had taken up the statistical problem of deriving the distribution of the correlation coefficient. His paper, submitted at the outbreak of war in 1914, appeared in *Biometrika* in June 1915. But he found there had been in 1913 a difference of opinion between Major Darwin and Professor Pearson about how the hereditary situation would affect the value of the correlation coefficient in the hypothetical case of a stable environment. Fisher wrote to Darwin in August 1915, vindicating Darwin's position and urging that Pearson's judgement should be challenged. Darwin, however, had accepted Pearson's strictures in 1913 and at once had publicly retracted his statement. He would not reopen the controversy, since he felt uncertain of his facts and wished to save the Eugenics Society from becoming embroiled in a personal feud. He did not forbid Fisher to write but asked, "Wherefore? It would make Pearson your enemy, I fear, and that should not be forgotten." And he made it clear that although he would like to see himself proved right, he, as president, could not accept an article for the *Eugenics Review* unless he could get assurances that Fisher was "on the right tack." This step precluded for the time being the possibility of Fisher's publishing. The interchange, however, showed each the quality of the other: Fisher appreciated Darwin's scientific perception and his lack of self-seeking, and Darwin appreciated Fisher's scientific understanding and his immediate impulse to correct what he felt to be an abuse of science and of justice. In addition, the need was indicated for a thorough investigation of the area of the controversy, and Fisher began the work which was to lead to his paper on the correlation between relatives [C P 9], eventually published in 1918.

□ □ □

Fisher wanted to work out whether the correlation coefficients for the continuous variables in man, which had been estimated for various physical measurements and between various types of relatives, were consistent with a Mendelian scheme of inheritance. It was a statistical problem: the welding together of Mendelism and biometry.

Charles Darwin had seen the small differences of continuous variation as the raw material of adaptive change. Galton had shown such variation to be heritable. Yule and others had pointed out that although in apparent contrast to the sharp differences from whose segregation Mendel's rules were inferred, the basic mechanism of this continuous variation need not be dissimilar, provided it was assumed that the expression of the character depended on the simultaneous action of many genes whose effects were additive. This Fisher also had perceived, as we have seen, at least by 1911. Pearson, however, disputed that the correlations observed between human relatives could be interpreted successfully on this basis. In particular, in the restricted case he had considered in 1903 [2] (where the character was supposed to be determined by a number of equally important Mendelian factors, the dominant and recessive phases being present in equal numbers and the different factors combining their effects by simple addition), Pearson had found that the expected correlation coefficients worked out uniformly too low. Yule [3] (1906), working on the same problem, had shown that by assuming absence of dominance, better agreement between theoretical and observed values could be obtained, but he had not been able to distinguish between the effects of dominance and environment in reducing the correlations.

Fisher set out in characteristic fashion to consider a more general case than had been studied previously. From the simple general case he was able to branch out to consider the complexities introduced by assortative mating, the interdependence of factors, genetic linkage, and multiple allelomorphs. The resultant paper [C P 9] was to lay the foundations of what is now called "biometrical genetics," establishing its methodology and fundamental concepts. Fisher himself only once again wrote on this subject, in connection with a particular set of data collected by O. Tedin, the analysis of which was given in a joint paper in 1932 [*C P* 96, with F. R. Immer and O. Tedin]. Perhaps once the basic relationship between the genes and the biometrical measures had been established, he found the investigation of particular genes more intriguing as well as more readily available to study. He undertook the first exercise in biometrical genetics because he was interested in human heredity, and this was the approach by which he could elucidate its mechanism.

Technically the paper [*C P* 9] was important in introducing the word *variance* into the statistical language and the concept of the analysis of variance components into its technology. To analyze human variability Fisher chose the squared standard deviation, or variance, because of its additive property: the variances contributed by independent causes of variability in the population sum to the total variance of the population. Many phenomena previously expressed in terms of correlation could be more clearly thought of in terms of variance components. By his distinctive naming of this measure Fisher intended that "the elementary ideas at the basis of the calculus of correlations" should be "clearly understood and easily expressed in ordinary language," while the ambiguities of current terminology were excluded.

His treatment resolved what had been controversial since the turn of the century. He showed not only that the correlation observed between relatives could be interpreted successfully on the supposition of Mendelian inheritance but that Mendelian inheritance must, in fact, lead to just the kind of correlations observed. Thus, he showed how assortative mating (homogomy) would increase correlations that Pearson, assuming complete dominance, had found to be too low. He showed how the correlations could be used to partition the variation into its heritable and nonheritable fractions, how the heritable fraction could itself be broken down into further fractions relatable to additive gene action, to dominance and to genic interaction, and how due allowance could be made for the correlation observed between spouses. Finally, he pointed out that the excess of the sib correlation over that found between parent and offspring, otherwise inexplicable, must follow from the phenomenon of dominance. After this paper there could be little doubt that the inheritance of continuous variation was entirely consistent with Mendelian principles.

The large effects due to dominance accounted for about 32% of the total variance. The proportion was found to depend on two ratios: that of the (unknown) gene frequencies for the pairs of factors involved and that of the (unknown) amount of dominance. As these two ratios were varied, the proportion could, theoretically, take any value between zero and unity. It was impossible to guess why the dominance ratio should have the value actually found, for, as Fisher then wrote, "We know practically nothing about the frequency distributions of these two ratios. The conditions under which Mendelian factors arise, disappear or become modified are unknown." In fact, the next major genetic problem Fisher was to tackle concerned precisely these questions, and by their elucidation he explained in the general case why dominance might be expected commonly to contribute nearly one third to the total variance in human measurements (*C P* 24).

□ □ □

In these war years Fisher reviewed a number of articles and books that condemned war as dysgenic. Most of these bore a pacifist complexion and condemned "militarism" in the same breath with war, as if the two were indistinguishable in their nature and consequences. One early review quotes the author's extreme claim that: "Militarism fights for, and supports, the lowest instincts of life; pacifism the highest" [6]. In more rational treatments, however, the same confusion was apparent, as in another phrase quoted by Fisher, "the consequences of war and militarism generally on the constitution of the race. . . ." [7]. To these Fisher responded by spelling out what "militarism" signified: training for war and the military virtues and spiritual tradition of a fighting race. These were, visibly, qualities that had enabled nations to conquer and to establish civilizations in the past. "Politically they enable a nation to sustain the burden of war, often to avert it, and at the worst to recover from it."

In contrast, war itself has dysgenic effects, and Fisher was concerned that racial repair after the war should not be neglected. In a short article in 1915 [8] he asserted that war was inflicting a real and selective injury on the nation:

A great body of men, selected from every part of the nation for three precious gifts, of health, courage and patriotism, have been subjected to an excessive mortality and . . . withdrawn from their share in reproduction.

Turning to practical means of minimizing the evil, he continued:

We are faced with two problems. How can wars be made less frequent? How can they be made less dysgenic? The first is simply a question of foreign policy . . . We rely on the general experience of the past in suggesting that that nation will suffer fewest wars which is so strong that it cannot be despised and so just that it cannot be suspected. Our policy must be to maintain the standards of national strength and honour so that we may preserve the confidence of our friends and the respect of our enemies. This much is so self-evident that it must be considered to be the aim of all sane politicians.

As to the second problem, conscription would make wars less dysgenic, yet it would still be necessary to select soldiers for their physical fitness, and thus war must always be dysgenic. Since wars were occasionally inevitable and inevitably dysgenic, should we not budget for their human cost? Victory in the current war would yield the liberty of our nation to pursue unhindered our own ideals. Such liberty may be used or wasted. It will be used if our nation is so organised in time of peace as to replace the men who have died in war by their own stock, bearing their own virile qualities. By so doing we shall save up in time of peace, for the inevitable cost of war.

He calculated that a moderate increase in reproduction of the men fighting the war would suffice. This might be achieved by recognizing in economic terms that men who had won distinctions and honors in war would serve their nation in peace by having larger families. Their pensions should be substantially increased for every child: "it is desirable that such men should be better off rather than worse off for each additional child they beget." Officers in the army and navy should be given financial inducements to offset their professional difficulties in marrying young and raising families. Similarly, among noncommissioned officers and men, the army system of allowing certain men to marry on the strength could be extended to a larger proportion of selected men.

This was a long-headed view of the problem, one curiously aloof from the current emotional reaction to the shock of losses during the first year or so of war. Personally, Fisher suffered like others: his brother Alwyn had recently been killed at the front, the ranks of his companions at college were already thinning. That loss, to him and to his whole generation, was irreparable. But Fisher felt it also for the following generations which had lost the companionship, talent, and leadership which would never be born. It is not surprising that the general reaction was, by whatever means, to make this war the last, that it became a crusade, The War to End Wars. It was perhaps surprising, in this atmosphere, that Fisher was contemplating the possibility of recurrence of war and hoping, over the long haul, to budget for its human cost.

Eugenics is an exercise in foresight, however, and the eugenist needs to look generations ahead. Quite a slight change of the differential in birth rates, like the modest increase Fisher proposed in the families of veterans, would have significant consequences for the nation. The dysgenic effects of war, being recognized, could be repaired. The racial changes taking place in peace were no less important, and dysgenic effects needed to be recognized and repaired in a similar manner.

In 1916 Paul Popenoe, editor of the *Journal of Heredity*, advanced a series of claims in support of the thesis that conditions of peace tended to produce an improvement in the race. In his review [9] Fisher took the opportunity to quote these and then, phrase by phrase, to deny them:

The whole passage shows a flagrant neglect of the well-established facts; that the poorer classes marry earlier and are more prolific than the rich and middle classes; that in the same class the most efficient and ambitious are the most tempted to postpone marriage and do in fact, have the fewest children and most clearly of all, that the dwellers in slums not only bear but rear a far larger number of children than those in less crowded districts. On any unbiassed judgement the conditions in the years preceding the war were very highly dysgenic.

Indeed, all the facts suggested that the distribution of the birthrate before the war was already draining the nation of its best abilities.

In an article on positive eugenics in 1917 (*C P* 8), Fisher considered the problem of devising practical means by which the middle classes might be encouraged to have more children. As he had proposed financial inducements for professional soldiers and veterans, so now he proposed financial measures to counteract the financial disadvantages suffered by parents, especially parents of large families, in comparison with their childless professional equals. He pointed out that it was among professional men and skilled artisans, where the low birthrate was most alarming, that organizations existed that could implement such measures. Professions such as medicine and the law had formed professional associations which set the academic and ethical standards for entry and maintained them, where necessary, by expulsion. Trade unions also exerted considerable influence on their membership. Such bodies might initiate programs of eugenic self-help within their membership. Believing that the root cause of the dysgenic selection was the difficulty of giving to children of the professional classes an adequate start in life, he suggested that professional bodies might, for example, endow public school scholarships such as were already offered to the sons of clergymen, extend mutual interchange of professional services such as was offered between the families of doctors, and provide special facilities for the long professional training of children of members.

Whereas in 1913 he had called on individual eugenists to sacrifice social success at the call of nobler instincts, by 1917 he was, in effect, calling on childless members of the whole profession, as a contribution to the future of their profession, to assume a share of the sacrifice made by members who were parents. "Such work can only be done by the professional bodies themselves," he insisted, but he was sure a range of possibilities would suggest itself to the bodies concerned, to suit their own organizations.

In reviewing various eugenic schemes, he reiterated his conviction, both on moral and utilitarian grounds, that eugenic policies should not involve direct interference from outside. Regarding one scheme that hinted at external regulation of marriage and procreation he commented dryly [10], "We feel that eugenic ends would be as successfully attained without undermining the independence and self-reliance of the citizen." Condemning another, because "only direct interference is contemplated," he wrote [11]:

There is nothing against which eugenists should guard more carefully than this narrow interpretation of the possible methods by which the race might be improved. Every change of law or custom is in some measure eugenic or dysgenic and it is the duty of eugenists not to strive for some fundamental remodelling of society . . . but to seize every opportunity of emphasizing the eugenic aspect of any proposed measure, and of creating a public opinion which will make obviously dysgenic legislation impossible.

His own suggestions for eugenic policy aimed to modify external circumstances so that the motivation of an independent and self-reliant citizen to have children would be increased. By decreasing the penalties of parenthood, he hoped to increase the balance of pleasure in his equation "Pleasure is Nature's bribe to persuade a conscientious man to obey its instinct."

□ □ □

In 1915 Fisher published a short article on the evolution of sexual selection [C P 6]. As happens not infrequently in his genetical work, he began with some observations made by Charles Darwin. Having dismissed Wallace's objections to Darwin's theory, he considered how sexual selection would occur in practice. The argument is qualitative: that animals have more or less conspicuous features (like large tail feathers or bright colour) and that these are in some cases a real guide to the health of the individual, that, therefore, a female who selects a suitor on account of such a feature profits her line, and that a tendency to prefer suitors in which the feature is well developed will, in consequence of natural selection, become firmly established among female instincts. Sexual selection is thus initiated.

Sexual selection will result in the increasing perfection of the feature itself, since males in which it is best developed will be preferred as mates. At the same time, female preference for it will become more pronounced. Even though the feature ceases to be a good indicator of natural superiority, the taste for it will increase through sexual selection until it has become so harmful as to cancel this advantage; only when natural selection kills the excess of individuals produced by the effect of sexual selection will the female taste begin to diminish and an equilibrium be established. Finally, the equilibrium may be broken by the rise of other features of interest and importance, and this will be followed by a gradual decay both of the original feature itself and the taste for it in the opposite sex.

Man being subject to the same process, Fisher considered how his greater mental and moral development led to even more interesting evolutionary possibilities. In the human mind the conception of desirable physical features had given place to a conception of beauty: a synthesis had been achieved. A more comprehensive synthesis was foreshadowed, for considerations of character entered into human judgements, which, fused with the idea of beauty, would give rise to the conception of a single valuation and appreciation of the total human personality. Sexual-social selection in man was correspondingly complex.

In thus discussing man, Fisher was developing the idea in his college talk on evolution and society that mental and moral characteristics have evolved under the influence of natural and sexual-social selection in a manner wholly analogous with the evolution of physical characteristics and that conscience was an extension, made possible by the extraordinary development of the human mind, of instinctive preferences and tastes similar to those discernable in other animals, even though the instinct in man might be a complex of intellectual, moral, or esthetic judgements.

In 1922, [C P 28], in a talk on the evolution of human conscience in civilized communities, he took the argument a step further. Evolved instinctive preferences were of unequal importance biologically for the race; the more important the choice biologically, the more strongly would moral feeling about it be developed. Sex and procreation were primary, for, as Fisher had said in 1912, "The solution which commends itself to nature and which is of interest to us as that which will be adopted in the future, is characterised . . . solely by fertility and the power of survival." Hence the strength of religious feelings and taboos primarily in connection with sex and procreation.

Fisher perceived the consequences of nature's criterion of fertility, reflected in changes in the moral attitude adopted toward family limitation during the rise and fall of civilizations. Among Greeks and early Norsemen, infanticide had been practiced at first from a high conscientious and moral aim, as a duty to the tribe, but, over time, the moral attitude had changed completely. The early preaching of Mohammed marked a revulsion against the practice among Arabs, because of an evolution of conscience. Extraordinarily strong moral opinion had grown up through the centuries. In all civilizations the prevailing religion condemned infanticide. The reason for the changed opinion was, Fisher said, that those to whom infanticide was repugnant left children in greater numbers and those who murdered their infants disappeared from the face of the earth.

The same changes of moral attitude had occurred in regard to abortion (Fisher cited a series of Greek and Roman writers to illustrate this dramatic reversal of feeling). In this case the early Christian church had adopted and given its sanction to the new morality. As for the modern method of family limitation by contraception, he predicted that if it were widely practiced it, in turn, would have the important selective effect of producing conscientious opposition to contraception among later generations:

The greater the economic pressure to which they are exposed, the more severe will be the selection and the more fiercely and clearly will their new morality be branded upon their conscience. Moral forces of the same intensity as those which destroyed the

Pagan Empire, or which in the seventh century flung the illiterate Beduin as conquerors across half the world, are developed gradually through centuries of hardship, degradation and temptation. In such a period as that which now lies before Europe is prepared the seedbed of a new religion.

Thus he suggested, before an audience which included Marie Stopes and Dr. Dunlop and Margaret Sanger (the leading protagonists of contraception in Great Britain and the United States), that encouragement of the practice of contraception would eventually defeat itself. Meanwhile, he deplored the fact that contraception was being practiced, mainly by educated people, with the immediate dysgenic effect of increasing the disparity in reproduction between different classes.

□ □ □

By mid-1916 Fisher had completed the article which he entitled "The correlation to be expected between relatives on the supposition of Mendelian inheritance." [*C P* 9]. He mentioned it when he wrote to Professor Pearson, on June 26, 1916 [14]:

. . . I have recently completed an article on Mendelism and Biometry which will probably be of interest to you. I find on analysis that the human data is as far as it goes, not inconsistent with Mendelism. But the argument is rather complex.

The difficulty, of course, was to achieve publication of such an article. In view of the paper's statistical treatment of human heredity, *Biometrika* would have been the appropriate journal for publication, and at the time Fisher may have believed that Pearson might be tempted by a good statistical argument to contemplate the possibility of Mendelian inheritance. If this paragraph of his letter was intended to bait Pearson's appetite, however, it failed; Pearson did not mention the article in his reply.

Fisher's paper was communicated to the Royal Society by W. C. Dampier Whetham, and its referees were the leading authorities in the two main subjects concerned: K. Pearson, biometrician, and R. C. Punnett, geneticist. But Mendelism and biometrics had long gone their separate ways, and the antagonism between the camps had become impermeable to reason. The referees may have agreed on nothing else in life, but they were united in rejecting Fisher's paper; each discarded it as a contribution to his own subject, while suggesting that it might conceivably be of interest to the other. It was too statistical for the one, too genetical for the other. The communication was withdrawn in January 1917 [1].

Two years after its writing the article was still unpublished. There was nowhere else to go, unless a man were prepared to pay for publication, and Fisher, at Bradfield, could not afford £50 to foist his jewel on an unreceptive public. Darwin, however, determined that the article should be published. The Eugenics Education Society, he said, wished to sponsor the publication. He found it could be done through the Royal Society of Edinburgh. The paper [C P 9] was submitted in June 1918 and read in July. It seems there was some doubt about publication; Darwin had the article set up in type for the *Eugenics Review* before the matter was settled, and it was published in October in the *Transactions of the Royal Society of Edinburgh*. Fisher also presented the results of this investigation in simplified form to the Eugenics Society [C P 10, 1918].

Five weeks later came Armistice Day, and Fisher unhesitatingly gave in his notice to Bradfield College. He was committed to his work there until the end of the school year, so that he had eight months in which to decide his future. Apart from the miscellaneous reviewing and some notes published in the *Eugenics Review,* he had two important scientific papers to his credit since his college days: the first had appeared in *Biometrika* and gave the exact distribution of the correlation coefficient, and he had since been condemned for it in the same journal; the second had just appeared in Edinburgh by the kindness of Major Darwin, in the teeth of authoritative rejection elsewhere. Fisher was unknown beyond a narrow circle; he had no independent income and no influential friends. He would have to take what was offered; but there should be openings, he thought, in university research abroad, and he could still turn to farming, though finance would be difficult and though, in fact, some of the gilt had worn off the farming gingerbread.

Early in 1919 Fisher applied for a job offered at Cairo University. Major Darwin was told and immediately made private inquiries about the cost of living and the conditions in Cairo and advised Fisher that the salary would hardly do for a man without an independent income. In March Cairo turned him down. He applied for a post in New Zealand; but the matter was not decided by the end of the school year. He inquired about work in the Agriculture School at Cambridge in vain. He applied for a fellowship at Caius College, but that year the fellowship was awarded to another man. (Fisher did, in fact, win a nonresident fellowship in 1920, when his crisis was over, and he rejoiced in the renewed contact with Cambridge men.)

He was still considering farming. Major Darwin advised against it. "In my experience," he wrote in March,

I have known failures among amateur farmers and I have known no successes. The farmer may be ignorant of many things but I believe he has a great store of knowledge, partly the result of experience and partly of tradition, which can be gleaned from no books, hence the failure of those who start without it.

He suggested looking for statistical jobs in England, getting to know men like F. Y. Edgeworth at Oxford or taking advice of acquaintances like G. U. Yule in Cambridge or J. F. Tocher in Aberdeen. Meanwhile, George Fisher was in touch with the auctioneering fraternity and was able to locate farming properties coming on to the market and to advise his son about their valuation. Ron and Eileen visited one such farm and began calculating how to raise the money. In April, Darwin wrote again, still unconvinced: "I have felt strongly what you say about leisure but to get leisure needs both absence of work and £.s.d. and unless I am utterly wrong neither of these will come from work on the land."

On one of his visits to Cambridge, Fisher met Horace Brown, the botanist, to whom he had been introduced by Major Darwin during the war and for whom he had done some work in connection with framing a mathematical model for the concentration curve of a diffusion column before it reached a steady state. Brown kindly recommended him for the New Zealand post. He also introduced him to Dr. E. J. Russell, who was visiting Cambridge looking for staff for Rothamsted Experimental Station. In August there followed an offer from Rothamsted: would Fisher come for 6 months or a year to find out whether the accumulated records could yield more information by a thorough statistical analysis than had been possible in the absence of a trained statistician?

Within a few days Fisher also received an offer from Professor Pearson at the Galton Laboratory. Fisher's interests had always been in the very subjects that were of interest at the Galton Laboratory, and for 5 years he had been in communication with Pearson, yet during those years he had been rather consistently snubbed. Now Pearson made him an offer on terms which would constrain him to teach and to publish only what Pearson himself approved. It seemed that the lover was at last to be admitted to his lady's court—on condition that he first submit to castration. Fisher rejected the security and prestige of a post at the Galton Laboratory and took up the temporary job as sole statistician in a small agricultural research station in the country.

It was a task after his own heart, offering the sort of challenge that had qualified his ambitions during the war, for the survival of the post itself depended on whether he could prove the usefulness of his peculiar contribution to agriculture: of statistics to practical research men. He was to be proved not as a farmer but in connection with farming and on the basis of his performance in applications of a subject he had already made his own: the statistics of small samples.

3
Mathematical Statistics

It is difficult now to imagine the field of mathematical statistics as it existed in the first decades of the twentieth century. By modern standards the terms of discourse appear crude and archaic and the discussion extremely confused. Admittedly, there existed sound bases in the theory already developed, if they could be distinguished. The whole field was like an unexplored archeological site, its structure hardly perceptible above the accretions of rubble, its treasures scattered through the literature; to assess its worth, one would need to dig and sift the soil beneath a modern home, for it was an occupied site. Under Karl Pearson at the Galton Laboratory at University College, London, there existed a great center of energetic endeavor from which flowed a profusion of statistical measures and tables by which to describe and distinguish the masses of observations being collected in the field. However, the principles that could justify the use of these measures had not been closely examined.

With many immediate and practical problems to grapple with in an expanding field, Pearson may readily be excused if his energies and his influence were not simultaneously mobilized to achieve the resolution of fundamental problems. The problems, moreover, were difficult both philosophically and mathematically. Even today there are differences of opinion among mathematical statisticians on some of the old issues. Once Fisher had attacked them, it was no longer possible to ignore these problems. More clearly than others, he recognized the fundamental issues and set himself the task of finding solutions. In the process he defined the aims and scope of modern statistics and introduced many of the important concepts, together with much of the vocabulary we use to frame them.

Reasons are not far to seek for this profusion of labors or for the confusion of thought which these early labors often concealed. Historically, mathematical statistics came into being as the convergence of two lines of investigation: probability theory and the theory of national populations, called statistics. The fusion of these two branches was imperfect, so that it was not clear to what extent the one could supplement the other in the new line of statistical work opened up in evolutionary biology following Darwin's work.

Probability theory had been the province of mathematicians. Laws of probability had been recognized as early as the sixteenth century by Cardano and Galileo, and the theory had been formalized and developed considerably during the seventeenth and eighteenth centuries in connection with gambling, then a favorite pastime among the aristocrats of Western Europe. In games of chance it is possible to calculate the probability of any particular outcome of the situation which has been established by the rules of play, because the probabilities of each event are known in advance; then, by combining the probabilities of all the events leading to the outcome in question, one can deduce its probability. With this axiomatic approach, the application of probability theory was an exercise in deduction and combinatorial mathematics, undertaken, in this instance, because the theoretical results had practical implications in gaming.

Some attempts were made in the reverse process, to find a way of inferring from the actual results what was the nature of the underlying but unknown situation. This, of course, is what the scientist usually needs to do: his observations record what was the outcome in certain instances and he asks what are the rules of play in Nature's game, which could give these results. It was in the science of astronomy that two main considerations in scientific inference were first appreciated: how, in estimating the quantities of interest, to make a just allowance for errors of observation, and, given some hypothesis, how to test whether it is consonant with the observations.

Methods to deal with the first of these problems had culminated in the method of least squares, fully developed by K. F. Gauss in the early nineteenth century as a method of estimation. The method was invaluable in combining observations. In surveying, for example, if the locations of four geographical points were considered and the distances between each pair of points were measured, then all these observations could be combined by the method of least squares to make the best estimate of the distance between any chosen pair. The method was, in fact, developed by Gauss in connection with both celestial and terrestrial surveying and came to be widely used in both fields.

Tests of the significance of discrepancies between hypothesis and observations had originated with a qualitative argument Galileo used in considering whether the comets were or were not, as the Ptolemaic system required, sublunary bodies. He considered whether the actual observations would plausibly arise by chance if the Ptolemaic hypothesis were, in fact, true, and he concluded that they would be very unlikely to occur in that case. He was led, therefore, to reject the hypothesis. D. Bernouilli first applied a quantitative argument of the same kind to astronomical data. Before considering the essay question set by the French Academy as to why the planetary orbits were nearly but not exactly in the same plane, he questioned whether the orbits might not reasonably be considered to be a random arrangement. To test this hypothesis, that the planetary orbits owe the degree of their coplanarity to chance alone, he calculated the dispersion of the points of intersection with a unit sphere of the poles of the planetary orbits, and thence the probability that points clustered as closely as or more closely than these were on the surface of the sphere would occur by chance. He concluded that the probability was too low to be considered plausible and that the planetary orbits could not be considered as random. Having demonstrated the reality of the phenomenon of nonrandomness by the test of significance, he felt justified in considering the reasons for the degree of coplanarity observed.

Identical problems of inference were recognized by the "statists," or statisticians, in making generalizations about social and political fact. They also were concerned with making allowance for the variability of their observations and testing their hypotheses about national populations. Indeed, it was in this context that the first recognizably modern test of significance was made, early in the eighteenth century. Dr. J. Arbuthnot, physician to Queen Anne, observed that for 17 consecutive years the number of male births registered in London had exceeded the number of female births. He calculated that if the two sexes were equally probable at birth, results as discrepant as those observed would occur by chance only once in 2^{17} times. He concluded that this probability was so small that the hypothesis must be rejected: he could with an easy mind assert that Divine Providence intervened in favor of the male sex.

Like astronomers, statisticians had to deal with variable material. Just as the astronomer wanted to locate the position of a star within certain error limits, so the statistician wanted to locate the central value for the population and to describe the distribution of individual variations about this point. Like the astronomer, he wanted to achieve this with relatively few observations and yet with sufficient accuracy to be a guide to social or political action. It was the Belgian astronomer Adolphe Quetelet [12] who developed the concept of the "average man," having found, for example, with physical measurements made on large numbers of army recruits that the individual values for height or girth were distributed about their mean in a symmetrical fashion, in a Gaussian or *normal* distribution.

Francis Galton extended this approach by considering the relation that existed between different physical characters, like that between the height and span of the same individual or between the heights of fathers and their sons. By scrutinizing the two-way tables he had made of his measurements, he found the marginal distributions were normal while the contours of equal density within the table were elliptical, the narrowness of the ellipse about its long axis indicating the intensity of the relation between the two characters. This property enabled him to establish the *correlation coefficient* as a measure of the intensity of their relationship, and the heights of father and son were said to follow a bivariate normal distribution.

Galton's interest in the bivariate case was stimulated by his wish to elucidate the role of heredity in causing and controlling the variability of the race. Since the correlation coefficient was a measure of the extent to which the same causes were at work in both father and son, he argued that it measured specifically hereditary causes, at least when it was derived from a sufficiently large number of individuals. Galton extended his work to less quantifiable human characteristics and found that for various measures of temperament and intellect the parental correlation was roughly the same as for physical characteristics, that is, nearly 50%.

Galton talked to younger men, in particular W. F. R. Weldon and K. Pearson, about these biometrical problems. Weldon had already begun biometrical work when, in 1890, he became professor of biology at University College, and a close friend of Pearson, professor of applied mathematics and mechanics there. That year was the turning point of Pearson's life, for he also was moved to turn his attention to biometrical problems, through Weldon's interest and the stimulus of Galton's book *Natural Inheritance* (1889).

Pearson took up the challenge with enthusiasm. In the outcome, however, he was somewhat distracted from developing Galton's work on the bivariate normal distribution. Many of the distributions of the data brought to him and collected by him in connection with social and biological observations showed departures from normality of greater or lesser extent, and his interest turned to the mathematical description of skewed and otherwise nonnormal univariate distributions. From this work emerged the Pearsonian system of frequency curves; these were all derived from a single differential equation and were characterized by differences in sign or value of the constants of that equation. To determine which of a dozen types of Pearsonian curve to fit to data, the first four moments of the empirical distribution were calculated; the values of the standardized third and fourth moments then determined the appropriate curve to fit.

What one gained on the roundabouts by this fresh initiative, however, one lost on the swings, for in Pearson's laboratory the special properties of the normal distribution were to some extent lost sight of. There were good theoretical reasons for according the normal distribution a central place in describing the distribution of errors arising in scientific experiments. Typically, in such experiments it is true that the overall experimental error is the resultant of numerous contributory errors. In an agricultural experiment, for instance, carried out to determine the increase in yield produced by a certain fertilizer, differences in fertility of the experimental plots, climatic differences, bird damage, errors in measuring and applying the fertilizer and in gathering and weighing the crop, and any number of other variables all contributed increments to the overall error. Now, the central limit theorem of Laplace had shown that almost irrespective of what the distribution of the individual incremental errors might be, the distribution of the aggregate would tend to the normal distribution as the number of contributing causes increased. Although this did not mean that one could expect to find exactly normal error distributions, it did mean (and observation of the distributions actually occurring confirmed this) that most error distributions met in practice would deviate about the normal rather than about some other central distribution.

Fisher was to take the apparently rash step of assuming general error normality and, as it turned out, was able to make tremendous advances which had very general practical applicability, because most of the results achieved on the assumption of normality were rather insensitive to the assumption itself.

□ □ □

A second cause of confusion originated in a long-standing conflict among mathematicians concerning the use of probability theory as an instrument of induction. No philosophical difficulty arose in classical probability; it was an axiomatic system in which the probabilities of the results were deduced from a situation known in advance. In the mid-eighteenth century, however, the Rev. Thomas Bayes had considered an approach to the problem of induction, for the first time using the known laws of probability to make inferences from an experimental sample about the population from which it was drawn. In order to apply the method to the wide class of real problems in which the sampling rule is not known, he had introduced a questionable postulate. He did not publish this work, and the validity of Bayes' postulate has so often since been doubted that it is tempting to speculate whether Bayes himself may not have been the first to admit doubts on the subject and have been thereby restrained from publication. Fisher, who did not accept the validity of Bayes' postulate, certainly believed so; in the introduction to *Statistical Methods for Research Workers*, for example, he wrote:

Bayes' celebrated essay . . . was published posthumously, and we do not know what views Bayes would have expressed had he lived to publish on the subject. We do know that the reason for his hesitation to publish was his dissatisfaction with the postulate associated with "Bayes' Theorem." While we must reject this postulate, we should also recognize Bayes' greatness in perceiving the problem to be solved, in illustrating the possibility of its experimental solution, and finally in realizing more clearly than many subsequent writers the weakness of the axiomatic method.

Bayes' essay was published in 1763, soon after his death, and his argument came to be accepted by the most influential mathematicians of the early nineteenth century.

In essence the argument runs as follows: it is desired to know the probability of an event from the results of experiment; if a statement can be made a priori concerning the probability of the event, then classical probability theory applies; if not, one cannot deduce a probability statement from the results of the experiment, unless Bayes' postulate is invoked. Suppose, for example, two urns are known to contain mixtures of black and white balls in different proportions, urn *A* having 90% and urn *B* 10% of black balls. A sample of five balls is drawn from one of the urns and all five are black. The question is, what is the probability that the sample was drawn from urn *A*? The answer can be calculated only if the chance of drawing from the two urns is specified in advance, for example, if it had been determined in advance that the sample should always be drawn from urn *A*, the probability of urn *A* would be unity irrespective of the sample. Bayes' theorem showed how this probability can be calculated from the sample for any other *specified* chance of sampling from the two urns. Bayes' postulate concerns the case in which the chance of the draw cannot be specified in advance and proposes that, in that case, the prior probabilities of the two urns should be assumed to be equal. The argument from Bayes' postulate came to be called the argument from inverse probability.

Bayes' theorem could be applied not only in assigning a probability to an event, such as the urn from which the sample was drawn, but also to the value of a parameter, such as the velocity of light. To invoke the argument it was proposed to attribute a uniform distribution to the value of the parameter in the a priori statement. This noncommittal statement seemed innocent enough; it simply stated the fact that, over the relevant range, information a priori gave every value of the parameter an equal chance of occurring. A problem arises, however, in applying this argument, because it is effectively impossible to make a statement of ignorance in general: if the value of the parameter, say, θ, has a uniform distribution a priori, then θ^2 (of which one is equally ignorant) and other functions of θ have particular nonuniform distributions. In consequence, one is led to mathematically inconsistent statements.

The possibility of deriving a posteriori probability distributions of parameters such as the velocity of light was extremely attractive, and no one could see how to bridge the gap between probability theory and scientific inference without invoking the inverse probability argument. In its favor was the most persuasive fact that the method gave results that were consonant with reason; for example, on the assumption that errors were normally distributed, it yielded the method of least squares. Thus, despite recurrent doubts, it gained admission into statistical practice. Even today the logical status of Bayes' inference is uncertain; it has remained a ghost to haunt mathematical statisticians, being neither incorporated into the living body of accepted theory nor finally laid to rest.

Laplace admitted the principle of inverse probability into the foundations of his exposition. Since he also developed mathematical analysis to a high state of sophistication, his free use of prior probabilities tended to obscure the logical issue. Gauss may well have had reservations on the subject, but he used the argument from inverse probability as one means by which to deduce and justify the least-squares procedure. His procedure was thus sometimes perceived as being dependent on inverse probability. In the latter half of the nineteenth century, the criticisms of men like Boole, Venn, and Chrystal brought about a change of scientific opinion about inverse probability such that the subject fell into disrepute.

Confusion about the bases of inference was compounded by confused terminology and consequent confusion of thought. There was a tendency to write and think of the frequency distribution which had been fitted to the data as if it were in fact the probability distribution of the population, to speak of a sample drawn from the population as if it were the whole population and of the statistics estimated from the sample as if they were identically the population values.

In practice, this confusion seemed unimportant. Statisticians were concerned with massive numbers of observations, and, in samples of the size which Pearson used, statistics such as the sample average did, indeed, approximate to the population values. The situation was very different with small samples, and the confusion of terminology became a hindrance to thought in this connection that was difficult to overcome.

□ □ □

When Fisher first became interested in the subject, the problems in relation to small samples were neglected, partly because such samples were considered unimportant and partly because the problems to be faced in dealing with them were very difficult.

In dealing with large samples it was possible to exploit the consequences of Laplace's central limit theorem. Statistics like the sample average or the sample standard deviation, when calculated from sufficiently large samples, were known to be practically normally distributed. If the distribution of a statistic could be treated as normal, then the deviation of any value from the center of its distribution could be tabulated in terms of the standard deviation of that normal distribution. One knew, for instance, that values which deviated by more than two standard deviations in either direction would occur just about once in twenty times. To make rough inferences from large samples, therefore, it was very often sufficient to determine the approximate standard deviation of the distribution of the statistic of interest (often referred to as the standard error of the statistic). This was not true of small samples.

The estimates derived from small samples are like the separate stars in a galaxy: they cluster about the central value but cannot pinpoint it. The accuracy of any estimate depends on the shape of the cluster, for, obviously, if the stars are widely diffused or strung out in a long tail, then the position of one among them will be relatively uninformative about the true center, whereas if they are arranged symmetrically close about the center, the position of one of them will fix the center within fairly close limits. Now, the distributions of various statistics in common use were not known so that the statistics were imprecise to an unknown degree. This made them virtually useless in experimental work, where one had not sufficient numbers of observations to assume the normality of the distribution of errors in the estimated sample statistics.

A further problem became apparent as soon as a statistic was known that had a distribution so skewed and changeable that even with a very large sample its error could not be calculated on the assumption of its normality. The distribution of an estimated correlation coefficient proved to be of this sort. In particular, the standard error associated with this estimate depends critically on the true value, which is unknown, and any error in estimating the true value will be compounded in estimating its error.

There remained the question of what statistics ought to be used for estimating a given characteristic of a distribution. In estimating the spread of a normal distribution, for example, one could use the average squared deviation from the sample average or the average absolute deviation. Which measure was the better? When one considered specifically the small samples of experimental values it became particularly important to make the best possible use of limited data, and for this purpose it was necessary to formulate criteria for selection among the possible statistics so that the best among them might be employed in statistical practice.

All these problems were obscured, of course, by the confident assumptions of the time: among them, that small samples were not important, and that the method of moments was appropriate to the fitting of any frequency distribution. In following Fisher's work, one wonders no less that he perceived the problems so clearly and set himself, against the current of the times for many years, to seek solutions, than that he eventually succeeded in so large a measure. There is a clue in his very first paper, as if, even as he set sail westward to seek a new route to the Indies, he had grasped intuitively the global nature of the earth; he did not use the well-known eastern route, because he could see the possibility of better success by another course.

The first paper [C P 1], published in 1912, concerns the problem of estimation. The current practice of fitting a theoretical curve to an empirical distribution by the method of moments contained an element of arbitrariness, for the method was chosen without theoretical justification. Why use these moments? Fisher suggested an alternative. In the Bayesian formula, the posterior distribution is given by multiplying together the prior probability of the parameter and the probability, for any given value of the parameter, of obtaining the actual data observed. The latter function Fisher later called the *likelihood*. In the 1912 paper he was able to show that progress could be made by studying the likelihood itself without reference to the prior probability and, in particular, that the estimates obtained by maximizing the likelihood have desirable properties. The method of maximum likelihood provides a direct solution to the problem of fitting frequency curves of known form, without introducing any arbitrary assumptions.

This discovery made it possible for him to contemplate wider problems and thereby doubtless clarified his goals in statistical research. How far he could go by the method of maximum likelihood he did not yet know, but it was certainly an approach well worth exploring. First, however, he was to explore the distributions of sample statistics, whose relative efficiencies he was now able to compare using his new ideas of likelihood.

□ □ □

The investigation of the distributions of sampling statistics was begun when Student's [13] paper "On the probable error of the mean" appeared in *Biometrika* in 1908. Fisher found this paper significant to a degree that has sometimes puzzled his successors, but it had virtue as the catalyst which precipitated Fisher's own work, so that its action was both magnified and hidden in his results.

Student's paper was unusual pre-eminently because it attempted to deal with small samples. Its author saw that as long as only large samples were contemplated, the range of the subject of statistics was severely limited. Until statistical methods were developed capable of dealing with small samples, the subject could not contribute to the wide range of applications in field and laboratory experimentation. Yet all the emphasis was laid on large-sample work. In this environment Student chose to concern himself with small samples; in a shop devoted to the sawing of house timbers, he chose to consider woodcarving and invented and fashioned a chisel to do the work.

With large samples the estimated standard deviation could be assumed to be essentially exact. With small samples it was subject to error that could not be ignored. In particular, when the standard error of a sample average was calculated, allowance had to be made for the fact that it too was an estimate subject to error. The fewer the observations, the greater the allowance that must be made for error in the estimate. With large samples, an observed average was known to fall with 95% probability within limits about two standard deviations on either side of the true value. Student found that with as many as thirty observations his value was very little different from two, but with smaller numbers it increased until with only two observations it was greater than twelve.

It did not matter that Student's mathematics were inelegant: a mathematician worth his salt could refine the argument and prove the empirical result—Fisher himself did so in 1912. What was significant was that an experimental scientist, who of necessity usually had only small samples from field or laboratory, had been so exercised about the precision of his estimates that he had explored the actual distribution of the standard deviation and calculated a function of the data that made accurate allowance for the errors of estimation. Fisher was ready to accept that the needs of experimental scientists are the challenges the statistical profession must meet, to recognize the practical importance of determining the precision of estimates from small samples, and to pursue the strategy indicated by Student's work by gaining a knowledge of the sampling distributions of the other needed statistics. As he put it later [*Statistical Methods for Research Workers*], "Once the true nature of the problem was indicated, a large number of sampling problems were within reach of mathematical solution."

The pseudonym Student had been adopted by W. S. Gosset, a chemist turned statistician almost in spite of himself by the necessities of his work at Guinness's Brewery in Dublin. He had visited the Galton Laboratory in the year 1906–1907 to study under Pearson and had made time while there to pursue his investigation of small-sample statistics and to prepare the first papers for publication. They show his flair for recognizing the essentials of a problem.

Guinness's Brewery had large farming interests, especially in growing barley for beer; in consequence, Gosset became involved in agricultural experimentation as well as with laboratory tests. In this connection, when he was in England, he used to visit the School of Agriculture at Cambridge and there he met F.J.M. Stratton, the astronomer, who was Fisher's tutor.

When Fisher brought to Stratton the matter of a discrepancy he had found between a formula published by Student and his own result, Stratton wisely suggested that his young pupil should write directly to the author. The introduction was made, and Fisher followed up the initial correspondence by sending to its author the mathematical proof of the formula for the frequency distribution of the t-statistic (originally called z as in the letter below). Gosset was sufficiently impressed to send the proof to Pearson with the following covering letter. [14]

12th September 1912   Woodlands, Monkstown
                                                                           Co. Dublin

Dear Pearson,

I am enclosing a letter which gives a proof of my formulae for the frequency distribution of $z(=x/s)$, where x is the distance of the mean of n observations from the general mean and s is the S.D. of the n observations. Would you mind looking at it for me; I don't feel at home in more than three dimensions even if I could understand it otherwise.

The question arose because this man's tutor is a Caius man whom I have met when I visit my agricultural friends at Cambridge and as he is an astronomer he has applied what you may call Airy * to their statistics and I have fallen upon him for being out of date. Well, this chap Fisher produced a paper giving 'A new criterion of probability' or something of the sort. A neat but as far as I could understand it, quite unpractical and unserviceable way of looking at things. (I understood it but its gone out of my head as you shall hear. I have lost it). By means of this he thought he proved that the proper formula for the S.D. is

$$\sqrt{\frac{(x-m)^2}{n}} \qquad \text{vice} \qquad \sqrt{\frac{(x-m)^2}{n-1}}$$

This, Stratton, the tutor, made him send me and with some exertion I mastered it, spotted the fallacy (as I believe) and wrote him a letter showing, I hope, an intelligent interest in the matter and incidentally making a blunder. To this he replied with two foolscap pages covered with mathematics of the deepest dye in which he proved, by using n-dimensions that the formula was, after all

$$\sqrt{\frac{(x-m)^2}{n-1}}$$

*Airy, G. B., author of the well-known textbook *On the Algebraical and Numerical Theory of Errors of Observations and the Combination of Observations* (3rd ed., 1879) London.

and of course exposing my mistake. I couldn't understand his stuff and wrote and said I was going to study it when I had time. I actually took it up to the Lakes with me—and lost it!

Now he sends this to me. It seemed to me that if it's all right perhaps you might like to put the proof in a note. It's so nice and mathematical that it might appeal to some people. In any case I should be glad of your opinion of it. . . .

Yours very sincerely,

W. S. Gosset

Nothing more was heard about Fisher's proof. Gosset's letter, however, not only establishes its existence but reveals the development of Fisher's thought at this time. In his paper "On an absolute criterion for fitting frequency curves" [C P 1], Fisher had derived the maximum likelihood estimate for the variance, and this had the divisor n instead of the $n - 1$ used by Gosset. (The latter had probably been adopted into use indirectly from the work of Gauss.) Having been persuaded to write his first letter to Gosset about the discrepancy, Fisher evidently began to consider the problem from an entirely different—and no less original—point of view, in terms of the configuration of the sample in n-dimensional space. The geometry would at once suggest that using the sample mean instead of the population mean was equivalent to reducing the dimensionality of the sample space by one, and therefore required the divisor $n - 1$ in the variance estimate. By adopting this formula in his second letter, on the basis of the geometric argument, Fisher showed his recognition, even at this early date, of the concept of degrees of freedom. Moreover, it is implicit in the geometric formulation that the sum of squares of the n *dependent* deviations from the sample mean must be distributed as the sum of squares of $n - 1$ *independent* deviations, and thus it yielded Student's distribution as given by Fisher in his third letter to Gosset.

□ □ □

A second paper of Student, published in 1908 after his year in London, gave the correct solution for the "Probable Error of a Correlation Coefficient" between independent variates. In 1914 Fisher produced the mathematical proof of Student's empirical solution, generalized his results, and sent the paper to the editor of *Biometrika,* giving the exact distribution of the correlation coefficient. Pearson responded in September that he was very interested in the paper but that it needed supplementing and polishing before publication. He also felt the need to check H.E. Soper's approximations, already published in *Biometrika* (1913), against the values given by Fisher's exact formula and to get curves drawn or tables of values calculated to give the changing shape of the distribution for different numbers of observations and values of the correlation, so that the probable error might be calculated appropriate to different sample sizes.

Fisher's interest in these projects was marginal. In the published paper [*C P* 4] he pointed out the occasions when Soper's approximations became useful, and he included some tabulated values of the variance and standardized third and fourth moments, calculated at Pearson's request, with a caution against their use in estimating error:

Tables are appended for inspection rather than for reference, which show the nature and extent of these changes in the form of the curves.

Pearson, however, saw in the proposed calculations the answer to practical needs and work on them began at the Galton Laboratory. In a long letter of January 30, 1915, having discussed the difficulties remaining before a test of significance could be applied to actual estimates of the correlation coefficient, he concluded:

I am still inclined to tabulate the ordinates of the frequency curves for r, starting, say, with $\rho = 0, .1, .2, .3, \ldots , .9$, and $n = 4, 8, 12, 16, \ldots , 40$. That is only 100 curves. I am quite aware that this will be very laborious, and can only be done by a trained calculator, who devotes his life to this sort of work, but I think it is the best thing to do . . . These tables shall be taken in hand as soon as opportunity offers (my best man for this is at present at work on a table of the incomplete Γ-function), unless you really want to do them yourself. I do not feel—subject to your reply—that the t-values you give will satisfy practical purposes, even had they been arranged for interpolation. I will publish, if you wish, your tables of σ^2, β_1 and β_2 but I think they lay us open to the criticism that they do not serve to answer the practical problems from which the whole discussion has arisen.

Fisher certainly did not "really want" to undertake the calculations—he had no calculating machine, little time and no assistance. Besides, it was not his style to assault with brute force the strongly held place. He had, in fact, his own approach and kept his faith in it, despite Pearson's expressed doubts.

The paper [*C P* 4] appeared in May 1915. As Gosset had mentioned in his letter to Pearson earlier, Fisher had used the concepts of n-dimensional geometry in deriving the proof of Student's distribution. It was doubtless this earlier exercise that enabled him, by the same approach, to come to grips with the distribution of the correlation coefficient and to solve it, as he did, within a week, a problem that had rebuffed all previous attempts. The geometric representation obviously had an esthetic appeal to him; as he remarked in this paper, "The five quantities [the first and second moments of the bivariate sample] defined above have, in fact, an exceedingly beautiful interpretation in generalised space, which we may now examine." He was to use the same approach repeatedly to provide simple solutions to other distribution problems that had proved intractable by purely algebraic methods.

Let the reader stand on this magic carpet and he is carried immediately into strange and intricate realms (e.g. *C P* 12), whirled through spherical planes, to catch a glimpse, perhaps, of a generalized octahedron with the highlight on a central section bounded by six squares and eight equilateral triangles which in a twinkling becomes eccentric when the squares become rectangles and the triangles are not all equal. He is shown meanings in the hardly comprehended figures he sees, until he is brought to earth, rather breathless and disoriented, precisely at the wished-for destination.

Fisher could see them with his mind's eye, the abstract and complex forms with their cognate meanings, unconfused. Indeed, his geometric insight was a dominant characteristic of his mathematical thought. He was a quick and accurate computer and had algebraic facility, but his imaginative mastery of spatial concepts was quite out of the ordinary. Doubtless the tendency to represent mathematical facts geometrically within the mind was innate: the toddler had somehow grasped the inverse binary system before he concluded that half of a sixteenth "must be a thirty-toof," and the preparatory schoolboy had found some means of producing the answers to mathematical problems without written calculations. Doubtless, his aptitude was strengthened by the exercise imposed on him at Harrow when he received tuition without aid of pen and paper. The exercise was also self-imposed and habitual, for his geometric vision did not come without effort; like an athlete he kept himself in training, and there were occasions throughout his life when he taxed his imagination to grapple with problems which seemed—and perhaps were— beyond its range. But he enjoyed the exercise, and that in itself is the best clue we have to his exceptional power in this respect. It gave him the joy of challenge, effort, and mastery.

In 1915 statisticians showed as little interest in Fisher's approach as they had in Gosset's beginnings. They did not experience the charm of the geometrical representation on their own minds, nor did they take up the challenge of the sampling distributions of other statistics derived from normal populations. Years later Fisher was still alone in sensing their importance. When he drew attention to the work still to be done in 1922 [*C P* 18], he offered pure mathematicians both a challenge and an appeal for help. Concerning the development of complete and self-contained tests of goodness of fit, he promised that "Problems of distribution of great mathematical difficulty have to be faced in this direction." He listed the existent work on sampling distributions and concluded, "The brevity of this list is emphasized by the absence of investigation of other important statistics, such as the regression coefficients, multiple correlations and the correlation ratio." In the same paper (*C P* 18), he wrote:

> For my own part I should gladly have withheld publication until a rigorously complete proof could have been formulated; but the number and variety of the new results which the method discloses press for publication, and at the same time I am not insensible of the advantage which accrues to Applied Mathematics from the co-operation of the Pure Mathematicians, and this co-operation is not infrequently called for by the very imperfections of writers on Applied Mathematics.

His appeal met with no response. Gradually he continued the work himself alone, deriving the distributions of the intraclass correlation coefficient [*C P* 14, 1921], the regression coefficient [*C P* 20, 1922], the partial correlation coefficient [*C P* 35, 1924], and the multiple correlation coefficient (*C P* 61, 1928). (The correlation ratio became irrelevant, the problem being completely taken care of by the analysis of variance.)

□ □ □

Meanwhile, work was going forward on the problem raised by Pearson in 1915 of determining the error of estimates of the correlation coefficient. As Figure 1a [p. 81] shows, the distribution of the sample correlation takes different and often highly skewed shapes, depending on the value of the true correlation, and the standard error approach cannot properly be used. Since it would serve for some of the more symmetrical curves, Pearson sought a criterion to decide how many observations would be necessary to bring the curves to approximate normality, and he initiated work at the Galton Laboratory to calculate the ordinates and moments of the various curves. However, even for large samples, when the correlation approached unity the curves were extremely distorted and cramped against this boundary value, and the changes in shape of the curve for proximate correlation values were relatively large. In such cases the probable error of the estimate could well fall outside the boundary in a theoretically impossible area. As Pearson remarked, it "would not serve to answer the practical problems from which the whole discussion has arisen."

To Fisher, the obvious thing to do was to find some transformation of the correlation curves that removed the boundary and allowed the curves to extend in a symmetrical shape from $-\infty$ to $+\infty$. He suggested such a transformation in the 1915 paper [*C P* 4] and evidently continued work on the idea, for he had a paper on the subject ready for Pearson in May 1916.

Nothing was published at the time, and we see from the following correspondence [14] that Pearson was preoccupied with his own approach to the problem and with other affairs.

May 10th, 1916   9 Horton Crescent
                                        Rugby

Dear Professor Pearson,

I am afraid I have been very slow about my paper on the probability integral. I have got it written out now, quite shortly as I have cut down on the unnecessary Mathematics. I think you will like the method of calculation in the last section, but of course I don't know what has been done. I have run out the first 27 values of R for $\theta = 90°$ and $120°$ in no time, but its no good multiplying repeatedly by a long number without a machine. A table by degrees from $90° - 180°$ should meet all requirements, and would save all double entry.

Yours very truly,

R. A. Fisher

May 13, 1916   7 Well Road, Hampstead, N.W.

Dear Mr. Fisher,

I ought to have written to you before and told you that the *whole* of the correlation business has come out quite excellently. We have calculated all the frequency distributions from 3 to 25, 50, 75, 100 and 400. We have got good formulae for the moments, the odd ones are all given in terms of complete elliptic integrals and the even as you know in terms of cot α. We have calculated in most cases the β_1 and β_2 and the modal values and these constants ought to be completely tabled in another month. By 25 my curves give the frequency very satisfactorily, but even when $n = 400$, for high values of ρ the normal curve is really not good enough. Soper's paper really failed because he took *the range* as a fixed quantum. To fit the frequency is far better, if you simply fit from β_1 and β_2 without regard to the range, the *appreciable* frequency then always lies inside the theoretical range. Many rather new points have arisen from the whole investigation, especially some, I think, new mathematical identities.

I do not know whether and when the results will be published. There is great uncertainty about the future of *Biometrika;* the war has cut off the bulk of our continental subscribers who were the mainstay of the journal—far more important in a way than the English subscribers; and, as I alone am now responsible for the journal, and the deficit was very heavy last year, it will probably have to be suspended or transferred to other proprietors, possibly American. It had been paying its way quite well before the war and readers had no doubt thought it was run by the Cambridge Press, but they have only issued it for the proprietors on commission and decline to give aid when hard times come. I only mention this because it is quite uncertain whether I can publish your paper, and you may care to issue it *at once* elsewhere. Many thanks for your friendly notice of *Biometrika* in the Eugenics Education Society's journal. I don't think it will convert that body!

Yours very sincerely,

Karl Pearson

May 15, 1916   9 Horton Crescent,
Rugby

Dear Professor Pearson,

Your news about *Biometrika* is extremely serious. It seems incredible that such a publication should be in jeopardy for lack of foreign support. It would be a most terrible loss as well as an appalling indignity, if this country cannot support such an important and valuable school of research. Could not the Universities be induced to tide over the War? They have a particular obligation to support such a work, and although I suppose they are hard hit, their honour is at stake in such a matter. I hope you will not allow it to pass into other hands without allowing the widest publicity to the injury which would be done to English learning.

As regards my own paper, parts of it will evidently have to be rewritten with fuller knowledge of what you have done. I should be very glad if you could send me copies of proofs of such of this work, as is complete, if possible with numerical tables. Also my own paper when you have done with it. I could probably have worked more profitably, if I had been in closer touch with the Laboratory, although such collaboration is never easy.

I remain,
Yours very truly,

R. A. Fisher

May 18, 1916   Department of Applied Statistics
University College, London, W.C.

Dear Mr. Fisher,

Many thanks for your kindly letter about *Biometrika*. . . . As regards our paper it is difficult to determine where, when or how it will be printed as matters are difficult at present with regard to all publication of scientific work. When it comes to type I will certainly send you proofs. With regard to your own paper it seems to me that it would be very nice to have a table of the probability integral, but it would undoubtedly mean on your lines a large amount of stiff work. I think it could be found for $n = 3$ to 25 and $\rho = 0$ to 1 by .05 with two or three months work (r from -1 to $+1$ by .05), but it would be a fairly long job, and I am not sure it could not be deduced as accurately from our completed tables of *ordinates* of the like cases. These ordinates enable the areas to be found either by quadrature, or by plotting and the integraph. My original idea was to use the latter method, but if you will work a table of the areas it would undoubtedly be preferable. Our work here has been *very* slow as I have only one man left on the staff and four women, and the man may be called up at any moment. Further we have been doing war work since last January, so that other things can only come in odd moments.

Yours very sincerely,

Karl Pearson.

Wartime difficulties, in fact, prevented the publication, and a year elapsed before Pearson's results were published in a "Cooperative Study" [15]. There is no record that Fisher saw the proofs, though he may have received them in part or "late in the day," as E. S. Pearson has suggested [14]. When he saw the paper, he was shocked to find that it contained a section that criticized his use of a Bayesian prior distribution for the correlation coefficient, a usage he did not accept and had, in fact, carefully avoided. Apparently, the 1912 paper [C P 1] and the 1915 paper [C P 4] together had failed to make his position clear, and the authors of the cooperative study had not felt the need to ask his confirmation of their interpretation of his work before condemning it in print. Unfortunately, their preferred approach was less efficient; still more unfortunately, their criticism was as unexpected as it was unjust, and it gave an impression of something less than scrupulous regard for a new and therefore vulnerable reputation.

Four years later Fisher was to set the record straight [C P 14]:

The writers of the cooperative study apparently imagine that my method depends upon "Bayes' theorem," or upon an assumption that our experience of parental correlation is equally distributed on the r scale (and therefore not so on the scale of any of the innumerable functions of r, such as z, which might equally be used to measure correlation), and consequently alter my method by adopting what they consider to be a better *a priori* assumption as to the distribution of ρ. This they enforce with such rigour that a sample which expresses the value of 0.6000 has its message so modified in transmission that it is finally reported as 0.462, at a distance of 0.002 only above that value which is assumed *a priori* to be most probable!

By the time he wrote these words, Fisher was ready not only to restate his concept of likelihood in such a way as to deny explicitly any dependence on Bayes' theorem but also to present the transformation

$$r = \tanh z$$

to reduce the skewed curves of the correlation coefficients to practically normal shape, for which the error distribution was readily available. The transformation is really remarkable. As shown in Figure 1a, the distribution of the correlation coefficient r has a different shape and a different spread, depending on the value of the true unknown correlation ρ; in contrast, the transformed distribution z, as shown in Figure 1b, is very nearly normal with very nearly constant and known standard deviation, whatever the value of the true correlation. By using the transformed statistic, a distribution which had been quite unmanageable and only to be employed by a very tedious series of incomplete and inexact approximations, was at a stroke transformed into a simple, familiar case.

The practical tasks of testing the significance of differences between correlation coefficients and of the difference of a coefficient from some theoretical value were thus immensely simplified; at the same time, the misleading nature was made apparent of the 'probable error' of statistics such as the sample correlation coefficient whose distributions are far from normal. The authors of the cooperative study could have saved their labors, for the z-transformation gave approximate integrals needed for calculating probabilities in place of the tabulated ordinates.

Like many good ideas, the z-transformation only gradually revealed its quality, even to its originator. Fisher had actually suggested it, quite tentatively, in the 1915 paper ($C\,P$ 4), as a transformation "aimed at reducing the asymmetry of the curves or at approximate constancy of the standard deviation." For this purpose it then seemed "not a little attractive, but so far as I have examined it, it does not tend to simplify the analysis." Some years later he used the same transformation when dealing with a practical problem concerning the correlation between pairs of twins ($C\,P$ 11, 1919). He was impressed by the fact that the curves for the intraclass correlations, although different from the interclass correlations previously obtained, could be rendered approximately normal by the same transformation and that the discrepancies between the sampling curves derived by the two methods were greater than the departure of either from the normal form. The advantage of the transformation became more apparent; indeed, it provided a very pretty solution to the distribution problem, and he went on soon after to write the paper ($C\,P$ 14, 1921) in which also he derived the distribution of the intraclass correlation coefficient. At the same time, he became aware of new contexts in

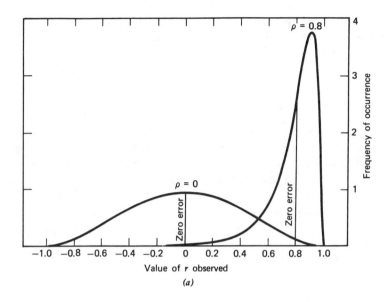

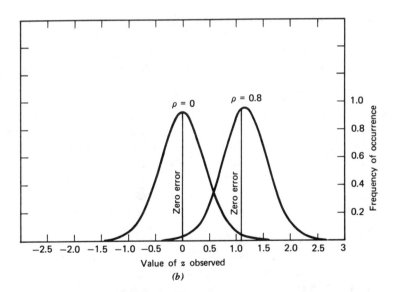

Figure 1. (a) *Distributions of the estimated correlation coefficient* r *for 8 pairs of observations* ($\rho = 0$ *and* $\rho = .8$) *and* (b) *The corresponding curves for* z. (*From* Statistical Methods for Research Workers § *35*).

which the transformation could be usefully employed. It was to play an important role in the evolution of the Analysis of Variance.

□ □ □

For Fisher, the publication without warning of the criticisms against him in the cooperative study made an object lesson he was not likely to forget. He was easily angered at what he conceived to be injustice, and it was not the first time that Pearson had "corrected" results that were, in Fisher's view, not in error. He had felt indignant on behalf of Major Darwin in such a case, when both the matter and the manner of Pearson's rebuke had seemed misconceived. That Darwin had accepted the correction and believed he must have deserved it because it came from so high an authority made the injustice the more cruelly flagrant in Fisher's eyes, and he was only persuaded to remain silent by Darwin's evident shrinking from controversy. Now the same thing had happened to himself.

This experience had its influence on Fisher's decision in the summer of 1919 to refuse an appointment at the Galton Laboratory. The post he was offered held the promise of opportunities for research in both his chosen fields of genetics and statistics; there he would enjoy the security and the facilities of a strong and prestigious department and exert an extended influence through his teaching and publications. On the retirement of Pearson (who was then 62) he might hope to succeed to the chair. These real advantages he jettisoned because he recognized that nothing would be taught or published at the Galton Laboratory without Pearson's approval. Fisher required the liberty to do his own work in his own way and, by it, to win his own reputation.

His ideas had not been noticeably successful with Pearson. Apart from the matter of the cooperative study in 1917, his paper on the probable error of the correlation coefficient had been refused in 1916, as was a note prompted by some work by K. Smith, published in *Biometrika*. A revised note on the same subject was rejected in 1918 on the grounds that Pearson personally disagreed with it. And, if Fisher did not know that Pearson was partially responsible for rejection of his paper on the correlation between relatives by the Royal Society in 1916, he did have Pearson's acknowledgement of a printed copy in 1918, which read: "Many thanks for your memoir which I hope to find time for. I am afraid I am not a believer in cumulative Mendelian factors as being the solution of the heredity puzzle."

Fisher's refusal was a disappointment to Pearson. Without agreeing with all Fisher's ideas, he had recognized his ability and made a bid for it. The fact that Fisher felt strong enough to refuse, and chose to do so, showed him to be a rival rather than a supporter. His independence may have seemed a personal threat, and, if so, one can understand Pearson's posture in defence of "his" territory against Fisher's incursions, which was even then leading them into a lifelong conflict.

Nevertheless, in 1920 Fisher sent his new paper on the probable error of the correlation coefficient first to Pearson. It was returned with the following explanation in a letter dated August 21, 1920 [14].

Only in passing through Town today did I find your communication of August 3rd. . . .
As there has been a delay of three weeks already, and as I fear if I could give full attention to your paper, which I cannot at the present time, I should be unlikely to publish it in its present form, or without a reply to your criticisms which would involve also a criticism of your work of 1912—I would prefer you published elsewhere. Under present printing and financial conditions, I am regretfully compelled to exclude all that I think is erroneous on my own judgement, because I cannot afford controversy.

Major Darwin's sympathy with Fisher under this reverse was heartwarming. Having seen Pearson's letter to Fisher, he wrote

As someone said, one must not treat Pearson like anybody else. I think he means to be civil but it is an astounding attitude to take up. To allow nothing to be published which does not back him up or which he personally does not have time to read! and pitch into—it is going too far.

It was going too far for Fisher's tolerance: it is said that he vowed never again to submit a paper to Pearson's *Biometrika,* and he never did. Darwin suggested the *Journal of the Royal Statistical Society* ("If they don't print mathematics they had better shut up") and offered to make informal inquiries within the society to find out their reaction. He was informed that the statistical society could not take the paper "because they have to cater for an audience many of whom could not understand it and they therefore have to limit the number of highly technical articles." His informant, Dr. M. Greenwood, was willing to send the paper on to *Metron,* as Darwin suggested. Thus it was at Darwin's instance that the paper ($C\ P\ 14$) eventually appeared in 1921 in the first volume of the new journal, edited by the eugenist and statistician, Corrado Gini, in Italy.

Soon after the break with Pearson, Fisher found that he was required to blow up the roadblock of Pearsonian teaching regarding applications of the χ^2 (chi-square) test of goodness of fit. This test had been introduced by Pearson in 1900, and it provided a useful and objective measure of the discrepancies between observed frequencies and those expected on some hypothesis under test. For example, on the hypothesis that a distribution was normal with a *known* mean and *known* standard deviation, the frequency with which the observations would be expected to fall into various classes was known and could be compared with the observed frequencies. The statistical significance of Pearson's criterion, calculated from these discrepancies, could be determined to an adequate approximation by reference to a χ^2 table. In this case there was a single constraint: that the sum of the expected frequencies should equal the total number of the observations. Pearson recognized this constraint by using his χ^2 table as if there were one less than the actual number of the frequency classes.

If the mean and the standard deviation of the hypothetical normal distribution were not known but calculated from the observations, however, expected frequencies would be constrained to agree with the observed values in these characteristics also, and a closer agreement between hypothesis and observations would be achieved than with independently established expectations. Similarly, in using contingency tables, for example, a 2×2 table in which were recorded the numbers of living and dead among the victims of typhoid fever who had or had not been inoculated, not only was the expected total constrained to equal the observed total, but the marginal totals were similarly constrained, so that the four cell frequencies were subject to three linear constraints.

The relevance of these additional constraints was not clearly recognized, and the χ^2 test was being used without modification in those cases, a practice that could give grossly misleading results. (A χ^2 of 3.84, which with one degree of freedom is significant at 5% probability level, with three degrees of freedom has a probability of 28%.) It was recognized that the situation was ambiguous, however, because by a different argument one could get different results. The reason for the discrepancy became clear when Fisher pointed out the effect of the constraints, in terms of the number of ways in which observed frequencies could be discrepant with the expected frequencies, which he characterized as the number of *degrees of freedom*. He showed how this number is reduced by one with the introduction of each additional linear constraint, and that when the table of χ^2 is entered with the appropriately modified number of degrees of freedom, the χ^2 test is valid in the cases that had been troublesome.

Fisher set down his argument, proved the general case, characteristically, by invoking the geometry of generalized space, and showed by examples how, with the concept of modified degrees of freedom, the χ^2 test could be applied in cases in which χ^2 had formerly given discrepant results or which had seemed to Pearson to require special treatment. A. L. Bowley, professor of economics at the London School of Economics (and a neighbor and bridge companion of Fisher in Harpenden), was kind enough to communicate the paper [C P 19] to the *Journal of the Royal Statistical Society* with the personal recommendation:

22 May 1921

I enclose a paper by Mr. R. A. Fisher, Chief Statistician at Rothamsted Agricultural Station and Fellow of Caius College, Cambridge, which I hope will be printed in the Journal. In my opinion it is of considerable importance and originality. It gives a very brief proof of Pearson's formula. . . . The original proof is lengthy and troublesome and buried in the *Phil. Mag.;* and it gives for the first time the correct application of that formula to contingency tables and other groups to which Pearson's analysis did not apply but to which his formula has been erroneously applied. I write this because Mr. Fisher's goods are not in the window and the importance of the paper might be missed in a light reading. Mr. Yule should be asked if he objects to the publication of his figures at the end of the paper.

In the event, G. Udny Yule submitted a separate paper showing how Fisher's modified test dispersed the difficulties he had experienced when testing the significance of effects from contingency tables. Both articles appeared in the *J.R.S.S.* in January 1922.

Nobody could mistake the butt of Fisher's attack, and the matter could hardly have been more calculated to upset Pearson. Pearson had introduced the χ^2 test more than 20 years before and from that time had been its chief interpreter. Now he was told, in effect, that he did not understand the primary principles of its application, that he had misled his followers and spread his own confusion, even in cases where he had managed to get the right answer, perhaps *"as an approximation, by means of very indirect reasoning."* There was, moreover, a sting in the tail of this peculiarly unpalatable document. Pearson had spent 30 years fitting frequency curves by means of their first four moments. He had elaborated the Pearsonian system of such frequency curves and had tested the goodness of fit of empirical data to the curves by the unmodified χ^2 test. Fisher now pointed out, in his final summary [C P 19], that the distribution of χ^2 was applicable not only to the contingency tables discussed in the body of the paper but

to all cases in which the frequencies observed are connected with those expected by a number of linear relations beyond their restriction to the same total frequency. In taking the goodness of fit of a frequency curve fitted by means of four moments the number of degrees of freedom has been reduced by four and since the four moments are linear functions of the class frequencies we shall take n' to be four less than the number of cells.

No wonder that Darwin hardly waited for the new year before inquiring anxiously, "Has Pearson gone for you yet?" No wonder there was a fluttering in the statistical dovecotes, as those who knew waited to see young David crushed under the massive weight of Pearson's familiar rhetoric. And Yule perhaps recalled how a previous paper of his own had exposed him to 150 pages of withering contempt from the same source 9 years before. In July the giant appeared on his own hill, the pages of *Biometrika* [16], and reiterated his position firmly before recognizing the existence of his insignificant and nameless adversary:

The above redescription of what seem to me very elementary considerations would be unnecessary had not a recent writer in the *Journal of the Royal Statistical Society* appeared to have wholly ignored them. He considers I have made serious blunders in not limiting my degrees of freedom by the number of moments I have taken.

(That shot had gone home!)

I hold that such a view is entirely erroneous and that the writer has done no service to the science of statistics by giving it broadcast circulation in the pages of the *Journal of the Royal Statistical Society.*

The argument continued with heavy sarcasm, which rose to a grand prophetic close:

I trust my critic will pardon me for comparing him to Don Quixote tilting at a windmill. He must either destroy himself or the whole theory of probable errors, for they are invariably based on using sample values for those of the sampled population unknown to us. For example, here is an argument for Don Quixote of the simplest nature. . . . I think this will illustrate what I mean by Don Quixote and the windmill.

Apart from its rhetorical flamboyance, the paper did show that there were points on which confusion might still exist. Fisher may have wanted also to take up Pearson's parting challenge; he therefore submitted a further paper on the subject of the χ^2 test. But when giants contend, it is well for lesser men to keep clear of their feet; Pearson had not named Fisher in his rebuttal but he had reiterated the name of the *Journal of the Royal Statistical Society,* and the editors of that journal stood clear lest he turn and rend them. They rejected Fisher's paper without offering any explanation to the author.

Stung by this rejection, Fisher was inclined immediately to resign from the Royal Statistical Society. He was dissuaded by Darwin's wise words:

29 January 1923

I think the Stats. have treated you badly but I hope you will think twice before resigning. The fault lies with at most two or three individuals even if more nominally consent. These men will go in time and the affair is quite forgotten. If you now protest to the Council or resign, you will get the reputation, justly or unjustly, of being very touchy and easily put out. That reputation will not die out easily. Therefore you will lose by any action. The dignified course is that which makes you appear to say "I don't care a damn what you do or say." Forgive me for writing thus plainly.

Fisher decided to seek an explanation. He felt there was a matter of principle involved. Since the editors had published the first paper on χ^2, the matter was a public issue, and their duty was to present such additional evidence as might enable the members of the society to make a just estimate of the case. His paper presented such evidence and answered criticisms. Darwin assured him that he had searched the paper and had found no discourtesy in its manner. If the editors thought the paper faulty in form or substance they ought to tell him; instead, they suppressed his note and refused to discuss the matter. Fisher suspected that they had no solid objections to offer against the paper, that they had simply bolted for cover in the face of Pearson's anger.

The editors, it appeared, had acted unanimously, and Fisher knew that in practice the council must support the united action of the editors. When, after several months of negotiations, he failed to get any satisfaction from the editors, he made the only gesture of protest open to him: he resigned. His reputation suffered, even as Darwin had foretold.

Meanwhile, the original paper had aroused considerable scientific interest and discussion. A. L. Bowley had reservations about Fisher's argument of another sort than Pearson's, and Fisher took the opportunity, in answering Bowley's paper in the new journal *Economica* [C P 31, 1922], to clarify certain points. Later, I. J. Brownlee, having performed a number of experiments that confirmed Fisher's application of χ^2, found that in the case of a coin-tossing experiment, a comparison of the results with areas of a normal curve gave values of χ^2 much too high. Fisher was delighted to have the chance so offered to discuss the matter in full in the *Journal of the Royal*

Statistical Society [*C P* 34, 1924]. He presented the classes of situations in which χ^2 is abnormally distributed. When the hypothesis is not, in fact, true and the values of χ^2 are found to be excessive, then they are only performing their prime function in indicating the inexactitude of the hypothesis to be tested. (This was the case with Brownlee's application.) If the method of fitting is inconsistent or the estimation inefficient, then certain consequences follow regarding the distribution of χ^2 and the validity of the χ^2 test. These cases were explored in detail. Thus in this paper the substance of the χ^2 controversy was wrapped up.

Opinion continued divided. Some statisticians continued to follow Pearson in opposition to the modified test, some considered the affair a highly technical piece of hairsplitting, and a growing number adopted Fisher's test.

A postscript was added in 1926—a *coup de grace* so deftly executed that to Darwin it seemed "wicked". E. S. Pearson had joined his father at the Galton Laboratory and had published in *Biometrika* the results of an experiment designed to test Bayes' theorem. These results comprised some 12,000 fourfold tables observed under approximately random sampling conditions. From them Fisher [*C P* 49] calculated the actual average value of χ^2 which he had proved earlier should theoretically be unity and which Pearson still maintained should be 3. In every case the average was close to unity, in no case near to 3. Indeed, Fisher pointed out that the general average, with an expected value of 1.02941 ± 0.01276, was "embarrassingly close to unity" at 1.00001, and from this he inferred that the sampling conditions had not been exactly random. There was no reply.

The work on distribution theory and tests of significance was in a sense preparatory to the work on estimation. Fisher was considering these problems concurrently in the early 1920s. Here the method of maximum likelihood came into its own.

The first paper on estimation (*C P* 12, 1920) compared current methods of determining the accuracy of an observation from a normal distribution. The two estimates of error most commonly used were the mean deviation m and the sample standard deviation s. The paper was, in particular, a mathematical examination as to which of these two statistics gives the better estimate of the true standard deviation of a normally distributed population of observations. It was a new idea that the merits of these two estimates could be compared with precision, and in considering them, Fisher had in mind applications of the same method to the wider question of the selection of the best estimates in general. In an introductory paragraph, he said, "The case is of interest in itself and it will be found that the method here outlined is illuminating in all similar cases when the same quantity may be ascertained by more than one statistical formula." His comparison showed that the sample standard deviation *s* is a better measure of σ than the mean deviation and, further, that it could not be improved by taking any other measure of spread. He showed that the sample standard deviation *s* has the unique property which he later called *sufficiency*. For any *given* value of the sample standard deviation *s*, the distribution of the mean deviation *m* is independent of the true error value of σ. Thus the whole of the information the sample contains respecting the true standard deviation σ is summed up in the sample standard deviation *s*.

The discovery of the property of sufficiency in the case of the standard deviation made it possible to hope that other sufficient statistics could be found by the method of maximum likelihood, and even that that method would in all cases provide sufficient statistics. If that were so, then the problem of estimation would be completely solved, for the property picks out one particular method of estimation as uniquely superior to all possible alternatives. By 1922 Fisher had gone far in examining the properties of the likelihood function and the properties of the estimates arrived at by maximizing that function. He presented his findings in a massive paper "On the mathematical foundations of theoretical statistics" [*C P* 18].

It was a *tour de force*. Although he apologized for its incompleteness, for example, in that he had not found any way to prove whether sufficient statistics would always exist, nonetheless it was conceptually a complete consideration of the asymptotic case in which the statistics are estimated from large numbers of observations. It stands as such a monument in the development of statistics that we pause to consider, without entering into mathematical detail, a few of the salient features of this 1922 paper.

Many of the ideas, of course, were not new. Gauss had used maximum likelihood to get least squares estimates, using, however, a Bayesian justification. Fisher had delivered it as if by Caesarian section from its Bayesian context and presented it as an individual in its own right. The requirement of consistency which Fisher introduced was no more than the common-sense requirement that the sample statistic should approach closer and closer to the true value as the sample size was increased. Fisher added the criteria of sufficiency and efficiency (the later meaning that the estimate has minimum possible variance in large samples). He acknowledged a debt to Student for the first work on sampling distributions and to Pearson, in particular for his development of the χ^2 test of significance and in general for the benefits his work had brought to the subject of statistics:

We must confine ourselves to those forms we know how to handle or for which any tables which may be necessary have been constructed. Evidently these are considerations the nature of which may change greatly during the work of a single generation. We may instance the development by Pearson of a very extensive system of skew curves, the elaboration of a method of calculating their parameters and the preparation of the necessary tables, a body of work which has enormously extended the power of modern statistical practice and which has been, by pertinacity and inspiration alike, practically the work of a single man.

Whatever its precedents, the totality of the 1922 paper was something new. The aim was clarity and precision, and the paper opened with a section of definitions. There were a number of innovations: the terms *consistency, sufficiency, efficiency, intrinsic accuracy, isostatistical regions, validity,* and *likelihood* were Fisher's expressions, new-minted as technical definitions of distinct ideas. In the text he commented that there had been a confusion of terminology because the same names—(mean, standard deviation, and so forth)—had been applied both to the true value, which we should like to know but can only estimate, and to the particular value at which we happen to arrive by our methods of estimation. He disentangled the two concepts, defining on the one hand the *hypothetical infinite population* whose *parameters* we should like to know and on the other the *sample* whose *statistics* are our estimates of the parameter values. The distinction was necessary at a time when, as he wrote: "It appears to be widely thought, or rather felt, that in a subject in which all results are liable to greater or smaller errors, definition of ideas or concepts is, if not impossible, not a practical necessity." To him it was just that: a practical necessity. The statistician who permitted vagueness to rule abdicated his professional responsibility for the refinement of precise methods of statistical analysis. Only by clear conceptions could one recognize the sorts of errors to be considered inherent in the data and distinguish them from errors arising from a failure to utilize all the information in the data. It is the first time *information* enters the technical vocabulary although Fisher owns no responsibility for "information theory" as it developed in the 1950s. He began with the notion that there was a precise *amount of information* in a sample and the proposition that it was the job of the statistician to ensure a minimal *loss of information* in the reduction of the data. Ideally, estimation could be *exhaustive,* as when sufficient statistics were available.

On the question of sufficiency, Fisher showed in the 1922 paper that the method of maximum likelihood gives sufficient statistics where they exist, for the method leads to the results that satisfy what was later called the "factorisation criterion" which is, in fact, "the mathematical expression of the condition of sufficiency," as Fisher pointed out at the time. Thus when sufficient statistics exist, the method is ideal.

Sufficient statistics do not always exist—and it was to be some years before Fisher was to know that they do not always exist—but he already had a second string to his bow, less satisfactory because it only applied to the asymptotic case (to large samples). He was able to show that for such samples the method of maximum likelihood always provides efficient estimates and was thus of immediate practical utility in a very wide range of problems. It provides methods of estimation for nonnormal (and correlated normal) material and quantal material, equivalent to those that had long been available for normally, independently distributed material by the Gaussian method of least squares, which is a special case of maximum likelihood.

In contrast, an examination of the method of moments revealed that it possessed low efficiency in large areas of the Pearsonian system of frequency curves, and only for a very limited area, close to the normal point, does it approach efficiency. Thus, by using the method of moments, statisticians were in effect throwing away a proportion of their carefully gathered data: if they were lucky as little as 20% but if they were unlucky nearly all of the value of their observations could be squandered in the process of estimation. This was particularly likely to be true in the estimation of the parameters of the nonnormal curves, for which the method of moments had been most employed in Pearson's statistical laboratory, since that method is incompetent to estimate any markedly nonnormal distribution.

Much of the 1922 paper was taken up with the examination of the efficiency of the method of moments, but the examination itself was really only an excursus from the main line of argument, justified by the fact that at the time no analytically competent discussion of the matter existed. In a similar spirit, Fisher allowed himself—even in a paper as theoretical as its title, "The mathematical foundations of theoretical statistics," suggests—to dwell on the details of certain other applications of the theory.

It was characteristic of Fisher that these illustrations should be allowed prominence. His concern for clear and precise notions in the theory of statistics was not that of a mathematical aesthete who cultivates purity for its own sake but rather the care of a craftsman that his tools be fit to deal with the materials he handles in his daily contact with the physical world. Practical considerations weighed with him, and he considered mathematics as the kit of tools of his trade, a way of handling practical problems in scientific experimentation.

In 1925 [C P 42], Fisher elaborated and expanded the results obtained in 1922. He showed that among all efficient statistics, that derived by maximum likelihood loses less information asymptotically than any other. However, by considering their distributions, he showed that some statistics that were efficient for large samples could nevertheless have low efficiencies for small samples. But for maximum likelihood estimates from small samples, the loss of information sustained when there were no sufficient statistics could be made good (provided a simple condition holds) by the use of ancillary statistics of the likelihood function.

Thus were the foundations of theoretical statistics laid, on which a large part of the subject still rests, a great and elaborate structure raised by many hands in the succeeding half-century. It is now almost impossible to realize that between the first publication concerning the method of maximum likelihood and its acceptance a decade or more later, Fisher was alone in consciously attempting to make good the foundations of the subject. It is easier to guess why he continued alone, against the tide: being aware of a great practical need, he could envisage some of the consequences of success. At the same time his sense of the need for clarification was sharpened both through the confusion which continued to plague the literature in the absence of firm theoretical foundations, and through his personal experience of experimental problems at Rothamsted.

4

Rothamsted Experimental Station

Rothamsted Experimental Station had its beginnings in the 1830s when the development of the science of organic chemistry was exciting the foremost chemists in Europe and its applications foreshadowed remarkable extensions in the science of physiology. Justus von Liebig, the celebrated professor of chemistry at Giessen University since 1824, first drew attention in England to the importance of organic chemistry for the subject of plant and animal nutrition. In an address delivered to the British Association at Liverpool in 1837 he urged British scientists to study organic chemistry in order to understand life processes. His message was so well received that he was invited to prepare a report on the subject for the association and he responded in 1840 by dedicating to the British Association a volume that was to have an enormous influence during the rest of the century, *Organic Chemistry in its Application to Agriculture and Physiology.*

Though Liebig was the most illustrious, and the most controversial, he was by no means the only able European chemist concerned about nutrition. J. B. Boussingault, professor of chemistry at the University of Paris, had in 1834 set up an institute at his farm at Pechelbronn in Alsace for the study, in particular, of the nitrogen cycle. In England in the 1830s, it was the work of

C. G. Daubeney, professor of chemistry at Oxford from 1822, that brought to an end a barren period in the development of agricultural science. Daubeney gained a broad understanding of problems in the geological and chemical history of the earth by his extensive travels in Europe, combined with his researches at Oxford on natural mineral waters.

One of Daubeney's students, John Bennett Lawes, left Oxford in 1834 to return, with his widowed mother, to the family manor of Rothamsted which lies about 25 miles north of London at Harpenden in Hertfordshire. Having spread bone meal on his heavy Hertfordshire clay for three consecutive years without benefit, he was challenged to find out why bone products, known to be effective phosphatic manures on sand, peat, chalk, limestone and light loams, failed on heavy loams and clays. In 1837 he turned a barn into a chemical laboratory and spent the next few years trying applications of bone meal, burnt bones, and various types of mineral phosphates with different quantities of sulphuric and other acids. He found the treatment with an appropriate quantity of acid gave very satisfactory results. In fact, it supplied the acidity which was already present in the lighter soils and which effected the chemical change in the phosphates from the inert dicalcic form to the active monocalcic form, called superphosphate.

In 1842 Lawes brought out a patent for the manufacture of superphosphate as an agricultural fertilizer. He set up a factory, and the business prospered. Superphosphate was immediately welcomed by farmers because of its notable success in increasing the yield of root crops, one of the props of the four-course rotation and a necessary winter foodstuff for sheep and cattle.

Bone products were always in short supply, but mineral phosphates soon became available in quantity, for example, by Daubeney's interest in utilizing the great Estremadura deposits in Spain, which stimulated the building of a railway to carry them to the coast. Lawes had a virtual monopoly for the production of superphosphate as an agricultural fertilizer, because his patent referred to all mineral phosphates yielding phosphoric acid. When his patent was challenged in 1850 and again in 1853, he did not attempt to defend the clauses referring to bone products but only the manufacture of superphosphate from mineral phosphates. His patent was upheld, fortunately for the laboratory at Rothamsted, which Lawes maintained and later endowed from the profits from his business.

In 1843 the laboratory had been established with the coming of the trained chemist, J. A. Gilbert, and with the layout of the first continuous field experiments, with wheat on Broadbalk Field and turnips on Barnfield. These experiments and others with barley and clover were given their final form in 1852, and grass plots were laid out in Rothamsted Park in 1856. Work on superphosphate ceased, and the inquiry broadened to embrace the whole range of problems in plant nutrition.

Rothamsted profited by the complementary characters of its founders: Lawes was a sound policy maker with administrative ability, acumen and the wide-ranging interests of the country squire; Gilbert had skill and meticulous care in chemical analysis and began the invaluable records, chemical, botanical and meteorological, which were maintained and extended during the rest of the century under his supervision.

The Lawes Agricultural Trust was formed in 1889 for the endowment of a continuing research station, but the work of Rothamsted continued to the end of the century under the personal guidance of its founders. Men of the village, taken on as boys, had learned specialized tasks in the routine of the station and grown old in its service. By the time of the death of Lawes in 1900 and Gilbert in 1901 Rothamsted was a marvellously efficient small institution but so conservative as to be scientifically moribund. Its revitalization became the task of A. D. (later Sir Daniel) Hall, then director at Wye College, when he accepted the position of director at Rothamsted on the death of Gilbert. During the 10 years of his directorship he raised money to bring in new men to lead new departments of soil science, bacteriology, and botany, thus beginning the infusion of youthful trained scientists, which his successor continued.

In these years two factors gave further impetus to the work at Rothamsted. First, it was the period when agricultural science gained academic standing. New university farms and schools of agriculture, like Wye College (associated with the University of London), provided a new class of trained scientists to agricultural research. Second, as the agricultural depression deepened, it became evident that the whole agricultural industry was in need of planned assistance. In 1909 the Chancellor of the Exchequer, David Lloyd George, brought in an act to expand the resources of the countryside and coastal regions by proper scientific development of afforestation, agriculture, and fisheries, under the guidance of properly equipped experimental forests and experimental farms. Rothamsted became a founding member, the most important of the network of institutions working together under the scheme, and it was a large beneficiary of the development fund. This meant an adequate financial basis for an expanding public responsibility. At the same time Hall became one of the commissioners who administered the development fund, and in 1912 he resigned from Rothamsted. He was succeeded as director by E. J. (later Sir John) Russell, who some years earlier had followed him from Wye College to Rothamsted.

During World War I agricultural scientists were called upon to take an increasing part in dealing with problems of food production. Russell not only managed to keep the Rothamsted Experimental Station going but to get a building completed to replace the structure that had collapsed in 1912, thus providing some increased and improved accommodation. When the war ended he was in a position to begin reconstruction and expansion. The physics department, begun in 1913, resumed on the return of B. A. (later Sir Bernard) Keen from active service. The department of bacteriology opened under H. G. (later Sir Gerard) Thornton, newly appointed after demobilization in 1919. The entomology and organic chemistry departments, begun in response to needs revealed during the war to study problems of infestation and insecticides, were strengthened, as were the original departments of chemistry and botany. The department of protozoology was recreated after the deaths in France of its two staff members. Research in plant pathology, which had previously been carried out by W. B. Brierley (mycologist) at Kew and by A. D. Imms (entomologist) at Manchester, was moved to Rothamsted under the same men after the war. Imperial College set up a laboratory of plant physiology at Rothamsted, and in 1923, the Bee Section of the Ministry of Agriculture was also transferred to Rothamsted.

When Fisher joined the staff at Rothamsted in October 1919, he found himself among a dozen or so specialists in various agricultural sciences, many of them graduates of Cambridge University. Among them he enjoyed the sort of fraternal companionship he had lacked since his college days. His scientific curiosity found satisfaction and stimulus as he came to know more of the work going on in the various departments. He settled down very happily under Russell's informal and mild direction, unimpeded by bureaucratic controls or irrelevant demands on his time. Indeed, Rothamsted offered him an environment almost ideally suited to his temperament, and he was never happier professionally than during the years he spent there.

□ □ □

Fisher's appointment was quite temporary and experimental in spirit. Sir John Russell [7] has recorded how he came to embark on the new venture. He had applied to both Cambridge and Oxford Universities for a young mathematician "who would be prepared to examine our data and elicit further information that we had missed." Neither could help. In Cambridge, however,

A member of the Committee, Dr. Horace Brown, introduced me to Ronald Aylmer Fisher . . . . when I first saw him in 1919 he was out of a job. Before deciding anything I wrote to his tutor at Caius College, whom I knew personally, asking about his mathematical ability. The answer was that he could have been a first class mathematician had he "stuck to the ropes" but he would not. That looked like the type of man we wanted, so I invited him to join us. I had only £200, and suggested he should stay as long as he thought that should suffice, and after studying our records he should tell me whether they were suitable for proper statistical examination and might be expected to yield more information than we had extracted. He reported to me weekly at tea at my house and always favourably. It took me a very short time to realize that he was more than a man of great ability, he was in fact a genius who must be retained. So I set about obtaining the necessary grant.

Initially, Fisher took lodgings in Harpenden. That year he had earned only £150 so far; he could not have afforded to move house, even if he would. It seemed best in any case to leave the family at Great House Cottage, where they were happily settled, until he saw how things turned out at Rothamsted. When the appointment was made permanent at the end of the first year, he looked for a house nearer his work; meanwhile there were sometimes weekend visits with his family and many lonely evenings without them.

Dr. W. A. Roach, a colleague working on insecticides at Rothamsted, recalls this time:

In those days he was always willing to walk and talk well into the small hours. This was most fortunate for me because I slept badly. The magic word "Arab" would start him on a fascinating discourse on Arab culture which only stopped when I was physically tired enough to hope for sleep and said "Goodnight, Fisher," on his doorstep. His discourse was not quite uninterrupted. The sight of a nearby pig never failed to cause him to leave off, usually in the middle of a sentence, scratch the pig and utter endearing remarks to it; then he would complete the unfinished sentence and continue walking and talking.

Every now and then Fisher paid me the compliment of "thinking aloud" to me about some problem exercising his mind at the time and I would say: "Oh! Come off it, Fisher, I can't understand that, talk down to my level." And he would talk in language intelligible to one of the meanest intelligence. I am proud to remember occasions when my practical farming experience was of value to him, in particular when he was examining the Broadbalk yields.

But greater minds than mine had difficulty in keeping up with Fisher. Once when making the fourth at bridge, while we were waiting for Fisher to arrive, Professor (later Sir Arthur) Bowley moaned about a difficult paper Fisher had asked him to vet and said, ". . . And I know full well that when I have taken a lot of trouble to make suggestions, he'll ignore them;" to which Gosset ("Student") responded: "You know, Bowley, when I come to Fisher's favourite sentence—'It is, therefore, obvious that . . .'—I know I'm in for hard work till the early hours before I get to the next line."

In 1920 the Fishers moved into a pair of cottages off Watling Street near the village of Markyate. Gudruna and her daughter were with them still, for a short period; Ron's sister Phyllis joined them after the death of her father, and Katie was born, a beautiful little sister for George. For a time Ron rode to and from Markyate on horseback. Roach recalls,

He took lessons from a Harpenden man who told me how he enjoyed the lessons and how much he learned from Fisher on a number of subjects unconnected with horsemanship. He was reticent about how much he imparted to Fisher. One day Fisher, smiling more beamingly, even, than usual, told us that as he, on his horse, emerged from the village the children were in the habit of gathering on a bank by the roadside and crying: "Hellaoo Jeeesus!"

In those days when children "collected" beards, he was pleased when some boy let out an exultant whoop as he passed—"Red beaver!"—to realize he was a rare scoop.

For the most part both Ron and Eileen relied on their bicycles for transportation, and it was on bicycles they set out for the seaside in the summer of 1921, to enjoy the first holiday of their married life, each carrying one of the children on a back carrier. It was to be a tragic event.

On the beach one day the toddler George dropped a handful of pebbles into the baby's open mouth. Eileen quickly drew out several from the back of the mouth. Ron took Katie up by the heels and shook out another, but it became evident that one had entered the bronchus. The baby wheezed; her sleep that night was restless, and she had fits of screaming, and choking. In the morning Eileen took her to a doctor in Bridport. He was unable to find the stone and sent them to Dorchester for an X-ray, which also failed to show anything. While they were still in the hospital late that afternoon, Katie suffered a frightening fit of choking, and Eileen demanded surgery. The operation failed to find the stone. A second operation the following morning revealed it lodged in the right bronchus. It was removed, and for a time Katie's breathing was completely free from noise and effort. But bronchopneumonia supervened. By evening Katie's temperature had soared, and she died at about three in the morning.

Eileen was desolated. For months afterward her tears came often, uncontrollably. Six months later she was taken to visit friends who had a new baby and had to retire, overwhelmed with a paroxism of grief. Fifty years later the thought of Katie brought tears to her eyes; it was a hurt, an emptiness, nothing would quite recover.

The work of the primitive cottages kept her occupied. She worked in the garden; pumped water from the garden well and filtered and boiled it, for it was unclean; chopped firewood; boiled the great kettle for baths in the tub set before the kitchen fire, just as she had at Great House Cottage. In the evenings, when supper was cleared and George in bed, the women took up their handiwork, as before, and reading and conversation filled the hours. It was as if Fisher had forgotten. But when, later. he came to look at the consequences of a child's death occurring early in a family, he understood, of his own experience and Eileen's, how such a death might raise desire for more children and produce the overcompensation he found in the data.

Swedish data collected by A.B. Hill showed a correlation between large family size and higher child mortality, and the higher mortality was attributed to the poorer socio-economic conditions of the larger families. Fisher, however, observed that child deaths in these families were concentrated at the beginning of the family and affected especially the first or second child. Thus the death preceded the large family; it could not be the consequence, but, he argued, it was the most likely cause of the eventually large size of the family. He later confirmed that such overcompensation for the early loss of children does take place. Victims of acholuric jaundice, who frequently lost children by miscarriage or infant mortality, were found actually to have more numerous living offspring than their normal sisters.

Soon the Fishers found a house in Harpenden, and on October 7, 1922, George's third birthday, they loaded their household goods on a cart, tethered the goats behind, and made their way to Milton Lodge. Eileen was expecting.

The household must have seemed shockingly unconventional to the residents, for Harpenden was a small town dominated by middle-class inhabitants, some retired, others with comfortable and dignified daily work in London or near-by Luton. Milton Road was the hard-core of Victorian Harpenden: large detached houses, with gardeners tending well-kept gardens and nannies in charge of decorous children. In contrast, soon the neighborhood became aware of a noisy group of young Fishers, barefoot in summer, whose games might spread from the unkempt front garden to the pavement and gutter outside.

Sir Bernard Keen recalls how Fisher himself appeared to his colleagues at Rothamsted:

Our early impressions of Fisher were of a shaggy man: copious brown hair, moustache and beard, the jacket fortified with leather patches on elbows and cuffs, the shapeless trousers, and the heavy but serviceable country boots; he was not interested in dress. But his quizzical eyes behind the thick pebble glasses, the cultured and pleasing voice, the elegant turn of phrase, devastating or urbane, his bubbly pipe, stuffed with peculiarly pungent tobacco, are vivid memories.

Initially Fisher made little scientific impact on the Rothamsted staff. He was absorbed in working over the voluminous past records, and in the mathematical development of statistics for small samples. But we quickly recognized, and enjoyed, his readiness to discuss, authoritatively, widely differing subjects.

Such was the man who in 1919 settled down to the desk he was given in the Library at Rothamsted, handy to the records, to study the long sequence of yields of wheat from Broadbalk.

There were a lot of data, all the yields, analyses, and supplementary information recorded throughout the years during which the 13 parallel plots of Broadbalk had received each its allotted kind and quantity of farmyard manure or artificial fertilizers. Lawes and Gilbert had extracted much information by personal observations at the time. Hall had taken 5- or 10-year averages to find general results and had compared results obtained in wet seasons with dry seasons. Could the data yield more?

In fact, the data yielded unexpected fruits, as one can see from the series of articles emanating from Rothamsted Statistical Department during the 1920s under the general title, "Studies in Crop Variation." From these studies we may see how, concurrently with his work on estimation, between 1919 and 1925, Fisher was led to the analysis of variance procedure and to the principles of experimental design. Within 7 years he had solved all major problems and placed in the hands of the experimenter both the techniques for conducting experiments and the mathematical and arithmetical procedures for making sense of the results.

How should we explain the fecundity of this period? Fisher was, of course, in his early thirties, when maturity often lends weight to the brilliance of youth; but also (and the same pattern recurs later in his genetical work), a period of theoretical consolidation preceded the opportunity of experimental applications and, when applications became possible, the experimental situation itself stimulated new advances. Fisher's studies of distribution theory and his work on estimation so far had been undertaken independently of empirical results. In contrast, the "Studies in Crop Variation" [C P 15, 32, 37, 57, 78] show how theoretical concepts were mobilized to meet the needs of particular applications and how these applications in turn promoted new developments in theory, new applications, and new analytical techniques.

The first thing Fisher wished to do with data on the yields of wheat from Broadbalk was to analyze the components of variation of the yields. In doing so, he employed ideas originating in his work on the correlation of relatives [C P 9], later to enter into the analysis of variance: the additive property of the variance itself, and the relationship between correlation and variance.

It tends to be forgotten in these days that before 1920, problems that today would be dealt with by the analysis of variance, were usually thought of as problems in correlation. If there were n families and k siblings in each, one could calculate the intraclass correlation among siblings by averaging correlations. But since with k siblings, there would be $k(k-1)/2$ pairs to average over, the direct method was tedious. J. A. Harris (1913) had proposed an abbreviated method of calculation for the intraclass correlation, which simply involved the identity between the sum of squares of deviations from the average over all families and the total of the corresponding sums of squares between and within families. Fisher used this method to estimate not correlations but the contributions of component variances to the total variance.

It is no accident that in *Statistical Methods*, he introduced the analysis of variance by way of intraclass correlation. He pointed out that when proper account is taken of the concept of degrees of freedom, and when r stands for the calculated intraclass correlation, the ratio of the between to within mean squares is

$$F^* = \frac{1 + (k - 1)r}{1 - r}$$

so that

$$z = \tfrac{1}{2} \log_e \frac{1 + (k - 1)r}{1 - r}$$

Thus since F (or half its logarithm z) is simply a function of r, all we can know about r is included in the analysis of variance table and, specifically, in F or z. Fisher noted that in the special case where $k = 2$,

$$z = \tfrac{1}{2} \log_e \frac{1 + r}{1 - r} = \tanh^{-1} r$$

and he was back to the transformation he had used earlier for the correlation coefficient (the particular case of assessing the correlation between twins, when $k = 2$, having been dealt with in $C P$ 11, 1919).

If Fisher could distinguish between independent variance components of the Broadbalk data, he could take advantage of the method and so estimate the relative values of the variances contributed by different causes and assign with more or less accuracy the several portions to their appropriate causes. The analysis was, however, complicated by changes in the yields, of unknown cause, which occurred over time.

Fisher made graphs of the actual yields of the plots during the 67 years (1852–1919) for which comparable data were available. As expected the graphs showed progressive deterioration over the years, but the changes in mean yield (see Figure 2) were by no means fully expressed by the simple deterioration. He was immediately struck by an unexpected feature they all displayed: the mean yield rose between 1852 and 1860 and, after a bad period in the seventies, reached a second maximum in the nineties. The winding snail-trail of the slow changes was common to all the Broadbalk plots from the

---

*The use of the symbol F for the ratio of mean squares was introduced later by Snedecor.

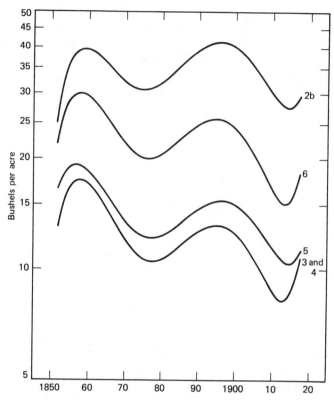

Figure 2. "Showing the course of the changes in mean yield in continuously manured plots of Broadbalk. The vertical ordinate is plotted on a logarithmic scale in order to show the proportionality of the slow changes in the several plots, and in order that the relative importance of soil deterioration may be compared." (From C P 15, 1921.)

richest to the poorest, whose yields, depending on the fertilizer applied, ranged from over 35 bushels per acre to a miserable 12 bushels per acre. The pattern was not found to be a feature of other fields, and it was not attributable to the weather. Whatever its cause, it was evident that the time trend must be eliminated from the data before the effect of known causes, like rainfall, could be accurately assessed.

In his reviewing Fisher [C P 7] had particularly noted a method for the elimination of time trends, suggested by Student. Student had proposed to represent the time trend by a polynomial. Changes over time could thus be eliminated, leaving only random variations of the observed values about a steady mean, and those could be analysed in the ordinary way.

The fitting of the polynomial expression in the manner suggested by Student was, however, impracticable in the context of the Broadbalk data. The labor involved in fitting perhaps half a dozen polynomial terms to the data for each of the 67 years and thirteen plots in itself was enough to cool the ardor of any investigator. Moreover, the estimates of the terms in the polynomial were mutually correlated. Since Fisher wished to allocate portions of variation to *independent* causes, he required uncorrelated, or orthogonal, values capable of distinguishing between the annual fluctuations, the progressive deterioration, and the slow changes. So he invented suitable orthogonal polynomials, functions of Student's polynomial terms but uncorrelated with each other. These provided an exactly equivalent result that was computationally manageable because the orthogonal polynomials could be calculated once and for all for the whole series of annual data.

In "Studies in Crop Variation, I. An examination of the yield of dressed grain from Broadbalk" [C P 15, 1921], Fisher shows how orthogonal polynomials may be obtained uniquely in succession. The estimated coefficients are uncorrelated with each other and are therefore independent of the number of terms taken. The coefficients are chosen to make the residual variance a minimum at each stage; consequently, the residual variance is reduced with every additional term taken. In the limit one could theoretically fit the mean and 66 terms to the 67 annual yields, making an exact fit to the data and accounting for the whole of the variance. If there were no time trend of any kind, the variances contributed by each polynomial term would, on the average, equal the fraction contributed by each individual observation to the total variance. In practice, the slow trends could be accounted for by a few terms only (Fisher chose five), and the residual variance of the $66 - 5 = 61$ higher-order terms could be used to estimate the variation about the fitted curve.

The concept of orthogonal polynomials thus led to the dividing up of the variance in proportion to the number of degrees of freedom each component enjoyed. There is no mention of "degrees of freedom" in the paper [C P 15], but the appropriate numbers are implied there in Fisher's Table II. The table is of interest in showing the stage of his development of the modern analysis of variance table in the autumn of 1920. In his words:

In Table II is shown the analysis of the total variance for each plot, divided as it may be ascribed (i) to annual causes (ii) to slow changes other than deterioration (iii) to deterioration; the sixth column shows the probability of larger values for the variance due to slow changes occurring fortuitously.

Below is displayed the entry for Plot 2b (farmyard manure) of this table.

Plot	Annual Causes	Slow Changes	Deteri- oration	Total	P for Slow Changes
2b	33.2 (61)	17.6 (4)	.4 (1)	51.1 (66)	.000,002

Fisher's table did not include the degrees of freedom, shown here in parentheses for each sum of squares, but it did show the probability for slow changes based on the appropriate number of degrees of freedom. The component due to "annual causes" (fluctuations in yield due to seasonal differences) was used as the error term with which to compare the values of the other variance components.

Before making this analysis, Fisher had to consider how many coefficients of the orthogonal polynomial he needed to represent the curve of the time trend sufficiently, without incurring the liabilities he saw in fitting excessively elaborate curves. He decided on five terms, and he checked the adequacy of the fit of the curve by inspecting the correlations between neighboring residuals. He noted that negative residual correlations are actually induced by the process of fitting an adequate curve, and therefore he compared the correlations of residuals with those expected to be induced by the fitting process. His values were not significantly different from expectation, and he concluded that the time trend was effectively eliminated by fitting the five coefficients of the orthogonal polynomial. Fifty years later, the problem of how best to determine the adequacy of fit of a polynomial of any degree has not been satisfactorily solved, nor has Fisher's suggestion been further explored that autocorrelations might be the most informative functions of the residuals to use in answering the question.

We may note another new statistical maneuver here. In discussing the serial correlation of residuals on the dunged plot, with a view to establishing the appropriateness of the degree of the polynomial he had fitted, he made a z transformation of the correlation coefficients. This is undoubtedly the right thing to do, but it had not been done before. In introducing the z transformation about this time [C P 11, 1919; C P 14, 1921], he used it for a correlation calculated from a variate regressed on some other variate; here it is used to transform correlates of a variate regressed on its own previous values.

As Bowley had said, Fisher's goods were not in the shop window. The analysis of variance had no public debut, in 1921 or later when it achieved its mature form, but was slipped in as part of the entourage of the Rothamsted experiment whose interpretation was the object of the paper. Similarly, orthogonal polynomials were here introduced into statistics for the first time, without comment, and a new use was found for the z transformation. Indeed,

Fisher was rather ambiguous about the derivation of some of his figures (those in "the analysis of the total variance" of Table II are actually sums of squares), but he was very clear in discussing what the data could tell about the effects of manurial treatments and the causes of the slow changes whose reality he had established by his analysis.

The investigation of the slow changes reads like detective fiction. It was evident from the test of significance that the slow changes in the mean yield were not fortuitous. Fisher advanced various arguments suggesting that they were not produced by changes in the weather. He dug down into the records seeking some agent working slowly enough to produce the tidal effect and the local variations he had discerned within the wheatfield, and he pulled out— weeds. In the 1850s a great deal of hand weeding had been done, for example, in 1853, during a little over 100 days of weeding, the work done was equivalent to the continuous labor of two men with hoes and seven boys hand weeding on a field of 14 acres.

The dominant weeds were four perennials, but in the absence of boy labor in the 1880s, the slender foxtail grass, an annual, had become enormously abundant and had continued troublesome ever since. The Education Acts of 1876 and 1880 had cut off the regular hand labor of the boys. In 1886 and 1887 hand weeding was resorted to, but the two wet summers that followed prevented this operation, and the field again became exceedingly foul and was therefore partially fallowed in 1890 and 1891. During the 1890s parties of schoolgirls were employed at Easter and Whitsun, on Saturdays and in the evenings. Fisher noted that "Sir John Lawes took much interest in the work, giving prizes to those who collected the greatest quantity." Thus the fluctuations of wheat yield reflected the pattern of fluctuations in the care shown for the condition of the field.

□ □ □

"Studies in Crop Variation, I" is a very pretty piece of analysis, full of originality and painstaking caution in checking the steps of the analysis. Substantial as it was, however, Fisher regarded it as preliminary; his paper closed with the words:

It is believed that the deviations from the smooth curves, which have been freed, for the most part, from the effects of exhaustion and weeds, form statistically homogeneous material for the study of meteorological effects.

By September 1923 Fisher had made this study. He reported his findings in a massive paper [*C P* 37, 1924), conceived as "Studies in Crop Variation, III:

The Influence of Rainfall on the Yield of Wheat at Rothamsted," but actually published not in the series in the *Journal of Agricultural Science* but in the *Philosophical Transactions of the Royal Society,* in view of the fundamental importance and very general interest of its contents. Not only was the topic of the influence of weather on crops currently popular among agricultural scientists, but its consideration introduced the whole complex of problems with regard to the use of multiple regression coefficients, a potentially powerful tool in a much more general scientific context. These were investigated in the paper before the particular analysis was made.

To start with, Fisher pointed out the difficulties and dangers of statistical analysis of meteorological data. First, he adverted to the difficulty of defining "weather," and the excessively large number of variates which could be regarded as constituting weather:

If we wished to analyse the sequence no more closely than by monthly averages, we should still have 12 values for rainfall, and 12 more for minimum and maximum temperature, dewpoint, grass minimum, solar maximum and soil temperature, nor would it be unreasonable to include . . . "Hours of Bright Sunshine" and averages for the direction and force of the wind. The number of meteorological elements might be made to exceed even the longest series of crop records available.

in which case "the number of unknowns would exceed the number of equations for them to satisfy." Second, he considered the dangers of selecting, according to common practice, those variates that in a particular body of data looked as if they had large effects, and to make this point he determined the null distribution of the multiple correlation coefficient. Third, he noted that these difficulties were enhanced by paucity of data, which further increased the chance the investigators would be led astray by spurious correlations.

As if the technical difficulties were not enough, he went on to discuss the practical difficulties. The calculation of the coefficients for rainfall alone was practically impossible, if the year were subdivided into discrete periods short enough to be meaningful. "Whereas the evaluation of determinants of 6 rows and columns is sufficiently rapid, disproportionate labour is involved in increasing the 6 to 12, and if 52 unknowns were attempted the labour involved would become fantastic." But even if the coefficients could have been calculated (as 50 years later they could have been on an electronic computer), the result would still have been imperfect, since the effect of rain does not fall in a series of discrete periods but may be expected to change continuously during the year. Fisher showed the task to be impregnable to direct attack.

Only then did he reveal that he had the key to the back door. He proposed to fit what would today be called a "transfer function."* If the value of the yield is a composite of all the effects of rainfall the crop has experienced during the year's growth—its memory, as it were, of the experiences of a lifetime—then, for a given rainfall, the yield records the weight the crop has attached to the experience of rain at different times, and the transfer function is the curve representing the continuously changing value of that weight throughout the year in terms of the additional bushels of wheat at harvest for every additional inch of rain. Transfer functions for three of the plots of Broadbalk, shown in Figure 3, illustrate the variety of responses of differently fertilized plots. Thus, though all show yields reduced by rain in January, an additional inch of January rain might result in a loss per acre of only 0.3 bushels on Plot 5 (2% of the average yield), of 0.9 bushels on Plot 10 (as high as 4.6% of average yield), and of 1.0 bushels on Plot 2b (2.8% of average yield).

Fisher proposed that the curve of the transfer function might be fitted by an orthogonal polynomial and that, since the weight changes relatively slowly, only a few coefficients of the polynomial would serve to represent it. Since the rainfall and the yield could also be represented by low-order orthogonal polynomials, it was possible, by what appears a virtuoso performance in juggling with three such expressions, to derive the actual values of the coefficients of the transfer function.

Before undertaking the calculations Fisher considered their practicability. A lot of work was involved, but he was able to show that the numerical manipulations did not require unreasonably heavy computation. By employing a process of repeated summation, the calculations became arithmetically straightforward and easy to check. He showed the manner in which the summation could be made and the further computational advantage to be gained by the symmetry properties which he introduced into the tabulation by taking the midpoint of the series as its origin.

His own experience at Rothamsted gave an edge to his ingenuity in avoiding unnecessarily laborious or complicated arithmetical methods. At first he had a desk calculating machine and a boy assistant. Later he was given a separate alcove in the library with a computing assistant. After some years the department came into possession of its own room and he had a statistical assistant, but he was always constrained to be thrifty with computation. Moreover he was constantly made aware of the need, in practice, for a means

---

*The expression "transfer function" entered the literature through applications in electrical engineering. It has an important place in recent statistical work with time series analysis. In economics, the similar idea of "distribution lags" was introduced by Irvine Fisher in 1925, and the advance of least squares fitting of distribution lags by polynomials is usually attributed to Shirley Almon (1965).

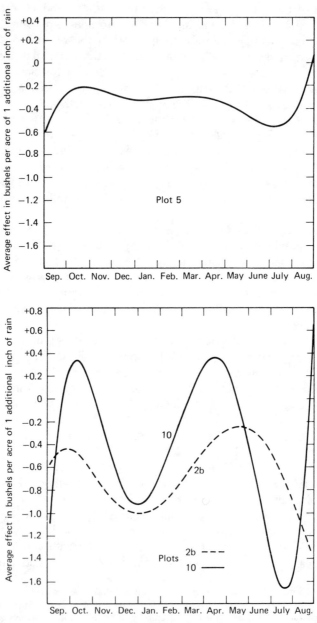

Figure 3. *"Transfer functions" for effects of rain on Broadbalk plots: Plot 5: Mineral manures only, mean yield 14.18 bushels per acre. Plot 10: Double dressing ammonium salts only, mean yield 19.5 bushels per acre. Plot 2b: Dung, mean yield 34.55 bushels per acre. (From C P 37, 1924).*

of checking arithmetical results, especially when the same sort of calculations were made frequently in routine work. Necessity was the mother of invention: one did not lightly undertake calculations that might take months to complete, such as are today undertaken unreflectingly because they can be performed in a matter of minutes on an electronic computer: one looked for ways to simplify and economize. It was good discipline.

Next Fisher devoted about one-third of the paper to devising and applying a series of checks on the validity of his analysis. He found no evidence of inconsistency. The transfer function was acceptable, and he felt justified in calculating the values, as planned, and finally discussing the effect of rainfall on yield under the different conditions of soil fertility represented by the individual plots of Broadbalk.

The main effect discerned was the influence of excessive rain in removing soluble nitrates. The average effect of increased rain was always harmful, October rain being least so and January rain generally the most damaging, particularly on the plots like Plot 10 that stood to lose their working capital of nitrates. (The poorest plots, like Plot 5, had little to lose and the richest, like Plot 2b, felt the loss less severely.) This finding had practical implications, for it suggested that nitrogenous fertilizers, at least, should be applied in the spring rather than in the autumn, if the results from Broadbalk were characteristic. The paper concludes with consideration of the applicability of the results to other soils than that of Broadbalk and of the usefulness of the analysis generally in predicting yields from rainfall records.

□ □ □

In "Studies of Crop Variation I" (*C P* 15), Fisher had split the total sum of squares of deviations from the mean into a number of independent components and made a valid estimate of the component variances by associating with each contributing sum of squares its appropriate share of the total number of degrees of freedom. In "Studies in Crop Variation II, The manurial response of different potato varieties," with M. A. Mackenzie [*C P* 32, 1923], the analysis of variance table was made explicit. Fisher himself was to describe the table as "merely a convenient way of arranging the arithmetic." Most experimenters would prefer to omit the word "merely," seeing how, by its simplicity and comprehensiveness, the orderly display brings them logical clarification about the comparisons possible with particular data, their error, and their precise meaning in the context of the experimental layout.

There was a great deal of confusion on these points at the time. A great number of coefficients were proposed to express different sorts of comparisons within tabulated data. These the analysis of variance served to unify and clarify. It became clear that one set of principles applied to all cases, so that estimating interclass and intraclass correlations, regressions and the correlation ratio, and the making of other special tests were subsumed by the analysis of variance procedure.

Setting up the table in one case, moreover, suggested its use in other contexts. Once it had been used for the fitting of the time trends in the Broadbalk data, it was ready to hand for fitting regression lines and for making oneway comparisons between and within groups, for example, between plots treated with different fertilizers. The analysis of variance table itself might suggest that not only one-way but two-way classifications were possible, for example, varieties and fertilizers could be compared simultaneously on the same plots; when that comparison in turn was tabulated, it would be obvious that the separate sum of squares and degrees of freedom associated with their interaction could be isolated. Similarly, when the experiment was replicated in separate blocks of land, the analysis of variance table displayed the basis on which elimination of block comparisons should be made. And when the new complex experimental designs were formulated, the form of the analysis of variance table was invaluable in clarifying the more complex analytical situations.

Having said this, one has to confess that Fisher's first published analysis of variance table is not correct. The tabulation itself did not prevent him from making a mistake in analyzing data from a complex experiment, though it did set out clearly what was already clear in his mind and did help him afterwards to expose the situation for others.

The experiment was made to test the differences, if they existed, between the manurial responses of twelve different varieties of potato. The field was divided, first, into two equal parts, one half receiving farmyard manure and a basal dressing and the other a basal dressing only. Each half was then subdivided into 36 plots, and the twelve varieties of potato were planted in triplicate, in a chessboard arrangement, on each half. Finally, each plot was split into three patches or subplots; one patch received only the basal dressing, the other two received in addition either sulphate or muriate of potash.

In his original analysis Fisher treated this arrangement as six manurial treatments (the three potash dressings with and without manure) applied to twelve varieties which were (nearly) all replicated three times.

Fisher first assumed (with apologies for oversimplification) that the yield of a given variety with a given manurial treatment was the sum of two quantities, one depending on the variety and the other on the manure. Then the sums of squares of all the deviations from the general mean might be divided into two parts; one measured the variation between parallel plots within triplicates and the other the variation between the means of the triplicates differently treated. Then, Fisher wrote, "If all the plots are undifferentiated, as if the numbers had been mixed up and written down in random order," the average value of each of the two parts would be proportional to the number of degrees of freedom in the variation of which it was composed.

Carrying the analysis further, he distinguished between the effects of variety, manure, and the interaction term, allotting to each their appropriate number of degrees of freedom and their definite fraction of the sum of squares of deviations. He set out the table of the analysis of variance in its now familiar form, with columns for the sums of squares, the degrees of freedom, and the mean squares of the contributing components and the residual error. His analysis of variance table then appeared as shown in Table 1:

Table 1   Analysis of variance table [from *C P 32*, with M. A. Mackenzie, 1923]

Variation due to	Degrees of Freedom	Sum of Squares	Mean Squares	Standard Deviation
Manuring	5	6,158	1231.6	35.09
Variety	11	2,843	258.5	16.07
Deviations from summation formula	55	981	17.84	4.22
Variations between parallel plots	141	1,758	12.47	3.53
Total	212	11,740	—	—

The comparison of particular interest was the interaction term expressing the difference in manurial response of the different varieties (called deviations from summation formula in Table 1).

The assessment of whether differences associated with the various "sources" really occurred was made by comparing the appropriate mean square with that which measured error (variations between parallel plots). The exact distribution of this ratio (now called the F ratio) was not known when the paper was written. Fisher therefore used an approximation based on the fact that the logarithm of an estimated variance is very approximately normally distributed with a variance that depends only on its number of degrees of freedom. Thus the logarithm of the ratio of two mean squares was distributed approximately as a *difference* of the two normally distributed variables. By the time that *Statistical Methods* appeared in 1925, he had tabulated the significance points of the exact distribution of the logarithm of this ratio (z). No significant difference was detected in the response of different varieties to different manures.

The corrected analysis published in *Statistical Methods* two years later is based on only one half of the field. In this analysis the effect of farmyard manure is not considered—the two halves of the field could not, in fact, be compared, for the error difference between the areas involved was confounded with the difference between their treatments. Furthermore, Fisher realized that he had what is now called a split-plot arrangement in which two different variances must be estimated, one the variance between patches within a plot, the other the variance between plots. The first was appropriate for testing for possible differences and interactions involving potash dressings, the second for possible differences and interactions involving varieties of potato.

It is also interesting that at this early stage he was dissatisfied with the analysis of variance procedure on another score. It entailed an oversimplification of the experimental situation. This consideration has not much troubled his successors who employ it without any such scruples. In fact, Fisher stated [*C P* 32, with M. A. Mackenzie]:

The above test is only given as an illustration of the method; the summation formula for combining the effects of variety and manurial treatment is evidently quite unsuitable for the purpose. No one would expect to obtain from a low yielding variety the same actual increase in yield which a high yielding variety would give . . . a far more natural assumption is that the yield should be a product of two factors one depending on the variety and one on the manure.

With the possibility of transformation so much a part of Fisher's everyday thoughts, he might have been expected to consider transforming the data, but in fact, he went on in the latter part of the paper to derive the appropriate nonlinear analysis. He suggested two methods, the first by successive approximation and the second by use of eigenvectors and eigenvalues. The former method has been rediscovered independently, first by H. Hotelling [18, 1933] and later by H. Wold [19, 1966] under the initials NIPALS, as a technique of economic analysis.

5

Tests of Significance

Between 1922 and 1924, Fisher was to be led to the realization that "many recently solved problems of distribution involve only a single family of distributions, that of χ^2, z and t." A multitude of cases formerly treated as unrelated were subsumed under a single, simplifying principle. Understanding of the relationships was accessible to him because of the insights he gained by representing the sample of n observations as a point in n-dimensional space. It seems fitting that Gosset, who had in 1912 provided the stimulus that prompted him to make the geometrical representation in the first place, should have been drawn into close friendship and collaboration as the statistical consequences unfolded.

In 1918 Fisher wrote appreciatively of Gosset's use of the difference correlation method. Gosset [20] answered at great length, and, characteristically, he concluded:

Well, I expect you've about had enough of this. I am sending it *via* University College as I cannot lay hands on our former correspondence and the only clue to your whereabouts is the postmark which seems to be Reading. If I ought to address you as Major or Professor pray forgive me.

Despite the lack of an address, Gosset's letter was safely sent on to Fisher. Fisher's correspondence suffered a sorrier fate, historically speaking: his letters to Gosset have nearly all been lost. Happily, Fisher kept letters from Gosset; later, possibly in preparing the obituary notice "Student" in 1938

[C P 165], he reviewed and added a summary to the collected correspondence [20]. Thus, it is possible to reconstruct something of the vital warmth and the similarity of scientific outlook that characterized the relationship of these two versatile men of science.

In his reply to Gosset in 1918 Fisher admitted he was neither major nor professor, but a schoolteacher looking for a job at the end of the year, and he mentioned some statistical work he was doing on Spencer Pickering's orchard data from Woburn. Immediately Gosset's sympathy was aroused:

Your fruit tree work must be interesting: as a gardener I am of course familiar with alternate seasons of apples and pears and (as a brewer) of hops. I don't know whether you are looking for a job in that line but I hear that Russell intends to get a statistician soon, when he gets the money I think, and it might be worth your while to keep your ears open to news from Harpenden.

Naturally, Gosset was the first to hear 8 months later that Fisher had been appointed to the post at Rothamsted and he was glad, perceiving that "there should be lots of interesting work to be done there and they might easily have got someone there who would have been worse than useless." He was happy to answer Fisher's request for advice on two subjects: the calculating machines he could recommend and information on home brewing.

In March 1922 Fisher sent Gosset some offprints, including the controversial χ^2 article [C P 19]. In a covering letter he suggested that the subject of biometry deserved a separate forum from statistics. Apparently, he had in mind the formation of a society such as Pearson alone could at that time have initiated, an extension of the informal Biometric Club at University College, in which the whole subject comprising the joint interest of biologists and statisticians could be considered. He knew he was the last man Pearson would listen to, and he hoped the idea might be better received if it came from Gosset, who was not only a leader in small-sample work but a former student and continuing friend of Pearson. Thus Gosset was sympathetic to both, but as his reply shows, he retained an independent judgement.

When I am next over I will see how the land lies at University College. Of course if the 'Biometers' are to be any use they should include the leading practitioners, but I rather fancy that Pearson's idea is that it is a sort of University College Club. Besides which, as you say, he is perhaps a little intolerant of criticism, most of us tend to that I fancy as we grow older.

At the time Fisher's suggestion came to nothing. The idea was, in fact, realized only after World War II, when the International Biometric Society was formed and national biometrical societies sprang up in many countries almost simultaneously. By that time the meanings of the words "biometrics" and "statistics" had changed; instead of excluding Fisher's concept of their subject, as they had in 1922, each had become a tributary to it.

Gosset realized that Fisher's work was coming close to the problems Gosset faced. On April 3, 1922 he inquired about the distribution of an estimated regression coefficient, "a problem to which he presumably received the solution by return," as Fisher noted later in summarizing the correspondence. On April 12th Gosset inquired about the distribution of the partial correlation and regression coefficients; Fisher noted, "this also was probably quickly answered, for on May 5th he refers to the solution."

The letter of May 5 shows that Gosset was both pleased to discover how Fisher was applying Student's test in regression work and to partial correlations and surprised at his suggesting its use to test the significance of the difference between two means from different-sized samples. The letter also shows Gosset's characteristic self-deprecation as a mathematician, for, having received Fisher's big paper "On the Mathematical Foundations of Theoretical Statistics," [*C P* 18] he wrote:

I fear that I can't conscientiously claim to understand it, but I take it for granted that you know what you are talking about and thankfully use the results!

It's not so much the mathematics, I can often say "well, of course that's beyond me, but we'll take it as correct" but when I come to "Evidently" I know that means two hours hard work at least before I can see why.

His feelings about Fisher's use of "evidently" and similar terms were to become a favorite joke in his correspondence with Fisher.

In the same letter, Gosset's reference to "tabulating your integral," shows that Fisher's extensions of the use of Student's tables had already given rise to the proposal of a new tabulation and also of a new method of calculating these tables, which was to result in 1926 in the publication of Student's "New tables for testing the significance of observations" [21], that is, the publication of the now familiar values of Student's *t*. Two changes were introduced in this tabulation. First, the number of observations, n', for which the table was entered was changed to the number of degrees of freedom, $n = n' - 1$, which was the more appropriate number. Second, the quantity *t* now tabulated was applicable directly to the ratio of a normally distributed quantity to its estimated standard error. It was, therefore, a more natural quantity to use than that originally tabulated by Student (that is, $t/\sqrt{n}$); moreover, it lent itself much more readily to tabulation.

In calculating the tabulated values in 1908 Gosset [13] had used a method that had the disadvantage that the formula includes an extra term for every two units added to *n*, which for large *n* becomes cumbersome. The original tabulation did not extend beyond $n = 10$. The table had later been calculated by the same method for larger *n* [22] but, in view of the difficulties, it is not surprising that when Gosset checked the two tables in 1923 he declared them to be "both perfectly rotten."

Fisher proposed a new method of attack. Using the value of t instead of $t/\sqrt{n}$, he perceived that the probabilities could be arrived at by an expansion in inverse powers of n. The expansion is asymptotic, so that as n is increased the number of nonnegligible terms of the expansion is reduced, and as n tends to infinity the leading term in t^2 alone remains. Student's formula became more and more complicated with increasing n, whereas Fisher's method became more and more simple, and even for small n, the leading term in t^2 required only four or five correcting terms.

In May 1922 Gosset and Fisher had not met, but Gosset was hoping "sometime" to visit Rothamsted; in August he proposed "next month"; on September 8 he wrote:

I have put off writing to you until I could make plans, though every post seemed to bring a paper from you, for which I thank you.

If it suits you, my sister and I will motor over on Friday 15th, sleep at Harpenden and return next day.

Meeting each other in the flesh confirmed all their favorable preconceptions, and they had a chance to talk with each other about the new tests of significance. Gosset could hardly comprehend that his table could ever be as important as Fisher insisted it was. After the visit he sent Fisher a copy of Student's original tables, "as you are the only man that's ever likely to use them!"

On October 12 Gosset wrote "I haven't had time to do anything with the type VII [Student's] curve—apples at home and business in the Brewery— but I hope to get on to it soon." A month later both men were evidently at work on the new tables, for he wrote on November 7:

. . . I calculated all the values for $t = 1$ from $n = 2$ to $n = 30$ to seven places. . . . Last night I checked your values for $x = 1$ (discovering a slight slip) from your correction formulae and calculated the same values to seven places. As I used the sum of at least four numbers of 7 places they also have an error in the 7th place due to approximations, but the correspondence is quite wonderfully close down to about $n = 10$.

One imagines Fisher working his motor Millionaire at Rothamsted, a large machine on which one turned a crank to set the number and inserted a plunger to start its noisy operation at each step; he provided the correction term formulae and a table of values calculated from his expansion. One imagines Gosset putting his hand-operated Baby Triumphator into his rucksack and carrying it home to work on the tables in the evenings, calculating t by his formula, checking with Fisher's results, and recalculating doubtful values. One imagines E. M. Somerfield, Gosset's first regular statistical assistant, left without the Triumphator and, as Gosset wrote to Fisher,

borrowing from all and sundry. Yesterday I found him with the machine which Noah used when quantity surveying before his voyage. The story goes that he subsequently bartered it for a barrel of porter with the original Guinness. Anyhow he doesn't seem to have been able to keep it dry and Somerfield wasn't strong enough to turn the handle.

Apart from Fisher's calculations and some assistance in checking, Gosset was to recalculate the whole table himself, under the restriction that the hand-calculating machine was in demand at the brewery during half the year. At the end of February he warned Fisher that "people are getting querulous about the machine and I really cannot spare daylight to work on it at the Brewery so I fear that I shan't do much more till next winter." Only in mid-October was he able to resume work: "The tabulating season having now commenced, I took a calculating machine home on Saturday and began work last night."

Meanwhile, in July 1922, shortly before Gosset's first visit to Harpenden, Fisher's paper had appeared on "The goodness of fit of regression formulae, and the distribution of regression coefficients" [C P 20], in which, in section 6, it was shown that the significance of the coefficients of regression formulae—linear or curvilinear, simple or multiple—could be treated exactly by Student's test. Though in May Gosset had seemed convinced by Fisher's argument to this effect, when he read this paper again after visiting Fisher, he became bothered about it; even as he was setting about the calculation of the new t table, he was putting the problem to Fisher with a pertinacity that refused to be quieted, until he could be convinced that this use of his table was, in fact, correct.

Gosset's difficulty was one that has troubled other statisticians in other contexts. It is perhaps the first instance in which Fisher's concept of the "relevant subset" becomes critical. In making the regression of y on x and estimating the significance of deviations about the regression line Fisher had proved that the distribution of the ratio of a regression coefficient b to its standard error followed the t distribution. It was not obvious to Gosset that the resulting test was legitimate, because the sampling distribution of the x's themselves was not taken into account. Fisher, however, argued that it was only the distribution of y's relative to the fixed sample of values of x actually obtained, not to the population of the x's, that had relevance and formed what he was later to call the "relevant subset."

Fisher's paper on tests of significance of regression formulae (*C P* 20) involved Student's *t* in a new and wider relationship which was not at first evident. Having shown earlier that year how χ^2 could be used correctly to test the goodness of fit of frequencies, it was natural to follow up this work by an investigation of the more difficult problem of the goodness of fit of regression lines. In this case it was found that the χ^2 distribution supplies only an approximation and that the true distribution is a Pearsonian type VI curve, which was referred to in the paper as a "modified χ^2." It was, in fact, the distribution of *F* (as it was later to be called), which Fisher was to introduce to the literature, in terms of $\frac{1}{2} \log_e F$, as the *z* distribution.

In March 1923, Gosset consulted Fisher about a problem of Beaven who wanted a formula for the error of a chessboard design. Although Gosset told Beaven he did not think anything could be done for him, he wrote to Fisher for confirmation. Fisher at once set out the calculations for Beaven's problem in the form of an analysis of variance table and, after some correspondence, the problem was sorted out.

The particular interest of this interchange appears in a letter Fisher wrote at its conclusion, from which it is evident that he was already attempting to get out a table of *F* using all the complex relationships of the functions involved: the connections between Student's curve, the normal (Gaussian) and the χ^2 (from Elderton's tables); the mathematical identities linking the probability of *F* with the beta function and the beta function with the binomial distribution; and the identities linking χ^2 with the gamma function and the gamma function with the Poisson series. It seems that he was not, at this stage, using the logarithmic transformation of *F* to *z*.

2nd May 1923

Rothamsted Experimental Station
Harpenden, Herts.

Dear Gosset,

I am glad the error estimate is straight now. A great beauty of splitting the sum of squares into fragments is that each fragment has independent sampling errors appropriate to the number of degrees of freedom. This greatly simplifies tests of significance; for instead of calculating, say, intraclass correlations, with some misgivings as to cross relationships, and performing my transformations and corrections appropriate to such correlations, one only has to make a direct comparison.

I have recently got out the formula involving n_1 and n_2 for such a comparison; it is no more difficult to evaluate than the formula for ρ of Elderton's table, but of course it would be more difficult to tabulate as there are two values of *n*, and the table is triply entered. It is interesting that your Type VII takes the place of the normal curve; n_1 corresponding with Elderton's n' and n_2 with your n. In fact the formula is a partial summation of a binomial, whereas Elderton's is a partial summation of a Poisson Series.

$$n_2 \to \infty$$

Limit

----------------------Student's Curve ----------------------Gaussian	Limit $n_1 \to 2$
----------------Student Elderton complex----------------Elderton	

To turn the limiting Elderton ($n' = 2$) into a Gaussian, we use χ instead of χ^2 but otherwise there is no transformation, and the notation can be uniform.

Yours sincerely,

R. A. Fisher

"Elderton" signifies χ^2 whose values Elderton had tabulated. The "Student-Elderton complex" is the double series of F curves.

By the summer of 1924 Fisher was able to summarize the relationships and the uses of the new function, $z = \frac{1}{2} \log_e F$, when he presented a paper [C P 36] at the International Congress of Mathematics in Toronto. Tables had not yet been calculated.

A year earlier Fisher had suggested the possibility of quoting Gosset's original table from *Biometrika* [13] in the book he was preparing (*Statistical Methods for Research Workers*, 1925). Gosset assured him this would not be necessary; he expected to finish the new tabulation that winter. Besides, "I imagine that they have the copyright and would be inclined to enforce it against *anyone*. The journal doesn't now pay its way though it did before the war and they are bound to make people buy it if they possibly can." In November Gosset had come to the conclusion that he must offer the new table first to Pearson, in case he was prepared to publish it in *Biometrika*. Fisher made no objection, and Gosset went to Pearson with his offering:

23/XI/23

Holly House,
Blackrock,
Dublin

Dear Fisher,

Your interpolation formula works like a charm though why the Dickens you chose those particular values of n I can't think. I can only go on "Watsonin'!"

I was over in London on my way to and from home the weekend before last and dropped in on K.P. I broached the subject of a joint table with an introductory note by you and he was prepared to consider it: he would I think have taken it but for a most unfortunate occurrence. While we were talking he mentioned that he was bringing out a new edition of *Tables for Biometricians* and I said 'Oh there are one or two mistakes in my small table in it', referring to a discovery of some in the odd numbers of .2 mentioned in the introduction to my second table in *Biometrika*. 'Well we'll put that right'. So I went away to get the corrections for him. That involved getting a 5th place for $n = 9$ and I couldn't check the 4th place. I left my attempt and went home and when I came back on my way to Dublin I found that he agreed with me and that the new table was wrong. On further investigation, both tables were found to be perfectly rotten. All .1 and .2 wrong in 4th place, mostly it is true by .0001 only, and quite a number of other ones. The fact is that I was even more ignorant when I made the first table than I am now and thought I was going to be accurate to 4 places by taking 5 in the working! and the second was of course constructed on the same lines though not by me. I ought to have checked it myself, but must have been pretty casual about it. Anyhow the old man is just about fed up with me as a computer and wouldn't even let me correct my own table. I don't blame him either.

Whether he will have anything to do with our table I don't know, I rather doubt it, but personally I feel I could hardly put it before him unless you are prepared to do quite a lot of checking either yourself or per Miss Mackenzie. Just as well you didn't take that table from *Biometrika!* It has been rather a miserable fortnight finding out what an ass I made of myself and from the point of view of the new table wholly wasted. However, I begin work again tomorrow.

Yrs. v. sincerely,

W. S. Gosset

Soon Gosset had "somewhat rehabilitated" himself by discovering a mistake in *their* version, which they had persisted in after comparison with his. He reported that "K.P. again wrote that he would be glad to consider our table." Fisher, busy writing *Statistical Methods* while his assistant was busy completing her thesis, must have jibbed at the prospect of more work on the table, for Gosset wrote on December 6, 1923, "It seems rather a shame to burden you with checking after what you say, but I think I may fairly put your own tables up to you." By this time *Statistical Methods* was becoming a reality, and Fisher was concerned to establish his right to use the t tables for his book, even if Pearson should publish them. In the end, however, the tabulations of χ^2, t, and z which appeared in *Statistical Methods* were Fisher's own.

On June 20, 1924 Gosset sent the completed tables, explaining the calculation and checking of the frame, the interpolations and their checking, and the special precautions taken with doubtful figures to the 7th or 8th place of decimals.

It is now up to you to write an account of the tables, and please don't let too much be clear or obvious, I'd like to understand myself as much as possible what I have been doing for the last two years. . . .

P.S. If you could let me have your account quite early next month I can probably take it to K.P. when I next get over. I'll get it typed as he is finding it more and more difficult to read manuscript.

Fisher sent Gosset his contribution on July 17, within the month, but too late for Gosset to take to Pearson.

I enclose the two notes I mentioned, the first of which is an attempt to give some idea of the multitude of uses to which your table may be put, and the second is a formal statement of the approximation formula. The first is larger than I had intended, and to make it at all complete should be larger still, but I shall not have time to make it so, as I am sailing for Canada on the 25th, and will not be back till early September.

The first note [*C P* 43] is a remarkable document. In it a proof was given for Student's results which, in 1908, had been partly intuitive, since on two points the demonstration was incomplete: (*i*) he had showed only that the distribution obtained for s^2 agrees with the assumed form in the first four moments, and (*ii*) he had demonstrated that s^2 is not correlated with $(\bar{x} - m)^2$, but not that the distributions are entirely independent. By use of n-dimensional geometry, Fisher showed that the joint distribution of $\bar{x}$ and s can be split into two factors, one of which is the distribution of $\bar{x}$ and the other the distribution of s. It follows that the two distributions must be wholly independent. Further, since the distribution of t is a ratio consisting of a normal deviate divided by an independently distributed χ, the same t distribution would be applicable to any other statistic that could be shown to be of this form. It was, therefore, appropriate for testing the significance of the difference between two means, of regression coefficients and curvilinear regressions.

In the closing section of the paper Fisher defined a "wider class of distributions which is related to 'Student's' distribution in the same manner as that of χ^2 is related to the normal distribution." That is, he dealt with the distribution of F given in 1922 (*C P* 20) and with its wider applications.

□ □ □

Publication of Gosset's table and the accompanying notes was curiously delayed. Ten months after receiving Fisher's contributions Gosset wrote to say he had "at last taken the table to K.P. together with your notes: I am not at all sure that he won't publish it." At that time he had only just written his own note to accompany the table and sent it on for Fisher to check and have typed.

After seeing Pearson again, Gosset reported:

K.P. is very anxious to publish your note about the use of the table, but doesn't like the binomial approximation which he considers requires a proof of convergence. It was in vain that I pointed out that, converging or diverging, the proof of the pudding lies (to me, doubtless not to you) in the fact that you get about seven places the same with $n = 21$ up to $t = 6$.

Anyhow, he returns both and I send them herewith. . . .

Fisher probably submitted his papers, and perhaps the tables, at once to *Metron*. (The decision against joint authorship of all the material under a single title was not made until later.) Meanwhile, Gosset, on his own initiative, tried to persuade Pearson to put Fisher's article on applications of Student's t in the next issue of *Biometrika*. When Gosset asked about it toward the end of September Pearson said he was still expecting to receive the paper. Thereafter the situation changed rapidly. On October 8, Gosset returned to Fisher a copy of Fisher's note, together with a new version of his own, obviously intended for Fisher to send on to *Metron* for publication with the tables. All their material on Student's t appeared in *Metron* a few months later.

□ □ □

In "Applications of 'Student's' Distribution" (*C P* 43), Fisher did his best to respond to Gosset's plea "not to let too much be clear or obvious," that is, he did explain himself in algebraic terms for those who, like Gosset, were dazzled but not illuminated by the geometry. Still, what was in his mind and what unambiguously defined the relationship between the various tests of significance was the geometrical representation.

This representation did much more than throw light on the t test. So far as Fisher was concerned, it solved for normally distributed errors all the distribution problems for what has come to be called the *general linear* model.* Special cases of this model cover all the standard distribution problems in linear, curvilinear, and multiple regression. All corresponding analysis of variance models are also covered. This is true whether the data come from randomized block designs, Latin squares, factorials, balanced incomplete blocks, or any other arrangement occurring by design or by chance. In addition, it solved the distribution problems for inter- and intraclass correlation coefficients, for the multiple correlation coefficient, partial correlation coefficients, and the

---

*Linear here refers only to the manner in which the *constants* or *parameters* appear in the model. For example, *curvilinear* regression problems, in which a response *y* plots as a curve against some variable *x*, are expressed in terms of a model that is *linear* in the parameters.

correlation ratio. Finally, it showed clearly the relationships between all these. Although all the geometrical results and the various relationships could be stated and proved algebraically, algebra did not offer the percipience of vision invoked by n-dimensional geometry, by which the results could be immediately *seen* rather than laboriously derived.

The reader unfamiliar with n-dimensional geometry can gain some idea of the power of the method by looking at a simple special case in which there are just $n = 3$ observations. Suppose an experiment is conducted in which y is the observation (say, the size of the kick of a frog's leg) when an electric current x is applied. Suppose it were known that if a relationship existed it would be a proportional one, so that $y = \beta x + \epsilon$ when β is the unknown constant of proportionality and ϵ is the experimental error. An investigator might wish to know whether his data exhibited a *statistically significant relationship* between y and x and he might wish to *estimate* the constant of proportionality β. In attempting the choice of appropriate statistical methods for the analysis of his data, he might wonder whether the method of *least squares* or perhaps *maximum likelihood* might be used to estimate the constant β. He might further speculate that the *t-test* or perhaps the *correlation coefficient* or even the *analysis of variance* with an appropriate *F or z test* might be used to assess significance. He could be forgiven if he was not clear as to how all these concepts were related. Once he grasped Fisher's use of geometry, however, all these ideas and their relationships would become clear at once.

If just three trials are run with three levels of the current (x_1, x_2, x_3), giving three responses (y_1, y_2, y_3), then we can write:

$$
\begin{aligned}
y_1 &= \beta x_1 + \epsilon_1 \\
y_2 &= \beta x_2 + \epsilon_2 \\
y_3 &= \beta x_3 + \epsilon_3
\end{aligned}
\quad \text{or} \quad
\begin{pmatrix} y_1 \\ y_2 \\ y_3 \end{pmatrix} = \beta \begin{pmatrix} x_1 \\ x_2 \\ x_3 \end{pmatrix} + \begin{pmatrix} \epsilon_1 \\ \epsilon_2 \\ \epsilon_3 \end{pmatrix}
\quad \text{or} \quad \mathbf{y} = \beta \mathbf{x} + \boldsymbol{\epsilon}
$$

where

$$
\mathbf{y} = \begin{pmatrix} y_1 \\ y_2 \\ y_3 \end{pmatrix} \quad \mathbf{x} = \begin{pmatrix} x_1 \\ x_2 \\ x_3 \end{pmatrix} \quad \text{and} \quad \boldsymbol{\epsilon} = \begin{pmatrix} \epsilon_1 \\ \epsilon_2 \\ \epsilon_3 \end{pmatrix}
$$

are called vectors and can be represented in three-dimensional space. For example, suppose the results were $y_1 = 2$, $y_2 = 1$, $y_3 = 4$, then the vector $\mathbf{y}$ could be represented by a line drawn from the origin to a point which had coordinates $(2, 1, 4)$ on the three orthogonal axes of the three-dimensional space. These axes are labeled (1), (2), (3) in Figure 4a, which shows the vector $\mathbf{y}$ and the corresponding vector $\mathbf{x}$ on the same diagram. Suppose that b is the maximum likelihood estimate of β and $\hat{y}_1$, $\hat{y}_2$, $\hat{y}_3$ the calculated values obtained by substituting b for β, so that

$$\begin{aligned}\hat{y}_1 &= bx_1 \\ \hat{y}_2 &= bx_2 \\ \hat{y}_3 &= bx_3\end{aligned} \quad \text{or} \quad \begin{pmatrix}\hat{y}_1 \\ \hat{y}_2 \\ \hat{y}_3\end{pmatrix} = b\begin{pmatrix}x_1 \\ x_2 \\ x_3\end{pmatrix} \quad \text{or} \quad \hat{\mathbf{y}} = b\mathbf{x}$$

Least Squares. For normally distributed data, it may be shown that the maximum likelihood estimate b is given by Gauss's method of least squares, that is, by minimizing the sum of squared discrepancies between observed and calculated values

$$S^2 = (y_1 - \hat{y}_1)^2 + (y_2 - \hat{y}_2)^2 + (y_3 - \hat{y}_3)^2$$

Since $\hat{\mathbf{y}} = b\mathbf{x}$, the vector $\hat{\mathbf{y}}$ is just some multiple of $\mathbf{x}$ lying along the vector $\mathbf{x}$. Geometrically, the expression for S means the distance between the point $\mathbf{y}$ and the point $\hat{\mathbf{y}}$. Evidently, it is minimized when the vector $\mathbf{y} - \hat{\mathbf{y}}$ makes a right angle with $\mathbf{x}$. Thus $\hat{\mathbf{y}}$ and hence b is obtained by dropping a perpendicular from $\mathbf{y}$ onto $\mathbf{x}$, as is illustrated in Figure 4a. In the example $\hat{\mathbf{y}} = 2.5\mathbf{x}$ so that for the data shown $b = 2.5$.

The Significance Test. If the null hypothesis, that there was no proportional relationship, were true, then β would equal zero. In that case $y_1 = \epsilon_1$, $y_2 = \epsilon_2$, $y_3 = \epsilon_3$. But if ϵ_1, ϵ_2, ϵ_3 are normally and independently distributed with constant variance, then it may be shown that the error vector $\mathbf{y} = \mathbf{\epsilon}$ has an equal chance of lying in any direction in the space.

The test of significance of b thus turns out to be a determination of how often the direction of $\mathbf{y}$ would have fallen as close or closer to the direction of $\mathbf{x}$ by chance, if there had been no relationship. To make the calculation we must imagine, as Fisher did, a sphere drawn about the origin with radius equal to the length of the $\mathbf{y}$ vector and imagine a cone produced by rotating $\mathbf{y}$ about $\mathbf{x}$ with fixed angle θ cutting off a cap from the sphere as in Fig. 4b. Then (area of cap)/(area of sphere) is equal to the chance of lying so close. This chance is clearly a function only of the angle θ.

Thus the test of significance could be made by calculating the angle θ and comparing it with a table of critical angles which subtend caps including 5%, 1%, and so on of the total area of the sphere. It is not easy for the layman to visualize more than three dimensions, but in fact, the same argument applies in 4-, 5-, . . ., n-dimensional space and thus the appropriate critical angle can be worked out for any number of observations. In practice the angle itself is not used but some suitable function of it; as will later be apparent, the familiar t, F and z statistics as well as the sample correlation coefficient r are all functions of such an angle.

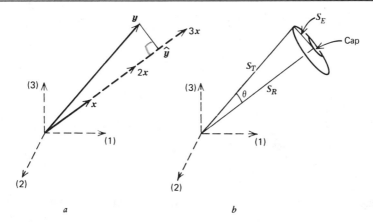

Figure 4. Geometrical representation of a sample illustrative of least squares and the accompanying test of significance.

Historically, various aspects of linear statistical problems such as this had been stumbled on from time to time, but their relationships were not usually recognized, and confusion was increased because different names were given to the same problem arising in different contexts. We use the simple case described above to illustrate how Fisher's geometrical argument and the corresponding analysis of variance made some of these relationships clear.

The Analysis of Variance Table. The squared length of the vector **y** in Fig. 4*b* is called the *total* sums of squares

$$S_T^2 = y_1^2 + y_2^2 + y_3^2$$

and may be split into parts: the *regression* sum of squares associated with the part of the data explained by the conjectured model

$$S_R^2 = \hat{y}_1^2 + \hat{y}_2^2 + \hat{y}_3^2$$

and the part associated with *error*

$$S_E^2 = (y_1 - \hat{y}_1)^2 + (y_2 - \hat{y}_2)^2 + (y_3 - \hat{y}_3)^2.$$

Since S_T, S_R, and S_E are lengths of vectors, the analysis

$$S_T^2 = S_R^2 + S_E^2$$

follows from Pythagoras' theorem.

Now, the experimental data vector **y** is free to move in $n = 3$ dimensions and hence has three degrees of freedom. But $\hat{y}$ must lie along the vector **x** and so has only $n_R = 1$ degree of freedom, whereas the remaining vector

$y - \hat{y}$ must be at right angles to **x** and so has $n_E = 2$ degrees of freedom. These basic facts for this (and any other) linear problem Fisher set out in the form of an analysis of variance table:

Source of Variation	Sums of Squares	Degrees of Freedom	Mean Square
Model (Regression) S_R^2		$n_R \ (= 1)$	S_R^2/n_R
About model (Error) S_E^2		$n_E \ (= 2)$	S_E^2/n_E
Total	S_T^2	$n_T \ (= 3)$	

***z* test, *F* test and *t* test.** If there is no relationship, both the mean squares are independent estimates of the error variance σ^2. However, if the mean square associated with the model relationship is much larger than that associated with error, doubt is cast on the null hypothesis.

Consider the geometric nature of the ratio F of the mean squares

$$F = (S_R^2/n_R)/(S_E^2/n_E) = (n_E/n_R)(S_R/S_E)^2$$

$$= (n_E/n_R) \cot^2 \theta$$

where $\cot \theta$, the cotangent of the angle θ, is the ratio of the length of the base of the right-angled triangle to the side opposite to the angle θ. Thus F is just a function of the angle θ, and consideration of whether F is too large to be explained by chance is precisely equivalent to consideration of whether the angle θ is too small. The difference in the logs of the root mean squares, which Fisher preferred to use, is

$$z = \tfrac{1}{2} \log F = \tfrac{1}{2} \log \{(n_E/n_R) \cot^2 \theta\}$$

which is, of course, again equivalent to the use of the angle θ.

Another way to approach the significance test would be to compare the estimate b of the slope with its standard error (calculated from the data). Fisher showed [*C P* 4, *C P* 43] that the ratio of the two quantities is distributed as Student's t. In fact, it turns out that

$$t = \sqrt{F} = \sqrt{(n_E/n_R)} \cot \theta$$

so that, once again, we are back to a precisely equivalent test which depends only on the angle θ.

Correlation Coefficient. Leaving the example of the frog's kick, suppose now that $y = Y - \bar{Y}$, $x = X - \bar{X}$ were deviations from averages of data (Y, X), measuring, for example, heights of fathers and eldest sons. Then

$$r = \Sigma xy / \sqrt{\Sigma x^2 \Sigma y^2}$$

is the correlation coefficient. Fisher had observed in 1915 that, using a well-known result in coordinate geometry, the expression on the right defines the cosine of the angle between the vectors **y** and **x** so that,

$$r = \cos \theta$$

and we are back again to considering the distribution of the angle θ between the vectors. It turns out that after reducing the degrees of freedom by one to allow for the estimated means, the same argument as before can be applied and leads to the distribution of the correlation coefficient.

Multiple Regression, Correlation, and Curvilinear Regression. The argument is subject to a vast generalization, which Fisher made in his paper on the application of Student's distribution [C P 43]. The example used above for illustration was a simple linear regression, in which the effect on the response of a single variable was under study. In the example, the problem of inference turned on the size of the angle θ between the vectors **x** and **y**. An exactly parallel argument applies, however, when there are several x's, so that the model is

$$y = \beta_1 x_1 + \beta_2 x_2 + \cdots + \beta_p x_p + \epsilon$$

The p sets of x's now define a p-dimensional plane (or hyperplane) in the space, and the significance test is necessarily made in terms of a function of the single angle θ between the observation y and its projection $\hat{y}$ *on this plane.* The argument also extends to quadratic and cubic regression, for we may always set $x_2 = x^2, x_3 = x^3$, etc. It also accommodates multiple correlation.

Indeed, the applicability is even more than this. Returning to the original model $y = \beta x + \epsilon$, suppose all the x's are *unity;* then the model is

$$y = \beta + \epsilon$$

and this is precisely the model that applies to Student's original problem in testing whether the mean β of a set of differences, represented now by the y's, could be zero. The unities corresponding to x are nowadays called indicator variables and, using such indicator variables, it is possible to use the argument to justify the significance test for all cases of the analysis of variance, whatever the design (one-way classification, Latin square, randomized block, incomplete block, etc.). Not only was Fisher thus able to derive the distribution of the various statistics when the null hypothesis was true, he could also derive the various noncentral distributions that applied when it was false. These were later to become of great importance in deriving the "power" of tests according to the Neyman Pearson theory.

It may be noted that for justification of the test of significance itself, it is required only that on the null hypothesis the vector **y** should be equally likely to lie in any direction. The assumption of identically normally distributed independent errors is *sufficient* to ensure this but is not necessary. Randomization would introduce a degree of symmetry into the error distribution that might justify the assumption of spherical symmetry of the error distribution as an approximation, *without the direct assumption of normality and independence* of the errors. This is perhaps one of the considerations that made the geometrical argument particularly attractive to Fisher and one which led to misunderstandings. Later writers would prefer a route in which, on the normal theory assumptions, (i) the distribution of each of the sums of squares in the analysis of variance was shown to be a χ^2 with the appropriate number of degrees of freedom and (ii) the component sums of squares were shown to be independent. Years later it was overlooked that in his 1926 paper Fisher had also covered this second route, and mistaken attempts were made to tie together the "loose ends" which Fisher was supposed to have left.

To understand this second approach, let us consider again the example with $n = 3$. If the null hypothesis is true then $\mathbf{y} = \boldsymbol{\epsilon}$, that is, the vector $\mathbf{y}$ is entirely the result of experimental errors. If these are normally and independently distributed with variance σ^2, then the probability density is constant on spheres centered at the origin in the space of Figure 5a, with coordinates labeled 1, 2, 3. Suppose now that we make a rotation to new axes, labeled $1'$, $2'$, $3'$ in Figure 5b, such that the new axis $1'$ lies along the vector $\mathbf{x}$. Then the distribution with respect to the new axes is the same as it was with respect to the old. Thus if the results were due to chance, the component $\hat{y}$ would be normally distributed about zero with variance σ^2. Its squared length

$$\hat{y}_1^2 \;+\; \hat{y}_2^2 \;+\; \hat{y}_3^2 \;=\; S_R^2$$

would be distributed as the square of a single normal variate with variance σ^2,

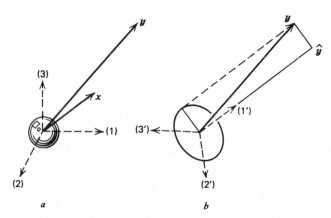

Figure 5. Derivation of distributions by a rotation of axes.

that is, as $\chi^2 \sigma^2$ with one degree of freedom. Also, the error component $\mathbf{y} - \hat{\mathbf{y}}$ which has squared length

$$(y_1 - \hat{y}_1)^2 + (y_2 - \hat{y}_2)^2 + (y_3 - \hat{y}_3)^2 = S_E^2$$

would be distributed as the sum of squares of two independent normal deviates, that is, as $\chi^2 \sigma^2$ with two degrees of freedom, independently of S_R.

In the first part of "Applications of Student's distribution" [C P 43], in discussing the significance of regression problems, Fisher commented:

It is perhaps worth while to give, at length, an algebraical method of proof, since analogous cases have hitherto been demonstrated only geometrically, by means of a construction in Euclidian hyperspace, and the validity of such methods of proof may not be universally admitted.

The derivation follows, which is an algebraic statement exactly representing the geometric argument presented above and generalizing it to cover more than one **x** vector. Thus the geometric proof, which was not universally understood, was converted to an algebraic proof none could dispute.

□ □ □

The first edition of *Statistical Methods for Research Workers* grew out of Fisher's experience and was moulded by the manner in which his own thought had developed. It was written between the summers of 1923 and 1924, and, as with many of his papers, he was still thinking through some of the problems as he wrote. Like many of his papers also, it ended with hints of the budding off of more new ideas which would be developed separately. The chapter headings read as chapters of his own intellectual growth in statistics: Introductory, Diagrams, Distributions, Tests of goodness of fit and homogeneity, Tests of significance of means and regression coefficients, The correlation coefficient, Intraclass correlations and the analysis of variance, Further applications of the analysis of variance, The principles of statistical estimation.

The objective of the book was shaped on the anvil of Fisher's scientific interest under the hammer of empirical problems. He conceived of statistics as a tool for research workers and shaped it here to their ends; the book consists largely of illustrations of statistical methods by means of numerical examples. The philosophy of this approach is clear from the Preface:

For several years the author has been working in somewhat intimate co-operation with a number of biological research departments; the present book is in every sense the product of this circumstance. Daily contact with the statistical problems which present themselves to the laboratory worker has stimulated the purely mathematical re-searches upon which are based the methods here presented. Little experience is suffi-cient to show that the traditional machinery of statistical processes is wholly unsuited to the needs of practical research. Not only does it take a cannon to shoot a sparrow, but it misses the sparrow! The elaborate mechanism built on the theory of infinitely large samples is not accurate enough for simple laboratory data. Only by systematically tackling small sample problems on their merits does it seem possible to apply accurate tests to practical data. Such at least has been the aim of this book.

I owe more than I can say to Mr. W. S. Gosset, Mr. E. Somerfield and Miss W. A. Mackenzie, who have read the proofs and made valuable suggestions. Many small but troublesome errors have been removed; I shall be grateful to readers who will notify me of any further errors and ambiguities they may detect.

Rothamsted Experimental Station
February 1925.

The book did not receive a single good review. The reviewers, faced with a volume so new in philosophy, scope, and method, were unimpressed with its virtues. They criticized the predominance in it of Fisher's own work and the absence of apology or warning that of "the methods favoured by the author and the conclusions he has reached some . . . are still in the controversial stage." [23]. They disliked his treatment of other writers on statistics: "The author is very economical in his references to earlier workers." [23]; and

Even if the statement that Professor Pearson's treatment of χ^2 contained a "serious er-ror" had not been disputable, and therefore improper in a work addressed to ele-mentary students, it would have reminded anyone of Macaulay's remark on a similar occasion, "Just so we have heard a baby mounted on the shoulders of his father cry out 'How much taller I am than Papa'." [24]

They disliked the absence of mathematical proofs, without which they felt themselves to be no longer on terra firma and feared that the research worker would be led astray by this attempt "to provide a short-cut to advanced results by avoiding a thorough grounding in the elements" [25]. And they regarded the whole treatment as of very narrow interest, dealing as it did with small samples. Student welcomed "the first book on statistics which deals mainly with this special technique" [26]. Others suggested "research workers whose statistical series necessarily consist of small samples" [23] might think the book worth looking at—with due caution, for Fisher "has given considerably more attention to the particular methods applicable to small samples than authors of most textbooks have deemed necessary." [24]. They objected to the inclusion of genetical illustrations and the technical vocabulary of biology as much as they did to the statistical difficulty of the book. On the whole, "If he feared that he was likely to fall between two stools, to produce a book neither full enough to satisfy those interested in its statistical algebra nor sufficiently simple to please those who dislike algebra, we think that Mr. Fisher's fears are justified by the result." [24].

Fisher was beginning to expect this sort of reception for his original work. He was never inured to it, but it was recognizably a fact of life. In 1947, speaking about a scientific career, he was able to express without bitterness the expectation he had formed of what recognition a young scientist should expect [C P 216]:

A scientific career is peculiar in some ways. Its *raison d'etre* is the increase of natural knowledge. Occasionally, therefore, an increase of natural knowledge occurs. But this is tactless, and feelings are hurt. For in some small degree it is inevitable that views previously expounded are shown to be either obsolete or false. Most people, I think, can recognize this and take it in good part if what they have been teaching for ten years or so comes to need a little revision; but some undoubtedly take it hard, as a blow to their *amour propre,* or even as an invasion of the territory they had come to think of as exclusively their own, and they must react with the same ferocity as we can see in the robins and chaffinches these spring days when they resent an intrusion into their little territories. I do not think anything can be done about it. It is inherent in the nature of our profession; but a young scientist may be warned and advised that when he has a jewel to offer for the enrichment of mankind some certainly will wish to turn and rend him.

Another point to remember has to do with recognition, which the young dearly desire. A ballet dancer gets her ovation on the spot, while she is still warm from her efforts; the wit gets his laugh across the table; but a scientist must wait about five years for *his* laugh. Recognition in science, to the man who has something to give, is, I should guess, more just and more certain than in most occupations but it does take time. And when it comes it will probably come from abroad.

In 1925 a letter arrived from abroad (from Prof. J. Glover at Michigan) welcoming the book with a warmth of appreciation that touched Fisher like a ray of sunshine. It was a token of the appreciation that was to come a little later most heartily from abroad. In contrast, it took time for his ideas to take root even at Rothamsted, where, despite some collaborative efforts, statistics still seemed irrelevant to most of the research workers. Nevertheless, Fisher was establishing himself and the role of statistics at Rothamsted, and the great changes that were, in consequence, to take place there within a few years were already inevitable.

□ □ □

Fisher made his first impact at Rothamsted as a personality, casually over a cup of tea. Regular afternoon tea had been instituted in 1906 when Miss W.

Brenchley had joined the scientific staff as its first woman member. Russell [17] recalls:

No one in those days knew what to do with a woman worker in a laboratory; it was felt, however, that she must have tea, and so from the day of her arrival a tray of tea and a tin of Bath Oliver biscuits appeared each afternoon at four o'clock precisely; and the scientific staff, then numbering five, was invited to partake thereof.

In the 1920s tea was served in the sample house or, in fine weather, on a trestle table set with urn and teacups outside the sample house. Fisher found the staff tea a particularly agreeable institution. He was a notable figure, with his shabby clothes and shaggy head. His spectacles lent strange gleams to his blue eyes, which seemed amused and smiling; the deep crow's feet round them suggested laughter, though he had acquired them also through his poor eyesight, screwing up his eyes habitually in the attempt to bring objects into focus. There he stood, pipe in hand, wafting the fumes of a pungent tobacco from the center of sustained conversation.

A snapshot (Plate 5) shows him leaning forward beside the table, one foot planted on the bench, an elbow on his knee, and his head held at a characteristic angle—jutting forward, as if in pursuit of his train of thought. He is shown holding a greatly cherished pipe, the bowl in one hand, the stem in the other. When the stem was in his mouth, a hand either held the bowl or was ready to catch it should it fall, as it frequently did. A hole burned in the bowl of this pipe was first plugged with blotting paper but, when this was found to be anything but tar-proof, Miss M. D. Glynne carried the pipe off to plug the hole with *pise de terre,* a material on which she was researching at the time.

He did not raise his voice, but his clear diction and incisive slow delivery enabled most of the staff members at tea to share his thoughts. This could be embarrassing, for he entered freely on the "delicate" subjects of eugenics. It was observed on such an occasion that the conversation among the rather strait-laced and serious young women within earshot became more intense, their faces extra serious, while an occasional blush betrayed how conscious they were of the remarks they were trying to ignore. Disconcerting, sometimes outrageous, but coherent, well-informed, and thoughtful, his conversation made an impression as notable as his appearance.

He was usually present at the regular colloquia at which Rothamsted staff described their work for their colleagues, which Russell had instituted so that the different departments should not lose contact with each other as the institution grew and the work diversified. On such occasions he was no mere auditor but a noticeable participant.

Duty demanded his appearance at various committee meetings. He did not enjoy committee work, but he was prompt to contribute whenever his calculations could help. Thus when the heads of all departments at Rothamsted were called together to hammer out some new means, in view of the growing staff, of allocating to the different departments an equitable share of the funds available for the coming year, they expected a difficult and tedious session. Fisher came in ready to propose a straightforward formula based on the average expenditure per man-month in each department, a figure that could be easily ascertained from departmental records for past years. The calculations were quickly done on the cumbersome Millionaire calculator and the result was adopted to everyone's satisfaction.

When the Federated Superannuation System for Universities (F.S.S.U.) was introduced for the staff, Fisher similarly calculated the benefits from the two options available under that scheme, one essentially a life insurance policy, the other a kind of deferred annuity. It was proposed that the former should be adopted at Rothamsted. Fisher disagreed. Finding that after only a few years the alternative would have the greater value and thereafter would continue to increase its lead, until at maturity it was worth twice as much as the proposed policy, he insisted on his right to choose for himself. He elected to take a different line than the rest of the Rothamsted staff. For himself, he preferred to take the risk of an early death and to abandon the much better provision for widow and family offered by the scheme proposed for Rothamsted in order to take the long-term gamble.

His whole life was a gamble of the same sort. A prudent man in his position would have welcomed additional security for his growing family, for their one security lay in his person, and he earned only enough to support the family from day to day without luxuries. A cautious man would not have returned his paycheck for the quarter as Fisher did in 1923, as a way of asserting his claim that the figure fell short of what he had been led to expect and that he was not prepared to accept the smaller amount. Happily, his faith in his value to Rothamsted was justified: the Lawes Agricultural Trust found it worthwhile during the following year to make the adjustments in his grade and seniority necessary to meet his expectations, even though these had arisen through a misunderstanding at the time of his permanent appointment to the staff in 1920.

Where a hesitant man might have waited for a formal invitation, he went where his enthusiasm led and made himself welcome. In 1922 he attended a meteorological meeting at which he officially represented Russell; in some mysterious way he was thereafter involved at the Meteorological Office, attending informal planning sessions and becoming established over time as a regular consultant. No one realized until he had left Rothamsted that he had no official status in this work, which had apparently originated through the helpfulness of his spontaneous participation at that one early meeting. Similarly, though a head of department, he seems to have been admitted quite unofficially to the gatherings of the junior staff club at Rothamsted. And he was certainly admitted, despite the shocked protests of midwife or doctor, at his wife's bedside at the birth of his children.

With Fisher in the party, unprecedented events tended to occur. At home, he tried to see if a baby a few days old, unlike older infants, could support its own weight with its hands, as its arboreal ancestors must have been able to do, by hanging on to his finger; the baby refused even to try. Perhaps he might have had done better to offer tangled hair in which the clutching fingers might have enmeshed themselves, as a more realistic reconstruction of primitive conditions. But what other father would cheerfully have dismissed his wife's fears that the tiny baby might fall (though it might be only a few inches on to a soft bed) in order to satisfy himself on a neglected point of evolutionary interest?

Already, quite soon after he had come to Rothamsted, his presence had transformed one commonplace tea time to an historic event. It happened one afternoon when he drew a cup of tea from the urn and offered it to the lady beside him, Dr. B. Muriel Bristol, an algologist. She declined it, stating that she preferred a cup into which the milk had been poured first. "Nonsense," returned Fisher, smiling, "Surely it makes no difference." But she maintained, with emphasis, that of course it did. From just behind, a voice suggested, "Let's test her." It was William Roach, who was not long afterward to marry Miss Bristol. Immediately, they embarked on the preliminaries of the experiment, Roach assisting with the cups and exulting that Miss Bristol divined correctly more than enough of those cups into which tea had been poured first to prove her case.

Miss Bristol's personal triumph was never recorded, and perhaps Fisher was not satisfied at that moment with the extempore experimental procedure. One can be sure, however, that even as he conceived and carried out the experiment beside the trestle table, and the onlookers, no doubt, took sides as to its outcome, he was thinking through the questions it raised: How many cups should be used in the test? Should they be paired? In what order should the cups be presented? What should be done about chance variations in the temperature, sweetness, and so on? What conclusions could be drawn from a perfect score or from one with one or more errors?

Probably this was the first time he had run such an experiment, for it was characteristic of him, having conceived an idea in one context, to revert to that context in expounding the idea later, rather than to select a new example from innumerable possibilities. And, of course, when he came to write *The Design of Experiments* (1935) more than a dozen years later, the "lady with the tea-cups" took her rightful place at the initiation of the subject; Fisher opened Chapter II, and the body of the discussion, with the proposition:

A lady declares that by tasting a cup of tea made with milk she can discriminate whether the milk or the tea infusion was first added to the cup. We will consider the problem of designing an experiment by means of which this assertion can be tested.

In the subsequent pages he considered the questions relevant to designing this particular test as a prime example, for the same questions arise, in some form, in all experimental designs.

□ □ □

Fisher was soon well liked and respected, though regarded with considerable awe of his mathematical powers, with a tinge of fear at his occasional rages, and with a hint of embarrassed amusement at his apparent unawareness of the claims of social conformity. He was congenial company and full of interest in new ideas. It was not long before he was consulted about the analysis of current experimental results and, even more gratifyingly, was asked to advise about an experiment which was only then being planned.

H. G. Thornton, head of the bacteriology department, was one of his earliest friends at Rothamsted—a man of his own age, cultured and very able, whose comfortable appearance and ineffable courtesy contrasted with Fisher's casual ways; their friendship was one of the special kind that Fisher enjoyed with a few lifelong comrades in whom he felt a whole-hearted trust and sympathy. Thornton early sought him out to advise on a method of testing the reliability of a plating system he had invented for making bacterial counts; it was the beginning of a series of investigations in which they consulted and worked together (some statistical considerations and consequences of which appear in *C P* 22 with H. G. Thornton, and *C P* 18, Section 12, describing the first nonlinear experimental design, 1922).

Fisher welcomed inquiries. Putting aside whatever he was engaged on, he would give his full attention to the case at issue, mastering the technical and biological factors involved and using them to guide him as he created statistical methods to suit the problem. Each case came to him as a unique problem. There was not in those days a wide repertoire of statistical methods on which he could draw, so that he was compelled to formulate individual solutions to suit individual situations. This also was his inclination and his philosophy; problems have individual features, so that it is profitable to consider them in their own right, both from the experimental point of view and for the sake of the development of statistics. Later he was to deplore how often his own methods were applied thoughtlessly, as cookbook solutions, when they were inappropriate or, at least, less informative than other methods might have been. Then and later he was concerned with the unique physical reality he confronted. He had to be inventive and ingenious to tailor the statistical methods to the experimental conditions; indeed, a number of pretty mathematical developments were occasioned in this way.

It was a two-way street: inquirers received helpful answers to their questions, and Fisher visualized, through the particular examples, the conditions within which the new statistics would have to work.

Believing that mathematical imagination was merely one direction in which any active intelligence might be developed, Fisher took his inquirers into his confidence regarding the statistical implications and possibilities of their work, explaining simply and clearly what he was doing and showing great patience with biologists whose grounding in mathematics made it difficult for them to comprehend the statistical argument. To him it seemed essential, if he were to do his statistical work efficiently, that he should grasp the experimental situation; it was equally essential that the experimenter, for his own sake, should grasp the statistical principles employed in his work. This exercise in the clarification of the statistical ideas for the experimenter and of the experimental ideas for the statistician brought immediate benefits to both parties, for they thereby achieved really cooperative research. At the same time, the cooperation resulted in a gradually widening and deepening understanding of the role of statistics in research.

Fisher did not expect the experimenter to do his own calculations, and consultation usually ended with his offer to work out the results when they became available. It might be months before the data were collected, yet on one such occasion Roach recalls that Fisher was able, without prompting, to give an accurate and succinct account of the proposals made long previously and to proceed immediately to calculate the results. His account shows also the quickness and flexibility of mind with which Fisher met various problems and the simplicity with which he recognized his own error.

I can see him now, alternately, nose close to my data as he short-sightedly read some figures through his thick-lensed spectacles, stroking his beard with his left hand; then moving left, working the Millionaire calculator with his left hand while he stroked his beard with his right hand; then back to the first position to record the figures obtained and read the next batch of data. Already his beard had that backward-curving tip. I noticed him dividing a large figure by 971 by short division (with the Millionaire calculator at his side) more quickly than I would have divided by 9.

On my way home I realized I need not have bothered Fisher because I could have got what I wanted from my figures algebraically. This I did but got the wrong answer. Having reviewed these last thing before going to sleep in the early hours, I woke with the dawn and wrote out what seemed to me my clearest solution and later took the resulting foolscap and said: "What's wrong with this, Fisher?" He looked at the figures for about as long as I'm taking to write this sentence and said, smilingly, "Where did I slip up? Yes, of course, the formula I used assumed a convergent series; yours is not." In a few minutes he worked out a new formula and got an answer quite close to mine.

Fisher not only gained a reputation for his extraordinarily rapid grasp of the essentials even of complex situations and arguments but also for his speed of response. Having grasped the problem, he could often propose a solution at once, or work one out within minutes, perhaps jotting a few squiggles on the foolscap pad before him. They were not always simple problems.

An example from the early 1920s was a question set him by Thornton. Thornton was making direct counts of bacterial cells in a soil suspension viewed under the microscope. Since the soil particles and bacteria were drawn toward the edge of the drop as it dried, they were unevenly distributed on the slide. To get meaningful counts, therefore, he had devised a standard of comparison. Shaking the soil suspension with a previously counted suspension of blue indigo particles that were of the same average size as the bacteria and were similarly affected when the drops were dried, he then counted the numbers of both red-stained bacteria and blue indigo particles. Since the number of indigo particles added per gram of soil was known, he could then estimate the actual number of bacteria from the ratio of the number of bacteria to the number of indigo particles observed. But the problem obtruded itself: how reliable was this ratio? How could one find its variance? He put the question to Fisher, and Fisher worked out the method in his head on his way across Harpenden Common from Milton Lodge; it can have taken no more than 10 minutes. In celebration of this feat of mental calculation, the method was thereafter known in the bacteriology department as the "Common" method.

When Fisher had computing assistants or voluntary workers in the department, working with him for a period of months, he expected that they would undertake statistical work of their own or in collaboration with him and that their work would be published. He found his first computing assistant, Miss Mackenzie, an apt pupil. The joint paper already referred to (C P 32) resulted in 1923, and Miss Mackenzie went on to prepare a thesis for her master's degree. The oral examination was held the following February. Gosset, being asked to act with Fisher as referee for the University of London, responded: "I suppose they appointed me because the Thesis was about barley, so of course a brewer was required, otherwise it seems to me rather irregular. I fear that some of Miss Mackenzie's mathematics may be too 'obvious' for me!" The referees met in Harpenden and Gosset stayed the weekend with Fisher: "I travel light so if you will let me know your address I will come straight up with my pack on my back." They seem to have enjoyed writing the report together, for Gosset afterward admitted he was "consumed with curiosity to know what the University thought of our report on Miss Mackenzie." As they had recommended, she was awarded her master's degree.

It was at Gosset's suggestion that Fisher received his first voluntary worker. On his initial visit to Harpenden he arranged for his assistant in Dublin, E. M. Somerfield, to study under Fisher for a few months. The method of instruction—giving him field data to work through—was never in doubt. It was how Fisher and Gosset themselves had learned the inwardness of statistical research. Gosset remarked "What you suggest about running Somerfield over any figures coming in from your various departments is exactly what I want." He was a little apprehensive on one point: "You will I fear find that you will have to be even more elementary with Somerfield than with me, but that will be rather good for you as your fault is that you consider us all to be mathematicians."

The visit proved highly satisfactory to all parties. Three months were extended to five, and all three men were involved in the discussion of their common problems. Soon after Somerfield's return, Gosset confessed that his assistant now understood Fisher's writings better than Gosset himself. Another result of the visit, which Gosset had hardly dared hope for, was that Somerfield was permitted to publish the work done at Rothamsted. It is likely that Fisher had something to do with this, for he was strongly opposed to the conservative attitude of the firm toward scientific publication by its employees. Not only should scientific information be made public, but scientists had a right to public recognition of their work. Guinness's permitted the publication but still insisted that Somerfield publish under a pseudonym; 'Alumnus' was chosen.

There were plenty of problems brought by the research workers at Rothamsted, once the word got around that Fisher could find answers. The phrase, "Fisher can always allow for it" became common usage and in 1928 was immortalized when Bernard Keen, then assistant director at Rothamsted, wrote a song on this theme for the annual Christmas party. To this party were invited all grades of staff at Rothamsted, together with their husbands and wives, and the special entertainment of the evening consisted of a play or revue, full of topical allusions and mild leg-pulls, written and acted entirely by members of the staff. The song, sung to the tune of "Wrap Me up in My Tarpaulin Jacket," was an immediate hit, hailed with delighted calls for an encore. Years later it was chosen, with one or two features from other parties, for the farewell gathering to Sir John Russell on his retirement in 1943 and again when he visited Rothamsted in his 90th year. Keen had caught something of Fisher's impact at Rothamsted, never to be forgotten by those who had experienced the early days when a difficulty such as lost results for a single plot in a field experiment or a mistake in manuring meant scrapping the whole experiment unless Fisher could "allow for" the error, which he generally could.

The crops on the field plots lay dying
At Woburn, and Rothamsted too.
And Russell will soon come home crying:
'Oh! dear, dear, oh! What can we do?'

Chorus: *Why! Fisher can always allow for it,*
> *All formulae bend to his will.*
> *He'll turn to his staff and say 'Now for it!*
> *Put the whole blinking lot through the mill.'*

What matter the yields have all gone to pot
And half the manures were not sown.
What matter if Miller's* pigs ate the lot,
And foraged Broadbalk on their own.

Chorus: *For, Fisher can always. . . .*

Then Wishart and Irwin and Otelling,
And Florrie and Dunkley as well,
Their breasts with modest pride swelling,
Said, "Shall we do likewise? We shall.

Chorus: *If Fisher can always allow for it*
> *Oh, why on earth, why shouldn't we?*
> *And as he's bagged chi-squared we'll bow to it*
> *And make up our own formulae."*

Statistics, you see, is a wondrous cult
For a non-mathematical mind,
Which wants but the final, or end result—
As to how its attained is quite blind.

Chorus: *And when Fisher says he's allowed for it,*
> *And you take his word, humble and meek,*
> *I wonder if when he allows for it*
> *He's just got his tongue in his cheek!*

---

*Miller was manager of Rothamsted Farm. Dunkley was the boy originally hired to assist Fisher in 1920 and through the years, had made himself invaluable to the department. Until her marriage in 1925, Miss Mackenzie was the only other permanent member of the department. Otelling was a temporary voluntary worker. The qualified statistical assistants were a recent innovation, Wishart having been appointed in 1927 and Irwin coming in 1928. Prof. H. Hotelling was a later voluntary worker, during the academic year 1929 – 1930.

6

The Design of Experiments

The whole art and practice of scientific experimentation is comprised in the skillful interrogation of Nature. Observation has provided the scientist with a picture of Nature in some aspect, which has all the imperfections of a voluntary statement. He wishes to check his interpretation of this statement by asking specific questions aimed at establishing causal relationships. His questions, in the form of experimental operations, are necessarily particular, and he must rely on the consistency of Nature in making general deductions from her response in a particular instance or in predicting the outcome to be anticipated from similar operations on other occasions. His aim is to draw valid conclusions of determinate precision and generality from the evidence he elicits.

Far from behaving consistently, however, Nature appears vacillating, coy, and ambiguous in her answers. She responds to the form of the question as it is set out in the field and not necessarily to the question in the experimenter's mind; she does not interpret for him; she gives no gratuitous information; and she is a stickler for accuracy. In consequence, the experimenter who wants to compare two manurial treatments wastes his labor if, dividing his field into two equal parts, he dresses each half with one of his manures, grows a crop, and compares the yields from the two halves. The form of his question was: what is the difference between the yield of plot A under the first treatment and that of plot B under the second? He has not asked whether plot A would yield the same as plot B under uniform treatment, and he cannot distinguish plot effects from treatment effects, for Nature has recorded, as requested, not only the contribution of the manurial differences to the plot yields but also the contributions of differences in soil fertility, texture, drainage, aspect, microflora, and innumerable other variables. He should not complain, then, if in this test her answer is that plot *A* produces 10% more than plot *B*, but in a different test, when the field is differently divided, in a different season, or with a different variety, she gives an answer quite different in magnitude and perhaps even in direction. She is not being capricious; she is simply answering the question as it is put to her. For the experimenter, however, her local accuracy of response opens up a Pandora's box of perennial problems of variation in the field. It was Fisher's contribution to grasp the emergent Hope at the bottom of it all: if the logic of Nature makes a logical interrogation necessary in statistical terms, it also makes it possible.

Nature is really an exceptionally fine witness. The experimenter can rely on her to answer to the questions in precisely the form in which they are asked; question and answer are logically complementary. This means that there are principles the experimenter may apply to the logical structure of his interrogation, by which he may formulate his questions in a manner which elicits the required information. In the assurance that his analysis will yield valid conclusions, he may *design* his experiment. His philosophy of experimentation becomes a marriage of design and analysis.

When Fisher went to Rothamsted in 1919, his first task was to see what could be learned by a careful statistical analysis of the records of experimental and observational data that had been collected there over the years, without the benefit of the understanding of statistical planning that was to come later. There could hardly have been a better school than this for his instruction in the design of experiments. The responsibility he consciously accepted was to provide a clear and probing summary of the evidence. The rigour of his analysis, however, highlighted structural defects in the experimental layout, and he became aware of the gaps in the record which halted the process of analysis or vitiated conclusions drawn from it. Thus the logic which should inform the design of experiments gradually became explicit through their analysis.

□ □ □

In some respects the analysis of experiments was already well understood. Suppose the experiment were a comparison of two manurial treatments, then if it could have been assumed that the field was homogeneous in fertility and all other things equal, any differences in yield between the halves differently treated would have been attributable directly to the difference of manurial treatment. In fact, it could not be assumed that variations within a single field were negligible; indeed, uniformity trials in the first decades of the century had impressed agricultural experimenters with the ubiquity and extent of soil heterogeneity. Two tactics were employed to deal with the unwanted variations: first, to eliminate them as far as possible and then to estimate the effect of the remaining variations on the differences under study. When the actual observed differences were large compared with the error that could be introduced by unwanted variations, the experimenter had confidence in the reality of any difference found experimentally, in proportion to the rarity of the occurrence of such a difference by chance alone. The outcome of the analysis was often stated in terms of a test of significance, having a conclusion in the form: *either* something has happened by chance which would happen only once in 20 (or in 50, or in 100) trials *or* there is a real difference between the treatments.

The yardstick against which experimental effects were measured was the estimate of field errors. What was not understood was what constituted a valid yardstick. At worst, one had no estimate of errors at all. Our experimenter who divided his field into halves and compared their performance under different treatments was in this case. Still, such experiments existed, and we have seen that Fisher's first analysis of variance table appeared in 1923 in the analysis of a complex experiment on exactly this plan: two halves of the experimental field received different basal manures, and so, irrespective of the superimposed complexity of plots and split-plots, no applicable estimate of error between the halves could be derived from the data nor any comparison made between them. That Fisher himself first made the analysis erroneously in a way which included this comparison not only illustrates the cloudiness of understanding that surrounded the subject but shows that Fisher had reason to emphasize, as a first principle of experimentation, the function of appropriate *replication* in providing an estimate of error.

An estimate of the difference between plots A and B that could occur by chance alone in our experimental field, could have been obtained by comparing the yields of the halves during a number of previous seasons when they had been treated alike, but it would have taken an inordinately long time to reach an adequate estimate by this means. Ten years seemed minimal. Since it would have been exceedingly inconvenient if every field trial had had to be preceded by even 10 uniformity trials, for the sole purpose of providing an estimate of error, it was necessary to devise a means of obtaining such an estimate from the actual yields of the trial itself. The method adopted was replication, as Fisher called it; by his naming of what was already a common experimental practice, he called attention to its functional importance in experimentation. The experimental field was divided into a number of plots of equal size, and each treatment was assigned to several of the plots scattered over the whole area. The differences between plots treated differently then provided comparisons between treatments, while the differences between plots treated alike provided an estimate of their error. It was a way of asking Nature at the same time to show the difference in yield under different treatments and to provide a yardstick against which to measure the significance of that difference.

□ □ □

In the early 1920s the number of replications of each treatment was usually three or four. It was an empirical compromise. Every additional replication added to the accuracy of the estimate of error, but if as many as 10 or 12 different varieties were to be compared, restrictions of experimental space and labor had to be considered. Moreover, in respect of the size of errors, a compact experiment was to be preferred to an extensive experiment which ran the risk of increased soil heterogeneity. The several treatments of an experiment were therefore often assigned to plots located in contiguous *blocks* of land. In the blocked experiment the comparison of treatments was made within the block and the differences in yield of pairs of plots within the block were combined with the differences of similar pairs replicated in the other blocks. Since the differences were all taken within the compact blocks, their real error was reduced, despite the fact that the blocks themselves might be allotted in any convenient manner and, perhaps, dispersed discontinuously in different fields or even on different farms.

The simplest and most versatile form of blocking was merely to select relatively homogeneous areas of equal size for the blocks and then arrange one replication of all treatments within each block. With a row crop, for instance, where a fertility trend across the rows was expected to affect yields differentially, the field might be divided into four broad bands, as blocks, running the whole length of the field; within the blocks, the plots would be narrow strips, again running the whole length of the field, for ease of cultivation. Commonly, however, fertility gradients were to be expected not only across but also along the rows, and with suitable crops it was preferable to divide the field in both directions.

An idealized block arrangement is illustrated in Figure 6. A fertility gradient is supposed to exist in a north-south direction, and the four blocks have been arranged to cut across the gradient in such a way as to make the plots within each block comparable. The comparisons between the treatments are therefore unaffected by the differences in fertility that are believed to exist between blocks.

A class of block designs of particular interest, which provided for the elimination of fertility gradients in two directions, was that which Fisher later designated the Latin square. Classically, in these arrangements the field was

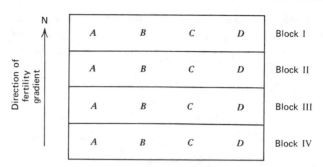

Figure 6. A systematic arrangement of 16 plots in four blocks, each with four treatments A, B, C, *and* D.

divided in both directions into the same number of plots, as in the 4 × 4 Latin square shown in Figure 7. Each row here constitutes a block of four treatments. In addition, the plots are so arranged that each column also contains one of each of the four treatments and is a block in its own right. The two-way blocking provided by the Latin squares removed the effects of soil heterogeneity in the two directions and so increased the sensitivity of the experiment. It was much used in testing differences between crop varieties or fertilizers, in which qualitative comparisons were to be made.

The arrangements illustrated in Figures 6 and 7 are systematic designs, that is, the arrangement of the treatments bears a regular relation to the arrangement of the blocks. The four treatments in Figure 6 are identically arranged within blocks and those of the Latin square are shifted one step across the columns for every step down the rows. Such simple systems were not often employed in practice, for obvious reasons: there was too much danger that fertility gradients might happen to run along the line of similar treatments, in an east-west direction in the block design or across the diagonal of the Latin square. Such gradients would naturally enhance or depress the apparent effects relative to each other and would alter the estimate of error, often increasing it. In an attempt to reduce the errors, alternative systematic designs had been introduced that avoided the obvious danger by using patterns in which plots containing similar treatments were never put closer to each other than was necessary. Such were the chessboard designs, sandwich designs, and Knut-Vik squares.

Unfortunately, all systematic designs contained the same flaw, although it was hidden in the arrangements usually adopted. As Fisher explained it once,

A	B	C	D
B	C	D	A
C	D	A	B
D	A	B	C

Figure 7.   A systematic 4 × 4 Latin square.

the experimenter games with the Devil and his object is not to be deceived. Let the Devil devise any arrangement whatsoever, in advance, for the soil fertility of the experimental plots; the experimenter must be ready to accommodate it by his experimental design. A systematic arrangement is prepared to deal with only a certain sort of devilish scheme. But the Devil may have chosen *any* arrangement, even that for which the systematic design is least appropriate; the experimenter may therefore lose, being deceived into attributing real effects to chance or supposing chance effects to be real, thinking really large errors to be small or small errors large. To play this game with the greatest chance of success, he cannot afford to exclude the possibility of any possible arrangement of soil fertilities, and his best strategy is to equalize the chance that any treatment shall fall on any plot by himself determining it by chance. Then if all the plots with a particular treatment have higher yields, it *may* still be due to the Devil's arrangement, but then and only then the experimenter knows how often his chance arrangement will coincide with the Devil's.

It is the same philosophy which makes a wise banker in games of chance prefer to use "fair" instruments, for only with them can he be sure of winning. The roulette wheel that is true will give him a certain margin of profit whatever the bets laid against him, for the chances have been arranged in the long run to favor him. The roulette wheel that is biased is not at all so safe, for, although his chance is increased with respect to some results of play, it is decreased for others, and he cannot be certain that the players will make the bets that profit him. He has to take all bets, as the experimenter has to accept all patterns of soil heterogeneity, and he stands to lose unduly from bets of a certain sort. A run of such bets will sometime inevitably happen and will break him as certainly as if the players suspected the direction of the bias of his wheel.

Fisher saw all this. It did not prevent him from enjoying the role of banker using a biased wheel at the Harpenden Fair in the summer of 1927. He set up his sideshow in among the gypsy vans and the stalls where women could be seen slinging toffee on a wall hook of their booth, where the steam organ of the merry-go-round lent its unforgettable rhythm to the clamor of men calling from the cocount shies, the peep-shows, and the swings. T. N. Hoblyn, then a voluntary worker in the statistics department, set up the wheel with him and during the evening helped him running it. He recalls:

It consisted of a homemade roulette wheel, mounted vertically on a pole so that all could see, and marked out in segments, to each of which was given a number showing the odds on it turning up when the wheel came to rest. We calibrated it beforehand at the Labs. and calculated the actual odds. Naturally, as it was a rather gimcrack affair, it was badly biassed. However, we cut the odds to about one-third, i.e., if the number was expected to turn up once in 100 spins, we gave 30 : 1 only and the result was an enormous success financially. I have a vivid memory of going back to the sideshow at about 10 o'clock in the evening, when the fair ground was lit with flares, to find R.A.F. with a large crowd around him, calling out the odds as to the manner born, flushed with success, and fairly raking in the money.

The experimenter with a systematic design had a biased wheel and paid the true odds every time.

With the diagnosis, Fisher proposed the cure. The condition on which a re-plicated experiment provides a *valid* estimate is that pairs of plots treated alike must on the average not be nearer together, or further apart, or in any rele-vant way distinguishable from pairs of plots treated differently; this can be done by arranging the plots within the agreed block pattern deliberately *at random*. Thus Fisher enunciated his own primary principle of experimental design: *randomization*. Chance alone should determine the allocation of treatments within the agreed block pattern; the throw of a die or the dealing of a card from a well-shuffled pack should decide the matter for each plot in a block, and the arrangement should be determined afresh for each block separately. Designs such as those illustrated in Figure 8 were the result when

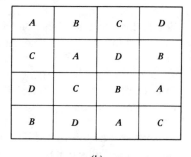

Block I	C	B	A	D
Block II	A	C	D	B
Block III	D	A	B	C
Block IV	A	B	C	D

A	B	C	D
C	A	D	B
D	C	B	A
B	D	A	C

(a)   (b)

Figure 8.   (a) *A randomized block design.* (b) *A randomized Latin square.*

chance determined the order of the treatments applied to plots in the block arrangement and in the Latin square. Each block of the randomized block design compromises, as before, one replication of all the treatments, and the randomized Latin square retains the double blocking arrangement, but, within these restrictions, the particular arrangement of treatments has been determined by a selective process that gives every possible arrangement an equal chance of being chosen.

□ □ □

As theoretical justification for the revolutionary proposal of randomization, Fisher argued in 1926 [*C P* 48] that

> The estimate of error is valid because, if we imagine a large number of different results obtained by different random arrangements, the ratio of the real to the estimated error, calculated afresh for each of these arrangements, will be actually distributed in the theoretical distribution by which the significance of the result is tested.

The theoretical distribution referred to is the distribution that obtains on the normal theory assumptions, so that Fisher here claims that the error ratio will, over all randomizations, have the distribution appropriate when observations have been drawn independently from a normally distributed population. This statement is not, in fact, strictly true, but it has since proved that for most data the distribution of the ratio generated by the randomization process does approximate adequately to the theoretical distribution.

It is uncertain just when Fisher made the intuitive leap by which he recognized the principle of randomization. In 1918 his analysis of variance of physical measurements of various human relatives was based on "normal theory" assumptions; the observations were assumed to have been drawn independently from a normally distributed population. Later, on analyzing the variance of field observations into the two components respectively within and between groups, he used the ratio of the variance s_B^2 of the group means to the variance s_W^2 within groups to assess departure from the null hypothesis. He obtained this distribution (of $F = s_B^2/s_W^2$) in 1922, although he often chose to work with half the logarithm of this ratio, $z = \frac{1}{2}\log F = \log s_B - \log s_W$).

Already, however, in 1923, in introducing the analysis of variance of field experiments, he had made it conditional not on "normal theory" assumptions but on randomization of the plot arrangement. Yet, in assessing his results, he used the same z test which was appropriate on "normal theory" assumptions. Evidently he saw that this "randomization" distribution of z would approximate the "normal theory" distribution of z. This conclusion was difficult to justify theoretically and was for years to cause trouble among statisticians.

At the time statisticians felt considerable concern about the normality or nonnormality of the distributions. We recall that Karl Pearson's series of frequency curves had been developed in order to fit nonnormal distributions and that they were then much in vogue. In contrast, Fisher seems from the first to have felt that moderate nonnormality was not an important factor. What he seems to have been much more concerned about was the latent assumption that invariably went with the "normal theory" assumptions, that the errors in the observations were independently distributed. In many, perhaps most practical cases, this assumption was clearly not justified. Observations of the fertility of adjacent plots, rain in successive hours of the day, or yield on successive milkings of a cow were not independent but highly correlated. Despite this, whatever assumptions were made about the specific distribution of the observations, analysis was traditionally based on the assumption of independence. For instance, the well-known formula for the standard error of the mean was only valid on the assumption of independence of the observations.

Almost certainly it was the evident lack of independence of field observations that led Fisher to seek a foundation for his analysis which did not involve this assumption. He knew that the effect of even a moderate lack of independence of the observations, unlike the effect of moderate nonnormality, could be disastrous to the analysis. Examples used to illustrate the analysis of variance in *Statistical Methods* were problems he had met in practice, and the first two-way table analyzed there is a case in point. The data record frequency of occurrence of showers of rain in successive hours of the day in different months of the year. Obviously, if it was raining between twelve and one o'clock, it was likely also to be raining between one and two. Thus the data for each month really represented a time series in which, because of the correlation of the observations, the ordinary "independent normal theory" analysis was invalid. Fisher recognized this and did not attempt to test the month-to-month effect for which the test was vitiated; instead, he used the example to illustrate how the analysis could be carried through to test the difference associated with hours of the day for which he could see the "independent normal theory" analysis was approximately correct.

Confronted by such correlated data, in the early 1920s Fisher must have been seeking an experimental arrangement by which, when the null hypothesis was true, the expected values of variance estimates made within and between groups should be equal. He found it in randomization, and when, in 1923, he introduced the analysis of variance, he did not take the expectation of variances for repeated samples from the same fixed experimental layout, assuming independently distributed observations, but took the expectation appropriate under randomization, that any observation was interchangeable with any other in the analytic expressions. On this basis, when the null hypothesis was true and there were no real differences between the group means, the variance calculated from the group means was on the average identical with that within the groups. The analysis of variance could be carried through because randomization ensured that this condition was met (as it would also have been met had independence assumptions been appropriate). He guessed rightly that this might make not only the mean values but the whole F distribution based on normal theory about right. Thus randomized experiments could be analyzed *as if* the observations were roughly normally distributed and independent.

No doubt these ideas were also supported on the theoretical side by Fisher's geometrical insight: he could picture the distribution of n results as a pattern in n-dimensional space, and he could see that randomization would produce a symmetry in that pattern rather like that produced by a kaleidoscope. This might approximate the spherical symmetry that would have been induced by normality. His confidence in the validity of his results must to some extent have rested on this insight.

A practical test was available, because many field trials had been run simply to test soil heterogeneity. In such uniformity trials the yields of different plots in a field under uniform treatment were recorded separately so that the data were equivalent to those from an experimental layout in which the null hypothesis was true. Fisher had used such results informally to check the effects on the estimated experimental error of various experimental arrangements which could be imagined as superimposed on the data. Similarly, in illustrating the value of randomization, replication, and blocking in *Statistical Methods* in 1925, he exemplified the estimation of error, using uniformity trial data, on the supposition that various treatments, in various arrangements, might have been superimposed on the data. Extending this notion, Eden and Yates [31] in 1933 ran a uniformity trial. Superimposing a series of different random arrangements on the data to simulate the application of treatments when the null hypothesis was known to be true, they demonstrated that the empirical distribution obtained from numerous randomizations was very similar to that appropriate to the "normal theory" z test.

In the absence of explicit mathematical proof, statisticians continued to be reluctant to accept randomization. If the analysis of variance were based on the theory of normal independent errors, it was inapplicable to the usual populations of the field; if it were not based on this theory, but by-passed the need for normal theory assumptions, it lacked theoretical justification.

Apart from theoretical objections, there was reluctance among the older experimental statisticians to abandon the practice of a lifetime and to learn the new philosophy and new techniques Fisher proposed. As an experimenter, his old friend Gosset was never thoroughly convinced of the necessity for randomization, and in this he was not alone. Randomization was a cornerstone of the new experimental methods, and these methods were not easily assimilated by the older generation. The first Ph.D. thesis* to come out of the statistics department at Rothamsted was the work of R. J. Kalamkar. Fisher characterized it as a useful series of papers illustrating the new methods. When it was submitted in 1932, however, the examination was delayed for lack of examiners. E. S. Beaven, the celebrated agriculturalist, refusing to act as examiner, wrote with indignation as a practitioner with 40 years of experience to denounce the new-fangled methods and the unprecedented demand they made for mathematical skills. G. U. Yule, a distinguished and able statistician, pleaded incompetence ("I simply cannot make head or tail of what the man is doing"). He apologized that he was too old a dog to learn new tricks and suggested that examiners should be found who were fully conversant with Fisher's work. In academic circles such men did not exist. Finally, J. F. Tocher of Aberdeen University accepted the task, because of his friendship for Fisher and his faith in his judgement, adding that he did not profess to be adequately prepared to give a critical appraisal of the work.

□ □ □

In these years there was no single place in the literature where the ideas Fisher was developing in connection with experimental design were gathered together and presented fully. The first brief general account of design appeared in the last nine pages of *Statistical Methods* in 1925. A fuller account appeared the following year [C P 48, 1926, "The arrangement of field experiments"], and this was elaborated in a joint paper with J. Wishart in 1930 [C P 85]. In these papers the principles of experimental design were enunciated and illustrated. Meanwhile new applications were rapidly being evolved and appeared piecemeal in discussion of experimental work in various agricultural journals.

   In 1930, Fisher taught courses in statistical methods at Imperial College (South Kensington) and also at the Chelsea Polytechnic. He spent the summer of 1931 at Ames, Iowa, giving seminars and lectures on the same topics. The notes he had prepared for these occasions formed the basis for the more comprehensive treatment of experimental design for which the need was now evident, and in 1935 *The Design of Experiments* was published. The volume not only introduced his ideas to the wider audience that was by that time

---

*Russell had arranged that work done at Rothamsted and vouched for by the head of department should be accepted by London University for examination as a thesis for Ph.D. or D.Sc. degrees.

clamoring for enlightenment about them, but also helped to clarify his position on some points of confusion.

In particular, we note here that in defending the common use of Student's *t* test in *The Design of Experiments,* Fisher introduced an aspect of randomization which had already been in his mind for some time. Student's *t* test is appropriate to the null hypothesis, that the two series are samples from the same normally distributed population. Theoretical statisticians had begun to stress the element of normality as though it were a serious limitation to the test. Fisher raised the question of whether we should obtain a materially different result by testing the wider hypothesis which merely asserts that two series come from the same population, without specifying that it is normally distributed. He proceeded to show that the results of a randomized experiment could be used to test the wider hypothesis and, by comparison with the *t* test, to check the appropriateness of the *t* test and the approximate normality of the observations. "It seems to have escaped recognition," he wrote, in a characteristic introduction of a new idea,

It seems to have escaped recognition that the physical act of randomization, which, as has been shown, is necessary for the validity of any test of significance, affords the means, in respect of any particular body of data, of examining the wider hypothesis in which no normality of distribution is implied.

As an example, he used the data from an experiment carried out by Charles Darwin in which the heights of maize seedlings grown from self-fertilized and cross-fertilized stock were compared. Pairs comprising one each of the contrasted seeds were planted in each of 15 flower pots, so that each self-fertilized plant was pitted against a cross-fertilized plant treated under conditions made as equal as possible. Darwin did not, in fact, randomize the allocation of seeds within the pots, as could have been done by tossing a coin to decide whether the seed were to be planted on the right-hand or on the left-hand side of each pot; but, Fisher said, had he done so, the randomization test would have applied to the results. With randomization, if the null hypothesis were true, the differences in height between pairs would be such as arose between pairs from identical populations and would be positive or negative as the physical act of randomization had chanced to determine. The differences actually obtained would thus form a sample from a population of 2^{15} numbers (or 2^k where k is the number of pairs) that could be generated by giving each difference alternatively a positive or negative sign; the average difference of a sample of 15 would be one case among the 2^{15} average differences that might have so arisen. The test of significance could thus consist of counting how many of the average differences which (given that the null hypothesis was true) could have occurred under different randomizations would have been greater than that actually observed. In Darwin's experiment the proportion was found to be 2.63%. Fisher compared this with the t test, which yields the result appropriate for independent normally distributed observations, and he showed that in this particular case there was very close correspondence between the results.

In the following year [C P 141, 1936], Fisher again used the randomization distribution in the same sort of way as he had for the paired t test but now checking the appropriateness of the unpaired t test which he had made on craniometric data. In that paper he went so far as to state that the only justification for using the normal theory test was that it was a good approximation to the randomization test.

Actually the statistician does not carry out this very simple and very tedious process [of calculating the randomisation set] but his conclusions have no justification beyond the fact that they agree with those which could have been arrived at by this elementary method.

By an extension of the theoretical considerations by which in 1926 he had justified the practice of randomization, the randomized experiment was shown to induce a distribution relative to the results which could be made explicit and, since it existed independently of normal theory, could be used as a check on the appropriateness of that theory.

□ □ □

The article on "The arrangement of field experiments" (C P 48, 1926), was not prepared with a view to meeting the criteria of mathematical proof but of introducing agricultural experimenters to the principles of experimental design, together with some examples of their application. It appeared in the *Journal of the Ministry of Agriculture* under a title apparently without reference to statistics, and it was addressed to men running experiments. At this time systematic designs were much in vogue. One important object of the paper was, therefore, to point out the advantage of randomized as opposed to systematic designs, in terms which would appeal to the practical man. Fisher summed up his argument for randomized designs as follows:

It is particularly to be noted that those [systematic] methods of arrangement, at which experimenters have consciously aimed, and which reduce the real errors, will appear from their (falsely) estimated standard errors to be not more but less accurate than if a random arrangement had been applied; whereas, if the experimenter is sufficiently unlucky, as must often be the case, to *increase* by his systematic arrangement the real errors, then the (falsely) estimated standard error will now be smaller, and will indicate that the experiment is not less but more accurate. Opinion will differ as to which event is, in the long run, the more unfortunate; it is evident that in both cases quite misleading conclusions will be drawn from the experiment.

In addition to emphasizing the unfortunate consequences of failure to obtain a valid estimate of error, he also made it clear that, to a large extent, experimenters could have their cake and eat it. Randomization would be conducted within the pattern of whatever block arrangement was considered likely to give the most precise results. The replication of treatments in block arrangements would still be carried out with random allocation of treatments to the plots within a block; Latin squares, retaining all the advantage of their blocking pattern in enabling the elimination of soil heterogeneity in two directions, could still be chosen randomly. Only the allocation of treatments within the block was to be affected by randomization. Whether in this case the real errors happened to be increased or decreased by the arrangement, the analysis would reflect the fact truly; it would not, like the analysis of a systematic design, give an inverted image of the real in the estimated effect on the precision of the experiment. The method of analysis was made appropriate to the condition of randomization in the experimental design. On this condition, the analysis of variance could be carried through and the z test applied. Information in the experimental data could not leak out in the analysis because it was held in the same watertight system of logic.

It is a little surprising, perhaps, that this paper, in 1926, should have been stimulated by a previous article in the same journal on the subject of "Field Experiments: How They are Made and What They Are," written by E. J. Russell [32], treating the subject in traditional terms, and advocating systematic designs. It is a measure of the climate of the times that Russell, an experienced research scientist who, as director of Rothamsted Experimental Station, had had the wisdom to appoint Fisher statistician for the better analysis of the Rothamsted experiments, did not defer to the views of his statistician when he wrote on how experiments were made. Design was, in effect, regarded as an empirical exercise attempted by the experimenter; it was not yet the domain of statisticians.

Nevertheless, at Rothamsted this domain had begun to be infiltrated by Fisher's influence. His statistical ideas, even the controversial idea of the randomization of experimental plots, that year had found their way into the conduct of some actual experiments; a change of attitude was occurring among the research workers as one or another of them sought help and found Fisher ready to discuss the experimental situation thoughtfully and offer good advice. These changes took place gradually and, in retrospect, Fisher's progress to dominance in the design of experiments seems smooth. Even the drastic changes in farm operations, undertaken to cope with the increasing complexity of field experiments in the later 1920s, were made with far less trouble than the administration expected, and, in this case, the tradition of scrupulous care over details in the classical experiments laid down by Lawes and Gilbert was no doubt an important contributing cause.

Two factors cooperated to effect the gradual acceptance among the research workers of an extended role for statistics in their experimentation. Rothamsted was a research institution, and in such places there was a growing sense of the need for statistical help in the interpretation of experimental results. Equally important, the particular statistician employed at Rothamsted was Fisher, and he took a broad view of his responsibilities. Biological workers were shy of mathematics and certainly to begin with felt that the statistician should only be called in after the data had been collected. When they went to Fisher, however, they found that, even as he interpreted their current results, they were being led to consider the lessons in design to be learned from the experiment already done and planning how best to proceed in order to elicit more informative results in the future. Their experiences demonstrated the benefits of discussion with Fisher before doing the experiment. It became evident that a statistician, like a physician, needs to be called into consultation early. He can sometimes manage to retrieve results from a sick experiment. He can carry out a *post mortem* on one that died. But his preferred task must be to assure the grounds of health from the beginning, so that the experiment shall have a good chance of leading a productive life and weathering the accidents to which it is exposed.

As we have seen, randomization did not mean haphazard scattering of the treatments on the plots but a random assignment within the chosen block pattern. With a Latin square a great deal of system was built into the design, for it had the restriction that each treatment appeared once in each row and in each column. The question then arose, how, under this one restriction, was the design to be chosen "at random"? Fisher's first task was to discover what should constitute the physical process of randomization. With Latin squares he felt that it was necessary to know all the possible arrangements in order to establish a means of selecting Latin squares in the statistical laboratory that would give every arrangement an equal chance of selection.

In 1924 he was considering the problem. He sought advice from P. A. MacMahon, who had enumerated the 4 × 4 and 5 × 5 squares and whose book on *Combinatorial Analysis* was the best recent account of the subject. A letter to MacMahon in July that year shows the caution with which he picked his steps across what was something of a philosophical bog to reach a firm method which should realize the ideal of imposing a random order. He wrote, in part,

It is possible that your method of solution will solve for me at least the first of the outstanding questions:

(1) What experimental technique of filling up the square will give each solution an equal chance of appearing? . . .

(2) Is such a technique necessary or sufficient for the statistical validity of an agricultural experiment? This question involves points which it is not easy to reduce to a purely mathematical form. An example will show you the kind of lines along which I have been working.

And he set out and discussed a field example.

By the end of the month the work had advanced. Fisher had evolved his own method of solution by direct enumeration instead of MacMahon's algebraic method, and he had by this method enumerated the 5 × 5 squares, finding 56 symmetrical pairs of reduced squares where MacMahon had discovered only 52.

In January 1925 Fisher had enumerated the 6 × 6 squares. He hoped MacMahon would think the method he had invented a valid one, although, as he wrote, "It is scarcely applicable to a 7 × 7 as there must be about 10,000 groups in that case"; this would mean about 250,000,000 reduced squares. Fortunately the most useful experimental layouts were no larger, and the enumeration of the 7 × 7 and larger squares could wait.

There are those today who might question whether this work was strictly necessary from the practical point of view. They might feel that it ought to be sufficient in practice to randomize rows, columns, and treatments piecemeal; or one could argue that it would be sufficient to make a complete enumeration of *one* of the possible sets rather than all of them. For Fisher, it was not enough: the aim was to select at random from all possible squares, and this would not be achieved without knowing what the possibilities were. Fisher chose to make the complete enumeration; combinatorial analysis was rather a special study, yet he was prepared to go beyond current knowledge in this field to get the results he wanted.

Clearly, he found it a fascinating task; he enjoyed pitting his wits against a purely logical puzzle and was delighted by the ingenuity required in the unravelling of the various possibilities. He was also intrigued by the individual character which some of the sets displayed. There was to be, in addition, considerable experimental spin-off from the purely theoretical work, for the combinatorial properties of the square arrangements soon came into play in the design of factorial arrangements and in the exploitation of Graeco-Latin squares and incomplete block designs. The Graeco-Latin squares are Latin square arrangements with rows, columns, and treatments (*A, B, C*) balanced (orthogonal) both with each other and with another variable, say, varieties (α, β, γ), whose contribution to the variance can therefore be estimated separately from other causes. Fisher's work led to some interesting results in combinatorial analysis, such as the demonstration of the truth of Euler's belief that there are no 6 × 6 Graeco-Latin squares.

The sheer fun of the chase played its part in his motivation; he was fascinated by the intricacy of numerical relationships and the beauty of numbers in all permutations and combinations. It was for fun that he invoked combinatorial theory at home in the painting of the "bricks." He acquired wooden blocks for his children. Their dimensions were 6″ × 3″ × 2″ so that they had distinct faces, edges, and ends. Each surface was painted in one of six colors, but the convention was adopted of having a white surface always opposite to a black, a green surface opposite a blue, and a red surface opposite a yellow. Within this restriction there are 96 different combinations of colors possible on the surfaces of the brick. Starting with 24 bricks he painted them so that no two were identical; then doubled and redoubled the supply until all 96 possibilities were realized. Such is the play of the combinatorial mathematician! And the bricks proved an inexhaustible source of fun for the children.

By the time he had enumerated the 6 × 6 squares, he could add a post script to MacMahon, "You will be interested to know that the Latin Square has been a great success agriculturally." Although in 1924 he had not yet been invited to design experiments at Rothamsted itself, he had been brought into consultation at Rothamsted with the Forestry Commission, in connection with an experimental layout in the forest nursery at Bagshot and had been able to introduce his new (randomized) Latin square design there that year. Plate 6 shows an experimental forest the Forestry Commission laid out in a 5 × 5 Latin square (of Fisher's) at Bettgelert in Wales in 1929. When the photograph was taken 16 years later the different effects of altitude on the different varieties of trees are visible a mile away.

In "The arrangement of field experiments" (*C P* 48, 1926), when Fisher discussed variety trials and qualitative comparisons of fertilizers, he proposed that by far the most efficient arrangement, as judged by experiments on uniformity trial data, "is that which the writer has named the Latin Square" (meaning randomized). Having by then solved the technical and mathematical problems of obtaining randomized Latin squares, he was mainly concerned that the experimenters should have ready access to them, without getting entangled needlessly with the theory. As he says,

The actual laboratory technique for obtaining a Latin Square of this random type will not be of very general interest, since it differs for 5 × 5 and 6 × 6 squares, these being by far the most useful sizes. They may be obtained quite rapidly, and the Statistical

Laboratory at Rothamsted is prepared to supply them, or other types of randomised arrangements, to intending experimenters; this procedure is considered the more desirable since it is only too probable that new principles will, at their inception, be, in some detail or other, misunderstood and misapplied; a consequence for which their originator, who has made himself responsible for explaining them, cannot be held entirely free from blame.

Thus he launched his randomized designs with an engineer aboard, as it were, to cope with problems arising with the vessel's new principles of design.

The offer from the statistical department did, in time, bring in a number of inquiries and requests from agricultural workers, running experiments under such varied conditions as were encountered with cotton in the Sudan, rubber in Malaya, rice in India, and tea in Ceylon. Inquirers were duly supplied with the cards of random arrangements, together with such advice and references to the literature as might be helpful, and often also with the offer to analyze the experimental results at Rothamsted.

The extending influence of Rothamsted itself brought new contacts to the statistical department. In 1923 Sir John Russell had visited the Sudan in connection with the Gezira scheme for growing cotton and, realizing then how isolated some research workers were from agricultural research news from Britain, had made the suggestion which led to the establishment of Commonwealth Agricultural Bureaux and to much better communication between workers in different regions. His later visits to Canada, Palestine, Australia, and New Zealand in 1928 and to South Africa in 1929 strengthened the ties between Rothamsted and agricultural workers in these countries. And always, the hospitable attitude at Rothamsted attracted visitors to the station.

Visitors discovered in Fisher a source of willing discussion and counsel about their problems. Hoblyn, sitting almost at Fisher's elbow in order to listen in, recalls:

He always gave everyone his full attention, whatever he was doing, and I was full of admiration for his virtuosity and the wisdom of his advice. It must be remembered that these were in the very early days of randomisation and that his ideas were often highly controversial.

After returning to their own place, men who had met Fisher at Rothamsted often wrote to him to discuss current work and to introduce colleagues. Further visits to Rothamsted resulted, often with the object this time of getting specifically statistical advice and experience.

As the new analysis and the new designs found a market, men were sent to Rothamsted to learn the elements of the new methods; consequently, the number of voluntary workers attached to the statistics department rose steeply: in 1928 there were three, in 1929 nine, and in 1930 thirteen, staying mostly for short periods but varying from 3 weeks to 3 years. So the influence of Fisher's ideas began to be felt abroad in the practical conduct of experimentation. The rapidity with which they spread may be gauged by the fact that only in 1926 was the first paper on design published and the first randomized experiments laid down at Rothamsted Farm.

☐ ☐ ☐

A number of field experiments were of the qualitative type for which Fisher felt the Latin squares were best, in which perhaps six varieties of potato or several types of nitrogenous fertilizer were to be compared. Most manurial and cultivational experiments, however, were not of this type. A single factor like the application of nitrogen might be affected by a number of other factors: the nitrogen might be applied early or late, in a light or heavy dressing, in the absence or presence of other manurial dressings. To investigate the results of nitrogen application the experimenter wanted to ask more than the single question, "which does best?" He wanted to ask a number of often dichotomous questions; because all were relevant to the same agricultural practice, he wanted to ask them simultaneously, for the effect of one factor might be enhanced at one level of another, as the heavy dressing of nitrate, for example, might prove valuable if applied early but damaging when applied late in the season.

This type of investigation presented experimenters with a dilemma. According to traditional ideas, if they were to make a precise comparison, they ought to vary the essential conditions only one at a time; they should isolate the individual factors of importance and test them singly. In practice, they could not do so in a meaningful way, since the answer to one question was often contingent on the answer to another. It was becoming clear that not only the simple additive effects but also the joint effects of different factors (their interactions) were important and that therefore experiments capable of discerning interactions had to be used, even if it meant seeming to abandon the idea of studying one factor at a time.

The answer Fisher produced in 1926 was the factorial design, so familiar today that its principles appear self-evident and its elegance and power are largely taken for granted in experimentation. Although one might have supposed the idea had come to him through his frustration in attempting to analyze experiments like that on Broadbalk (which now can be seen as an incomplete factorial layout), he later [C P 248, 1951] traced the idea to origins in his genetical thought: "The 'factorial' method of experimentation, now of lively concern so far afield as the psychologists, or the industrial chemists, derives its structure, and its name, from the simultaneous inheritance of Mendelian factors." Thus we should regard the 2^3 factorial cube shown in Figure 9b as derivative from Fisher's conception of an analogous cube representing not three treatments at two levels each but, perhaps, three genetic loci with two alleles at each. Indeed, in 1946 (see Figure 12, Chapter 13) he had occasion to represent the genetic basis of the rhesus blood groups by just such a factorial cube.

In 1926 it seemed a "remarkable consequence" of the simultaneous investigation of several factors that large and complex experiments could have a much higher efficiency than simple ones. As Fisher said,

No aphorism is more frequently repeated in connection with field trials, than that we must ask Nature few questions or, ideally, one question, at a time. The writer is convinced that this view is wholly mistaken. Nature, he suggests, will best respond to a logical and carefully thought out questionnaire; indeed, if we ask her a single question, she will often refuse to answer until some other topic has been discussed.

The factorial design is well exemplified by what must be the commonest of all manurial experiments, a minor classic, designed to test the application, singly or in combination, of nitrogen (N), phosphate (P), and potash (K). A basal manure has, we assume, been applied overall to the field. Each of the three factors is then to be applied at two levels (that is, its absence or presence as an addition to the basal dressing), indicated by a plus or minus sign under Treatments in Figure 9a, or the absence or presence of a lower case letter in respect of yields. Each of the eight combinations shown constitutes a single 'treatment' for comparison with the other combinations. In practice the 8 combinations in this factorial design would be replicated several times, using a randomized block layout. The analysis outlined would thus be applied to the average yields at each of the 8 conditions indicated by lower case letter combinations.

In Figure 9b the four treatments at the corners of the back face differ from those on the front face only in respect of the application of nitrogen; there are thus four contrasts measuring the effect of nitrogen alone. The average or main effect of nitrogen is thus

$$N = \tfrac{1}{4}\{(n-1) + (nk - k) + (np - p) + (npk - pk)\}$$

as indicated in Figure 9c. Comparison of the top with the bottom of the cube shows that the same is true for contrasts involving phosphate alone and comparison of the left-hand side with the right-hand side shows the same for contrasts involving potash alone.

If, now, we compare treatment combinations on the top right of the cube with those on the top left $(npk + pk) - (np + p)$ the contrast measures the effect of K in the presence of P. Similarly, comparing the bottom right with the

	N	P	K	Yield
1.	−	−	−	1
2.	+	−	−	n
3.	−	+	−	p
4.	+	+	−	np
5.	−	−	+	k
6.	+	−	+	nk
7.	−	+	+	pk
8.	+	+	+	npk

a. Treatments

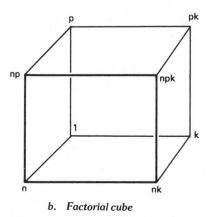

b. Factorial cube

Effects

Main Effects
$$4N = (n - 1) + (nk - k) + (np - p) + (npk - pk)$$
$$4P = (p - 1) + (np - n) + (pk - k) + (npk - nk)$$
$$4K = (k - 1) + (nk - n) + (pk - p) + (npk - np)$$

Inter-actions
$$4NP = (np + npk) - (p + pk) - (n + nk) + (k + 1)$$
$$4NK = (nk + npk) - (k + pk) - (n + np) + (p + 1)$$
$$4PK = (pk + npk) - (p + np) - (k + nk) + (n + 1)$$
$$4NPK = (npk + n + p + k) - (1 + pk + nk + np)$$

c. Treatment contrasts

Figure 9. Treatments and treatment contrasts of a 2^3 factorial, represented as a cube.

bottom left $(nk + k) - (n + 1)$ we have a second measure of the effect of K, this time in the absence of P. Thus the extent to which the effect of K is different as a result of the presence of P—the PK interaction—is the difference between these two measures $(npk + pk) - (np + p) - (nk + k) + (n + 1)$. This provides us with a contrast involving the four treatments on the southeast diagonal with four treatments lying on the northwest diagonal. By the same argument the diagonals across each of three pairs of parallel faces provide us with a contrast of four treatment combinations with some other four treatment combinations and yield, respectively, the 3 two-factor interactions. A little thought will show that the test of the PK interaction in the presence of N (front face of cube) with the PK interaction in the absence of N (back face of cube), on the four diagonals across the cube, will yield a measure of the three-factor interaction, NPK, measuring, for instance, whether the PK interaction is changed by the presence of N. Thus in the full factorial design, every observation is used for every comparison made; from eight observations we obtain four differences each for the seven treatment contrasts, and each of these comparisons is affected only by such soil heterogeneity as exists between an average of four plots contrasted with another average of four.

The tremendous advantages of the factorial design are apparent. In the first place, if the variables N, P, and K were so obliging as to behave independently, we have obtained with eight plots an estimate of the effect of each of these factors, as accurate as could have been obtained in three such layouts testing one factor at a time. In the second place, we have a bonus whenever, as is usually the case, the factors do not in fact behave independently. From the factorial we have estimates, of equal precision with those of the main effects, of all the interactions NP, NK, PK and NPK. No experiment testing one factor at a time provides any of this information. Each observation is made to work for us seven times over in estimating the three main effects and four interactions.

As already mentioned, the set of eight treatments in the 2^3 factorial would not be run singly but in replicates. The replication would normally take the form of a randomized arrangement of compact blocks, each containing the eight treatment combinations. In addition to the other advantages of the factorial design, the randomized block factorial has a further very considerable advantage over studies of one factor at a time. The conclusions drawn from the single-factor comparisons in the factorial design are given a much wider inductive basis. In the experiment discussed above, we know the main effect of nitrogen not only at some fixed level of the other plant nutrients but under a variety of conditions of phosphate and potash: we have estimated the interactions. The conclusions may be further generalized by replicating the factorial blocks on somewhat different material; blocks could be run on fields at Rothamsted having different histories of past cultivation, or blocks on the clay soil of Rothamsted and the sandy soil of Woburn could be compared to see to what extent the response to nitrogen varies with variations in the soil texture. Evidence of consistency of response over a variety of other variable conditions gives added force to the particular conclusions drawn from the factorial experiment.

□ □ □

At Rothamsted Fisher was early brought into consultation with people who had to run experiments with more than one factor at a time. In 1923 his paper with Mackenzie (*C P* 32) presented an analysis of such a complex experiment, run by Thomas Eden, a crop ecologist at Rothamsted, in which it was intended to investigate in particular the interaction between two variables: different types of potash manure and different varieties of potatoes. In that case although three variables were introduced, they were tested on areas of different size: basal manures were compared by half-fields, potato varieties on plots within the fields, and potash manures in rows within the plots. In consequence, the comparisons were not all equally precise, and three estimates of error would have been required to test the three effects. Moreover, the progressive fragmentation of the field required in order to introduce three factors in this way resulted in rows consisting of only seven plants each, rather too small a sample to be reliable. The poor design of this experiment prompted Gosset to suggest that Fisher should start designing experiments. Fruits of this idea first appeared in the field, and in print, 3 years later.

In 1926 T. Eden ran a full factorial design in a randomized block arrangement to test manurial effects on winter oats. This seems to be the first of the Rothamsted experiments carried out unambiguously to Fisher's specifications, and it was the experiment he used, even before the results were available, to illustrate the advantages of the new factorial designs when he wrote "The arrangement of field experiments" (*C P* 48). The results were presented in a joint paper with Eden in 1927 (*C P* 57). They were highly satisfactory. The discussion opens with a reference to the enviable accuracy of the results. The authors point out that while randomization ensured the validity of the conclusions, at the same time, contrary to common opinion, a randomized experiment had proved it could be extremely accurate. This they attributed to three aspects of the design: eightfold replication had been used in place of the usual triplicates and quadruplicates; the choice of combinations of treatments was such that each plot yield recorded contributed to the accuracy of all the comparisons desired (a full factorial); and the effects of soil heterogeneity were adequately eliminated by the block arrangement. This statement was not a needless reiteration of well-recognized principles of experimentation; at the time randomization was highly controversial, adequate replication was considered extravagant, factorial designs were quite new, and the size and shape of plots and blocks had not been seriously considered.

Eden and Fisher had by that time been working for some years in cooperation, and both had been learning from the sharing of their own experience. Eden ran the field experiments, and Fisher analyzed the data; as they discussed the results and prepared the reports on them, they saw that improvements in design were possible. As he came to know Fisher, Eden was willing to try out the new ideas, as is apparent from a joint article published in 1929 [C P 78] on a series of trials on potatoes, which had been run in the years 1925–1927.

In the first year two experiments were run, one to study qualitative and the other quantitative factors. The designs were the sort of experiments commonly laid out at this time. Both the qualitative experiment, a Latin square, and the quantitative experiment, a block arrangement, were run in fourfold replication only; both were systematic arrangements; and the block design was an incomplete factorial whose blocks run right across the field.

The experimental layout in 1926 shows some changes: a systematic 4×4 Latin square design was again run for the qualitative test. In the quantitative test there were still four blocks but they no longer ran across the field but quartered it in two directions, making more compact blocks. More important, each block included a complete factorial of 16 treatments, and the plots were arranged at random in each block. The results of this experiment, though improved, were rather inaccurate. More replications in smaller blocks were needed, but it was impossible to arrange for more replications of the same design because of difficulty in harvesting the increased number of plots.

The problem was overcome by amalgamating the qualitative with the quantitative experiment; instead of the two separate experiments run in 1926 with a total of 80 plots (the Latin square with 4×4 plots and the block arrangement with 16×4 plots), they ran a single experiment with 81 plots. There were now nine replications of a randomized block factorial. The blocks were smaller than those in 1926 by the omission of one level of treatment of both nitrogen and potash, but the much improved accuracy was ample compensation. Both the qualitative and quantitative comparisons were included in all combinations. In both qualitative and quantitative aspects the results showed for the first time statistically significant differences; the nature of the differences in response associated with type of potash were revealed, and there was even a perceptible interaction between the type of potash and the quantity of nitrogen applied, an interaction between qualitative and quantitative factors.

□ □ □

In the 1926 paper (*C P* 48), Fisher was setting up a very powerful set of tools for experimental design, appropriate to the investigation of quantitative problems such as the choice of the appropriate mixture of nitrogen, phosphate, and potash. He was advocating the use of the highly efficient factorial experiments to generate a set of "treatments" to be contrasted, and this could be done accurately by building them into blocks. It soon appeared to him, however, that this arrangement, which guaranteed efficiency by the use of the factorial design, which guaranteed the elimination of unnecessary differences in soil fertility by blocking, which guaranteed a valid estimate of error by randomization, even this fell short of perfection. Already he perceived the innate conflict between the principles that justified the block designs and the factorial designs, respectively.

The factorial design had to be set within the block design. The larger the factorial block, the more factors that could be packed in, the more efficient the factorial would be. But if all the factor combinations were to be put in a single block, the block size would get very large, and the error would increase as comparisons were made between plots scattered more widely over a larger area. Although it would be tempting, for instance, to test all of six relevant factors together each at two levels in a 2^6 factorial, a single block of $2^6 = 64$ plots would be liable to large effects of soil heterogeneity. It was obvious that a tug of war must develop between the claims for the precision of the small block and the efficiency of the large factorial.

The factorial arrangement itself suggested a way out of the impasse, for it made independent estimates of the effects of each of the treatments and interactions, any one of which corresponded to a contrast between a certain selection of the treatment combinations and the remainder. Information on some of the higher-order interactions was likely to be of relatively little interest. If one were prepared to sacrifice this information, one could split a block in two by selecting those treatments for one block whose contrast with the treatments in the other block was normally used to measure the interaction considered dispensible.

In the factorial test of nitrogen, phosphate, and potash, for example, interest might center on the three main effects. Although it might be felt to be essential to discover any interactions that existed between pairs of factors, the three-factor interaction might confidently be expected to be negligible. Then, Fisher suggested, the single block of eight treatments could be split into two smaller blocks, each comprising that group of treatments that was normally contrasted with the other in estimating the NPK interaction. From Figure 9c that contrast is seen to be (npk + n + p + k) with (1 + np + nk + pk). The new blocks would, therefore, be random arrangements of these groups of treatments, as illustrated in Figure 10. The NPK interaction would thereby be "confounded" as Fisher called it with the difference between the two new blocks and would be eliminated as such in the analysis. All the other six comparisons would be unaffected, for they would remain unconfounded, two of the four comparisons in each case residing in each of the new blocks. Fisher mentioned this strategy at the very end of the 1926 paper, almost as an afterthought, as a means of reducing the size of the factorial block and thus improving the accuracy of the experiment without sacrificing any of the advantages of the factorial design.

Elaborating on these ideas, Fisher came up with a great variety of remarkable and useful confounding schemes. His skill in combinatorial manipulations here found new scope. The 2^6 factorial on 64 plots, for example, now need not be run in a single block but could be laid out in as many as eight blocks of eight plots each, while ensuring that none of the main effects or two-factor interactions were confounded with block differences.

Having suggested the possibility of confounding, Fisher soon saw that, in practice, estimates even of the confounded interaction terms need not be completely lost in a randomized block design. They might be partially confounded, different comparisons being sacrificed in different blocks, so that all the comparisons appeared unconfounded somewhere in the experiment as a whole, though with fewer replications than other effects. The origin of the 3^3 factorial makes a particularly beautiful example. A project was being discussed in connection with the manurial requirements of young rubber plantations. Apparently inattentive, Fisher spent his time jotting his minute squiggles on the large foolscap pad he always carried with him, and before the meeting was over, he presented the design, tailor-made to their requirements.

It was desired to test three levels of dressing of each of the nutrients, nitrogen, phosphate, and potash; that meant 27 treatments in one replication. Fisher proposed three blocks of nine treatments in each replication, confounding with the block differences two of the eight differences measuring the three-factor interaction. Since there are four mutually orthogonal arrangements of the 3×3 Latin squares, it was possible, with four replications, to ar-

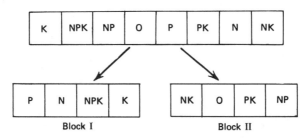

Figure 10. A 2^3 factorial block split by confounding NPK interaction with block differences.

range that any two interaction contrasts that were confounded in one replication would appear unconfounded in the other three replications. The loss of information was minimal.

Because he was a consultant in the planning of various experiments, Fisher found new uses for the principle of confounding and extended and generalized it in a large number of ways. As he said (in *Statistical Methods*), "the variety of the subject is, in fact, unlimited." It was possible to employ double confounding in cases where two sorts of heterogeneity exist, as in a Latin square; for instance, certain treatment effects could be confounded with rows and others with columns. It was possible to exploit the orthogonality in three dimensions of the Graeco-Latin squares for partial confounding in large complex experiments. Every experimental situation called for separate consideration, and many beautiful and elaborate designs were created for particular cases, so as to conserve all the advantages of the factorial arrangement and squeeze out of it every drop of information it could supply, while conforming to the block size appropriate to the field.

7

The Genetical Theory
of Natural Selection

An adequate and permanent home, close to his work at Rothamsted, gave Fisher space he needed to expand his extraprofessional activities. The first consideration was to provide accommodation for the projected large family. Milton Lodge was a ten-room Victorian building, a big house for the family of four who moved in in 1922: Ron, his wife, his sister Phyllis, and his son George. Fisher was looking to the future. Some years later, being accosted at the Eugenics Society by a well-known advocate of contraception who demanded how many children he had, his reply was "Three, so far!"

With the family he required a place at home where he might pursue his studies, uninterrupted by household affairs. A small morning room on the ground floor became his library; his desk was set under the window, overlooking the circular front drive surrounding a monkey-puzzle tree; the armchair and footstool, characteristic of all the rooms he used, were placed by the fireplace and, in the evenings, the adults often gathered there for their reading and conversation, Eileen with her knitting or mending beside him and Phyllis with embroidery in the daybed opposite. When Fisher was in the library no interruption and no disturbance by noises outside were permitted. It was here that he continued his investigations into genetical theory in parallel with his statistical work at Rothamsted.

A residue of his farming enterprise continued in the large back garden, with chickens and goats, orchard, and vegetable beds. Sawing horse and grindstone were set in the bricked yard alongside the house, and around the yard outbuildings were allotted for storing food and bedding for the animals, sheltering bicycles and firewood, and housing large items like the churn, the mangle, and the cross-cut saw. Here, as at Bradfield, Fisher began the task of double-trenching the garden beds. Having learned of work done at Rothamsted during the war on the preparation of household and garden waste for compost, he built his compost heap and treated it according to the newly discovered principles, adding the chemical mixture recommended to promote bacterial activity, which was by that time on sale under the brand name of Adco.

Friendship with D. M. Morland of the Ministry of Agriculture Bee Section at Rothamsted stimulated him to become a beekeeper. He was affected by the collaboration between Rothamsted and Long Ashton Research Station in two ways. Cider research was the primary interest at Long Ashton, and cider became the regular accompaniment to Fisher's evening meal. Moreover, following the discovery of the importance of the carbon/nitrogen ratio to fruiting, and the publication from Long Ashton in 1934 of detailed studies of this phenomenon in fruit trees grown under different conditions and worked on different root stocks, Fisher planted an espalier of his favorite apple varieties at home, to be cultivated in accordance with the resultant new theory and practice of pruning.

Geographically, Milton Lodge was well situated for Fisher's purposes. Milton Road lies parallel to the eastern edge of Harpenden Common, only a few hundred yards away, with the railway running between. The house was at the northern end, close to the railway station and the village, while at the southern end half a mile away there was in those days a dairy farm, and rutted country lanes tunnelled under the overarching hedgerow trees. This was the road they took to pasture the goats on the Common, the four-year-old George clutching the end of a chain and being led by the billy-goat, Marsyas, rather more rapidly than was convenient for his short legs.

Fisher was shocked to learn that by-laws forbad the keeping of pigs in this exclusive neighborhood, but they did not mention goats. Therefore, when a neighbour greeted him over their common fence and remarked that when the wind was in a certain direction the odor of goat became offensive, he was unrepentant. A year or two later, however, residents became aware of a change in the usual evening procession home from the Common. Marsyas had been tethered not far from the houses of East Common. The wind had, no doubt, been "in a certain direction," and someone had put an end to the offensive odor for ever: Marsyas was dead. Fisher trundled home the corpse in a wheelbarrow, while George trailed behind, all along Milton Road, keening his woe (and anger) for the sad end of Marsyas.

Harpenden Common was an inheritance from the medieval system under which villagers in England enjoyed the common use of a patch set aside for the purpose, and every man had his sheep, his goat, or even his cow pastured there. Fisher still looked upon it as a part of the rural economy, not to be wasted. He pastured his goats there, enjoyed the wild raspberries and blackberries gathered there, and was only sorry that in the twentieth century he could not send out a daughter as goose girl to spend long summer days enjoying it, for it was beautiful and full of interest to him and should, he felt, satisfy equally the girl's eye and mind and the goose's crop.

On the western side of the Common, as it narrows toward the village, lies Rothamsted Experimental Station, only a 15-minute walk from the home. Fisher could now return for lunch instead of taking sandwiches, and he found the walk across the Common pleasant and healthful. He did not have to conform to strict hours of work and usually had a leisurely breakfast before putting on his boots, picking up a coat and an ash stick, and walking out; in the evening he was equally unhurried about coming home.

His pace seemed leisurely. Though he covered the ground fast, his walking pace seemed easy, in rhythm with the slow swing of his stick. His speech was slow, although unhesitating. He read much, but the reading was taken slowly, usually aloud, and he gave each word due time and emphasis. With each phrase or the rounding out of a sentence he paused, slowly nodding his head and stroking his beard, like a wine lover glass in hand who savors every drop. The reading, too, was interrupted for leisurely reflection and conversation. His physical deliberation was, however, comparable with that of a tight-rope walker, maintaining the poise of a highly nervous psychophysical system. The very control of his movements betrayed nervous energy: his exactitude in folding and refolding the newspaper, his manner of tilting his chair and maintaining its balance, perhaps on a single leg, and the nice timing of his fingertips drumming on the table.

Behind this deliberation of action was a mind equally deliberate. He did not skim his reading; he knew precisely what had been said, not its rough outlines; and he remembered it. He did not have a vague notion of what he wished to say; he had something clear in his mind and spoke his thoughts precisely, as near as humanly possible. But there was nothing elephantine about this deliberation. His mind was constantly strung to concert pitch, feeding richly but never satisfied, sparking continually but never discharged. The most negligible observation might start a train of thought on any subject under the sun; for instance, the sight of his son's arithmetic primer started him thinking of the appearance and the uses of the Sieve of Eratosthenes. Sometimes, as in that case, the thoughts led to a published article or note [C P 76, 1928]; more often they were explored in conversation, and sometimes they were stored away in mental notes or longhand jottings until, perhaps years later, they were found relevant to another problem. Thus the enumeration of the 6 × 6 Latin squares, done in 1924, was only published in 1934 (C P 110, with F. Yates) because the question had become important in experimental design and the incomplete enumeration published by Jacob in 1929 had called for correction. His mind was exposed on all sides, like the leaves of the sundew plant, sensitive to the slightest impact, ready to engulf and digest every intellectual morsel that lighted on it.

Stimulated on all sides, his mind gave him no rest. Habitually he thought, read, and conversed late into the night, his interest generating the energy needed to sustain it; even when he attempted it deliberately, he could hardly weary his mind into relaxation. For years he was afflicted by sometimes crippling headaches, yet his mind persisted in thinking at speed, clearly and comprehensively.

Under this intellectual pressure Fisher's temper was as irritable as his intellect. He would "fly off the handle" for no apparent reason and was fiercely intolerant of petty inconveniences. His anger was unpredictable: he might tolerate a real nuisance for years or blow up over a trifle. There was usually an element of frustration in the precipitating situation which, being expressed, left no personal rancor. Sir Bernard Keen, then assistant director at Rothamsted, recalls a typical instance of Fisher's complex character:

He button-holed me to unload his anger about some administrative decision by Russell, and immediately followed this with a generous tribute to his wisdom in supporting concentration on natural selection research. It was merely Fisher's way of saying that wise men are sometimes foolish.

His irritation was not always directed toward others; he sometimes frustrated himself, his absent mindedness about things being the complement of his preoccupation with ideas. He was very upset when he missed a train connection through his own fault and therefore missed a Eugenics Society dinner at which he had hoped to introduce his wife to Mrs. Darwin. He and Eileen spent the evening reading "Othello" together, because he needed to purge the emotions of his disappointment by acting out a grander tragedy. On another occasion, he had been in Ireland collecting measurements of triplet children and had attended a dinner on his return through London. Traveling home with the data and the dinner jacket in his suitcase, he left the case on the train. Even as the train moved out of the station he realized his loss and reported it, but the case was never recovered. His anger fell on the dilatory porter who would not telephone in time to recover the case at the next station. Another unhappy episode occurred toward the close of a long series of breeding experiments, when he slew with his own hand the key mouse, a male whose progeny should have clinched the genetical argument. That night he could settle down to nothing, was restless, wretched beyond anger. He picked up Descartes and then resorted to Marcus Aurelius, whose stoicism, if anything, might fortify his own to accept what he had done.

□ □ □

Under the pressure of intellectual activity at this period was produced a magnificent burgeoning in print. From 1922 to 1935, a period chosen to cover the development of the ideas and the actual writing of his three major books, he was also writing articles at the rate of somewhat more than eight in each year. The articles covered a remarkably wide field. In 1926, for instance, titles included "On the capillary forces in an ideal soil" (pure physics), "The arrangement of field experiments," "Periodical health surveys," "The variability of species," "Eugenics: can it solve the problem of the decay of civilizations," and "Expansion of 'Student's' integral in powers of n^{-1}." In 1924, when Professor Bowley requested all of Fisher's offprints for the London School of Economics, he expressed surprise on receiving them, rather like Queen Victoria receiving, as requested, all the works of the author of *Alice in Wonderland*. Bowley had framed his comment in the terms of a statistician; the articles were so numerous, he said, that they ought to obey the law of large numbers, yet he could not predict from past performance what would be the size or shape or subject of Fisher's next paper. As Fisher's audience grew, those who received predominantly statistical or agricultural or genetical papers separated out; soon there were three distinct lists. Some names appeared on all three, of course, but it became reasonable, on receiving a request for offprints from a stranger, to ascertain the interests of his correspondent.

Fisher loved conversation, and his wide knowledge and insatiable curiosity made him a fascinating conversationalist. He would discourse speculatively on matters not yet settled in his own mind to his satisfaction; he would expound his thoughts in process of formulation; he would probe the foundations of another man's argument to make sure he had missed nothing; and he took a gamin delight in pricking a windbag. He regarded self-identified "authorities" as fair game, and, in conversation, his challenge was often delightfully urbane. Sir Bernard Keen recalls one teatime at Rothamsted, when a Sikh visitor was dilating to a group on Sikh religion, dogmas, and history, that Fisher joined the group and then by gentle leading questions showed he knew far more of the matter than the Sikh himself.

William Roach gives his own zestful account of an encounter with Sir Arthur Keith.

One afternoon when setting off from the Lab. I met Fisher just arriving and said, "Aren't you going to the lecture?"—"What lecture?"—"Keith on the Rhodesian skull."—"Now, I wonder what attitude he will take," says Fisher, walking by my side. "Good! You are coming, but what about letting Mrs. Fisher know?" But Fisher went on with his really learned discourse uninterrupted by ticket getting, and settled comfortably in the railway carriage, oblivious of pained looks of occupants trying vainly to read, but who gave up in desperation and just listened to Fisher.

As soon as Sir Arthur Keith had finished his lecture, Fisher went up to him and soon had him in knots; another chipped in and was pulverised by Fisher more quickly than I thought possible. Never have I seen a lecturer spring through a theatre door as quickly as Sir Arthur Keith made his exit to safety. Fisher, turning from the other victim, looked surprised but made me the most willing audience for the further development of his thesis.

Soon I interrupted with, "What about some food, Fisher?"—"Yes, let's eat," said Fisher and continued his peroration. We boarded a bus, Fisher sitting just inside the door and I at the other end of the seat for three. As the conductor was still vigorously ringing the bell a red-faced lady scrambled on the bus and breathlessly asked numerous questions, to which the conductor was still saying, "Yes, Lady, we go . . . Yes, Lady, I'll put you down . . ." (and Fisher still developing his thesis to me and a puzzled lot of fares) when the voluble lady backed into the bus and sat on Fisher's lap. She sprang up as if electrocuted, saying "I beg your pardon," and Fisher interrupted his discourse sufficiently to say, "Not at all," lifting his hat gracefully and giving his usual beaming smile. Still apologizing profusely and backing away from Fisher, she sat full weight on my lap. Whereupon she levitated white-faced to the other end of the bus, while Fisher proceeded with his discourse. This was only interrupted when I had guided him to my favourite Soho restaurant and we each had a foaming pint of beer before us. Fisher put down what was left of his and beamed and I started laughing. "What are you laughing at?" said Fisher. "The occurrence on the bus."—"What occurrence?" And I had to tell him. Whereupon he laughed so uproariously as to be the centre of interest of almost the entire assembled company.

T. N. Hoblyn, Fisher's student a few years later, found Fisher's enthusiasm and zest for life as infectious as ever:

He was always very good company, enjoyed the good things of life and was one of the best informed men that I have ever come across on a wide range of subjects. One night driving back from London after a meeting of the Genetical Society, he discoursed the whole way to Harpenden on the roads of England. (After that I had to read G. K. Chesterton.)

There is a story that he and I were discovered sitting in my car in the middle of Broadbalk field in the early hours, singing at the tops of our voices. I have always denied the truth of this, though I do remember playing pontoon with him at the farm manager's house until 2 a.m. and that the car had to be towed out of the field by the farm tractor next morning.

One afternoon at tea in the laboratory Fisher remarked with a twinkle in his eye that his son George wanted to keep white mice "but on this we do not see eye to eye." Nevertheless, some months later, in February 1924, he announced that it was his birthday and that George had presented him with a pair of white mice with the offer to look after them for his father. In fact, Eileen had conspired to get George his two white mice, and, fortunately, she was not sent two pink-eyed albinos: one of the mice had dark eyes. Some months later when Roach teased him about his love of white mice, Fisher said he was still not clear about the eye-color factor.

From that time onward he always had some feature of his mouse stocks under genetical study. From those two original white mice were born a generation, all showing the wild-type agouti coloration. The following generation began the process of segregating out the variegated coat color factors inherent in the grandparents. In addition to the white and agouti, there appeared pied and self coats in black and brown. There were 70 cages of descendants by the time the stocks were moved from the attic at Milton Lodge to the Galton Laboratory in 1933. Yet larger numbers of their descendants were transferred 10 years later to the genetics department at Cambridge where, finally, Fisher boasted a birthrate only "somewhat less than that of the human population of Eire; they have just over a thousand born each week and we have just under a thousand."

In 1924 he was trying his prentice hand with living genetical material, making cages and collecting wood shavings from the woodyard over his back fence, planning menus, setting up households, and opening the first of his notebooks to record the appearance of the first pregnancy with a "p" beside the date. When "pp" appeared the litter was imminent. Eileen and George helped in the mouse room, and in the summer of 1924, when Fisher left for Canada, Eileen took over the recording and in George's company put up new matings according to instructions. Soon they were referring to her own unborn child as "Peepee," as the family were to do ever after.

The mouse-room arrangements were quite amateurish. It proved less simple than it might seem to keep experimental mice inside a cage and to prevent the incursion of wild mice from outside. The wooden cages were soon patched with metal corners and reinforcements. The breeding program could be disturbed in other unexpected ways. Checks and accidents were not infrequent, and they led to intimate observation of the social behavior of the mouse colony, as we see from Fisher's observations in 1932 [27] when he was moved to comment on a plan to exterminate rodents by destroying females only. The plan had been criticized, but his experience suggested that there might be indirect effects of the policy that could invalidate the criticisms.

Dr. Fisher said . . . For the purposes of his experiments it had been necessary that the successive litters of each doe should be sired always by different bucks. At first he had tried waiting until the litter had been born before changing the buck. But this course had been quickly found to be disastrous. Mice appeared to have a strong sense of territory, and the doe usually flew at the intruder, who, for his part, defended himself very meekly, and displayed every anxiety to escape. If, after they had settled down, they were left for the night, it was found in the morning that the litter had been slaughtered, and was being quietly eaten by both adults. It was considerably safer to introduce the doe and her litter into the buck's cage, for his infanticidal instinct appeared to be conditioned chiefly by finding himself on alien territory.

In the light of these facts the practice had next been adopted of changing the bucks before the litter was born, in fact as soon as the pregnancy would be detected, so as to give the buck introduced as long as possible to settle down, and "feel at home" in his new surroundings. It was particularly interesting that this plan worked well for old bucks, but generally failed when a young buck was introduced into the experiment. The infanticidal instinct was, in fact, very finely adapted to the biological needs of the male mouse. If a buck which had been capable of engendering for only ten days, were confronted with a litter which had been begotten certainly as long as nineteen days ago, calculation might show him that he could not be its father. In practice he might not do the calculation, but he did slaughter the litter.

Wild bucks certainly possessed the same instinct. Until recently, before the old cages had been replaced, the first sign of an incursion by a wild buck had been always the loss of one or more litters already born and often, what was equally annoying, the replacement of the next experimental litter by one of obviously wild parentage.

He grew fond of his mice as he grew familiar with them. He studied genetical factors, but for him the mouse was never merely the vehicle for some factor he was investigating; it was a creature that had evolved under natural selection, whose instincts, territorial and otherwise, were quite as interesting as the colors of its coat.

Fisher had a great faith in the sort of intimacy that daily observation and handling forged between mice and men; it gave the investigator constantly renewed opportunities to notice new facts, and it made the mice accustomed to handling so that their lives were less disturbed by it; conveniently, they became tamer and more tractable. When new styles of cage were devised which allowed of the rearing of mice "untouched by human hand," he resisted the idea as being quite unsuitable for genetic research. The physical handling of living material was, he insisted, an essential educational experience for the geneticist. By the habitual practice of observation in the mouse room, students would be put in the way of noticing things that might lead to new discoveries and to the investigation of new problems in a way that would be impossible if they were given merely theoretical training.

□ □ □

Fisher himself could not undertake any large-scale genetical investigation, but he managed to initiate and carry through an inquiry on human triplets between 1923 and 1927 [C P 70 1928]. The object of the study was, in the first instance, to obtain reliable data on the physical measurements of twins, because in his analysis of twin measurements from Thorndike's data (*C P* 11, 1919 and *C P* 27, 1922) he had found that the twins represented an apparently homogeneous group; their measurements did not show distinct classes of 'fraternal' and 'identical' pairs but a single class of twins correlated to an intermediate degree. It seemed either that Thorndike's figures were unreliable or that the current theory of twinning was in need of revision.

Since multiple births were not registered as such in Great Britain, Fisher could not identify twins from the register. It occurred to him that a Royal Bounty was awarded for triplets, and he found he could get names and addresses of these triplets through records in the charge of the Secretary of His Majesty's Privy Purse. He gained some financial support for the inquiry from the British Association and from Major Darwin. To give the inquiry a personal touch, his sister and his wife wrote by hand the letters to the parents of triplets, soliciting their cooperation. Then, as triplets came of age to be measured, Fisher visited their homes and collected measurements and genealogical records. Other friends helped in distant districts, and one notices among his acknowledgements the names of his first two voluntary workers at Rothamsted: E. M. Somerfield, after his return to Guinness's, collected data in the Dublin area and L. H. C. Tippett, at the Shirley Institute in Cheshire, did the same in the Midlands and North.

The precautions taken in this inquiry show the problems in the collection and comparison of twin data of which Fisher had become aware. First, previous measurements had included twins of various ages. Since it was impossible to allow for the growth factor, Fisher decided to measure all individuals at a set age (6 ½ years). Second, the existing collections were often selective; recognizing that "only data obtained on a carefully planned and uniform basis are capable of yielding valid conclusions," Fisher planned to measure all triplets born between set dates (Oct. 8, 1917— Sept. 28, 1920) in those families where two or more of the triplets survived to be measured. Further, in genealogical inquiries the tendency to record striking cases was to be resisted. He therefore requested all the parents of triplets to return records of births among their relatives of stipulated degree, and he rejected all incomplete records and extraneous information. Finally, the precision of previous measurements was uncertain. As others had done, Fisher made measurements as accurate as possible by standardizing the position of the body for measurement. In addition, he arranged

for the majority of cases measured, and especially for those of which the author had the greatest suspicion (namely, his own) that not only should the measurer satisfy himself that an accurate measurement had been taken, but also that independent duplicate measurements of the same value should be obtained in order to yield by comparison an independent estimate of the accuracy of measurement.

During the course of the inquiry, other workers resolved the original question of the genesis of twins. By 1927, twins were confidently believed to be of the two classes already identified, either fraternal (dizygotic) or identical (monozygotic). G. Dahlberg had evolved a set of measurements by which he discriminated between the two sorts. Fisher had not made these particular measurements, but he was able to make the first estimate of the percentage of monozygotic pairs among English twins by comparison of his own data with correlations observed by Dahlberg. This gave him an estimate of 57% for the monozygotic pairs among his triplets. This estimate was confirmed by the somewhat lower and less precise figure 53.3% given by the proportions of the sexes. Finally, he calculated the minimum estimate: if the monozygotic pairs had absolutely identical measurements, the percentage of monozygotic pairs would be as low as 47%.

Using the difference between the minimum estimate and the best estimate he then checked his earlier conclusion [*C P* 9, 1918] that environmental factors play a negligible part in determining human measurements. If the proportion of monozygotic pairs were really 57%, then the correlation between these pairs was not 100% but something like 93% and the uncorrelated fraction must be accounted for by nongenetic effects. Fisher was surprised that the correlation was so imperfect but suggested, in explanation:

If, as seems probable from the mutual interaction of the parts in development, very minute initial variations of growth of particular organs may in some cases lead to appreciable ultimate differences in the measurements, these will be as great for twins as for unrelated children.

From the family records Fisher analyzed the evidence of inheritance of the twinning tendency. He showed how such records could best be used to show whether the tendency was transmitted through the mother or the father, and also whether it affected specifically the proportion of identical or of fraternal twins. (His limited data did not show it, but the genetic tendency does, in fact, affect the proportion of fraternal twins in a population but not the proportion of identical twins.) His analysis emphasized the importance of recording sex in order to distinguish between the two types of twinning tendency. It illustrated the importance of the fact, previously neglected in such analyses, that "the best controlled evidence of inheritance is the contrast in twin frequency between the different groups of relatives in the same collections."

□ □ □

The need was felt, especially among those few biologists who in the 1920s believed in evolution by means of natural selection, for a quantitative expression of changes in gene frequencies, which were to be expected over the generations in response to natural selection acting on the population. Fisher was not alone in taking up this work; after the World War I both J. B. S. Haldane in England and Sewall Wright in the United States undertook primarily statistical researches into the genetics of populations. By means of the statistical formulation a framework of expectations was built for evolutionary geneticists. On the basis of the assumptions made, conclusions could be stated with assurance; speculations could often be confirmed or negated by the statistical approach, and explanations offered for some of the unexplained phenomena of nature.

This was an area of biological study less readily accessible to the biologist than to the mathematician. In the preface to his *Genetical Theory of Natural Selection* (1930) Fisher described the contribution an imagination trained in mathematical concepts and procedures could make to biological science, and here he characterized the basis of his unique contribution. Quoting a remark by Eddington, "We need scarcely add that the contemplation of a wider domain than the actual leads to a far better understanding of the actual," he added,

For a mathematician the statement is almost a truism. For a biologist, speaking of his own subject, it would suggest an extraordinarily wide outlook. No practical biologist interested in sexual reproduction would be led to work out the detailed consequences experienced by organisms having three or more sexes; yet, what else should he do if he wishes to understand why the sexes are, in fact, always two? The ordinary mathematical procedure in dealing with any actual problem is, after abstracting what are believed to be the essential elements of the problem, to consider it as one of a system of possibilities infinitely wider than the actual, the essential relations of which may be apprehended by generalized reasoning, and subsumed in general formulae, which may be applied at will to any particular case considered. Even the word possibilities in this statement unduly limits the scope of the practical procedures in which he is trained; for he is early made familiar with the advantages of imaginary solutions, and can most readily think of a wave, or an alternating current, in terms of the square root of minus one.

Fisher's paper in 1918 [C P 9] on the correlation between relatives had raised a question concerning the values of the dominance ratio found by analysis of human measurements. In 1922 he published a consideration of this question [C P 24] which established the power of the mathematical approach through what is now called "population genetics." He treated the genetical situation in the same way that, as a mathematician, he approached the theory of gases, an analogy which had appealed to him from his college days and which he mentioned on several occasions in a genetical context. In introducing the 1922 paper he wrote:

The whole investigation may be compared to the analytical treatment of the Theory of Gases in which it is possible to make the most varied assumptions as to the accidental circumstances and even the essential nature of the individual molecules and yet to develop the general laws as to the behaviour of gases, leaving but a few fundamental constants to be determined.

His very general approach avoided the necessity of assuming, as others had felt obliged to do, which of two alternative characters (allelomorphs) was dominant, to what extent dominance occurred, what the relative magnitudes of the effects produced by different genes were, in what proportion the allelomorphs occurred in the population, whether two or more allelomorphs were involved in a particular locus, to what extent genes were linked, to what extent preferential mating occurred and even the intensity of selection and of environmental effects. "When factors are sufficiently numerous the most general assumptions as to their individual peculiarities lead to the same statistical results."

Taking a dimorphic factor with gene frequencies in any proportion $p:q$, he inquired in general what conditions of selection are required to maintain those gene frequencies in the next generation. He went on to consider what numerical relationships of the population influence the chance of survival and the rate of spread of a particular gene. He deduced quantitative expressions for the effect of chance extinction of genes on the variability of the population, the effect of a modest amount of selection (1% differential survival) and the effect of dominance. The mathematical elaboration is considerable, because complexities are successively introduced, but the simplicity and generality of the results are beautiful.

From statistical considerations alone a number of empirical results became explicable. For instance, gene frequencies were found to be maintained in a stable equilibrium if the heterozygote is at a selective advantage relative to both homozygotes; that is, when a mixed pair of genes, *Aa*, is more successful than either of the homozygous pairs, *AA* (dominant), or *aa* (recessive). One consequence is that such genes are maintained in the population; they accumulate while others are eliminated. This explained why such genes were commonly found. It explained instances of "heterozygous vigour" and, to some extent, the deleterious effects sometimes brought about by inbreeding. And it suggested how, under intense selection, the heterozygote's advantage might be so enhanced that balanced lethal systems were established, as in the celebrated cases of *Oenothera* and *Drosophila*, when, in general, neither homozygote is viable.

The rate of diminution of variability, in a population unaffected by selection or mutation, was shown to be almost inconceivably slow; even a very slight selective effect must overwhelm the effect of chance extinction. A number of important consequences flowed from these calculations. The rate of mutation needed to maintain the variability of a population was much lower than biologists generally supposed and, in the absence of selection, would be minute. Also, when a gene is represented in a large number of individuals, even if they form only a small proportion of the total population, chance will play little part in determining its survival. This observation negatived a suggestion put forward shortly before by the Hagedoorns (*On the relative value of the processes causing evolution* by A. L. and A. C. Hagedoorn, reviewed by Fisher, *C P* 17, 1921), a theory which was later revived by S. Wright under the designation "random drift."

Since selection was shown to remove preferentially those genes that have marked or important effects, the existence of multiple factors was explained and their importance to evolution emphasized:

It is therefore to be expected that large and easily recognized factors in natural organisms will be of little adaptive importance and that factors affecting important adaptations will be individually of very slight effect. We should thus expect that variation in organs of adaptive importance should be due to numerous factors which individually are difficult to detect.

Charles Darwin had been right to base his evolutionary theory on the selection of small continuous variations.

Finally, Fisher calculated the proportion of the variation due to dominance deviations under various conditions of mutation and selection. He showed that the assumption that leads to the level of the dominance ratio actually observed in man is one of "selection maintained in equilibrium by occasional mutation." Since other assumptions led to different results, this finding fixed the main parameters of the evolutionary situation with which the geneticist had to deal. In addition, if the equilibrium were maintained by occasional mutation, populations would differ in the variability they maintained.

In all cases it is worth noting that the rate of mutation required varies as the variance of the species, but diminishes as the number of individuals is increased. Thus a numerous species, with the same frequency of mutation, will maintain a higher variability than will a less numerous species; in connection with this fact we cannot fail to remember the dictum of Charles Darwin, that "wide-ranging, much diffused and common species vary most" (I. Chap. ii).

Thus the 1922 paper struck a first blow in favor of Darwin's neglected theory of natural selection.

□ □ □

At that time the prevailing feeling about Darwin's work was one of mixed reverence and scepticism. The evidence Darwin provided for evolution was accepted and admired, but his views on its mechanism were regarded as out-moded. Natural selection was thought of as being of minor importance only, capable at most of bringing about small adaptations. A contemporary witness, E. B. Ford, then doing research with Julian Huxley at Oxford, in describing the current beliefs, exhibits the lack of definition of ideas that fogged the issue:

It was held that in the main evolution was controlled by processes not really under-stood but at any rate vaguely mutational, a view to which the new science of genetics, or Mendelism, was believed somehow to contribute, insofar as it had any evolutionary significance at all, and this was doubted. Fisher and I, together with Julian Huxley, were at that time almost alone in this or any other country in preaching the importance of selection and in believing that the heritable variability upon which it could work was provided by Mendelism and in no other way. I ought to add that J. B. S. Haldane was of a similar opinion.

Today, the state of scientific opinion regarding Darwinism that prevailed dur-ing the first 30 years of the century appears a curious retrogressive eddy in the current of scientific advance. At the time, however, it flowed strongly. In op-posing it, statistical studies gave backbone to the biological evidence, first, by providing a quantitative framework for genetical hypotheses concerning evo-lution by natural selection; later, statistical methods were no less essential in planning and interpreting the biological observations that demonstrated selec-tive effects in nature.

Sharing as they did the same minority view as to the importance of natural selection, Fisher and Julian Huxley were much thrown together in the early 1920s. Huxley was at Oxford University, doing research on genetic physiology and using his influence among his colleagues to try to reopen the subject of natural selection for scientific discussion. In consequence, it was at Oxford that Fisher met the man who was to be the partner of his field work on natural selection, the biologist whose skills complemented his skill as a statis-tician, the companion of many hours, Edmund Brisco Ford.

On the surface there was little similarity between the meticulous Oxford bachelor and the *pater familias* with his casual ways and old, untidy clothes but they found each other sympathetic, enthusiastic about the same ideas, and mutually stimulating. In conversation Ford found himself inhibited by some people, and certainly many were inhibited by Fisher, but they hit it perfectly, and as W.J. Cory put it, they would "tire the sun with talking."

Ford tells the story of their first meeting:

Our meeting which took place in 1923 was typical of him. Like so many good things in my life it was due to Julian Huxley. Though I was only an undergraduate at Oxford at that time, he and I were researching together in genetic physiology in the earliest days of that subject. Meeting Fisher somewhere, he mentioned that he knew an undergraduate who had interesting ideas on genetics and evolution. Fisher was a Fellow of Caius—he was only 33 but he was already becoming famous. Other people in his position might possibly have asked briefly about me, a few might have even invited me to go to see them. Fisher's reaction was different. The Fellow of Caius took a train to Oxford to call on the undergraduate.

Characteristically it did not occur to him to let me know that he was coming so I was out when he arrived, and he settled down in my rooms to wait for me. On opening the door of the sitting-room on my return, I was surprised to find it full of smoke from pipe tobacco, a thing which disgusts me, and to see a stranger there, a smallish man with red hair, a rather fierce, pointed, red beard, and a very white face. The cast of his countenance slightly resembled that of King George V. As he got up and came towards me I noticed his eyes, hard and glittering like a snake's and seen through spectacles with lenses so thick that they resembled transparent pebbles. He took my hand in a firm, bony grip and, bending slightly forwards, gave me a momentary but most searching inspection. Then his face relaxed into a charming smile, the beginning of nearly forty years of friendship.

With this friendship began Fisher's involvement with field data relevant to the genetical theory he was developing at his desk. The first result was their demonstration of the undirected nature of variation. If the variation within a species were due to rare and random mutations, as they supposed, and not to mutations induced by the environment, then, as Fisher had shown in 1922 [*C P* 24], the variability of different species would be proportional to their relative abundance, and the greater variability of common species would be apparent even in local samples from a uniform environment. It was decided to compare the variability of pigmentation on the wings of various species of night-flying moths. Thirty-five of these species were found which could be compared using a single color chart. These species were classified both for range and abundance, and local samples were collected from which their variability was estimated. It was found (*C P* 52, 1926, with E. B. Ford) that variability was strongly correlated with abundance and probably correlated also with increased range, thus demonstrating the truth of Darwin's dictum and implying that the rare mutations by which variability was maintained were not adaptive but random changes.

These observations had a double aim and tested an additional evolutionary hypothesis, which was discussed in a separate paper (*C P* 59, 1928, with E. B. Ford). It was well known that in polymorphic moths the females but not usually the males are polymorphic, that in mimetic butterflies the females are the mimics, whereas the males usually have a nonmimetic form. One explanation might be that in these natural orders the female was inherently more variable than the male and that natural selection had enhanced the variants in certain cases, with the consequent evolution of the polymorphic or mimetic females. To test the general case, it was decided to test sex differences in the variability of coloring of the night-flying moths, which are neither mimetic nor polymorphic.

The females of the 35 species were found to be generally darker as well as more variable than the males, and it was therefore necessary to standardize the measures of variability for the different color classes. When this was done, the females in each abundance class still showed an indubitable excess of variability over the males. The observations were very suggestive of selection taking place to favor the protective coloration of the females at the expense of the paler males. The general evolutionary hypothesis, too, was confirmed; the greater variability in the wing color of the females was evidently of selective advantage, for, even at the initial stage of differentiation of wing coloration, represented by *Nocturna,* it had led to the evolution of inhibitors of variability in the males of the whole group tested.

The investigation of the night-flying moths was particularly gratifying, for it verified and demonstrated the theoretical points without delay. Most of the theory developed by Fisher in the 1920s that culminated in the publication of *Genetical Theory of Natural Selection* could not so easily be confirmed.

This work led Fisher in 1928 to embark on further investigations of variability and relative abundance through another friend of Huxley's in Oxford, the Rev. F. C. R. Jourdain, who had made measurements of length and breadth on his large collection of bird's eggs. Fisher wished to compare the variability in size of these eggs with the relative abundance of the species from which they came. It was a spare-time activity and took many years to complete. Having assembled 100 pairs of measurements from nearly all of 150-odd species of British nesting birds, he was able to calculate in what way the observed variances would require correction, in view of the differences between species on the average length and breadth of the eggs. Early in 1933 he welcomed Huxley back to England with the news that he was going on to the more difficult part of the enquiry, namely, the relative abundance of British nesting birds. This was difficult because of lack of information about the numbers of birds of the various species. Fisher's enquiries from the Natural History Museum in 1930 had been unable to elicit any statement of relative abundance. However, another Oxford contact, W. B. Alexander, kindly prepared, in consultation with other ornithologists, estimated orders of abundance within each group of birds, and early in 1936 Fisher wrote happily to Ford:

Of course the combination of such heterogeneous material is exceedingly difficult but it is quite certain that the variance both of length and breadth increases with abundance, as it did with your, I think, 35 species of moths.

Available information as to abundance is, I suppose, weaker than it was with moths and certainly variability of eggs is a very indirect approach to the variability of birds. In fact, the greater number of species is fully needed to bring out so small an effect clearly.

The analysis of these data, published in 1937 [*C P* 153], exhibited again, in quite a different class of animals, the truth of Darwin's dictum that widely diffused, common species vary more than more local or rare species.

□ □ □

In retrospect, Fisher admitted [*C P* 86, 1930] that

the mathematical treatment in 1922 left much to be desired, since on reflection it appeared that more searching questions could be asked, and especially the probable progress of new mutations could be traced statistically, making the distinction between the frequency of the new gene and that of the old.

He investigated these questions in the course of the writing of *The Genetical Theory of Natural Selection*. He was working on the relevant chapters in January and February 1929. Later in 1929, he received the manuscript of an investigation by Sewall Wright into the distribution of gene frequencies, from which it appeared that the value given in 1922 (*C P* 24) for the "time of relaxation" was wrong by a factor of two. Fisher acknowledged the correction in a separate article [*C P* 86, 1930], putting on record his acceptance of Wright's value but not of Wright's conclusion about random survival. Rather, "Both periods are in most species so enormous that they lead to the same conclusion, namely, that random survival . . . is a totally unimportant factor in the balance of forces by which the actual variability of species is determined." He then proceeded to develop "a more rigorous and comprehensive treatment of the subject" than he had managed in 1922.

The resulting paper was submitted in 1929 to the Royal Society, whose referees, not for the first time nor for the last, reported that a genetical paper by Fisher was unsuitable for publication by the society. The referees had a genuine difficulty in judging this paper, for Fisher reached conclusions which most geneticists would find unacceptable by the use of mathematical techniques which most would find incomprehensible. The paper was, therefore, rejected for the same qualities which in the event proved it fundamental in the study of the evolutionary spread of a major gene. When, after a very long delay, the decision and the paper were returned to Fisher, he immediately sent the work, unaltered, to the Royal Society of Edinburgh, who published it at the earliest possible moment, the manuscript being received on March 21, 1930, read on May 5 and issued in the journal later that year.

In this highly mathematical document Fisher utilized a very powerful method of approach, indicated but not employed in 1922. Using differential difference equations he established the relative probabilities of an event or a combination of events occurring in successive generations. He was thus able to estimate, for a population of given gene frequencies, the probabilities of the appearance in the next generation in 1, 2, 3, . . . *n* individuals, of genes at 1, 2, 3, or more genetic loci.

By this entirely novel method of handling the problem and through a brilliant manipulation of the functional equations, it became possible for the first time to examine cases in which one allelomorph is extremely rare and to show the relation between the number of genes maintained in a population and the rate of mutation.

Considering the probability of survival of individually rare mutations, it was shown that the vast majority of mutations are doomed to extinction. If they have no selective advantage, eventual extinction becomes almost a certainty, for only $1/2n$ survive (where *n* is the number of individuals breeding each generation). For mutations having a small selective disadvantage, the chance is still lower, yet finite. However, the value changes very rapidly in passing from a small negative to a small positive selective advantage. The probability of success increases fiftyfold in passing from $an = -1$ to $an = +1$ (where *a* is a minute selective advantage). Thus selective advantages of the order of the inverse of the population number make a very great difference to the chance of survival of the mutation. It is almost impossible to conceive that any mutation will long prove so exactly neutral selectively that its chance of success will be unaffected when measured on so sensitive a scale. Fisher concluded

The neutral zone of selective advantage is so narrow that changes in the environment and in the genetic constitution of the species, must cause this zone to be crossed and perhaps recrossed relatively rapidly in the course of evolutionary change, so that many possible gene substitutions may have a fluctuating history of advance and regression before the final balance of selective advantage is determined.

With larger values of selective advantage *a*, the probability of survival approximates $2a$. Thus a mutation enjoying a 1% selective advantage will have practically a 2% chance of establishing itself. Such a mutation could hardly recur in the population many more than one hundred times before in fact realizing that goal.

By the same argument it was possible to show also what proportion of the contributions to the genetic variance of the species was made by mutations of different selective advantage. It was found that, assuming mutations to be equally frequent at different levels of utility, the variance contributed for negative values of *an* exceeding 2 was nearly equal to the small value $1/2an$ but that it increased sharply in the immediate neighborhood of neutrality, passing through unity at $a = 0$. Since for higher values of selective advantage the supply of mutations might be expected to fall off, the contribution to the specific variance would probably reach a maximum for slightly favorable mutations. Thus the portion of the genetic variance to which evolutionary progress was to be ascribed seemed to be concentrated in groups of factors each determining a very minute selective advantage.

This result was quite contrary to the assumptions of geneticists who trusted in the pressure of mutations with large effects to drive the evolutionary machine forward without the intervention of selection. Even those few who believed in natural selection must have been surprised by the statistical result that demonstrated the necessary existence in the system of a sensitivity to selective differences beyond anything they could have hoped.

Discussing the matter in *The Genetical Theory*, Fisher showed that selective advantages of 1%, or even much less, were quite sufficient to explain evolutionary progress, and suggested that in nature the selective advantage might well be as high as 1%. At the time a value as high as 1% seemed excessive to most geneticists. Today, since ecological geneticists have found values of 40 to 60%, and sometimes much higher, to occur with polymorphic characters, the assumption of a selective advantage as high as 1% may seem unduly modest. Yet, as Fisher showed, most evolution must depend on genes with very small selective advantage. The higher values actually found, however, have radically altered our ideas about the speed of evolution possible in a rapidly changing environment.

□ □ □

In *The Genetical Theory of Natural Selection*, Fisher extended the discussion of problems of population genetics such as were treated in the papers of 1922 and 1930. He developed thoughts on sexual selection and on mimicry which had appeared in the literature over the years. He gave a brief account of his new theory of the evolution of dominance. He paid considerable attention to man; in fact, the last 96 pages, over one-third of the whole volume, are devoted to inheritance in man and the consequences of social selection in his societies. As with the earlier chapters, some of this material had appeared previously, some of it was new.

The extent of Fisher's originality and achievement in this volume is apparent from the summary of the earlier chapters, given by K. Mather [28]:

He formulated the Malthusian parameter, as he called it, for representing the fitness or reproductive value of populations, and he developed his fundamental theorem of natural selection, that "The rate of increase in fitness of any organism at any time is equal to its genetic variance in fitness at that time."—a finding that has been rediscovered (using much more elaborate mathematics) at least once in the succeeding thirty years. He showed that even the smallest genetic change can have no more than a half chance of being advantageous and that this chance falls off rapidly with the size of the effect the change produces; that the environment must be constantly deteriorating from the organism's point of view and that this deterioration offsets the action of selection in raising fitness; that any net gain in fitness is expressed as an increase in size of population; and that more numerous species carry relatively more genetic variation so that they have a greater prospect of adaptive change, and hence of survival, than their scarcer fellow species. He argued that since mutation maintains the variation in populations, its rate of occurrence will determine the speed of evolution just as selection determines the direction . . . He discussed sexual selection, developing the view that natural selection will tend to equalise the parental expenditure devoted to the two sexes rather than equalise the sex-ratio itself, and he considered the action of selection in Batesian and Mullerian mimicry. He also developed a comprehensive mathematical theory of gene survival and spread under selection, special aspects of which he extended in a number of papers over the next fifteen years.

Just as Fisher had brought together Mendelism and biometry in the paper on the correlation between relatives, and had displayed how these different instruments for the scientific representation of nature complemented each other, so in *The Genetical Theory* he brought genetics together with evolution by natural selection, rewriting Darwin's score in the symbols of population genetics. Here, for the first time, natural selection was discussed in its own right in the light of a statistical approach. In consequence, evolution could no longer be divorced, in the minds of serious thinkers, from the process of natural selection acting on the heritable variation conferred on a population by the Mendelian mode of inheritance. From the publication of *The Genetical Theory* in 1930 one can date the beginning of the returning tide of Darwinism, the neo-Darwinism of twentieth century evolutionary geneticists.

Although much of the thought involved in *The Genetical Theory of Natural Selection* dated back to the early 1920s, and some at least to his college days, it was not until the autumn of 1928 that Fisher began to write the book. Once he had embarked on it, however, he wrote with remarkable rapidity. Like his other genetical work, it was done at home with his wife. He would stride about the room, or mull over a pipe, as he dictated and she took down his words in longhand. Apart from the mathematical chapters IV and V, the whole book was written in this way, by dictation, between October 1928 and June 1929. In June the whole thing was in the hands of the publisher.

Fisher seemed to know exactly what he wished to say, holding the whole ordered argument in his head, and even his deliberation over the detailed expression of his thoughts did not often give pause to the pen of his amanuensis. Yet, once he had set a passage down on paper, he rarely changed a word or needed to rearrange the order or insert omissions.

His capacity to hold in mind the numerous details of a complex argument was remarkable, as was his precision in expressing what he meant. From much later in life, Dr. R. R. Race recalls a "staggering exhibition of his extraordinary powers of dictating." The occasion was the composition of a complex paper which involved holding in his head gene frequencies and the expected frequencies of possible crossovers.

All this he did and spoke in such lucid English that nothing had to be changed; it is recorded in *Nature*. It was quite unpremeditated: I said, "Shouldn't these results be published?" and he said, "Right," and marched into the secretary's room, sat down, put his feet on the table, and immediately and straightforwardly, without once going back on what he had said before, created a gem of a paper.

As each chapter of the book was completed, it was sent to Major Darwin. Darwin had been urging upon Fisher, repeatedly since the publication of the paper in 1922,

my wish that you should deal with the whole problem of selection mathematically: you will have a small audience but it will gradually be realized that many of the problems can be attacked in no other way.

Throughout the 1920s the two men had talked and corresponded over matters of natural selection, and the sharing of their knowledge and interest was a constant pleasure and stimulus.

Sometimes each thought the other had suggested some idea that they both liked. But Darwin often had doubts, and these stimulated Fisher to explanation and clarification of his views. In 1928 an epistle representing Darwin's doubts about Fisher's theory of the evolution of dominance was rewarded with a very long letter in return. The profound mutual trust of the two men is exhibited by Fisher's words near the beginning of this letter: "You are one of the very few people who will ever appreciate the consequences of my suggestion so I shall be especially particular that you shall understand me clearly about its framework." But if Darwin had particular doubts, he was very confident in general: "I look on my letters to you in the light of pins, the pinpricks to urge you on to your great work on the mathematical theory of inheritance," he wrote in 1927. When at last the work was on the stocks, Darwin was happy to read Fisher's manuscript critically (as Fisher a few years earlier had read and commented on Darwin's book on eugenics).

Fisher was concerned to gain Darwin's concurrence about the discussion in the first chapter, rather to assure himself of Darwin's peace of mind than his own, for it concerned the evolution of Charles Darwin's reasoning about inheritance and natural selection. In contrast, in some of the later chapters Fisher felt his temerity in putting forward some of his own ideas, and then he demanded not agreement but criticism. When Chapter III, on dominance theory, was ready to send, he wrote,

You will groan to hear that I am going the whole hog about dominance. Any example to the contrary is therefore badly needed. Has the homozygous dominant in blood groups ever been found? . . . Is the pink version of many blue flowers analogous to albino mutants in many mammals? . . . I want you to read Chapter III when you form an opinion on the whole-hoggism.

P.S. Thank you ever so much for real encouragement [about Chapter I].

Major Darwin read Chapter III and commented and received counternotes to his own and protested, "I feel a little alarmed that you take my remarks so seriously. If you think the matter over again and stick to your point, I shall be satisfied." In February Fisher sent the mathematical "Chapter IV which will have to be Chapters IV and V it has grown so confoundedly long. Do not try to read it, except the summary and any points which the summary makes you want to look up in more detail. I have made an abominable mess of the whole thing and failed to get out an adequate solution of nearly all the problems but I hope it may at least show what further work is needed." Darwin, without comprehending the mathematics, responded, as always, in the right spirit: "The impression I get from this chapter is that you have been digging in virgin soil and if you have not covered the whole surface it is because the ground is very, very stiff. In pioneer work of this kind no one can be expected to solve all the problems."

In March, Fisher sent the first of the chapters on man, remarking "I do not expect you to agree that I am necessarily right about man but only that I am approaching the subject in a rational spirit." In mid-June he sent the whole section on man simultaneously to Darwin and to the publisher, of whom he remarked to Darwin, "I am afraid he will have a shock when he reads the human chapters and I only hope you won't. I feel on a knife-edge between timidity and audacity and need all the wisdom I can collect if I am to keep my balance." So between timidity and audacity the work was finished.

On the whole audacity seems to have carried the day, the audacity of a research scientist laying himself, his creative ideas, on the line. He had doubts; he felt the inadequacy of his treatment of the subject; he felt the need of wisdom in presenting a balanced view of man. And yet the way to a better assurance and a more adequate treatment lay in exposing his reasoning for the facts of nature and the logic of fellow scientists to assault and possibly to overthrow. Some ground had been won and was held with fair confidence but, working always on the frontiers, he appropriated concepts at a high level of risk: it was no place for intellectual arrogance; indeed, we see from the correspondence with Darwin how far from arrogance he was. Yet he was prepared to gamble. A little later, when he offered for publication a genetical paper that contained some admittedly tentative and conjectural views, liable to be mistaken and condemned, he explained to Darwin, "Personally, I think we ought to go on guessing as intelligently as possible, and if it is an error, it seems one on the generous side to do some of it in public."

The volume was published in 1930 and received some good reviews but, as Fisher had predicted, it never sold well. Its influence on genetical science has been profound but curiously indirect, and it has remained unfamiliar to many geneticists. Yet it has stood the test of time. When a second edition was brought out more than a quarter of a century later no changes in principle were needed, and the changes Fisher chose to make affected mainly the chapter on dominance theory, for, by then, the original concept no longer needed apology.

Once again the reader does not find Fisher's goods displayed in the shop window. One difficulty has been that a number of the theorems are not fully explained or have incomplete proofs. For example, Fisher stated, without proof, that sexual recombination increases the rate of evolution at least twice over that in asexual organisms. Since then other geneticists (H. J. Muller; M. Kimura and J. F. Crow) have been led to different results. Eventually, however, W. Bodmer [29] proved Fisher's theory directly and was able to show that the formulations which had led to a different result were not complete, nor, it appears, as complete as Fisher's thought had been in 1930.

In the absence of a fuller exposition and perhaps because of the unfamiliarity of the text itself, some of the theories have been rediscovered independently by later workers. Such rediscoveries and reconfirmations suggest, to some who follow Fisher, that there may be more yet to be unearthed in the book. Fisher was, in Ford's words, "a geneticist of such prescience that the genius of his conclusions is still unfolding itself today."

In his preface to *The Genetical Theory of Natural Selection,* Fisher remarked: "The deductions respecting Man are strictly inseparable from the more general chapters but they have been placed together in a group commencing with Chapter VIII." Like him, we find it convenient to treat this aspect of his genetic work in a separate section, despite the fact that the same theory supported both applications.

In 1920 Fisher became an honorary business secretary of the Eugenics Education Society. Apart from the administrative work, with its close collaboration with Major Darwin, and the continuing series of reviews for the *Eugenics Review,* he contributed his own ideas at the meetings of the society, in particular, the ideas he had formed about the biological history of civilizations.

Through his extensive readings in the history of past civilizations he was now able to illustrate his thesis concerning their rise and fall in terms of the changes in moral attitude which took place during the course of a civilization [C P 28], the changes in intellectual achievement and emphasis [C P 29], and the concurrent changes in political and economic conditions at different stages of its rise and fall. All followed the general pattern to be expected if the selective forces in society were what they appeared to be both from internal evidence (contemporary records of the civilizations) and on analogy with the course of Western civilization. The biological history put its signature on every age. The consequences of the growth of social selection, in parallel with the increasing relative importance of wealth in a moneyed economy, could be distinguished in all areas of life and affected every level of society. If the complex of innate tendencies to sterility should continue to be favored by social promotion in the West, as it had been in previous civilizations, it seemed inevitable that Western civilization should founder on the same rock as its predecessors.

In 1926 Fisher drew together these ideas in an address given at the Annual General Meeting of the Eugenics Society on the subject, "Eugenics: can it solve the problem of decay of civilisations?" This paper was also presented in French at the Conference of the International Federation of Eugenics Organizations in Paris in July [C P 53, 1926]. In this talk he elaborated the thesis first put forward in 1912 [C P 3, 1914] that the differential birthrate, which had apparently come into existence following the establishment of a moneyed economy in known civilizations, was sufficient to explain the subsequent decay of the civilization.

Observing that a relatively populous, organized, cooperative state, by these very advantages, adds to its own wealth and knowledge, Fisher remarked how improbable it should be that any civilization, once organized and established, should ever fail, and yet how in all cases without exception, after a period of glory and domination accompanied by notable contributions to the sciences and the arts, civilizations *have* failed. Explanations offered for one or another failure seemed inconclusive in particular and inadequate in general. He argued that:

A physician observing a number of patients to sicken and die in similar though not identical conditions, and with similar though not identical symptoms would surely make an initial error if he did not seek for a single cause of the disorder. The complexity of the symptoms, and of the disturbances of the various organs of the body, should not lead him to assume that the original cause, or the appropriate remedial measures must be equally complex. Is this not because the physician assumes that the workings of the body, though immensely complex, are self-regulatory and capable of a normal corrective response to all ordinary disturbances; while only a small number of disturbances of an exceptional kind meet with no effective response and cause severe illness? Have we not equally a right to assume a self-regulatory power in human societies? If not, we should be led to think that such societies should break down under the influence of any of the innumerable accidents to which they are exposed. Human societies of various kinds have adapted themselves to every climate from the Arctic to the forests and deserts of the tropics. They share the territories of the most savage or the most poisonous animals, and often implacable human enemies. Social progress has not been arrested by the introduction of new weapons, of alcohol, or of opium or even of infanticide; yet these introductions might each of them seem to threaten the existence of the race. That civilized men, possessed of more effective appliances, with access to more knowledge, and organised for the most detailed co-operation, should prove themselves incapable of effective response to any disturbance of their social organisations, surely demands some very special explanation.

If we turn to our own civilisation in the hope of isolating a single cause which is capable continually and persistently of injuring the racial qualities of future generations, we are at once confronted with the phenomenon of the differential birthrate.

He argued that at the beginning of an indigenous civilization the population might be supposed to have stratified itself into various occupations roughly in accordance with innate abilities of the different families. With the development of aggregates larger than the individual family, however, the primitive military advantage of belonging to a numerous kindred becomes outweighed by the burden of supporting an increasing family on a limited domain. From that point on, the innate qualities of the race, in all classes of society, cannot but deteriorate. Up to this stage in the development of the society the social status of an individual is fairly stable and will continue for a while to be determined primarily by birth and only to a minor degree by the wealth he possesses or can accumulate. Social position will still have economic advantages. Progress in civilized arts also will continue subsequent to the epoch when racial deterioration has set in, and these improvements will include the more and more thorough utilization of such ability as is available, with a correspondingly increasingly rapid exhaustion of the innate abilities from which it is drawn.

Under these conditions the time must come when the breakdown of the social strata, long prepared for by intermarriage and social promotion, at last becomes evident. Historically, political movements were found to arise aimed at the attenuation of the privileges of birth and at the reorganization of the social strata on a basis, principally or entirely, of wealth. Fisher observed that this movement "in the reorganization of our own societies has already attained, in this respect, practically its full effect." He exemplified parallel developments in the past by reference to the revolution of 750 A.D. in Islam. The revolutionary movement "asserted the claims of the subject peoples and of the unorthodox sects of Islamism, and, in its success, it established a constitutional monarchy of which the first Vizir was a self-made banker with no pretensions to family." The dissolution of class barriers in a plutocratic society completes the vicious circle, by maintaining that stream of infertile promotions by which the fertility of the upper classes is continually forced down.

In Western civilization, he said, the earlier investigations gave no adequate idea of either the widespread character or the intensity of the differential birthrate. He summarized the most recent results, drawn from the figures of the 1911 census in the United Kingdom, showing an inverse relationship between fertility and social status. A steep and regular rise in fertility was evident from the professional occupations to the lowest grades of labourers.

And this phenomenon was more or less definitely established in every civilized country.

To counteract the differential birthrate at its roots, Fisher proposed abolition of the economic advantage of small families by instituting family allowances.

In order to nullify, or better to reverse, the action of social selection upon the innate causes of fertility, it would be necessary that the allowances should cover the full net cost of the children in every class, or, to put the matter more exactly, that a member of a large family should find himself in a position of similar social and economic advantage, to an only child of parents in the same station. This presupposes that the allowances shall increase, at least proportionately to the earnings of the father.

Family allowances had come to be considered in the 1920s for economic reasons. After World War I demobilization had been followed by unemployment, international financial crises, low wages, and desperate strikes. Families were hard hit, and laborers' families were often reduced virtually to destitution. In France employers had adopted a scheme of payment to their laborers which incorporated family allowances within the wage scheme, in addition to a low basic wage. It was a way of cutting the total cost of labor without undue hardship to any, and employers thereby succeeded in paying a modest living wage to all their employees. In contrast, in Great Britain a "living-wage" was supposed to cover the cost of a wife and three children. In their wage bills, therefore, British employers paid for hundreds of thousands of nonexistent wives and millions of nonexistent children, while the larger families (that is, half the child population) were still unprovided for from the workers' wages.

In 1924 Eleanor Rathbone, in *The Disinherited Family,* pointed out the inequalities of such a system. She insisted that the rearing of the next generation was a national concern and responsibility, which should not be charged exclusively to the parents. She proposed the institution of family allowances, preferably within professional or work groups, with the object of sharing the burden more evenly among the whole population. Fisher reviewed her book and contributed to the discussion of family endowment at the Eugenics Society that year [30] with considerable enthusiasm. The plan envisaged mainly social improvement; however, Fisher noted, it could be made to serve eugenic purposes, on one condition, namely that the family allowances were not made at a flat rate but proportional to the father's income, scaled to the cost actually incurred by the parents. The French scheme also was not designed for eugenic purposes but could be made so if family allowances proportionate to the basic wage were extended to the salaried occupations, where the effects of social selection were most serious.

Another political expression of economic concern for families, strongly backed by the Eugenics Society, culminated in the scheme of income-tax rebates adopted in the 1928 budget. In those days the lowest-paid of the taxpayers, roughly typified by the highly skilled artisan or elementary schoolteacher, represented a class of the greatest eugenic importance, being numerous and moderately endowed with valuable gifts. Fisher, explaining the proposals in advance for the readers of the *Eugenics Review,* and (after their adoption) considering future policy, had to point out that tax rebates were inherently inadequate. Abatement rarely rose above 1% of income, whereas he calculated that the actual expense of a child must be at least 10% of income for most parents. The future policy of the Eugenics Society must be to work for the institution of family allowances.

Ideally, all working adults might be called on to pay into a professional or national mutual insurance scheme against the cost of raising the next generation. Bachelors and childless as well as parents would then contribute perhaps 10% of their income, while parents drew from the fund an allowance of 10 or 12% of their income for each dependent child. The cost of children would then be spread evenly throughout the working life; the redistribution of wealth would result in an equalization of the standard of living of men doing the same job of work, irrespective of family responsibilities, and the force of social selection for infertility would be broken. It was a scheme by which economic relief, social justice, and eugenic benefit might all be provided without undermining the self-respect of individual beneficiaries or denying the economic principle, ideally, to give a monetary reward according to the social value of the service rendered.

Fisher explained his theory, presented such quantitative evidence as he had found in the literature and offered his proposals for scaled family allowances in *The Genetical Theory of Natural Selection* (1930). The subject received favorable comment in a number of reviews. It seemed possible that the book might influence public opinion and even help guide the framing of a national scheme of family allowances. Fisher wished to strike while the iron was hot, and letters discussing this part of the volume accompanied some of his gift copies, for example, to the Bishop of Birmingham (an old friend in the Eugenics Society).

In consequence, in 1931 Fisher was invited, together with J. Huxley and R. Ruggles Gates, to take part in a small conference on "Eugenics and the Church," organized by the Bishops of Winchester and Birmingham at Winchester, to consider what might be done within the Church of England to equalize the standard of living of clergymen of equal professional standing but with different family responsibilities. Subsequently, in 1932 the headmaster of Winchester College raised the question at the Headmasters' Conference, whether some graded family allowance system, or equalization pool, might be established within the teaching profession.

The campaign continued. For the first issue of the *Family Endowment Chronicle* in 1931, Fisher wrote an article on "The biological effects of family allowances" [*C P* 94]. In September the same year, as representative of the Eugenics Society of London, he carried the message of the retiring president of the federation, Major Darwin, to the Third International Congress of Eugenics at Cold Spring Harbor, New York. There a section was devoted to differential fertility which stimulated informal discussions on both family allowances and sterilization of the unfit. In April 1932 he was present at the family endowment conference at the Eugenics Society, where he spoke on "Family allowances in the contemporary economic situation" [*C P* 100]. In June he devoted the Herbert Spencer Memorial Lecture at Oxford to the subject of "The social selection of human infertility" [*C P* 99].

Meanwhile, as an active member of the Population Society, he was engaged with J. Huxley, J. Gray, J. B. S. Haldane, and others from the beginning of 1931 in planning the British Association Centenary Meeting on Population in 1932. When the time came for the meeting, he was in hospital, but Huxley read his paper. Feelings, on that occasion, evidently ran high, as Fisher learned from a friend's irate account:

Huxley had read your paper for you . . . At the end of the morning, MacBride got up and said that the opinions of Dr. Fisher were of no value, since he is not a trained biologist, and a mathematician's ideas are of no importance in the problem of population. At this there was applause. Baker loudly shouted "Shame!" and there was applause from another section. That MacBride is a damned, ignorant, pig-headed fool I knew. I knew also that he is quite unscrupulous. I did not know, however, that he had not learned how to behave. . . . Unfortunately, he is just one of a type of a whole class of biologists with neither the inclination nor the capacity for critical thought.

Perhaps Fisher was just as well off in bed that day, although he did know what to expect from MacBride and other opponents of natural selection and was much more concerned with the many who might give him a hearing than with the reactions of the few bigots.

In 1932 the Eugenics Society was persuaded to look into the eugenic effect of family allowances and set up a committee under Fisher's chairmanship to investigate and report. In 1934 the pamphlet on family allowances prepared by this committee was accepted and published by the society. The committee gave unanimous support for the principle of establishing family allowances proportional to parental income and adequate in amount to restore to parents the standard of living enjoyed by single men doing the same work. Nothing more came of the campaign.

It was only after World War II that family allowances were introduced in the United Kingdom, and then it was purely as a social measure. They came as a part of the social revolution represented by the Beveridge Report of 1943, on which so much postwar social legislation was based. Although Sir William (later Lord) Beveridge himself had in the late 1920s endorsed eugenic ideas, including scaled family allowances, social reform had a prior claim on his sympathies, and he limited his ambitions to what might be practicable under pressure of modern egalitarian sentiment. Naturally, when he consulted Fisher in preparing this part of the report, Fisher urged again, finally, and in vain, that the allowances proposed should not be made at a flat rate but proportional to income and adequate in amount to compensate for the cost incurred by parents, and he reiterated his argument in print [*C P* 198, 1943]. The family allowances that were adopted failed on both counts.

□ □ □

Through the 1920s Fisher grew ever more insistent that the Eugenics Society should become involved with scientific research. There was little money available, but the research committee did actually undertake some research into mental deficiency in cooperation with other institutions. This pointed up the further moral that not only research but sound research was required. The society needed more scientific members and officers whose influence could ensure that experiments were carefully planned and carried out so as to enable valid statements to be made at their conclusion. Without this sort of scientific commitment, the society, its anchor cut, would drift at the mercy of any eloquent appeal (by enthusiasts for birth control or prohibition or free love or compulsory sterilization of "the unfit" or state responsibility for children from "bad homes," for example). A danger Fisher was particularly aware of was that in the absence of scientific leadership, social scientists of an environmentalist persuasion—and theirs was a majority view in the Eugenics Society as elsewhere—could divert the efforts of the Eugenics Society from their proper study of human inheritance to serve a noneugenic social function.

In the later 1920s Fisher brought several of his friends into the Eugenics Society. The number of scientists on the committees increased, and as their influence began to be felt, they provoked some anxiety among the nonscientific officers so that a noticeable reaction set in against what was called the "Rothamsted lobby." A few years later the conflict came to a head. Convinced that the Eugenics Society should be reorganized so that it might be guided by a predominantly scientific organization, Fisher was pressing hard for cooption to the General Policies Committee, immediately, of a number of scientists. He had the full support of scientists, including Huxley, in this attempt. His candidates were Professor R. Ruggles Gates of London University;

Dr. Ward Cutler, Dr. H. G. Thornton, and Miss L. M. Crump all of Rothamsted; Mr. E. B. Ford of Oxford University; and Dr. J. Fraser Roberts of the mental hospital at Stoke Park near Bristol. The secretary of the Eugenics Society was a sociologist who shared the environmentalist view and distrusted the Rothamsted lobby. At his instigation a special committee with emergency powers was formed early in 1934, and this committee, going over the head of the council, assumed virtual control of the society. The council was rendered impotent, and men like Fisher, Ford, Fraser Roberts, and Thornton resigned from the society rather than lend their support to a movement whose policies they could no longer hope either to approve or amend. The scientific movement was routed and, with it, Fisher's hopes that the Eugenics Society would again serve genuinely eugenic ends. The report of the Committee on Family Allowances, a dead letter in a Eugenics Society without adequate scientific ballast, was one of the last tasks Fisher performed with the Eugenics Society he had served for more than 20 years.

□ □ □

Between 1929 and 1934 Fisher was also deeply involved with a campaign for the legalisation of sterilization on eugenic grounds. The Eugenics Society had formulated its policy with regard to such sterilization as early as 1926. It was an object close to Darwin's heart to see it adopted nationally, and in 1928 he prepared and published in the *Eugenics Review* a prototype for a Parliamentary bill. This was not presented on the floor of the House of Commons, for it was realized that first the public must be informed about the eugenic aspects and reassured about the medical and legal aspects of sterilization. An opportunity was presented by the publication in 1929 of the report of the Ministry of Health on mental health. This report showed an increase in the numbers of mental cases, both in schools and hospitals, sufficient to arouse alarmed comment in the press, and thus it opened the way for what grew into a campaign to have the occurrence, treatment, and control of mental deficiency subjected to investigation and the advisability considered of a national policy of voluntary sterilization.

On investigation of the data, there was no doubt that factors contributing to mental incompetence were heritable and that the numbers affected were increasing. Even taking account of improved ascertainment, since the 1911 census something like an increase of 50% was indicated in every class of patient. The increase was concentrated in the class of marginal ability, persons who frequently had been temporary patients in mental institutions and who were sufficiently subnormal to be usually unemployed and often unemployable. Ignorance and dirt, combined with poverty, led to the needless prevalence of malnutrition, accidents, and disease in their homes, and the death rate among the children was shockingly high. For the parents' sake, attempting to cope with a life already almost beyond their powers, for the sake of the children brought up in unfavorable conditions and endowed with generally subnormal abilities, and for the sake of posterity, these families would have been "better unborn," as the title of one contemporary pamphlet put it.

The Eugenics Society considered that it was insufficient merely to advise such persons to use contraceptives or even to provide and explain the use of contraceptive devices. It was quite unrealistic to advise or require celibacy. Sterilization offered a method of preventing procreation more foolproof than contraceptives and more conducive to normal social and sexual life than abstinence. It was exceptional in being both effective and free of undesirable side effects.

The Eugenics Society also made clear from the beginning that the operation should be unambiguously voluntary. They believed that if a policy of sterilization were undertaken on a voluntary basis and supported by safeguards in medical practice, it would prove its worth in the experience of the patients and become sufficiently widespread to make a real difference to the numbers of defectives being born; whereas, if it were imposed, it would be resisted and resented. Sterilization should be regarded as a right and not as a punishment. It should be performed in general hospitals and not in mental hospitals. Already, private patients enjoyed this right, but there were impediments that virtually denied the possibility in public hospitals. That it should be made readily available to all who wished it represented social justice and not discrimination. Every effort would need to be made to ensure that no stigma attached to the person who chose to avail himself of the opportunity; it would be, in fact, like other measures taken under medical advice, the choice of a socially responsible individual.

In proposing a policy of voluntary sterilization, the first difficulty to be overcome was legal. Current law was ambiguous, and new legislation would be required to clarify the position of the physicians performing the operation on eugenic grounds. In December 1929 a committee of the Eugenics Society was therefore formed to promote the legalization of sterilization. Fisher was among their number. In 1930 the committee published a pamphlet setting out fully and fairly the case for sterilization on eugenic grounds, and it was distributed to all local county authorities and members of Parliament. Further publicity was given to the subject through the lectures offered by the society whose speakers were invited to address local groups.

Public concern, aroused by the publication of O. E. Lewis's investigations (in the Ministry of Health's report), also stimulated considerable public discussion. Huxley presented the biological basis of the campaign in the pages of the *Daily Mail*. A number of eminent scientists, speaking in the B.B.C. series on "What I Would Do with the World," though not members of the Eugenics Society, made special reference to eugenics. At the British Association meetings that year no less than five members of the Eugenics Society spoke on closely related matters. The movement was gathering momentum.

In December 1930 the Eugenics Society published a second edition of their pamphlet. It included a draft of a Parliamentary bill, and this bill was introduced in Parliament the following year. It was, as expected, defeated, but the vote was only 2:1 against the measure and served to publicize the growing support for the legalization of sterilization. The new pamphlet also contained an appeal to the Minister of Health to set up a Royal Commission to inquire into the whole question of the incidence, treatment, and control of mental deficiency, with special reference to sterilization.

Although the minister was not inclined to act and the Mental Health Bill passed later that year was reported in the *Eugenics Review* to have "no aspects of eugenic interest," the British Medical Association (BMA) was moved to set up an investigating committee and invited some members of the Eugenics Society, including Fisher, to sit with the doctors on this committee. There were 21 members, most of whom were already publicly committed to their views before they met; their views were incompatible. Fisher's most important task was as a statistical interpreter. He found himself under some suspicion, an outsider on the largely medical body. It was uphill work persuading conservative members to consider the evidence in support of the eugenic effectiveness of sterlization. They pointed to Punnett's earlier calculations, made on the assumption that feebleness of mind was due to a single recessive gene, which showed that even after 990 generations without reproduction from homozygotes there would still be a small proportion of feebleminded in the population. Fisher pointed out that, on the same assumptions, even on the most conservative estimate, sterilization would reduce the incidence of feeblemindedness by 17% in the first generation, or more like 36% if assortative mating were taken into account. They suggested that sterilization would not be effective if the genetic basis were multifactorial; Fisher tried to convince them that, in fact, the effects of sterilization would in that case be much more quickly apparent than if a single recessive factor only were responsible.

It was surprising that so large a committee of persons who had already prejudged the issue succeeded in producing a report at all, but within 2 years they did report. Although the report confirmed previous studies in emphasizing the gravity of the increasing numbers, it failed to give a strong lead by united recommendations to deal with the problems. To relieve the congestion in mental institutions and to give patients a better chance of managing their life in the community, brief hospitalization was proposed, including courses of technical training and instruction in social hygiene. This policy of "socialization" did not envisage eugenic controls. There was a hint in the report that sterilization might be used as an adjunct to "socialization," but the point was not made that the effects of "socialization" without sterilization must be highly dysgenic; for patients who formerly would have been segregated in mental institutions would be returned to the community under the new policy.

The publication of the BMA report provoked citizen groups and individuals to petition the Minister of Health again for a Royal Commission. The Eugenics Society backed the request. As a result, that year a departmental committee of the Ministry of Health was set up to investigate. It consisted of eight members representing both sexes and much the same variety of opinion as had the BMA Committee. Fisher was coopted to this committee also. The smaller size of the committee, the tactfulness and patience of the chairman, Mr. L. G. Brock, and the determination among the members to give positive recommendations to the government, all contributed to the result: the publication in 1934 of a unanimous report by the departmental committee [Statistical Summary in *C P* 120, 1934]. Evidence was taken of the numbers of persons affected, the heritability of the mental conditions, the eugenic and social consequences to be anticipated of sterilization, and other measures. There was discussion of voluntary or compulsory controls and the whole gamut of medical, legal and ethical considerations. Some details were omitted from the report because of intransigent disagreement within the committee, but the main lines of a definite policy had the support of the whole.

The policy proposed was that which the Eugenics Society had long supported: the legalization of sterilization on eugenic grounds and the provision of accessible means of undergoing the operation, on a voluntary basis under medical advice, and with appropriate safeguards. In one respect the Brock report recommended measures beyond any which the Eugenics Society had dared propose. Not only sufferers from heritable diseases but also those who had reason to believe themselves to be carriers of such diseases were included in the recommendations. This meant that a much wider and more effective eugenic policy could be followed.

Fisher's work for sterilization ceased with the publication of the Brock report. No government action resulted. The current government would not risk raising a controversial issue toward the close of its term in office, and the subsequent government was involved with the constitutional crisis of Edward VIII. Public concern about mental health dwindled as more immediate issues appeared in the press.

Now, there were occasions throughout his life when Fisher gave an impression of being intolerant of opposition, impatient, contemptuous of fools. Certainly he had a hot temper and (making allowance for his severe headaches) he was sometimes pettily irritable and disagreeable. Such lapses need to be seen in perspective. An intemperate man would not have been asked to serve on the successive sterilization committees, nor would he (who disliked committee work) have chosen to serve in such a thankless capacity, repeatedly, over a period of 6 years, nor would his participation in the investigation and discussion of difficult and controversial matters have been likely to eventuate in unanimous and positive recommendations. For this worthwhile objective, Fisher submitted himself to listen with patience, to answer with sustained reasonableness, and to endure the tedium inevitable in negotiating points of disagreement and finding a form of words acceptable to all. It was his pertinacity, loyalty, reasonableness that characterized the tenor of his relationship with these committees, as it was usually with other groups with whom he worked.

Evidently he won the respect of the medical men, for in 1932 he was invited to become a member of the Medical Research Council's newly formed Committee on Human Heredity. He was glad to accept. It was a position from which he might be able to influence research programs both in the planning stages and in analysis of results. He was eager to be involved in research in human heredity for itself and for the sake of promoting eugenic legislation. His connection with the Medical Research Council grew more intimate after he became professor of eugenics at University College. His own bid for research funds, intended for research into mental deficiency, was accepted in 1934, and the serological unit was established which, throughout its life as a research unit, was supported by funds administered by the Medical Research Council.

□ □ □

One difficulty with any human studies was to get adequate data. In respect of the fertility of different sections of the population, a national survey was really required. The census figures from the 1911 census gave very helpful information, but the families then recorded had been born at latest at the turn of the century. The birth rate had fallen steeply since then, but because the 1921 Census did not include the relevant questions, the relative rates of reproduction of different classes could only be guessed. In *Genetical Theory* Fisher commented,

Although the general character of the facts may be said to be established, in most western countries with certainty, yet the actual quantitative data available are far less satisfactory than they might easily be made and this for two principal reasons. In the first place they are much out of date and relate, in some of the most important instances, to the rate at which births were occurring in different classes as much as 35 years ago; while as to what is now occurring we can only conjecture from the past course of events . . . In the second place, a completely satisfactory comparison between different occupational groups must involve the mortalities occurring in these groups and the natalities as affected, not only by the birth rate of married persons of given age, but also by age at marriage and frequency of celibacy . . . In all the examples given it will be seen that our knowledge falls very far short of this level.

Fisher became deeply concerned that the 1931 census should be used to get good recent figures for birthrate for different classes of occupation and that the registration of births should be amended in such a way as to be a continuation of the census record and include all the necessary information. At the time the registration of births could not be coordinated with census records because there was no consistent classification of "occupation of the father"; nor were there records at birth registration of previous births and deaths of siblings.

In 1927 the Eugenics Society began, through him, to bring these considerations to the notice of the Registrar General. Memoranda were prepared, correspondence and interviews took place with committees and individuals involved, objections had to be overcome and recalcitrants persuaded. By May 1929, Fisher was not without hope: "Vivian [the Registrar General] seems much inclined to let us down but I shall write to him again. I should not like him to feel comfortable about doing nothing." Because the possibility of census reform or even of a thorough-going reform of registration seemed far out of sight, he put forward the idea that somebody with research money to spend might be persuaded to put up one of their employees as a local Registrar, equip him with card indexes and clerical assistance, and set him to make a proper survey of one registration district. The advantage would be to provide an ideal framework for further bioeconomic enquiries. There was no objection to the plan in principle, but no sponsorship for it in practice.

In April 1930 the Registrar General was under pressure from other groups to have the census requirements of the Eugenics Society withdrawn and was prepared to promise reform of registration procedure at this price. After negotiating this with him, Fisher reported, "If we get the terms laid down, we shall do very well for the future at some immediate loss especially of the rather critical post-war period." But he noted that the new birth registration forms were due to be in use on January 1, 1931 and the Registrar General "will really have to get busy if he is to do what we want. I discussed all the points raised with him for nearly two hours in all and believe he will not jib utterly at any item. That is, he will do it if the pressure is as heavy as I thought it must be." In the outcome, the Eugenics Society did not even get the promised change in registration. The census over, the pressure on the Registrar was relieved, and there were always reasons for procrastination.

It was frustrating to be dependent on others to gather the data with which to work. By the close of the 1920s Fisher was restless in his wish for a chance to organize his own research in human genetics. He could see the opportunity offered by having appropriate census and registration forms for use together, so that population trends could be traced from year to year with up-to-date information. And he could see the opportunity offered by genetical science for the study of human heredity.

At the Galton Laboratory studies were made using the statistical methods developed by Galton and Pearson and the historical and comparative method associated with the name of Dr. Archibald Reid, but genetics was neglected. As early as 1924, when the Rockefeller Institute of Health was established in London, Fisher had prepared a notice for the *Eugenics Review:* "to bring to the attention of the Ministry of Health the urgent desirability of establishing a Chair of Human Heredity in relation to disease, with facilities for training advanced students in methods of research appropriate to this subject." Such methods, he suggested, should include the Mendelian analysis of the hereditary complex as well as the other methods mentioned above, and he pointed to the recent successful elucidation of the factorial basis of the isoagglutinins of human blood (A B O blood groups) as an example of what might be done, adding that "in the absence of an institution especially devoted to such researches, the scope of these methods is not at present widely understood."

In 1929, while he was fighting for better census and registration forms, he was attracted by the advertisement of a research professorship in social biology at the London School of Economics. The opportunity to direct his own research was obviously appealing, and he wrote to Major Darwin about it, wondering what assistance the professor would have. For his research program, he stated, he would want a geneticist and a research psychologist to work with him. Darwin protested he should "first catch his hare before fattening it," and used Charles Darwin to exemplify what a man can do without institutional support. He wanted Fisher to get all his mathematical contributions out of his head before becoming entangled with directing research. Fisher, in contrast, felt the limitation to the amount of theorizing any man can do and the great need for well-conducted experiments. He applied for the professorship, to which L. Hogben was appointed.

Fisher's letter setting out his position for Darwin is interesting in exhibiting his attitude to statistics as a tool valuable as it is applied to a valuable object. The worth of the hare, unfattened, was, he wrote, firstly a small increase in salary and secondly "the possibility that my work in mathematical statistics will be more valuable if applied to researches on man." The letter also displays the relationship that he perceived between theoretical and experimental research and the need he felt for institutional support which would give him an opportunity to "answer the problems and consolidate the conclusions at which he had arrived."

Would you agree with me that at about fifty, your father had decided that there was little more to be done for the subject out of his own head but that as a good theorist makes a good observer so still more in experimentation that there was a great need for well directed experimentation which should answer the problems and consolidate the conclusions at which he had arrived?

If this is so he was several generations in advance of his time and, in the absence of a ready supply of trained assistants and under the restriction of working at his private expense, he was unable, without being unwilling, to set a much needed example of what a director of research should be.

Write Fisher for Darwin and we get a clear notion of the restrictions that cramped Fisher in his genetical research and were beginning to gall him.

Throughout the 1920s he had been on the sidelines of genetics. In his boyhood he is said to have watched his brother Alwyn playing with his train set and not to have been allowed to touch it, and he had sat for hours at a stretch, happily observing, learning, and planning exactly what he would do if he could operate the system. Just so in the 1920s he had observed, learned, and planned. He had worked out the genetical theory. He had established the principles of experimental design. He had ideas which wanted testing, and he could not rely on others to test them or to do it in his way. Further developments, especially in human genetics, waited only for the opportunity he could not create, the opportunity for him to direct particular researches into human genetics. He did manage in 1929 to begin an experimental program on poultry, but this was, as before, a private venture, made possible by the cooperation of Rothamsted and by sponsorship from private sources and small research funds.

8

The Evolution
of Dominance

Original contributions to science are peculiarly personal. They arise not from the evidence but from the mind which seeks to give coherence to the evidence. Fisher's theory of dominance modification was no exception.

Other geneticists accepted dominance without worrying much how it came about. To Fisher, however, the question early became important for eugenics, as we see when he asked it in 1920, in a review of *Inbreeding and Outbreeding* by E. M. East and D. F. Jones [*C P* 13]: "One question, however, arises from the maize experiments, which East and Jones do not answer and on the answer to which depend all the human applications of their results. Why are the recessive factors in maize harmful?" The question continued to puzzle him until in 1928 he found a solution to satisfy him. But the solution only came to a mind prepared in advance to contemplate it, primed with the sort of construction of other evidences that induced it and strengthened by the statistical approach that justified the boldness of the conclusion. For it was Fisher's construction of earlier evidences that compelled him to add this piece of theory. For him it made sense of a troublesome phenomenon in terms of familiar and well-ascertained processes. To him, mulling over the problem, it became suddenly obvious that the relevant facts of evolution, natural selection and genetics converged upon this solution, and required it, as the arch requires the keystone which, once laid, itself confirms and stabilizes the fabric upon which it is based.

Fisher's construction of the evidence was, in fact, so different from that of his contemporaries that his proposal was met with incredulity and denial. It is against a stormy backcloth of adverse scientific opinion that we follow the glow of his excitement, kindled in the formulation of a theory which seemed good to him and which he carried at once confidently into the field, and the delight with which he backed the theory and saw it succeed in accounting for a series of challenging natural phenomena. He, like his rivals, put against it the hardest facts he knew and it held its own; although it was tried at all points through years of controversy, in logic and in experimental verification, even the original application, still debatable,has not been dismissed by scientific proof: it remains "not out."

□ □ □

Dominance was a feature of Mendel's discoveries which could not have been predicted of a particulate biparental scheme of inheritance. However, its logical independence was not recognized after the rediscovery of Mendel's work, and explanations of its existence did not attempt to show its logical necessity but only to fit it into the physical scheme of genetics.

Fisher's question was *why,* and his argument started from the observed incidence of dominance [C P 68, 1928]. First, the prevalent wild type was found to be almost invariably dominant to mutants occurring in experimental cultures or variants selected for domestic breeds. Second, the dominant was found to be noticeably superior to the homozygous recessives, which were usually deformed (and it was a startling fact that the largest class of mutations at that time identified in the fruit fly, *Drosophila melanogaster,* were actually lethal as homozygotes). Third, in the case of multiple allelomorphs, when more than one alternative was found at a particular genetic locus, the dominant was found to be dominant to all other allelomorphs while these usually showed no dominance among themselves.

In the 1920s the followers of Bateson continued to interpret dominance as the presence of a gene absent in the recessive, though this was becoming a cumbersome way of managing new facts. Some mutations, which should have been "absent," were found to be at least partially dominant, and these too were explained as being due to losses of genetic material, this time of "inhibitors" of the mutant effect. To explain the anomalous dominant characters in domestic poultry, Punnett had even considered the possibility that "something had been added" that prevented the inhibition of the mutant effect. But already the ground was cut from under the "presence or absence" thesis by the discovery of multiple allelomorphs. Since there could be no more than one sort of absence, there must in this case be more than one sort of presence, the one completely dominant to the others; in explaining this dominance the hypothesis was void.

A view which became more popular during the 1920s was that dominance expressed a quantitative difference in respect of some one physiological or biochemical function. Although Fisher agreed that there was evidence that in some cases the physiological effect was of this kind, he did not regard such inactivation as being, in itself, a sufficient cause of recessiveness. There were both dominant and recessive *Pied* mice, both dominant and recessive *White* fowls, and, in general, there was the coincidence of the wild-type normal with the dominant, which could scarcely be due in all cases to the coincidence of a greater physiological activity with a greater viability of the organism! In fact, the dominance of a character could not logically be attributed either to the presence or to the activity of the gene itself.

From an evolutionary point of view it was evident that dominance could not be a quality intrinsic to the gene, for the dominant was found consistently to be the prevalent wild type. If evolution had taken place by the preferential survival under natural selection of one or another of the alleles occurring through rare mutations at any locus, simply by the replacement of an inferior by a superior gene, then the allele for the dominant in one age must have been replaced during the evolutionary history of the species, and its successor must *then have acquired dominance* over its previously dominant rival. Without being itself modified in its homozygous effect, that is, almost certainly without being itself replaced, it had picked up in addition the property of dominance.

Some genetic factors examined experimentally had been shown to owe their characteristic manifestation not to a single gene but to a primary gene acting in conjunction with others that were capable of modifying its somatic expression. Thus the homozygous *hooded* rat or recessive *pied* mouse could, by selection in different directions, be modified in appearance until the pigmented areas covered practically the whole body or left it nearly white. Fisher argued that, in a similar fashion, if an organism heterozygous for some mutant factor had an appearance initially intermediate between the two homozygotes, any existent modifying genes would be subjected to natural selection. Those that tended toward the suppression of the effects of a deleterious gene in the heterozygote would be favored, and the result would be a gradual elimination in the expression of the advantageous gene, of differences between the heterozygous and the homozygous phase; the advantageous gene would become dominant through the action of the selected modifying genes, and the deleterious gene would become effectively recessive. This was the theory of the origin and nature of dominance that Fisher put forward in 1928 [C P 68].

It is worth emphasizing here the divergence of Fisher's perception of the situation from that of his contemporaries, his faith in the efficacy of natural selection, backed, as it was, by statistical investigations and, equally important, his attitude to genetic interaction.

Recalling the antipathy that existed between geneticists and biometricians, it is hardly surprising that Fisher differed from most geneticists in taking for granted the possibility of genetic interaction. If human intelligence resulted from the action of a number of genes, as he had assumed in 1911, then these genes could be regarded as modifying the manifestation of each other. Perhaps they ought to be so regarded, for they could be considered independent only so far as the outcome was attributable to purely additive effects of each of the genes. Although additive effects were sometimes assumed as a convenient fiction, offering mathematically the simplest case, it could not be supposed that the assumption would be often justified in nature. Therefore, he took into account the possibility of the interaction of genes when, in 1916, he was considering the multifactorial determination of various human measurements [C P 9]. In effect, his early interest in continuous variation in man, as a product of Mendelian process, made him early aware of the ubiquity of modifying factors.

Mendelian process implied essentially stable genes. He [33] therefore challenged W. E. Castle's interpretation in 1916 of the results of selection on pigmentation of hooded rats. Castle interpreted the changes in color as being due to changes at a single locus. Fisher responded:

As a scheme of inheritance Mendelism has the manifest advantage that it is established beyond question. . . . If the variations among the hooded rats be due to subsidiary Mendelian factors, there could scarcely be less than 100 of them. Professor Castle speaks of this hypothesis as definitely disproved and we should be glad to see the reasons for this conclusion stated more fully. In the figures which he gives there are indications, however slight, that a Mendelian explanation is possible.

His discussion shows, in effect, that acceptance of gene interaction led him to contemplate the involvement possibly of "innumerable and minute" genes at a time when contemporaries preferred to believe in very numerous mutations. At the same time many geneticists accepted the occurrence of large mutations, or saltations, affecting many different somatic features simultaneously. Here again Fisher disagreed.

When it was discovered that the frequency of crossing over between portions of chromosome was under genetic control, being promoted or inhibited by genes, he saw that close linkage would be developed by natural selection wherever it was advantageous. The concept of numerous interacting genes lent itself to the extension that, with close linkage, a whole group of such genes might be modified as a unit, so that the primary gene acted merely as a switch mechanism between two or more progressively divergent groups of modifying genes. In particular, this would provide a genetic basis for polymorphism, for in species consisting of two or more distinct forms, each form must be supposed to comprise many mutually harmonious modifications unsuitable to the other; these could have developed only if the genes had been inherited generally as a group, and been selected *inter se* for their contribution to the group effect.

Fisher first published these thoughts in 1927 [C P 59]. Punnett had proposed that the trimorphic female of the butterfly *Papilio polytes* had arisen by a single mutation (or saltation); to him it seemed that it must have done so, since the trimorphism had been shown to be controlled by only two genes, both limited in their obvious effects to the female sex and the one apparently necessary for the manifestation of the other. The female forms were quite distinct, mimics of three other species, and different also from the male forms.

Fisher illustrated his argument in favor of the controlling genes acting merely as a switch by the example of sex, the oldest and most prevalent dimorphism we know. Commonly, sex is associated with a whole chromosome, but in fish of the genus *Lebistes* (guppies) crossing-over of genetic material had been observed between the sex chromosomes, X and Y, of the heterogametic male. Fisher remarked that in this case one ought more properly to speak of the sex gene than the sex chromosome. The case was thus analogous to that of the mimetic butterfly. But, he added:

we should nevertheless at once recognize our folly if we argued that because the sex difference in *Lebistes* is apparently determined by a single factor, therefore a female fish of that genus, with the appropriate adaptations of her sex had arisen by a single saltation from a male of the same species! Or *vice versa*. . . . Since the reproduction of the species requires the cooperation of both sexes, we may be certain that the origin of the sex factor antedated the evolution of separate sexes, and has persisted, in its function of switch, unchanged during the whole course of the evolutionary development of these types. . . . The morphological contrast determined by the factor at a late stage may be quite unlike that which it determined at its first appearance.

That evolutionary changes do take place in such alternative forms through the selection of modifying genes and not by modification of the gene itself had been demonstrated by W. E. Castle's breeding of hooded rats [34]. Fisher cited a crucial experiment in which progeny of two contrasted lines (dark and light) had been bred back to the self line and their progeny in turn inbred to obtain an F_2 generation with segregation of the *hooded* gene. The *hooded* grandchildren did not resemble their very dark or very light *hooded* grandparents, as they would have if the gene itself had been modified. In each line, they exhibited the characteristic appearance of the original unmodified ancestors through the restoration to the *hooded* gene of the unselected modifying factors present in the self rats. The gene itself, then, could be taken as uninfluenced by selection, but its external effect could be influenced, apparently to any extent, by means of the selection of the modifying genes.

In continuous variables, in sex determination, mimicry and other polymorphisms, in apparently insignificant recessive color factors of *pied* mice, *hooded* rats, *Dutch* rabbits, the same story repeated itself. There always seemed to be an abundance of genes available, capable under selection of modifying the somatic expression of any gene. And, when it occurred to Fisher that dominance itself might be due to genes that modified the character of the heterozygote, he realized that something parallel to the effect of deliberate selection in hooded rats occurred spontaneously in mutant stocks of *Drosophila* in culture [*C P* 68]. Mutants in their pristine form usually had well-marked bodily effects, but several workers with *Drosophila* had remarked that after isolating a mutant in separate stock bottles for a time, the mutant had appeared to be less distinctive and more normal than it was at first; it seemed to have reverted somewhat toward the wild type. When such modified mutants were outcrossed to unrelated wild flies, however, and the mutant type recovered by inbreeding, the offspring showed a return toward the extreme condition at first observed. This clearly demonstrated that the mutant gene itself had not been altered and that the change in its expression must be due to modifying genes. Since such mutant stocks often showed from the first some degree of reduced viability and some variability in the intensity with which the mutant character was manifested, it was evident that modifying genes existed in the mutant stocks and had been subjected to natural selection, with the result that those producing the most normal expression of the mutant, being the most viable, had spread through the stock.

It was a relatively small step from the concept of the modification of the expression of a deleterious mutant in the homozygote to the realization that dominance itself had developed in consequence of the action of modifying factors on the expression of a deleterious mutant in the heterozygote. It was clear that a selective advantage accrued to the heterozygote, when modified to resemble the normal form, and it was clear that genetic effects were susceptible of such modification. The motive and the means for the evolution of dominance were in evidence.

The piece of the puzzle which was at first missing was the realization that the opportunity also existed because the known mutations had been constantly recurring during the evolutionary history of the species. Evolution was *not* wafted forward on a breeze of predominantly favorable mutations, as was assumed by current evolutionary theories; on the contrary, it had gone forward in the face of a blizzard of predominantly unfavorable mutations. Moreover, the species had met the same elements in the blizzard for hundreds of thousands, or even millions of years and this had left unmistakable traces in the incidence of dominance throughout the length and breadth of the animal and vegetable kingdoms.

But was the opportunity great enough? Here Fisher's approach through population genetics helped check the qualitative argument. In August 1928 he wrote to Major Darwin: "I'm sure that we have very much the same picture of evolution in our minds, but the picture in my mind has been changing of late, not in any way in principle, but by groping after approximate magnitudes in the proportions of the different parts." The rate of occurrence of mutations in *Drosophila* cultures gave a notion of how many mutants the wild population might contain at any time. Homologous mutations, like the albino in mammals, showed that this mutation at least had been occurring since an early stage in the differentiation of the species. Since individual mutations were rare, the mutations that had already occurred in laboratory stocks must be the common recurrent mutations of nature, importunate failures which had been constantly clamoring at the evolutionary gates for hundreds of thousands of generations and had been as constantly rejected; they must have occurred with an enormous total frequency during the evolutionary history of the race.

Statistical calculation showed that although individual mutations were rare and the homozygous mutant would be kept exceedingly rare by counterselection, the much more numerous heterozygote, though still rare, would be modified at a rate depending on the initial viability of the mutant. Toward the end of its modification, when it approached very close to the viability of the wild type, the numbers of both heterozygous and homozygous mutants would increase very rapidly and the frequency of the mutant gene would begin to depend on the viability of the mutant homozygote. The homozygote might then begin the long haul in the same direction the heterozygote had already gone, toward normality. With a persistent mutation in which even the homozygote had not too bad a chance of survival, the homozygote might follow in the footsteps of the heterozygote all the way and eventually become indistinguishable from the wild type. Such mutations might, in fact, leave no trace for genetic research to reveal.

Although mutations at the same locus would be individually too rare to occur together in the same individual frequently enough to develop any mutual dominance, each had recurred in the presence of the wild type times without number. Thus the lack of dominance among the multiple allelomorphs that were recessive to the dominant wild type was explained. At the same time the dominance of the prevalent wild type, the recessiveness of most deleterious mutants, and the incompleteness of dominance in a few cases all followed from the principle of modification of dominance by natural selection. Only the anomalous case of dominance in poultry remained inexplicable when Fisher wrote his first paper on the modification of dominance (*C P* 68). He expressed the opinion, then, that his theory "throws no light upon such a case as is presented in poultry, where a majority of the genes which distinguish domesticated races from the wild type appear to be nonlethal dominants."

□ □ □

A good new theory, like an advantageous mutant, has an immediate advantage in coping with its environment, and Fisher's theory quickly proved advantageous in explaining two instances, the one typical, the other exceptional.

In the summer of 1928 J. B. Hutchinson, a young cotton geneticist, came from Trinidad to spend the summer as a voluntary worker in the Statistics Laboratory at Rothamsted. He was able to tell Fisher about unpublished results of cotton breeding which provided a very striking confirmation of the evolution of dominance. It appeared that the Sea Island species fairly frequently produced a mutant called *crinkled dwarf* which was, in this species, completely recessive. However, when the Sea Island *crinkled dwarf* was crossed with other New World cottons in which the *crinkled dwarf* mutation was unknown, and the hybrids self-fertilized, there was no segregation of the progeny in a clear 3:1 ratio, as there should have been with complete dominance. In fact, the progeny consisted of every gradation of intermediate, with no clear dominance at all. Evidently, in contrast to the Sea Island, the Peruvian and Upland species in which the crosses had been made had not experienced *crinkled dwarf* as a recurrent mutation and therefore had not developed the dominance to it of the specific type. This was a far more direct demonstration than Fisher had at all hoped for; it was decisive evidence for his theory from an entirely unforeseen quarter. He was thrilled by this early support for his hypothesis.

The evidence from New World cottons was one of the confirmations of his theory Fisher used later the same year when he expounded dominance theory in *The Genetical Theory of Natural Selection,* and he was to use it again on various occasions during the next few years. In the second edition of his book, however, he withdrew this example and in place of its indirect confirmation used the direct proofs which were by then available. During the intervening 30 years, cotton breeding had revealed a situation in this case more complex than was convenient to use any longer as an introductory example; more than one distinct *crinkled dwarf* mutation had been found, and more than one distinct system of modifiers of the heterozygous appearance in different species. Thus, Fisher wrote,

While Hutchinson was undoubtedly right, in what is now generally agreed, that modifying factors are abundant, and that much modification has taken place in the phenotypic appearances of the different heterozygotes and homozygotes at this locus, yet these are more numerous than was at first thought, and the interpretation at first put on some of the facts is untenable.

In 1928, in contrast, the very simplicity of the interpretation of the first few results, on the basis of the new theory, and the difficulty of interpretation on any other theory, told strongly in favor of the evolution of dominance.

In the same summer, Major Darwin in private correspondence with Fisher raised the problem of poultry again in connection with a recent publication on the *rose* and *pea* comb factors. His letter was a part of the continuing reflective consideration, together with Fisher, of all things evolutionary. On August 7 Fisher replied and, even as he wrote, we see a solution of this problem comes to him and he begins to like it. Only at the end of a long letter does he turn to the curious case of dominance in poultry:

Is not the case of poultry queer? There must be eight or ten factors in domestic breeds, nonlethal and dominant to the apparently wild-like characters. I do not feel it personally as a difficulty to my theory of dominance because, on any view, one would want to know why poultry should behave differently from other beasts and birds, to say nothing of plants, and to this we have no clue. That species crosses have occurred is likely and, though all possible species have, I believe, single combs, they may, as you suggest, have genetically unlike single combs which on combination might give Rose and Pea. Is any form of unintentional human selection possible? Were hens only kept at one stage, constantly outcrossed with wild cocks and so only dominant novelties selected?

Yours sincerely,

R.A.F.

I believe this works. The primitive fancier would have to be always selecting heterozygotes from wild type birds in the same brood and would therefore be constantly increasing the contrast. Dominance of several of these fowl dominants is very variable in its completeness in different breeds. How's that!

Fisher hurried into print with "Two further notes on the origin of dominance," dated August 22, 1928 [C P 69], describing the situation so far uncovered in New World cottons and elaborating the scheme he envisaged of human selection of novelties among domestic poultry. He wrote:

I must now withdraw the statement that the theory of the evolution of dominance appears to throw no light upon the case of poultry; for that theory seems to be capable of doing its full duty of explaining both the rule and the exception.

Observing that mutant forms which in nature are kept rare by counterselection are often sheltered from competition under domestication and prized as novelties, he noted that the domestic hen was, in its own country, constantly liable to be mated by wild cocks: "This is frequent in India to the present day and must have been the prevalent condition especially in the early stages of domestication by jungle tribes." In the primitive conditions in which the domestic hens were as a rule covered by wild cocks, their broods would have been heterozygous for factors peculiar to the domestic flock, and any mutant novelties that could have been picked out from the broods would not, even initially, have been completely recessive; the selection of such novelties would then constantly favor the more distinctive appearance, the more dominant modifications of the mutant appearance.

In addition, a greatly accelerated rate of modification of dominance was to be expected in this case, where the whole of the population rather than a minute fraction of the population under selection consisted of heterozygotes.

It will be observed that this explanation covers both the high proportion of non-lethal dominants, and the high degree of dominance shown by them in the breeds of domestic poultry; it is moreover capable of direct experimental verification, for it does not involve the corresponding modification of the wild species. The crucial test would consist in crossing any one of these dominants, such as dominant *White,* continually back into a line of genuinely wild jungle fowl; the *White* gene should then, on the view developed above, lose its dominance to an appreciable extent in a few generations, and when the homozygous *White* was reconstituted in birds principally of wild ancestry, it should be appreciably different from the heterozygotes in the same stock. *White* would not, however, in my view be expected ever to become a complete recessive in such an experiment.

Immediately, he embarked on the preliminaries necessary for breeding experiments that could provide the experimental verification of his predictions.

The theory was rich with implications. Fisher concluded his "Two further notes" by reverting to the astonishing power of natural selection:

The work of adaptation has hitherto seemed to be the only one upon which natural selection is engaged, and nothing could be more difficult to measure than achievement in this respect. It has now been shown that the same agency, as a minute by-product of its activity, must also tend to modify dominance, and, if the recessiveness of each several mutation be referred to this cause, the vast number of reactions which must have been so modified gives a measure of its efficacy, which might have startled even a Weismann.

It was in the remarkable power attributed to natural selection that the theory met its first opposition. Was it reasonable to apply the same mathematical formula to selective intensities ranging from the very large to the very small? There was a feeling that although larger selective intensities would have a predictable influence, the smaller selective intensities would be subject to interference and so reduced to nullity. It seemed incredible, for example, to Sewall Wright [35], the population geneticist, that a small selective intensity of, say, 1/50,000 the magnitude of a larger one should produce the same effect in 50,000 times the time.

In answering Wright's criticism [C P 81, 1929], Fisher admitted that Wright's criticism was genuinely aimed at his theory; he had made the assumption that the smaller selective intensity would have its proportional effect in the long run and the assumption really might be doubted. But Fisher argued that if the conditions on which the stability of the gene ratios rest were liable to change, stability might turn to instability, and, in that transient state, a minute but steadily increasing selective intensity would be well fitted to tip the balance. And it was such factors, suffering the feeblest selection, that at any one time were the most numerous. So long as the slight selective pressure was consistently maintained, there was no reason to suppose that its minute size would make it inconsequent or its slow action make it ultimately ineffective.

Though their results differed numerically, Fisher and Wright could come to an agreement about the algebra. The important difference between them was in the interpretation of the numbers. Since both views were based on conjecture that could not be verified by direct experimentation, Fisher felt it was important that the point at issue should be clearly distinguished from points merely of mathematical formulation and that biologists should be free to judge whether they could risk accepting the biological assumption Fisher had made in formulating a consistent and coherent biological theory, or risk rejecting it and discarding the theory without further investigation.

Thus the theory blocked this particular attack: neither scored. But the play had personal overtones. Between Wright and Fisher there were a number of exchanges on the subject during the early 1930s, and the two men diverged at the same time on other points of evolutionary theory for the same reason, because they laid a different emphasis on the role of natural selection. The closely similar interests which had for years brought them into friendly contact from this time divided them.

□ □ □

In the autumn of 1928, when Fisher decided to "go the whole hog" in presenting the case for the evolution of dominance in *The Genetical Theory of Natural Selection,* he wrote to Major Darwin that "any examples to the contrary are therefore badly needed," and he put forward a whole series of questions that might be relevant: he wanted to test his idea and see how it stood up against all sorts of biological data. Happily, his papers in 1928 [*C P* 68, 69] produced a response; in January 1930, J. B. S. Haldane [36] came forward with "examples to the contrary" and proposed his own solution. Fisher joyfully took up the new challenge [*C P* 87, 1930], confident that his theory would perform well.

Haldane called attention to the peculiar dominance phenomena found by R. K. Nabours in the grouse locusts *Paratettix* and *Apotettix* and by O. Winge in the fish *Lebistes reticulatus*. The wild forms were all visibly polymorphic, and the polymorphism was in each case determined by genes or gene complexes that were very closely linked in inheritance. In respect of dominance, each species had a relatively common "universal recessive," together with a number of usually less common dominant forms, which, if allelomorphic, showed no mutual dominance but had heterozygotes combining the characteristics of the two dominant homozygotes. Thus in the presence of the polymorphism and close linkage the pattern of dominance was an inversion of that found with respect to a deleterious mutant.

Haldane had previously suggested with respect to grouse locusts what M. Demerec had later suggested with respect to *Lebistes,* that the close linkage was due to several chromosomes from the same parent being generally transmitted in a group to the same offspring. In 1930 Haldane added that such linkage between chromosomes could be accounted for by sectional translocations and that the dominant genotypes were themselves due to the duplication of such translocated segments. Thus, although conceived as an alternative to Fisher's, his proposal tacitly accepted the revolutionary idea that dominance had evolved.

Haldane's interpretation of the incidence of dominance in these cases depended on the quantitative physiological effect of duplicated genes, some of the genes in the duplication having a greater effect when three or four were present than when only two were found. Fisher's [*C P* 87] reaction to this proposal was:

This would indeed suffice to explain why the heterozygous duplication should differ from the universal recessive. But it gives no explanation of the completeness of dominance; it does not explain why the homozygous duplication should resemble the heterozygote. For this it is necessary to suppose, not only that some of the genes in the duplication have a greater effect when three are present, but also that none of them should exert a further effect when the three are increased to four; and this curious type of limitation must be observed in each of the possible duplications to which the different dominants are to be ascribed. Attractive as the suggestion is, therefore, in respect to linkage, it can not be regarded, as it stands, in any sense as an explanation of the dominance observed in these groups.

In developing his own case Fisher observed that the instance he had previously discussed, that of a deleterious mutant subjected to counterselection and only maintained at an extremely small frequency in the stock by persistent mutation, was very different from the example raised by Haldane, that of the dominance of the less numerous genotypes of polymorphic species. The latter genotypes could not be regarded as being subjected to counterselection but must, in any stable polymorphism, be maintained by a balance of selective actions, under which the frequency ratio of the contrasted genes settled down to a condition of equilibrium. As he had shown in 1922 [C P 24], the equilibrium would be stable if the heterozygote were at a selective advantage compared to both homozygotes.

Fisher had observed in 1927 [C P 59] that in mimetic butterflies the homozygous mimic enjoyed the advantage of its coloration probably only at the cost of infertility or impaired viability associated with the mimetic gene; the heterozygote, in contrast, enjoyed the advantage of the mimetic form (through the dominance of the polymorphic character) and at the same time some greater viability associated with the recessive. In the same way, the polymorphism in grouse locusts must, he thought, be due to a balance of selective actions, and he predicted that the homozygous dominants would be found to be at a selective disadvantage in respect of viability or rate of reproduction. He wrote at once to Nabours and was able to obtain figures from his breeding experiments and to check whether there had, in fact, been the predicted deficiency in numbers of homozygotes raised in culture. In January he visited Winge in Copenhagen and talked with him about *Lebistes*. The information from Nabours and from Winge confirmed Fisher's interpretation. Privately he was cock-a-hoop. Sending to Major Darwin a copy of his paper [C P 87] answering Haldane's challenge, he wrote on October 11, 1930:

> It must be nearly a year ago that I wrote you that Haldane had attacked dominance theory on the strength of dominance exhibited in grouse locusts and in the fish *Lebistes*. I thought at the time that his allies might betray him, and give unexpected support to the theory, as apparent exceptions are wont to do. So far they have come up to expectations nobly. Of course I need more data to make a complete case but I think this paper may serve to make sure that the necessary observations will be made.

Nabours' figures, analyzed in Fisher's paper [C P 87] showed, as predicted, a deficiency in the numbers of polymorphic homozygotes of the various types observed in culture as against the numbers expected; compared with heterozygotes, the average difference in viability amounted to about 7%. He concluded:

> The severest test of a theory is to build upon it a system of inferences, for if any rigorously logical inference is found to be untrue the theory fails. If on the contrary,

facts previously unsuspected are inferred from the theory, and found on trial to be true, the theory is undoubtedly strengthened.

His theory had passed the test.

The polymorphism in *Lebistes* occurs in sex-linked color factors in the Y chromosome of the male fish. Because there was some crossingover of genes from the X to the Y chromosome and vice versa, it had been possible to breed male fish with the color factors in both X and Y chromosomes. Since the homozygotes appeared no different from the usual heterozygotes, the color factors were called dominant; in the females, however, the colors were completely suppressed. From Winge Fisher learned that the X chromosome was in this case regarded as the "universal recessive" in wild specimens and that the genes for additional color patterns in the males were in nature more fully confined to the Y chromosome than they were found to be in the more "beautiful" domesticated breeds. Could Fisher's theory handle these facts? Again, the theory did its duty:

We have here a group of facts extremely suggestive of the interpretation which we should at once infer from the view that dominance has been determined by selective agencies, namely that the colour variants are advantageous in the male and disadvantageous in the female.

Having shown how, on this view, the observed facts would all be explained, he added: "It is difficult to imagine how the observed facts in *Lebistes* could more closely simulate those to be anticipated on the theory of selective modification of dominance." To the cases of polymorphism raised by Haldane, Fisher himself added that of garden snails in which the same features appeared, namely, a polymorphic species with a universal recessive and close linkage of the factors contributing to the various dominant polymorphisms.

The special case of polymorphism was, he found on consideration, peculiarly favorable to the evolution of dominance, for dominance modifiers should be particularly effective, both because the heterozygotes, being advantageous, formed a perceptible percentage of the wild population, with a consequent increase in the selective intensities, and because in the presence of close linkage the evolutionary modification of the tract of chromatin concerned would take place almost wholly in the heterozygous individuals. If, as Haldane suggested, the linkage were due to duplication, so much the better; one could infer that duplication, as such, had not been subjected to counterselection but, on the contrary, might be advantageous by assuring the linkage.

One of the points emerging from the discussion of polymorphism was that if the homozygous dominants were at a selective advantage in nature but at a disadvantage in respect of viability or fertility, it would be possible to make a direct experimental determination of the magnitude of the selective advantage of one color pattern over another in nature. Given the frequency of the different types in a natural population and an experimental determination of the relative viabilities, and perhaps fertilities, it should be possible to calculate the magnitude of the selective advantage not due to constitutional causes, which in nature favored the dominants over the recessives. With increasingly precise data on these points, therefore, it should be possible to put Fisher's interpretation of the dominance phenomena in grouse locusts, for instance, to a quantitative test.

Privately Fisher urged on Nabours the unparalleled opportunity he had in the genetic breeding program already established with his grouse locusts; by a combination of laboratory tests on relative viabilities with observation of the relative frequencies of the phenotypes in the field he might not only clear up the whole question of dominance but open up the investigation of selective intensities in nature. Nabours did, in fact, make collections in the field in 1933, and Fisher later analyzed the data [C P 167, 1939]. In culture it was found that heterozygotes enjoyed a *biological* advantage; for, in comparison, the average viability of the dominant homozygotes was 7% lower and that of recessive homozygote 5 – 6% lower. In their natural environment he found the ecological advantage of heterozygotes was much higher: double dominants suffered at least 40% elimination, presumably by selective predation; and statistical considerations suggested that homozygous dominants must suffer equally adverse selection. The *selective* advantage of the single dominant heterozygote over other dominant forms was surprising not only for its magnitude but for its exhibition of the sharp selective discrimination made by predators among individuals having very similar appearances.

Fisher adverted to these experimental possibilities in his paper in 1930 [C P 87] and himself began breeding snails at home in order to study the polymorphisms of background color and ring number of the shell of *Nemoralis hortensis*. It was at this time also that E. B. Ford conceived the long-term project he was to carry out along these lines, investigating selective intensities in nature in polymorphic species by the combination of laboratory breeding and field observation. This was to lead to remarkable results in the 1940s and 1950s, recorded in Ford's book, *Ecological Genetics* (1962), which was planned, even in 1928, to be the outcome after 25 years of this type of research.

Already in 1930, dominance theory was yielding prodigiously gratifying results. That year two young geneticists joined the fray, convinced by their own results that Fisher's theory worked. J. B. Hutchinson [37] offered an interpretation in terms of dominance modifying factors of a recently published collection of results concerning manifestations of the gene for polydactyly in various breed crosses in poultry. At the same time E. B. Ford [38] presented an entirely new and important theoretical point. Observing that in some cases one mutant gene had more than one somatic effect, he pointed out that the several characters were not equally dominant. Recessive characters were seen to be damaging to the organism, whereas dominant or semidominant characters controlled by the same gene suffered no perceptible disadvantage. The dominance of each character had apparently been modified independently in accordance with its selective advantage. Here, again, was decisive evidence springing from an unforeseen quarter; again, existing observations could be explained by the theory while they were certainly anomalous on any view that recessiveness of mutations was due to the inherent properties of the physiological system. The theory was becoming irresistible.

□ □ □

The explanation Fisher proposed in 1928 for the case of domestic poultry afforded another test of the general theory of dominance modification, for, with the tentative hypothesis, he had proposed a crucial experiment to check whether the dominance of certain characters had, in fact, been modified under domestication, in a manner consistent with his interpretation.

His proposal was to reverse the genetical history (as he envisaged it) of the domesticated fowl, inasmuch as it affected the dominant characters investigated (see Figure 11). Assuming these had become dominant by acquiring modifying genes peculiar to domestic flocks, if fowls carrying the dominant were crossed repeatedly back to wild jungle fowl, the domestic ancestry would be reduced in each generation by one half; thus one half of the genes acquired under domestication would be stripped away and replaced by their wild alleles. Within a few generations the characters visible in his heterozygous flocks would cease to be stable and well-defined and would become variable and lose their dominance to an appreciable extent. Chicks heterozygous for the "dominant," having in the fifth generation only 1/32 part domestic ancestry, would then be inbred. If the character were inherently dominant the progeny would segregate in the familiar Mendelian ratio, $3:1$ showing the dominant character. In that case dominance could not be attributed to modifying genes acquired under domestication. Fisher predicted instead that the homozygous form would be appreciably different from the heterozygote, giving the ratio $1:2:1$.

Having proposed this test, Fisher at once began to make preparations to carry it out himself. He inquired about possible sources of funds. He consulted experts in poultry genetics and with them drew up a list of seven "good and reliable" dominant characters. He obtained four Japanese Silky

SUGGESTED HISTORY

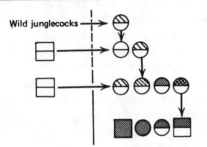

Wild junglecocks →

Domestic hen, heterozygous
for a novelty not recessive

Broods from which heterozygotes
for novelty were selected

Reaction of organism to novelty
is gradually enhanced by selection
of modifying factors

Domestic flock showing
novelty as highly dominant

EXPERIMENT: STAGE I   Reversing the history.

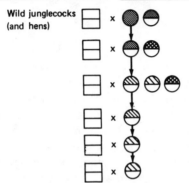

Wild junglecocks
(and hens)

Domestic breed hens (and cocks)

1929: 1st generation, 1/2 domestic.
Variable manifestation as some
modifying factors lost.

1930: 2nd generation, 1/4 domestic.
Variable and diminishing
manifestation

1931: 3rd generation, 1/8 domestic.

1932: 4th generation, 1/16 domestic.

1933: 5th generation, 1/32 domestic.

EXPERIMENT: STAGE II – Testing dominace in wild stock

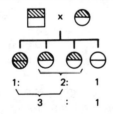

1934: Inbreed heterozygotes

Segregation in Mendelian ratios:
Either 1:2:1: ratio — No dominance:
homozygote distinct
Or   3:1   ratio — Novelties
retain original dominance: homozygotes
indistinguishable from heterozygotes

EXPERIMENT: STAGE III – Checking

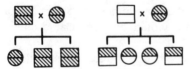

Prove identity of homozygotes and test
variability of homozygotes and
heterozygotes by breeding parallel
broods of each genotype

Figure 11.   Experimental program with poultry dominants.

pullets that displayed five of these characters: black internal pigment, feathered feet, rose comb, crest, and polydactyly. Two characters, dominant white and the sex-linked barred had to be introduced from other breeds. All were to be included in the experiment.

To obtain genuine wild jungle fowl, Fisher sought help of the London Zoological Society, and his first wild cock was actually borrowed from the London Zoo for the 1929 season. He rather hoped that the breeding might be arranged at the new Zoological Gardens then in preparation at Whipsnade, only 11 miles from Harpenden, but this hope came to nothing. The secretary of the Zoological Society, however, introduced him to J. Spedan Lewis whose private aviary at his country house contained game birds as well as a rare collection of owls. Spedan Lewis became much interested in the experiments; he set about finding jungle cocks of guaranteed ancestral purity, and he suggested the possibility that the experimental breeding might actually be done at his Leckford Farm, where he could supply equipment and expert breeding care. Though the latter plan failed, Spedan Lewis not only supplied the wildfowl for later seasons but made the experiments possible by his generous sponsorship throughout the 8 years during which they continued.

Thus it happened that in May 1929 Fisher collected his jungle cock from Spedan Lewis and brought it home in the company of his friend H. G. Thornton, the Rothamsted bacteriologist who had kindly offered to give the cock and the four Silky pullets a temporary home at his place near St. Albans. Meanwhile, Fisher negotiated with Rothamsted for the loan of a piece of land, arranged the purchase of hen houses and fencing, and hired a poultry keeper. At last the birds were transferred to the field at Rothamsted, and the experiment began to take shape.

From the progeny of the four Silkies, heterozygotes were obtained showing any or all of five characters. These were gradually to be segregated into lines displaying each dominant separately. In getting birds with several of the dominants together, the first object had been economy. In addition, however, the arrangement afforded the opportunity to observe linkage between the several genes and, in fact, one by-product of the experiments was the amendment of the chromosome mapping for the linked genes *crest, dominant white,* and *frizzle.*

By good fortune the Silky breed was a small fowl, suitable for breeding with the tiny jungle cocks; in contrast, there was some difficulty in introducing the two additional factors which occurred only in larger domestic breeds. Early in 1929 an attempt, which proved fruitless, was made to mate the wild cock with a barred Plymouth Rock hen that was evidently too big for mating to be successful; later in 1929, however, M. S. Pease in Cambridge presented a small *barred* pullet that was successfully mated and produced a brood that year. Similarly, broods for *dominant white* were eventually obtained in 1930 by reciprocal crosses using a small White Leghorn female and a male of the same breed which was sexually mature at less than 6 months old. The seven lines of dominants and the line of wild jungle fowl were thus established.

As the experiment continued, it became apparent that the wildfowl were not doing so well as the more domestic stock; it was hardly possible to keep up the numbers to the desired level. Fisher anxiously consulted Spedan Lewis about precautions in rearing and was advised that jungle fowl were notoriously delicate and difficult to breed. The delicacy of the wild stock meant that broods were small and that many birds successfully hatched did not survive to maturity. There was a large proportion of infertile eggs, practically all the first broods failing completely, and later broods had hardly time to make the necessary growth before winter. Since in Fisher's experiments one half of every generation would in any case not show the character under study, and thus had no experimental interest, the practical problem was grave. As the proportion of wild ancestry increased, the stock became ever more susceptible to the failures experienced with pure wild stock, and the danger of premature termination in every line increased. In fact, one line was thought to have been lost, and several of the others depended on the survival of single birds.

To the practical problem was added the technical difficulty that if birds survived in proportion to their domestic ancestry, the genetic constitution of the survivors would not be what had been planned, for an excess of domestic factors would be preserved in the lines. The replacement by wild of domestic genetic background would be retarded. When the progeny were inbred in the sixth generation to produce homozygous birds, the character of interest might still be obscured by modifying genes originating in the domestic breeds. Despite this, it might be necessary, in view of the feebleness of the wild lines, to reduce the program from five to four generations bred to wild stock before inbreeding. This reduction was actually made in three of the seven lines, with the result that, in particular, a gene modifying *feathered feet* was evidently segregating in that line until the end of the experiment.

□ □ □

Already in the first two generations, however, the heterozygous birds were showing the sort of changes in appearance Fisher had predicted: a variable manifestation of the character and decreasing dominance, that is, a less marked display. In the first generation of crested birds, for instance, the form of the crest was changed because of the change in the structure of the feathers, being now a tuft of somewhat erected and elongated feathers growing on the fore part of the skull, not at all resembling the little mop of loose feathers of the Silky breed. In the second generation there was noticeable variation in the length and erectness of the feathers. Later the length tended, if anything to decrease, but there were too few birds reared in later generations to detect any trend in the average. In the third generation there were 15 birds, of which only 4 were crested; in the fourth generation, 7, of which a single cock was crested. In the fifth generation, of three chicks, two survived and, by good fortune, one was a crested cock and the other a crested hen. These two were mated together in 1934.

In the *barred* line, the second generation showed exceedingly varied progeny. Characteristically, this gene is manifested by a transverse stripe across all the feathers. Yet one of the heterozygotes used for breeding appeared normal in all respects save for having a brown breast; his progeny proved that he carried the *barred* factor. The low viability and backwardness of the *barred* males in the first 3 years was very noticeable, and only one of the brood of 1932 attained adult plumage and was fit to mate in 1933. Good broods, however, were obtained, and in the following year also one cock was again fit to breed and was mated to five *barred* females with a view to producing homozygous cocks in 1934.

The crucial year was to be 1934, but one tantalizing preview occurred in 1933. *Feathered feet* had proved so quickly to be recessive that it was thought desirable to obtain homozygotes without delay, although the extent of feathering was still very variable. After only 3 years, some of the heterozygotes were being correctly classified by the minimal manifestation of a single feather on one foot at birth (and soon after birth the feathers were lost and the heterozygotes appeared normal); indeed, judging by the inequality of numbers classified as normal and feathered, some heterozygous birds were already being classified as normal. It was decided, therefore, that the fourth generation should be inbred in 1933. Fisher had predicted in 1928 that the homozygotes on reconstitution against a largely wild genetic background would appear appreciably different from the heterozygotes. From this cross, therefore, he looked for homozygotes with distinctly well-feathered feet. The outcome was beyond expectation. The homozygotes were immediately distinguishable from the heterozygotes: they had abundant feathering over an area that, instead of terminating at the base of the fourth toe, extended across the back of the foot up to the second or third joint; in addition, they showed a completely unexpected feature—a skeletal deformity involving the loss of one bone and a shortening of the end bone of the long toe. This deformity quickly proved a disadvantage in life: one bird died in the shell, three within a few weeks of hatching, one in the third month, and the survivor, always lame, broke her leg and had to be killed that winter. It was, therefore, not possible to prove the homozygosity of these birds by breeding from them; the experiment had to be repeated in following years. Nevertheless, it seemed certain that, far from being a harmless or dominant factor in the wild stock, *feathered feet* was a deleterious mutant almost completely recessive.

While, under domestication, the feathering of the feet had been enhanced and rendered more dominant by selection, the crippling effect of the skeletal changes had been suppressed even in the homozygote. This latter was precisely the sort of ultimate modification Fisher had predicted in 1928, in the case of a homozygous mutant following in the footsteps of the heterozygote and being rendered normal. Only the feathering of the feet betrayed the existence in the domestic breeds of a gene that, if unmodified, caused a crippling brachydactyly, but, when the genes modifying the homozygote were stripped away and replaced by their wild alleles, the original effect of the homozygous mutant was exposed. The later stage of the modification of the genetic effect on the skeleton had been reversed by the experimental breeding. Further breeding in the same line showed a reversal also of the earlier stage, the modification of the heterozygote which had rendered it recessive. A heterozygote was produced that exhibited skeletal changes of the same kind, though less extreme than the homozygous manifestation. Evidently, this heterozygote had been stripped of the modifying gene(s) which under domestication had rendered the brachydactyly recessive, and had been exposed in its primitive form as being intermediate between the two homozygotes.

□ □ □

Tension mounted in the spring of 1934 as the crucial year of the experiment was reached. The unexpected and important skeletal difference between the homozygous and heterozygous *feathered feet,* found in 1933 after only four generations crossed to the wild, set the stage for the following season. Heterozygotes of the fifth generation were to be mated, and there would be homozygous chicks in most lines. Fisher had predicted that one-third of those displaying the character, being homozygous, would be appreciably different from the heterozygotes. If the homozygotes were distinguishable, Fisher had proved his point. But no one knew whether the character would be sufficiently different to carry conviction, or whether the season would favor the tenuous breeding program. Would there be chicks? Would they live long enough to be classified?

Fisher, now working in London, visited the chicken pens at Rothamsted regularly every Saturday morning in the company of his children. As the season advanced he would inspect the embryonic chicks that had not survived to hatching. The fierce jungle cocks strutted with raised hackles against the fence of their pen making aggressive gestures, and the Fisher children eyed them askance, remembering a pretty-looking jet-black silky-feathered cock with a bright, large rose comb which their father had given them for the poultry yard at home. *He* had been sired by one of these jungle cocks and, inheriting the temper and the fighting spurs of the paternal line, had flown at those who dared to feed him with jabbing beak and plunging spurs. More than once he drew blood.

The first broods began to hatch, and Fisher visited the pens more often. He was writing to interested parties, urging them to visit and witness for themselves the outcome that summer: to Spedan Lewis, whose support had made the experiments possible, to Huxley and Ford at Oxford, to Haldane in London, to the American geneticist, T. H. Morgan, who was visiting England that summer, to the poultry geneticists at Cambridge, and to the representative of the Royal Society Grants Committee, Professor MacBride, who ought not only to learn by correspondence what the experiment was about but also to see for himself what had been done for their money. Expeditions were made to Harpenden to tour the chicken pens, to see the exhibition of transparencies (showing bone formation) and stuffed birds (showing black internal pigment) which had been preserved in previous years; the parties returned to Milton Lodge for lunch and more talk followed through the afternoon.

The first brood of *crested* stock consisted of three apparently normal birds. It occasioned no excitement, for the crests were not expected to develop for some months after birth. Then, at the end of April, the second brood hatched and on May 1 Fisher wrote to MacBride:

This last week a beautiful case has turned up. I had only two crested birds from last year and had not much hope of getting homozygotes this year but eleven hatched out in one brood a few days ago and the homozygotes, three of them, are immediately recognizable at hatching by a marked cerebral hernia which altogether alters the form of the head.

Cerebral hernia has long been recognized as a recessive and Davenport states that it is only found in crested breeds. There seems, however, to have been no idea that it is only a manifestation of the Crested mutation when homozygous. Once this is recognized, it is clear that the mutant Crested is more nearly recessive than dominant, for the homozygotes can be recognized at hatching whereas it will be four or five months before the crests appear.

I am rather afraid of the chicks with hernia dying, especially when they are old enough to be frightened when caught, so, if you could manage it, it might be worth while making your visit early, coming again later in the year if you could spare a few days. I don't know when Morgan is coming . . .

Fisher's luck held long enough to make the story convincing. Two further broods of *crested* birds hatched that season, and of 26 chicks born, 5 were herniated, close to the fraction of homozygotes (one-quarter) expected from the cross. Evidently the primitive reaction to the homozygous *crested* mutant, as observed in wild stock, was a bursting of the brain case between the anterior portions of the frontal bones, between which the brain protruded, forcing them apart. The heads of chicks that died young were preserved as transparencies and living chicks were photographed to show the phenomenon. Only one herniated chick died before a crest could become apparent, another died already crested at 82 days, still without having developed any bony roof to the skull; two, both crested, were killed by a weasle in September, and one hen, with an exceptionally large crest, survived to prove her homozygosity in the following season.

Fisher, reporting his results later that year [C P 135], concluded:

The mutation should not be described either as dominant or recessive, since the heterozygote is clearly distinguishable from both homozygotes. If the terms dominant and recessive are used at all in such a case they should only be applied to particular aspects of the mutant gene's effects. In the skulls, I have observed no difference between the heterozygous crested and the uncrested birds. Thus it might be said that in the rupture of the brain-case the mutant is recessive, being analogous in this respect to the more harmful effects of most other mutations; but that in the crest itself it is semidominant. . . . We may thus suppose that relatively severe counter-selection against cerebral hernia in the wild population is responsible for its recessiveness, while the mere erection and elongation of the feathers of the head have been less, if at all, harmful.

The different degree of dominance of the innocuous crest and the dangerous hernia, both due to the same gene, illustrated again the point E. B. Ford [38] had made in 1930.

One could not hope that other characters would prove as sensational as *crest,* especially because no one knew what form the distinction in the homozygote would take. This was particularly important in respect of the factors for pigmentation. Nevertheless, in 1934, not only factors affecting structure but the sex-linked color factor, *barred,* gave unambiguous results. The heterozygous *barred* males had done very badly and the fifth generation were illdeveloped; it was feared that they, and still more their homozygous *barred* progeny, would be unfit to breed. The first brood in 1934, of three chicks, however, contained two which were unmistakably homozygotes for *barred.* They were more different from the heterozygote than the heterozygote was from the wild type, lighter than either, with large white patches on the sides of the head and neck including the ears, with light wings, and with the dark lateral stripes on the body quite obliterated. In contrast, the heterozygotes had but a pale spot in the middle of the dark band which normally runs from the forehead down the back of the neck. In 1934 one could conclude that although the heterozygote was throughout life distinguishable from the wild type, the mutant was certainly not dominant but more nearly recessive against the genetic background of the wild jungle fowl.

By the end of the 1934 season Fisher had publishable results on three genes and had a promise of success also with *feathered feet.* In June 1934 he spoke to the Linnaean Society about the results with *polydactyly* [C P 116]. In November he reported to the Royal Society [C P 135] the results to date with *crest, barred,* and *polydactyly* and concluded that "any dominance, therefore, which is shown by them in breed crosses must be due to modification during the period of domestication." Later that year he made a preliminary report on crest and hernia to the Genetical Society.

□ □ □

These reports had a double task: first to persuade his audience to contemplate what was still an alien theoretical construction and then to present his results which in a particular case confirmed it. Fisher knew how difficult it would be to penetrate the ramparts of what was generally assumed to be known to be otherwise: He wanted to appeal, like Cromwell to his colleagues, "I beseech you, in the bowels of Christ, think it possible you may be mistaken."

He did his best to persuade his audience at the Linnaean Society [C P 116], as scientists, at least to give him a hearing:

[Mutations] are regarded as the actual causes of evolutionary change—but are they? That is the question I want to discuss, and to answering which I hope such facts as I have to give may contribute.

The question deserves our most careful consideration, so careful that I must ask my audience to assist me in a very difficult undertaking. It is not that I mean to put forward any complicated argument—on the contrary, all I shall do is to describe a simple, though prolonged experiment, and to exhibit some of the results. The really difficult thing that I ask is to avoid *assuming,* at least during our discussion, either that mutations do, or that they do not, govern evolutionary change—to maintain an open mind on this point while the evidence is being considered, and so to give the evidence a fair chance of carrying whatever conviction it may be entitled to.

In the polydactylous line, characterized by an extra hallux on the foot, there was very variable manifestation in the second generation, and some of the birds with normal feet, which were therefore not used for breeding in the polydactylous line but elsewhere in the experiment, actually produced polydactylous chicks, thus betraying recessive behavior of the "dominant." It was true that this character had not enjoyed the same unquestioned reputation as a dominant as had *crest,* for example; as early as 1904, Bateson had reported instances of five-toed birds bred from four-toed parents. Nevertheless, *polydactyly* had habitually been termed a dominant and had been entered on Fisher's list in 1928 as in this respect reputed to be "good and reliable." Then, in 1929, Punnett and Pease [39] had summarized the numerous progenies they and previous workers had obtained with *polydactyly* in different breed crosses, offering their own interpretation. To Fisher it seemed clear from these data that dominance had been much modified in the course of the formation of different breeds, for the mutant was shown to have behaved variously as a complete dominant or as a recessive or in an arbitrary manner which Hutchinson [37, 1931] had interpreted as being due to the independent segregation of the *polydactyly* gene and the genes modifying its dominance. This was also Fisher's view.

The diversity of display in Fisher's heterozygous stocks suggested that the unmodified homozygote should have a still more marked manifestation than the usual heterozygote. In the heterozygotes, one foot was often normal, the other (usually the left) having an extra claw, a fused extra bone and claw, or one or two separate extra bones, the maximum and commonest expression being on each foot an extra metatarsal bone bearing a hallux with two phalanges.

On inbreeding in this line, in 1934, of 31 chicks, 11 were normal and 20 polydactylous, and 6 of these 20 showed exceptional manifestations such as had not appeared in previous generations. These were the homozygotes. Characteristically they had four additional bones on each foot, that is, three instead of two extra phalanges, together with the extra metatarsal which was elongated backward in the homozygote; the individual bones, too, were larger than in the heterozygotes. No doubtful or intermediate bird appeared in the limited material of 1934: the homozygotes were distinct.

In addition to exhibits of transparencies and diagrams to illustrate his verbal classification of the variety of the bone structures, Fisher presented a graph of the frequency of occurrence of different numbers of bones for different genotypes; this, apart from any other distinctions that might be made, displayed a simple and absolute numerical difference between heterozygous and homozygous *polydactyly*. The heterozygotes had any number from the normal four bones to a maximum of ten in the two feet together; the homozygotes had between twelve and fourteen. Fisher concluded:

The case is, therefore, a typical one of absence of dominance. The heterozygote is intermediate, and usually distinguishable from both homozygotes, though occasionally it resembles the normal. Such slight departure from complete intermediacy as exists is in the direction of recessiveness.

In its dominance, therefore, this factor bears out entirely the anticipations with which the experiment was planned, as do also the factors for Crest and Hernia and for Barred.

The discussion [40] that followed Fisher's presentation gives a glimpse for posterity of that sturdy conservatism from which Fisher's introductory remarks had slid like water off a duck's back. Professor E. W. MacBride, embryologist:

Both Dr. Fisher and he recognized that whatever mutations might be they had no significance for evolution. But he differed entirely from Dr. Fisher in regarding 'natural selection' as an alternative cause for evolution. Natural selection, regarded in that light, was not only a truism but a dishonest truism . . . The curious and interesting mutations described by Dr. Fisher, instead of being ascribed to the action of mysterious 'genes' —and of the making of 'genes' there was no end, —could all be accounted for as results of one 'weakness of growth' of varying strength.

He was neither a Mendelian nor a believer in natural selection, and he offered a purely qualitative embryological interpretation of the facts. Professor R. R. Ruggles Gates, in contrast, welcomed the demonstration of a single Mendelian factor for polydactyly, with incomplete dominance. But

In his own view the great majority of specific differences were non-adaptational and the bulk of the raw material of evolution, on which natural selection might or might not act, was supplied by gene mutations—but Professor Fisher does not appear to distinguish between adaptational and non-adaptational characters. . . .

Most geneticists would have shared the doubts of Ruggles Gates about the efficacy of natural selection to determine specific differences, let alone to produce, as a minute byproduct of its activity, such mutual adaptations within the genetic complex as Fisher proposed.

Mr. Michael Pease, poultry geneticist from Cambridge, had been consulted early as to which characters in domestic poultry were "good and reliable" dominants. He had provided a bird for the experiment in 1929. He had visited Harpenden at least in 1930, 1932, and, at Fisher's urgent invitation, in 1934, that he might see the experiments himself. Now, he

much regretted that he had been unable to grasp what the exhibition in front of him was supposed to demonstrate . . . He had expected to see a 3 : 1 segregation in the one case and a 1 : 2 : 1 in the other. But the demonstration in front of them showed nothing of the sort; it lacked entirely a standard of comparison. . . .

For ordinary purposes it might be sufficient to say that a character was dominant; but it did not necessarily follow that dominance was 'perfect.' There were not many dominant genes in poultry for which it could be said with confidence that the homozygote could not be distinguished from the heterozygote in every circumstance and at every stage of the bird's growth.

A fog had descended on the facts which Pease, along with other poultry geneticists, had asserted with assurance in 1929: the once good and reliable "dominants" had lost their definition since they had come to demonstrate an unacceptable thesis. Fisher was tempted to point out the change of attitude, before discussing this criticism.

In reply to Mr. Pease, he would like to say that the particular purpose and method of the experiment had been in print, and personally known to Mr. Pease, from its inception six years before. Until the result of the tests was known he had not suggested that the comparison was in the least unsatisfactory.

However dubious his hearers might be about Mendelism or the efficacy of natural or human selection or the part played by modifying genes in the somatic manifestation of a gene, Fisher's results took a great deal of explaining away. The logic of the program, and the realization, after five generations, of the initial predictions, was irrefutable.

The experimental breeding was planned to continue in 1935 in order to confirm that the birds identified as homozygous were indeed so, and in the hope of producing parallel broods consisting exclusively of homozygotes and heterozygotes, in which the differences between these genotypes and the variability of each could be observed. But the 1935 season was disastrous: in over 50 sittings of eggs, only 15 chickens were raised. The experiment was extended to the 1936 season, which proved hardly better. Fisher did not succeed in breeding enough birds in any line to produce the parallel broods. The poor breeding seasons and the high proportion of wild ancestry in the experimental stock cooperated with breeding failures that were apparently due to some of the genes under study. In the sex-linked *barred* line, poor fertility was certainly associated with the gene for *barred*. There was a deficiency of homozygotes, only 5 of 37 chicks being *barred* where a ratio of $1:4$ was expected, and the single survivor from these broods only fathered two chicks. The homozygous *dominant white* hens produced no progeny at all.

Fisher referred to *dominant white* as *pile* [*C P* 161, 1938], observing that it was not dominant and that, acting on the wild constitution, it did not produce a white bird but an appearance that the breeders referred to as pile. The large feathers were white, but the wings of males were deep chestnut, while yellow and brown coloration was distributed on the heads and necks, and especially the breasts of females.

Even in the good season of 1934 this cross gave, in five broods, only 21 chickens, of which only two males and two females attained adult plumage. One male and one female were judged homozygous, for they had noticeably less yellow on head and neck than the heterozygotes. Of these two birds, the cock lived long enough to father one brood but died before the hatching of the six *pile* chickens which proved him to have been homozygous. Only at this stage was it realized that the homozygotes characteristically lack the speckling black dots on the wings and tail shown by the heterozygotes. Had this indication been noted, the cock should have been unambiguously distinct from heterozygotes. The observation was not made, however, and could not be repeated for the hen, and another homozygous hen produced later, failed to give any chicks. Nevertheless, the diminished pigmentation on the feathers and, in particular, the lack of the speckling distinguished the homozygotes clearly from the heterozygotes.

With the largely quantitative differences of the color factors it was a real advantage to observe a definite qualitative difference between homozygotes and heterozygotes, like the black speckling on the *pile* fowls. In the line for *black internal pigment* such a criterion was also observed, again almost too late. The intercross for *black internal pigment* had been made in 1933. In that year there were no wild pullets available because the wild line had bred so badly; consequently, the two surviving pigmented birds were mated together. Of the chicks only a single pigmented pair survived. The male was later proved to be homozygous, but the appearance of his down feathers as a chick was not observed. One of his chicks, however, was observed to be different in this respect, with extended pigmentation on the down of the head. Further breeding in 1936 produced 14 of 46 chicks with the darker down coloration and black internal pigmentation. That this appearance, segregating when the line was inbred and appearing only in this line, was, in fact, due to the homozygous mutant was confirmed by the observation that the feet, which in the heterozygote appeared green in the young chick, in these cases were much darkened so as to appear black rather than green. Several birds showing the down character were proved on breeding to be homozygous. Thus it appeared that *black internal pigment,* like the other genes, had an enhanced expression in the homozygote.

Rose comb was the only doubtful case. With this factor the variability in the heterozygote was difficult to gauge: in the males its manifestation was exceedingly variable; in the females and chicks its lack of development made any study of variability impossible. This factor, however, proved to be one of the most interesting in the experiment, for during the preparation of heads for the examination of the hernia, it was noticed that birds with rose combs had not only the comb but the underlying frontal bone widened between the orbits. The width in adult males was increased on the average by 25%. Fisher had not birds enough himself to measure any difference that occurred between heterozygotes and homozygotes; he suggested, however, that such measurements could be made on existent flocks of any breed in which both single and rose combs occurred. Since human selection had been made for the comb shape and not for the bony structure, one could, from the measurements, establish the unmodified dominance relations of the gene.

The observation of differences in skull width was actually made by A. E. Brandt, who spent the year 1935–1936 with Fisher, on a Rockefeller Experience Fellowship, and at the weekends worked with him on the chicken experiments. With *crested* and *white* birds also Brandt found ways to predict upon examination at 10 or 11 weeks, which birds would develop into heterozygotes. "In each case I did not reveal the basis of my prediction until we had tried it out and demonstrated better than 90% success."

For Fisher the particular interest of the results with *rose comb* was evolutionary. Reporting these results in 1938 [*C P* 161], he remarked:

It is a matter of some general interest that, of the four structural characters tested in this experiment, all, without exception, should affect the development of the skeleton. In my previous communication it was shown, as further breeding has confirmed, that the gene responsible for Crest, causes, when homozygous, an opening of the frontal bones leading to cerebal hernia. In the present paper it has appeared that the gene responsible for Feathered Feet causes, when homozygous, a pronounced brachydactyly, with the suppression of the terminal phalanx of the outermost toe. This is also partially brought about, in some genetic combinations, in the heterozygote. The fourth structural character studied has been polydactyly, the skeletal nature of which has never been in doubt. In view of these examples the remarkable conservatism of the skeleton in phylogeny should certainly not be ascribed to any lack of abundance of mutations affecting its structure, but rather to the persistent success of selection in preventing such mutations, as must constantly be occurring, from having any evolutionary effects, save in favourable and most exceptional circumstances.

Every one of the characters studied gave interesting and suggestive results. Apart from the inconclusive evidence for *rose comb,* by 1937 every character showed lack of dominance against the wild background, and in only one, that for *pile,* was the mutant even so dominant as to have a heterozygote appearing more like the homozygote than it was like the wild type. The most harmful manifestations of the mutants had been suppressed in the domestic breeds, rendering them, in this respect at least, completely recessive (as in the case of hernia) or even normal in the homozygote (as with the brachydactyly of feathered feet). At the same time the more innocuous manifestations of feathered feet, rose comb or crest, had under domestication acquired modifying genes that rendered the appearance to a considerable degree dominant. It seemed very probable that every one of the breed characters studied originated in a deleterious mutation: against the genetic background of the wild jungle fowl, four manifested themselves in skeletal changes, three of which were obviously disadvantageous, and of the three factors affecting pigmentation, *barred* and probably *pile* also was associated with infertility.

In 1938 [*C P* 161], when Fisher's final report on poultry was published, he could point to a case of the *reversal* of dominance which completed the story.

L. C. Dunn and W. Landauer [41] had reported results with the *rumpless* factor in poultry for which they offered their own interpretation but which to Fisher seemed a conclusive case of the modification of dominance through selection. Initially, the *rumpless* character was found to be completely dominant in the breed in which it occurred. On outcrossing into a different breed, however, the character had been reduced to an intermediate status; then, on inbreeding the hybrids and selecting the less rumpless manifestation, the character had been reduced to complete recessiveness, and finally even the homozygous mutant was showing signs of becoming modified towards the normal condition.

□ □ □

During the years of the poultry experiments, the theory of the evolution of dominance was being, in turn, denied, criticized, belittled, and finally ignored. As time went on positions shifted and the original problem was rather lost sight of. In 1932 Haldane, accepting the proposition that dominance had evolved under natural selection, proposed that selection took place not among the characters produced by multiple genes but among multiple allelomorphs. Fisher welcomed this as an extension of his theory, such as any theory might expect to receive as its sphere of application was explored; he agreed that selection among multiple allelomorphs would in certain cases make a significant contribution to the evolution of dominance. He accepted Haldane's suggestion as a supplement to his own, however, not as an alternative, for, generally speaking, selection of multiple genes must be the more powerful agency. Haldane, nevertheless, presented his theory as an alternative to Fisher's; it was accepted as such and it was generally preferred because direct selection did not require the efficacy of the very small selective intensities associated with selection of modifying genes.

Among geneticists today the concept of the evolution of dominance is almost unknown. Nevertheless, it has played a significant part in transforming scientific opinion in respect of evolution by natural selection. It has opened up the whole field of ecological genetics by which the efficacy of natural selection has been fully demonstrated.

In the science of genetics proper, the theory has been no less influential, for it has been the means by which the much wider principle of multifactorial action has been recognized. Thus, in the second edition of *The Genetical Theory of Natural Selection* (1958), Fisher introduced the chapter on the evolution of dominance in a changed manner,

The principle of this chapter was novel enough in 1930 to require a good deal of explanation or apology. In the light of modern knowledge, however, it is seen to be one only, but one of the most interesting, of the aspects of a much wider principle, namely that the effects by which any gene substitution is recognized depend on the results of interactions with, possibly, all other ingredients of the germ plasm and so may be altered or abolished by changes in these latter. Although this wider principle is now recognized, yet it was the need to give an evolutionary interpretation of the genetic facts regarding dominance that was the first occasion of its recognition.

Recognition of this principle has liberated genetical opinion to consider an expanded universe containing very numerous interacting genes. That universe today shows examples not only of genetically controlled linkage and crossover and complex systems controlled by a switch mechanism but also schemes of feed-forward and feed-backward control of biochemical processes, due to multifactorial genetic action, and of accurate or (deliberately) inaccurate transcription of the genetic code.

Not least important, the exploration of dominance modification has demonstrated the value of the methodology of research into inheritance, to which Fisher attributed Mendel's success and which he felt was Mendel's primary contribution to the subject, a methodology based on statistical logic. It was from the logic of Mendelism that Fisher inferred that genes must be very numerous indeed, that their interactions must be profoundly important; that systems of genes must have evolved, under the control of perhaps a single gene capable of switching the whole appearance of an organism to one or another of its polymorphic forms, and that in such systems the genes involved would commonly be closely linked. It was from statistical considerations that he deduced the probable history and relative rates of change at different stages in the modification of the somatic manifestation of the gene; from statistical arguments he predicted the *biological* advantage of the heterozygotes for the polymorphic color factors in grouse locusts, and he similarly predicted and found the *selective* advantage of the heterozygous polymorphic forms over the homozygotes. Genetical and evolutionary inferences came to the experimental proof when they were presented as part of the theory of the evolution of dominance.

9
The Role of
a Statistician

In preceding chapters it has been convenient to consider separately the several aspects of Fisher's work in such a manner as might exhibit the continuity in the development of his thought touching various branches of scientific research. This treatment of our subject has the further justification that Fisher's reputation grew during the 1920s and his influence became felt through different parts of his work, appealing to rather different specialists. Mathematicians were impressed by the advances made in distribution theory and estimation, and it was the substantial body of such work that won Fisher election to the Fellowship of the Royal Society, as a mathematician, in 1929. Research scientists came to the realization, more slowly but with growing enthusiasm, that his statistical applications offered them what they greatly needed, ways of dealing with small samples and of planning experiments with assurance that the conclusions drawn would be valid. Geneticists appreciated his incursions in their subject in varying degree, as they had more or less sympathy with his evolutionary interpretation of Mendelism and faith in his statistical handling of the subject. In 1928, in recognition of his work in population genetics, he was awarded the Weldon Memorial Medal, an award presented triennially for work in quantifying phenomena of evolution and heredity.

As his reputation grew, he used his influence to draw together his various audiences. His own philosophy did not admit the separate treatment of mathematical from agricultural or genetical ideas or of theoretical from practical problems, and he deplored the academic divorce of mathematical from biological studies. Mathematics simply provided a rigorously logical way of thinking about the real world, especially its teeming biological problems. Experimental facts and mathematical forms were the warp and woof of the philosophy of experimentation which he created, and they were inseparably interwoven in a single fabric for the making of scientific inferences. He was concerned that this philosophy should be appreciated in the education of statisticians and research scientists.

In this chapter, considering Fisher's middle years when his views on the nature and role of statistics became explicit, it seems appropriate to consider the interplay between mathematical and biological aspects of his work.

□ □ □

Superficially, his life appeared a medley of disparate activities; his undertakings in 1929 illustrate how his increasing prestige simply increased their number and deepened his commitment in every phase of his life. At home, at the beginning of that year, he had perhaps 30 or 40 cages of mice, and he was arranging to set up experiments also with snails and chickens. In the evenings he was busy dictating his book on natural selection and by February had written half the chapters. At the end of March he was scouring the village in search of a woman capable of taking charge at Milton Lodge during his wife's sixth confinement; he succeeded only on the eve of the birth, and being ejected from bed, he spent that night writing the summaries of chapters of the book that were already completed. In June he sent the complete manuscript to the publishers. The same month he learned that his sister, Phyllis, who had lived in his home during the previous 9 years was dying of disseminated cancer; she passed away 3 weeks later, at the age of 46.

At Rothamsted he had his regular work and his researches. In addition, he had a larger number of voluntary workers visiting the statistical laboratory than in any previous year and more consultative work both at Rothamsted and elsewhere. Inquiries reached him from all over the world about experimentation, especially about manurial and breeding trials; correspondence from cotton experimenters alone came that year from Gezira in the Sudan, Trinidad, Madras, and Tashkent.

His public roles were as multifarious as those of a quick-change artist. In May he was stirring the Registrar General to reform the census and registration forms; the next day he traveled to Zurich to speak at a Public Health congress about the possible uses of information on public record, if supplemented by certain family data, in the elucidation of human genetics and the guidance of eugenic policies. In June, he was re-elected to the Royal Statistical Society—Major Darwin urged him, with a solicitude that could not be withstood, to make his election to the Royal Society the occasion of peacemaking after his unnatural estrangement from the R.S.S., so that the old quarrel might at last be forgotten, and Fisher was not unwilling to be persuaded by his old friend. In July he played host to the Genetical Society when it met at Rothamsted; there was little genetical work at Rothamsted, but Fisher was attempting to interest both Rothamsted and the Genetical Society in work on natural selection and was interesting himself in statistical methods applicable to breeding work. J. B. Hutchinson, working that summer in the statistical laboratory, was doing notable work on the application of the method of maximum likelihood to the estimation of linkage. At the same time, Fisher was secretary of the British Association Tables Committee and was much concerned that the tables (published in 1931) should be of the most useful sort for men without desk calculators, both accurate and easy to employ. As a member of the Marine Biological Society, that year he analysed data on sex differences of gonad development in herrings. He attended a conference of empire meterologists to whom his recent publications on multiple correlations and correlation tables in comparative climatology [C P 61, and C P 65 with T. N. Hoblyn] were of particular interest. He was a business secretary of the Eugenics Education Society and represented that body in the discussions preliminary to the formation of an international population union and, at the Rome meeting on population, spoke on the differential birthrate. He touched the economics of child-rearing in an article on "The Overproduction of Food" [C P 82], and he was informally giving statistical advice to members of the Society for Psychic Research (indeed, his continuing assistance was rewarded, rather disconcertingly, some years later by his election, sceptic though he was, to the fellowship of that society).

□ □ □

At this stage of his career, Fisher's election to the Fellowship of the Royal Society, on the first occasion when his name was put forward as a candidate, came to him personally with all the happy sensations that scientific recognition of the highest sort could bring to a man who was ambitious not only for himself but for the subject with which he was identified. What it meant to him personally in encouragement and stimulus, one senses from his acknowledgement of Major Darwin's congratulations:

I knew you would be glad and your pleasure is as good to me almost as though my own father were still living. He lived long enough to see me fail in two occupations and to hear me *say* I was on my feet in research. That is nine years ago, and it has gone well. . . . Do you remember the help you gave me in getting my first Edinburgh paper accepted and introducing me to Horace Brown?

As a Fellow of the Royal Society, Fisher never forgot how much his election had meant to him, or lost consciousness of what it must mean to any man young enough to profit by it in his career. He believed election might sometimes open professional opportunities to the young scientist whose reputation was not yet fully established and most certainly would give him moral support and encouragement that would be especially welcome if his work had been (as the best work must be) original enough to challenge tradition. Therefore, he maintained that it was a mistake for the society to elect a large proportion of older men: men over 50, whose reputations were secure, whose original work was nearly completed, to whom election would have little practical value. Instead of setting a seal of approval (Forever Rest Satisfied) on what had been achieved, election to the fellowship should be an investment in promising talent and so actively promote the titular goal of the Royal Society of London for Improving Natural Knowledge.

Whenever the opportunity arose within the society, he argued for the election of the younger candidates and for a policy aimed at electing most Fellows before they reached the age of 45. When it was objected that this might result in some elections that would later prove to be mistakes, he replied that one might observe among the current Fellowship that election at a more advanced age had not obviated such mistakes.

In the early 1930s, when the Royal Society felt the need to increase its numbers and was considering how this might be done without lowering the honor of Fellowship, Fisher suggested establishing a class of Associate Fellows. So long as the Associates were as numerous as the full Fellows, he believed the honor of election to Associate Fellowship would be very considerable and the honor of election to full Fellowship would be enhanced. The plan would, incidentally, augment the funds of the Royal Society available for its publications. He suggested sponsorship of the more expensive forms of publication: "I think coloured plates and mathematical symbols are the most expensive, and prefer the former; but tables of original data have, in my mind, a special claim." Above all, the society might in this way give immediate recognition to some 500 young scientists whose work showed originality, and the prospect of such recognition to others long before they could hope for election to the Fellowship in the ordinary way. When the Royal Society decided simply to increase the number of the Fellowship, Fisher turned his hand to calculating the increase in the number of the annual elections appropriate to maintain the Fellowship at the increased number decided upon.

Fisher was gratified that he had been elected as a mathematician. He had not "stuck to the ropes" and proved himself in an orthodox fashion; the scientific standing of the subject of statistics, as he had shaped it, was enhanced by his election and he gained a voice in representing its importance to scientists in all fields, within the Royal Society, and in guiding its future.

□ □ □

He had long thought that conventional instruction in mathematics was inappropriate as training for mathematical statisticians, for although statistics uses the deductive arguments of mathematics, it relies immediately on inductive reasoning, a mode of thought of which mathematicians, as such, have little experience. As early as 1923, as a fellow of Caius College, he had broached with the Master the possibility of establishing a Chair of Mathematical Statistics at Cambridge. In 1926, when he was awarded his doctorate, he was urging that such a chair ought to be established immediately. Cambridge vacillated for years and, at last, in 1931 created a Lectureship in Mathematical Statistics, attached to the School of Agriculture. To this post they appointed John Wishart, who had been for 3 years Fisher's assistant at Rothamsted. There the matter rested for a quarter of a century. Evidently, statistics was considered a second-class academic discipline. Yet Fisher, foreseeing in the 1920s its importance as an academic subject, was already concerned not only that it should be taught but that it should be taught in its own terms, not as a branch of mathematics but as a complementary discipline.

Receiving congratulations on his election to the Royal Society, he expressed this concern to his former mathematics master at Harrow, C. H. P. Mayo, then in retirement in Devon:

It is great luck to be recognized by mathematicians because academically speaking statistics has scarcely been discovered yet, though the demand for it in research is becoming a little felt. I wish I knew what sort of organisation would suit the subject, for it would be deadly to it to be isolated as a self-contained study. I mean, to create Chairs and Readerships to train statisticians to occupy Chairs and Readerships. What we need is a fairly intensive mathematical training, together with *very wide* scientific interests, not so much in established knowledge as in the means of establishing it.

The demand for statistics in research *was* becoming a little felt. When Fisher addressed the Royal Dublin Society in January 1932 [C P 98], he referred to the need of the biologist for "assistance in the inductive process by which general theoretical conclusions are drawn from bodies of observational data."

How greatly this has become a practical need, in all fields in which biological work is being put upon a quantitative basis, will not easily be realized without personal contact with the scientific workers, agronomists, entomologists, botanists, geneticists, marine

biologists and many others, from Europe, Asia, Africa and America, who, in the first
five or ten years of their research experience, discover that their really urgent practical
problems are essentially statistical; and who, if they are fortunate enough to have the
opportunity, apply to attach themselves as voluntary workers to a statistical research
laboratory, such as the department for which I am responsible at Rothamsted. Diverse
as the qualifications of these men are, they all have this in common, that while they are
of a generally high intellectual capacity, they have received, in their university training,
no preparation whatever for the statistical problems which are bound to confront
them, as soon as they come into real touch with the questions they are set to investigate.

On this occasion, in Dublin, Fisher appealed against the classification of
mathematics as the most specialized and isolated of scientific studies, at the
opposite pole from the "messy" subject of biology. "Such a division is logically indefensible," he said, "and what is more important, it is becoming inconvenient."

It is the method of reasoning, and not the subject matter, that is distinctive of
mathematical thought. A mathematician, if he is of *any* use, is of use as an expert in
the process of reasoning, by which we pass from a theory to its logical consequences,
or from an observation to the inferences which must be drawn from it.

Gosset (who was in his audience) had pioneered the latter inductive
process of reasoning, and Fisher equated the importance of "the logical revolution effected by 'Student's' work" for human thought in general with the
first mathematically exact use of deductive reasoning by the Greek geometers
of the generation of Euclid. Clearly, he was hoping to appeal to mathematicians to perceive the opportunities open to them through biological applications. Speaking on a biological topic, "The bearing of genetics on theories of
evolution," he made it his thesis that Mendel's statistical logic was the complement of Darwin's biological insight and that the history of their ideas showed
how each had required the other in order fully to establish evolutionary
theory. In general, the inductive reasoning of statistical method had become a
necessary part of the biologist's intellectual equipment in research.
    The Royal Dublin Society had chosen Fisher's topic from a quartet of sug-
gested titles. Fisher had offered "The Theory of Estimation" in mathematics,
"Statistical method and experimental theory" in statistics, "Dominance" in
genetics, and in evolutionary theory the title actually chosen. All were subjects
currently exercising his thoughts, and we note that he might have presented
his thesis under any of the titles. All his work was Janus-faced and looked at
the bearing on the topic equally of experimental detail and of mathematical
reasoning. Mathematics was the obverse, biology the reverse of the same
coin, and he wanted to interest, and to train, both mathematicians and

biologists in handling the new currency of research. As he wrote to Arthur Vassal at this time,

Naturally I have had most of the fun myself as there are very few mathematicians with any serious interest in biology or working contact with biologists. . . . Why can you not cut out all the stocks and shares and quadratic equations from the compulsory mathematics of biologists and doctors and put in the elements, and the ideas of statistics?

Fisher knew just how ill-adapted were the courses taught in statistics to the needs of research scientists, for when he had acted as examiner from 1925 – 1927 for statistical questions set at Leeds University, he had not only seen but attempted to answer all the examination questions from a previous year. Current emphasis was on actuarial methods and economic theory, and the questions were ill-conceived as tests of a student's familiarity with statistical method.

For many years this type of examination question continued to be set, so that in 1949 Prof. J. Maclean, by then a very old friend of Fisher, wrote to him:

By chance I learned that you had set the question paper in 1945 for the London Higher Schools Statistical Method and I wonder if you would be interested to comment on the paper for this year, enclosed, or, more profitably, throw light on the general situation. Your paper was so utterly different from others I have seen.

In response, Fisher commented on each question in turn; few proved satisfactory. The general situation seemed a triumph for academic conservatism. Unchanged were the emphasis on economic problems, the frequent irrelevance of the problems set to testing a student's understanding of elementary statistical methods, and the ill-considered demands for long arithmetical computations and, occasionally, for answers impossible from the data provided.

As for biologists at the university, their statistical training also continued to be neglected. When Fisher moved to Cambridge in 1944, J. Gray, professor of zoology, consulted him, hoping to introduce a course in elementary statistics for biologists, to be examined in the mathematics paper of the Natural Sciences Tripos. Fisher suggested a syllabus for the course to suit the needs of biologists, and although Gray was enthusiastic and pressed on with the committee work to get changes made in the mathematical paper, even at that date his efforts were fruitless; no changes were made in the Tripos.

In contrast, already when he was at Rothamsted, Fisher was concerned that both mathematicians and biologists should receive an appropriate introduction to biometrical investigations. His department had some educational responsibility to the voluntary workers, and his practice shows the sort of experiences he thought most profitable for the young research worker. Voluntary workers came from a variety of fields with more or less practical experience. There could be no strict pattern of instruction. This suited Fisher, and he began from their existent knowledge and, if possible, with the problems in their own field.

Any voluntary worker with the department for a period of months was set to work on the analysis of actual data, often from the Rothamsted records, for Fisher took seriously the value of experience in computation. Advocating a university introduction to computing in 1924, he wrote:

It is not merely that the experience gained by working through a few examples under an experienced computer would save much time and energy but that the art of practical computation itself will only advance in proportion as it receives the attention of university teachers.

In effect he aimed to give workers a chance to familiarize themselves with tools of statistical craft as he had become familiar with them, and to evolve better ways of using them as he had done. Each had to consider the possibilities of a given set of data, to select and employ some tried statistical techniques, such as were described in *Statistical Methods,* and thus to learn their appropriateness or their inadequacy for his problem. He gained facility in the physical act of computation and learned to avoid needless or impracticable computation. Working with the figures from an actual experiment, he was in a position to discern the experimental peculiarities of his data and so refine his analysis. He grew more competent and more confident that he could follow the processes of reasoning and accomplish the necessary computations when confronted (alone in the field) with different data.

Fisher would discuss the data and indicate a possible line of attack. He could often offer practical tips about computation. And he was glad to discuss sticky points arising at any stage of the work. Generally, however, he left it to the students to demand help when they recognized the need, for this also afforded useful experience. Either they struggled through alone, discovering resources within themselves and gaining in self-confidence, or they discovered some limitation in their current understanding or in current statistical techniques and thus put themselves in a position to learn.

Another consequence of setting students to solve practical problems was to help overcome a natural tendency to gravitate to problems of a more theoretical nature and more within the scope of their previous experience: Fisher testified, in 1941,

I have frequently been impressed with the advantage a worker has gained, especially in self-confidence and resourcefulness, by being confronted *malgre lui* with problems of a so-called "applied" or practical nature, which in reality are problems requiring exploration and judgement rather than the application of ready-made formulae.

Fisher plunged his students into the dangers and challenges of the real world; for those with a mathematical background, in particular, it was the only way to achieve a meaningful application of their academic training to problems of inductive reasoning.

□ □ □

As he introduced statistical methods to students in person, so he introduced them to a wider audience in *Statistical Methods*. This is more like a manual for apprentices than like an academic textbook. Throughout the book, he urged his readers to work through the examples for themselves. Practical, technical, theoretical, and philosophical problems are discussed primarily through numerical examples. The examples are nearly all imperfect as illustrations of statistical theory, for Fisher was not only concerned with the statistical tools but also with the physical material to which they were applied. He introduced real experimental data as he had met it, in the raw, and worked it as a craftsman works a block of wood, to illustrate both the application of statistical methods and the regard for the structure of the problem which, like the grain of the wood, must guide the craftsman in his selection and application of the working tools.

This was the natural approach to an inductive as opposed to a deductive study. Understanding of the art itself arose not from the academic mathematical background of the statistician but from the actual data.

The success of the book showed that research scientists rapidly discovered its value. They needed information about statistical procedures that might be applicable to their data, and they were prepared to struggle with the difficulties and obscurities of a new range of thought, because it approached problems which were recognizably their own and promised hope of reaching a solution.

From its publication in 1925 the book sold well. A second edition appeared in 1928 and was sold out by early 1930. (From Nyasaland, Fisher received an urgent appeal from the Cotton Growing Corporation: they had searched everywhere but could not lay hands on a copy, which they needed for the current season's work. Could Fisher get them a copy? He could not, but he sent his own desk copy from the laboratory.) Later in 1930 the third edition appeared, and every 2 years thereafter for 20 years (with one omission during the war) further editions were brought out. In every edition there were additions, for new work had added to what Fisher characterized as "the little storehouse of tried and useful methods" and provided new examples of their application. Apart from his own contributions, he borrowed methods and examples from other workers, most often from the current work of his colleagues, past and present.

The imperfections of the book as a mathematical treatise troubled mathematicians who, as Fisher put it once, "would perhaps have welcomed a concise little elucidation of the mathematical connections, stated and assumed, in the text, and references to the sources of particular mathematical proofs." The book did not provide such material, nor was there a convenient way for the mathematician to discover it. The foundation papers to which the book referred were nearly all mixed up in respect of theory and practice, theoretical statements being put in ad hoc when they were immediately needed, perhaps on discovering that the point had not been made explicitly clear, and by no means in the orderly succession of a connected theoretical treatise. Mathematicians attempting to teach Fisher's methods were perplexed as to how they could present them as coherent mathematical theory.

Although he admitted this difficulty of the mathematician, Fisher had been quite deliberate in presenting *Statistical Methods* as he did, as statistical methods and not mathematical proofs. He intended it for research workers. It was difficult to explain the necessity for this difference of orientation of the statistician or research worker to the mathematician who dealt so differently with the same mathematical processes. It expressed a whole philosophy of experimentation, a whole way of understanding and learning that is foreign to most pure mathematicians.

When he received thoughtful criticisms of the form of the book, then in its 8th edition, from a professor of mathematics who asked for a more connected mathematical treatment, Fisher explained his reasons for retaining his original approach and concluded,

> In my view a statistician ought to strive above all to acquire versatility and resourcefulness, based on a repertoire of tried procedures, always keeping aware that the next case he wants to deal with seriously may not fit any particular recipe. Of course, I know that my book has been described as though it supplied such cut and dried recipes and has been criticized for not supplying the mathematical background by which they may be better understood; but . . . from my point of view, this is a misapprehension, based on the belief that the understanding required can be obtained from the mathematical background rather than, as I think, from the particular peculiarities of the actual body of data to be examined. This kind of understanding seems to be ignored to a frightful extent in mathematical discussions.

It was through his handling of bodies of data that Fisher himself discovered valuable applications for his mathematical training and ability. Through intimate participation in every phase of the making of scientific inferences from the real world, he gained an understanding of how mathematics could serve new purposes in scientific research. He knew from experience that every stage of the process was important in practice, and, working himself with each link of the chain he improved statistical understanding and technique at every stage. Drawing on mathematical theory in utilitarian fashion, he found new uses for old knowledge, and he thereby also increased theoretical understanding of the mathematics. It was his practice from the beginning at Rothamsted to interest himself in new features of the real world, and to the end of his life he continued to find in them the stimulus to new mathematical applications.

At the first stage of making scientific inferences—in seeking to apprehend the structure of the problem—the use of graphs was an invaluable aid. The process was so important a part of statistical method that Fisher devoted a whole chapter to the subject of diagrams in *Statistical Methods*. The first thing he had done with the Broadbalk data in 1919 was to plot the data, and it was the first thing he did on many subsequent occasions. But a graph may be drawn on many scales, and he pointed out the advantage in many cases to be gained by making a transformation in the scale of the observed frequencies, as he had done when he plotted the yields from Broadbalk on a logarithmic scale (see Figure 2, p. 102).

Not a few men must, over the years, have made their first acquaintance with Fisher when they plotted their data with him. The staff at Rothamsted and visitors from other research stations in the early days often knew only enough of statistics to take their problems to Fisher. Much the same applied to young Sergeant George Box (later to become his son-in-law and R. A. Fisher Professor of Statistics at the University of Wisconsin, Madison), at the time when he visited Fisher in Cambridge in 1944. Without training in statistics, he had found himself "the statistician" in a research laboratory in the army. Unable to deal with data from diverse skewed and truncated distributions, he got leave to visit Fisher with his problem. Fisher took him into the garden, and they worked together on a rustic seat in the orchard through the lovely summer day. Fisher suggested making a reciprocal transformation of the data; he read out the probit values while Box looked up the reciprocals and Fisher plotted the transformed values. It worked like a charm. The new diagram showed points clustered along parallel straight lines, revealing the simple linear relationship of the transformed variables. They talked about the experimental situation, and Fisher mentioned published work which might be helpful. He seemed entirely at leisure to spend the day with the unknown sergeant.

Such simple preliminaries as the drawing and inspection of graphs led on to a statistical analysis that involved computation. Fisher, like most biological research workers in the 1920s and most statisticians outside Karl Pearson's laboratory, had not a hive of staff humming at their desk calculators. It was beyond his resources, as it was beyond his ambition, to produce exhaustive tabulations of mathematical functions like the gamma function or the incomplete beta function. He made his calculations for immediate purposes. Nevertheless, if necessary, he was prepared to undertake very heavy computations. The actual labor involved, for example, in the computation of the orthogonal polynomials to describe the slow trends in the yields from Broadbalk, was immense; pages of foolscap were covered with numerals in Fisher's minute handwriting. It was not a task lightly undertaken, even with his Millionaire motor calculator. Characteristically, he did not merely do the computations but considered how best to do them, for simplicity and accuracy of computation. By the manner in which he tabulated the values, he managed to reduce the work to a series of summations and differences, thus making it possible for an investigator to hand over such work with straightforward instructions to his assistant and also to check the arithmetic easily, revealing such errors as inevitably creep into any considerable computation. It was his attention to these practical details that made the computations manageable despite their onerous nature.

In a similar fashion he set out the computation of the sums of squares for the analysis of variance table in such a way that the work is extremely straightforward and easy to check. Any statistical computation is likely to deal with large numbers and to occasion some errors. In routine laboratory procedures like the analysis of variance, the cumulative saving in time and expense of spirit, by having easy arithmetical checks, is literally incalculable. In a practical sense it made possible the use of the complex factorial designs that were subsequently developed on the basis of the analysis of variance procedure.

Naturally, where tables existed, Fisher was glad to use them. But tables varied in their usefulness, particularly with respect to the ease with which the desired value could be read and interpolations made. It was no doubt personal experience of difficulties with the existent tables of χ^2 and t that led to his making changes in their form which have now become universal. What was needed was comparison of probability values, and Fisher conceived the notion of turning the tables inside out and tabulating not values of the probability for given χ^2 or t but χ^2 or t for suitably chosen probability levels. (In the case of χ^2 in *Statistical Methods*, $P = .99, .98, .90, .80, .70, .50, .30, .20, .10, .05, .02,$ and $.01$.) By this means he produced concise and convenient tabulations of the desired quantities. This form also made possible a tabulation of z in manageable compass. The table of z is triply entered, for n_1, n_2, and the probability. By tabulating on separate sheets those percentage points (5%, 1%, 0.1%) of the distribution of z that were of immediate interest to the experimenter, all the relevant information (which might be culled from the thick volume of tables of the incomplete beta function) was made immediately accessible in a few pages. This form of tabulation Fisher introduced in the first edition of *Statistical Methods,* and one may judge of its success by the fact that in the 20 years following publication he gave his permission to about two hundred authors who wished to reproduce his tables of χ^2, t, and z in their own statistical texts.

The tabulation of z shows another characteristic feature of the tables later produced by Fisher and Yates (*Statistical Tables for Biological, Agricultural and Medical Research,* 1938). The higher tabulated values of n_1 and n_2 form a harmonic progression. Thus $n_1 = \ldots$. 6, 8, 12, 24, ∞. In consequence, approximately linear interpolation is possible if the reciprocals of n are taken.

In the later 1920s Fisher was secretary of the Tables Committee of the British Association (B. A.), and in considering what tables should be published and in what form, he reiterated his opinion that the usefulness of the tables would depend on whether they were made to serve the needs of workers who had not even a desk calculator. If there were tabulations with complicated formulae for interpolation, even a man with a machine might not find them convenient to use, let alone the man Fisher was most concerned with. In comparing the usefulness of alternatives he applied the most practical test: wishing to use for one of the B. A. tables an interpolation formula due to Jordan rather than the more familiar formula proposed, he challenged his friend and colleague on the committee, A. C. Aitken, to make a series of timed tests using the two methods.

The aim of tabulations was entirely practical, but the calculation of the tables drew on mathematical formulations of a highly theoretical kind. We have noticed how, working with Student's tables, he was led to introduce the $1/(n-1)$ correction and how, for the function t (though not for Student's original function), he could use the asymptotic expansion to derive t for the larger values of n. In deriving the percentage points of z, similarly, he exploited the possibilities of the identities linking the probabilities he sought with the beta function or with the gamma function.

In this sort of mathematical juggling, he constantly made transformations that resulted in brilliantly simplifying the functions to be tabulated: the transformation of the correlation coefficient was of this sort, as was the tabulation of percentage points of z, the difference of the logs of the standard deviations, rather than of the variance ratio, F, itself. In his introduction to the B. A. tables [C P 91], he provided a series of notes exhibiting how various functions are related, how they may be expressed as differential equations and how the equations may be solved.

□ □ □

Just as Fisher drew on mathematical concepts in improving methods of calculation and tabulation, he used them to meet other statistical needs. He was the first to use an asymptotic expansion to determine probabilities, for Student's *t* Table [*C P* 44]. He also discovered the appropriateness of such an expansion to describe the probabilities of gene frequencies [*C P* 86, 1930]. He found a new use for combinatorial analysis in selecting random Latin squares and, again, in determining the required partitions in deriving the coefficients of the *k* statistics [*C P* 74, 1929]. He applied group theory in deriving the complex designs with balanced incomplete blocks and in solving problems of confounding in factorial designs. It also had applications later in dealing with genetic segregation in polyploid plants. He used functional analysis in solving the problem of the relationships between gene frequencies from generation to generation [*C P* 86, 1930] as he had applied functional analysis earlier to a problem arising in an entirely theoretical context, in L. H. C. Tippett's work [*C P* 63, with L. H. C. Tippett].

Tippett was at the time pursuing a graduate course in statistics at University College, London, in preparation for work at the Shirley Institute of Cotton Research. Under Karl Pearson's direction, he was investigating the problem of finding the frequency distribution of the largest or smallest member of a sample. He had attempted to determine the distribution empirically by taking the first four moments of the distribution of the extreme value in samples up to 1000 drawn from a normal population. Then, during his last two or three terms at University College he spent two days a week at Rothamsted. Fisher learned of his work and became interested in the problem of extremes, purely as a mathematical problem and, in Tippett's words, "produced the paper on which he associated me as co-author."

The argument stems from the perception that

Since the extreme member of a sample of *mn* may be regarded as the extreme member of a sample *n* of the extreme members of samples of *m* and since, if a limiting form exists, both these distributions will tend to the limiting form as *m* is increased indefinitely, it follows that the limiting distribution must be such that the extreme member of the sample of *n* from such a distribution has itself a similar distribution.

Defining a function $P^n(x)$ as the probability that the greatest of a sample is less than *x*, it was possible to inquire what solutions for $P^n(x)$ would satisfy the functional equations implied by the above statement. The beauty of the formulation at once revealed itself, for it appeared that only three functions exist that satisfy the requirement. The extreme value problem reduced to determining for any particular sample of extreme values which of the three distributions was appropriate.

On taking up his employment at the Shirley Institute in 1925, Tippett found an immediate application for his theoretical work. F. T. Peirce had been studying the effect of specimen length on the strength of cotton yarn. Since each test measured, as it were, the weakest link in the tested fiber, the breaking strength of sampled fibers was distributed as an extreme value. Thus Tippett was able immediately to supply the solution to the problem in its mathematical aspect.

In later years this work has become fundamental to the development of what is called "reliability theory." In producing and testing the components of space craft, for example, for which very high reliability is required, the probability of failure can be assessed and reduced to the required level only through a knowledge of the distribution of the extreme values and by assuring that the whole of the distribution of failures shall fall within whatever minute proportion of the population shall be assigned.

One of the most rewarding mathematical formulations, and one which Fisher used repeatedly, especially in the derivation of exact sampling distributions of statistics in common use, was the representation of the mathematical situation in geometrical rather than algebraic terms. By representing a sample of n elements by a point in n-dimensional space, it was relatively easy to visualize quantities such as the t and χ^2 statistics calculated from the sample, in terms of angles and distances in that space. Fisher's interpretations looked like sleight of hand. Gosset remarked once of a problematical distribution: "I take it that whatever it is follows at once 'obviously' from a consideration of n-dimensional geometry"; and on another occasion he demanded whether a certain equation came out of n-dimensional space "or, what is much the same thing, your head." In fact, Fisher was applying a mathematical construction which, though not generally familiar, appealed to him as appropriate to a whole class of problems. In consequence, he was able to elucidate that class of problems, as no one else did.

All these distributions arose from the parent normal distribution and involved linear and quadratic forms whose mathematical behaviour is relatively well understood. There were, however, other statistics in common use, like the measures of skewness and kurtosis which involve third and fourth powers and are not susceptible to straightforward mathematical treatment. Their distributions were, and remain, unknown. In the absence of well-behaved functions to describe the distribution of such statistics, and therefore of any exact tests of departure from expectation, it was desirable, at least, to determine the elementary properties of the distributions. The most obvious way to do this was in terms of the moments; specifically, one might calculate the mean, standard deviation, and the standardized third and fourth moments of the statistics of interest.

Suppose, for example that the statistic of interest was Pearson's measure of skewness, which for a sample of N observations is given by

$$\sqrt{\hat{\beta}_1} = \frac{\Sigma(x - \bar{x})^3}{N} \bigg/ \left\{ \frac{\Sigma(x - \bar{x})^2}{N} \right\}^{3/2}$$

Comparison of $\sqrt{\hat{\beta}_1}$ with its standard error could give some idea of whether the hypothesis that the sampled population was normal ($\sqrt{\beta_1} = 0$) was tenable or not. When one considers that $\sqrt{\hat{\beta}_1}$ is itself the *ratio* of terms of considerable complexity it is obvious that the calculation even of the standard error of $\sqrt{\hat{\beta}_1}$ involves great difficulty.

Possibly, Fisher originally interested himself in the moment problem in connection with tests for nonnormality. He was not unconscious of the contemporary pressure to demonstrate at least approximate normality in order to justify the statistical tests he applied. And, if one had observations enough to test their normality, one first step might be to compare the measures of normality proposed (that is, the standardized third and fourth moments) with their standard errors on the hypothesis of normality. When he introduced the new sampling moment functions, the k statistics, in the place of the crude moments, he chose to illustrate their use in making this comparison [C P 74, 1929]. The example showed well not only the great simplification of the calculations but the numerical effects on tests of significance of using the actual distribution.

At the time it seemed the attempt to derive moments of moment statistics had reached an impasse. In 1897 W. F. Sheppard had developed the algebraic method of deriving such moments. It had since been used extensively by Pearson, and a number of workers had published extensions and refinements of the algebraic method in *Biometrika*. Despite the noble indifference shown by these workers to the horrific complexity of the algebraic expressions and to the interminable labor of their solution, they had nevertheless found it possible to attempt only the simplest cases. As Fisher said,

These results are subject to two somewhat serious limitations: the great complexity of the results attained detracts largely from the possibility either of a theoretical comprehension of their meaning, or of numerical applications; it has also led to great difficulties in the detection of errors, which have had on more than one occasion to be corrected by subsequent workers. Secondly, partly no doubt in consequence of this complexity, attention has been almost solely confined to the direct moments of single statistics, and the product moments, specifying the simultaneous distribution of two or more statistics, have been largely neglected.

In another context T. N. Thiele had already introduced the parameters which are now known as cumulants and had defined certain sampling statistics, which evidently proved unsatisfactory, for C. C. Craig [42], while he developed this work and gave a number of further results in 1929, nevertheless concluded:

It rather seems that the best hopes of effectively further simplifying the problem for statistical characteristics lie either in the discovery of a new kind of symmetric function of all the observations . . . or in the abandonment of the method of characterising frequency functions by symmetric functions of the observations altogether.

If algebraic formulation was to be made manageable, some other symmetric functions having simpler properties would have to be employed, but the situation was desperate when the experts could contemplate abandoning the chase.

As it happened, Fisher had already discovered the desired new kind of symmetric function. His paper [C P 74] appeared in 1929. It has formed the basis of nearly all subsequent work on the subject, for with the introduction of the k statistics, that part of the problem was removed which had been caused by the inappropriateness of the functions employed.

It was on a long train journey from London to Glasgow, during which he diverted himself with the problem, that Fisher seized on the new functions. He jotted down some notes on an envelope, which he subsequently lost, and, in writing the paper afterward, was conscious that he could not express with the clarity of his first vision the sequence of steps in the evaluation of the coefficients.

One aspect of the new functions which was immediately appealing was the fact that the expressions could be determined by an application of combinatorial analysis. The k statistics were nicely related to their population values (the cumulants κ) in a way that was not true of the relationship of sampling moments to their population values.

The equations which connect the moment functions of the sampling distribution of moment statistics [the k statistics] with the moment functions of the populations from which the samples are drawn [the cumulants] correspond in univariate problems to all the partitions of all the natural numbers, and in multivariate problems to all the partitions of all multipartite numbers.

In consequence of this perception, when John Wishart, Fisher's new assistant at Rothamsted, bogged down in an attempt to calculate the moments of multivariate populations, Fisher was able to suggest that he should use the nice combinatorial properties of the cumulants by considering, instead of the sample moments, the equivalent k statistics. By this means, Wishart completed his investigation before Fisher was ready to publish, and consequently the new functions were first introduced into the literature in Wishart's paper [43], with the promise that Fisher's treatment would follow shortly.

The starting point was the replacement of the population moments by functions of the moments, called cumulants. The moment generating function may be expanded in the form.

$$M = 1 + \mu_1' t + \mu_2' \frac{t^2}{2!} + \mu_3' \frac{t^3}{3!} + \ldots$$

The logarithm of M may be similarly expanded, and the cumulants were defined as the coefficients of t, $t^2/2!$, $t^3/3!$ . . . in that function:

$$K = \log M = \kappa_1 t + \kappa_2 \frac{t^2}{2!} + \kappa_3 \frac{t^3}{3!}$$

The κ's are thus determinate functions of the μ's and can be obtained explicitly. In particular,

$$\kappa_1 = \mu_1', \ \kappa_2 = \mu_2, \ \kappa_3 = \mu_3, \ \kappa_4 = \mu_4 - 3\mu_2^2$$

where the μ's without primes are moments about the *mean*. One basic advantage in using the cumulants in preference to the moments is that for independent random variables y and x, whereas the rth moment of $y + x$ is not in general equal to the rth moment of y plus the rth moment of x, this additive property does apply to the cumulant functions, so that, in general,

$$\kappa_r(x + y + z + \cdots) = \kappa_r(x) + \kappa_r(y) + \kappa_r(z) + \cdots$$

Now, these are all theoretical values; as μ_1' is the mean of the population and not of the sample, so κ_1 is a population value. Fisher was interested primarily in the behavior of moments of small samples. To this end he introduced the k statistics, which stand in relation to the cumulants as the sampling moments do to the population moments. The k statistics are functions of the sums of squares, sums of cubes, and so forth of the observations that have the property that the mean values of $k_1, k_2, k_3, \ldots$, are equal to the corresponding values of the cumulants $\kappa_1, \kappa_2, \kappa_3, \ldots$, and it turned out that this property was sufficient to define them.

Fisher was able to show that these ideas which so greatly simplified the study of moments of single distributions were readily generalized and so made manageable problems involving several variables, which had been too complicated to attempt with the earlier crude techniques.

Another new area to which Fisher was led by his Rothamsted experience is now called nonlinear design. An early problem that arose in protozoology concerned the estimation of the number of organisms per unit volume in a given sample. The dilution method used consisted of making a solution from the soil sample and diluting a unit volume successively with the same volume of water. After incubating the samples it was possible to tell which was sterile (contained no organisms) and which fertile (contained one or more organisms). A good deal of effort was expended in this work, and the question arose how best to choose the dilutions, that is, how to design the experiment. This is the earliest known example of nonlinear design.

Many statisticians today think of the study of nonlinear design as being a recent development, yet this example appeared as early as 1922 (*C P* 18), and, as W. G. Cochran has remarked [44, 1972], was actually the first design problem to which Fisher published a solution, before he had published anything on linear designs such as randomized blocks, Latin squares or factorials. Moreover, he considered it in a sophisticated way, taking into account factors to which later less practical investigators have tended to give scant attention.

Nonlinear design concerns the choice of experimental conditions when the observations are nonlinear functions of the parameters to be estimated. The inevitable mathematical consequence of nonlinearity is that the formulae defining the best design contain the parameter itself: the choice of experiment intended to estimate the parameter depends on the very parameter value which is unknown. In practice, then, the optimal design is only as good as the investigator's guess of the parameter value. Let us see how Fisher tackled this tautological conundrum.

Fisher first considered how to choose the experiment that would maximize his measure of information about the logarithm of the density per unit volume. This is essentially equivalent to making a design that minimizes the *percentage* error of this required density. He then considered how fast the information would fall off for nonoptimal dilutions made at higher and lower values. If there were no basis at all on which to guess, one would theoretically need to make an infinite number of dilutions, covering all possible values. Fisher's calculation showed, however, that eight twofold dilutions that straddled the best value would provide 97½% of the information supplied by this theoretical infinite series of twofold dilutions. In practice, some kind of guess could be made, although perhaps rather an inexact one. If it was thought that the guess might be out by a factor of 1000 (which is roughly 2^{10}), then by making 10 additional dilutions, 18 in all, the experimenter could be sure of covering the optimal value. Thus the problem of the uncertainty of the guessed value was tackled in a practical manner.

Furthermore, since calculation of the exact estimates of the density would be laborious and the Rothamsted laboratory were doing 38 such tests daily, Fisher provided a table that enabled the density to be estimated in less than 5 minutes. To do this he employed the method of moments. Having first calculated that in this case the method had an efficiency of 88%, he could afford to adopt the approximation. Indeed, the reason for the inclusion of this example in this paper, in which the inefficiency of the method of moments was exhibited at large, appears to be to show that a precise knowledge of the efficiency of any method is a prerequisite for its use. Inefficiency of known amount might be justified in particular circumstances where the time and expense of additional complicated calculations were balanced against time and expense of additional experiments.

□ □ □

Through the 1920s Fisher was exploring various means of inductive inference. He extended the range of tests of significance. On the basis of the likelihood function he was able to formulate a theory of estimation. First he obtained estimates using maximum likelihood; later he introduced ancillary statistics of the likelihood function and, finally, considered the whole of the likelihood function to describe the data.

Although, using the likelihood, he could state the relative plausibility of different values of some parameter of interest, he could not by this means obtain statements about the interval within which the parameter lay with a given probability. The logical status of statements of likelihood and of probability remained distinct. A solution to this difficulty presented itself to Fisher at the close of the decade. It emerged from his work on tests of significance. These, although valuable for their own purposes, are, of course, generally less informative than the method of maximum likelihood, for they provide only a dichotomy between hypotheses and not an estimate of the value of the parameter.

Fisher was preparing for publication a paper on the last of the distributions required for making common tests of significance (the general sampling distribution of the multiple correlation coefficient), together with some tabulated values (C P 61, 1928). He noticed, for the first time consciously, that in the test of significance the relationship between the estimate and the parameter is of a type he later characterized as pivotal. He argued that if one quantity was fixed, the distribution of the other was determined; in consequence, once the observations fixed the value of the observed statistic, the whole distribution of the unknown parameter was determined. J. O. Irwin, his assistant at Rothamsted at the time, was to present a paper to the British Association meeting in 1929, and Fisher suggested he might put the idea in his paper, which he did.*

*In *The Design of Experiments* Fisher attributes the first application of the idea to E. J. Maskell. (1930).

In 1930 (*C P* 84), Fisher presented the argument himself, using the example of the correlation coefficient. Calling it the fiducial argument, he offered it as a means by which true probability statements concerning continuous parameters may in some cases be inferred from the data. Since it is applicable only to statistics of continuous distribution (like the mean, standard deviation, regression, or correlation), its use was made possible through his solution already of a number of problems of distribution and tabulation of the distribution values. It seemed to him reasonable to suppose that arguments of the fiducial type must have been overlooked by earlier writers on probability because they did not have these distributions available to them. In 1930, however, he felt they "force themselves on our attention."

Soon after, it became clear to him that the fiducial argument could only be employed when sufficient statistics existed. Having defined the conditions under which sufficient or exhaustive estimation was possible [*C P* 108, 1934], he used the argument to derive fiducial distributions for a number of continuous parameters for which sufficient statistics were available [*C P* 108, 109, 124, 125]. By 1936 he felt the time was ripe for a general solution for the range of more complex cases of simultaneous estimation. Characteristically, he set out the general theoretical problem in the form of a practical example, although he was addressing a mathematical audience. The occasion was the opening lecture of the Harvard Tercentenary Conference in 1936 [*C P* 137]. Having given an outline of the history leading up to recent developments in the logic of inductive inference, he closed with the words:

My final word on this topic is a query, the answer to which so far is unknown, and which is, therefore, at present a challenge to our mathematical intuition. May I put the problem in this form?: —

The agricultural land of a predynastic Egyptian village is of unequal fertility. Given the height to which the Nile will rise, the fertility of every portion of it is known with exactitude, but the height of the flood affects different parts of the territory unequally. It is required to divide the area, between the several households of the village, so that the yields of the lots assigned to each shall be in a predetermined proportion, whatever may be the height to which the river rises.

If this problem is capable of a general solution, then . . . one of the primary problems of uncertain inference will have reached its complete solution. If not, there must remain some further puzzles to unravel.

In introducing this paper in *Contributions to Mathematical Statistics* (1951), he referred to the still unsolved problem:

The author submits that the existence or nonexistence of solutions, or, in general, the conditions of solubility, of the problem of the Nile . . . must supply the key to the nature of the inductive inferences possible in each type of problem.

Some years later it was shown that there is no general solution to the problem of the Nile. Solutions exist only for special forms of the distribution of fertility.

The origin of the fiducial argument is important in the context of its times. It appeared in 1930, as a consequence of Fisher's concern for the needs of research scientists, when he was not only providing tests of significance but deeply committed to improving both the theory and practice of estimation. It was a consequence also of his early rejection of Bayes' postulate, which, if he had not seen cause categorically to reject it, would have obviated the need to look for alternative routes to statements of probability. The paper (C P 84) is entitled "Inverse Probability," and the first controversy it caused was on account of the argument there presented against the currently popular acceptance of the argument from inverse probability. His reasoning on this subject was more fully developed [C P 95, 1932; C P 102, 1933; C P 109, 1934] in response to critics of the first paper, and in particular Harold Jeffreys, who took up the defence of inverse probability. Fisher's challenge forced the debaters into a more critical appreciation of the underlying assumptions and of the logical implications of those assumptions. By doing so he illuminated the issues, so initiating not only controversy but the gradual rationalisation of theories of inductive inference.

It is not irrelevant to the theme of this chapter that it was particularly in controversy over fiducial inference that a divergence of philosophy between Fisher and other influential mathematical statisticians became apparent. Men trained in mathematics but without Fisher's commitment to and feeling for the experimental situation were prominent among the statisticians now beginning to exploit Fisher's work. The tendency was for them to consider the statistical theory in academic situations susceptible to the deductive reasoning of pure mathematics rather than the practical situations requiring inductive insight. This tendency led to misunderstandings of Fisher's theoretical work and to a series of confrontations with him in the 1930s. In the realm of scientific inference fundamental differences were exposed that still divide different schools of statistical inference. There will be occasion to say more of this later.

Plate 1. R. A. Fisher at 26 months.

Plate 2. R. A. Fisher at about 9½ years.

Plate 3.   Steward at First International Eugenics Congress, 1912.

Plate 4. Passport photograph, 1924.

Plate 5. At tea outside the Sample House at Rothamsted.

Plate 6. A 5 × 5 *Latin square laid out at Bettgelert Forest in 1929 to study the effect of exposure on Sitka spruce, Norway spruce, Japanese larch, European larch,* Pinus contorta, *and Beech. Photograph taken about 1945, reproduced by permission of the Forestry Commission.*

BETTGELERT FOREST

LAYOUT

N

Elevation:					
1730–1800′	*B*	*A*	*E*	*D*	*C*
1530–1730′	*C*	*E*	*B*	*A*	*D*
1460–1590′	*A*	*C*	*D*	*E*	*B*
1340–1460:	*D*	*B*	*A*	*C*	*E*
1250–1340′	*E*	*D*	*C*	*B*	*A*

A. Sitka spruce
B. Japanese larch
C. Sitka spruce/Japanese larch 50/50
D. Sitka spruce/Pinus contorta 50/50
E. Norway spruce/European larch 50/50

Two rows of Beech planted on each side of the series.

Plate 7. Layout of Bettgelert Experiment.

Plate 8.   March 1929, Fellow of the Royal Society.

Plate 9.   Mrs. Fisher with Elizabeth, 1931.

Plate 10. R. A. Fisher, 1938, with sons George (aged 18) and Harry (14).

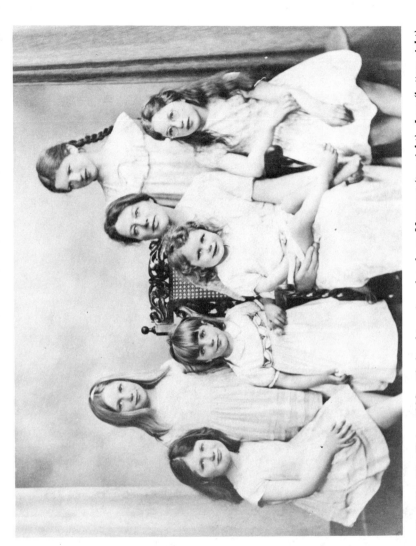

Plate 11. *Mrs. Fisher 1938, with daughters, in order of age, Margaret (top right), Joan (bottom right), Phyllis (top left), Elizabeth (bottom left), Rose standing beside her chair, and June in her lap.*

Plate 12. At the seaside, 1938.

Plate 13. Caught repairing mouse cages, Rothamsted, 1943, by T. N. Ollrycht; with Amy, the favorite and last of his experimental dogs.

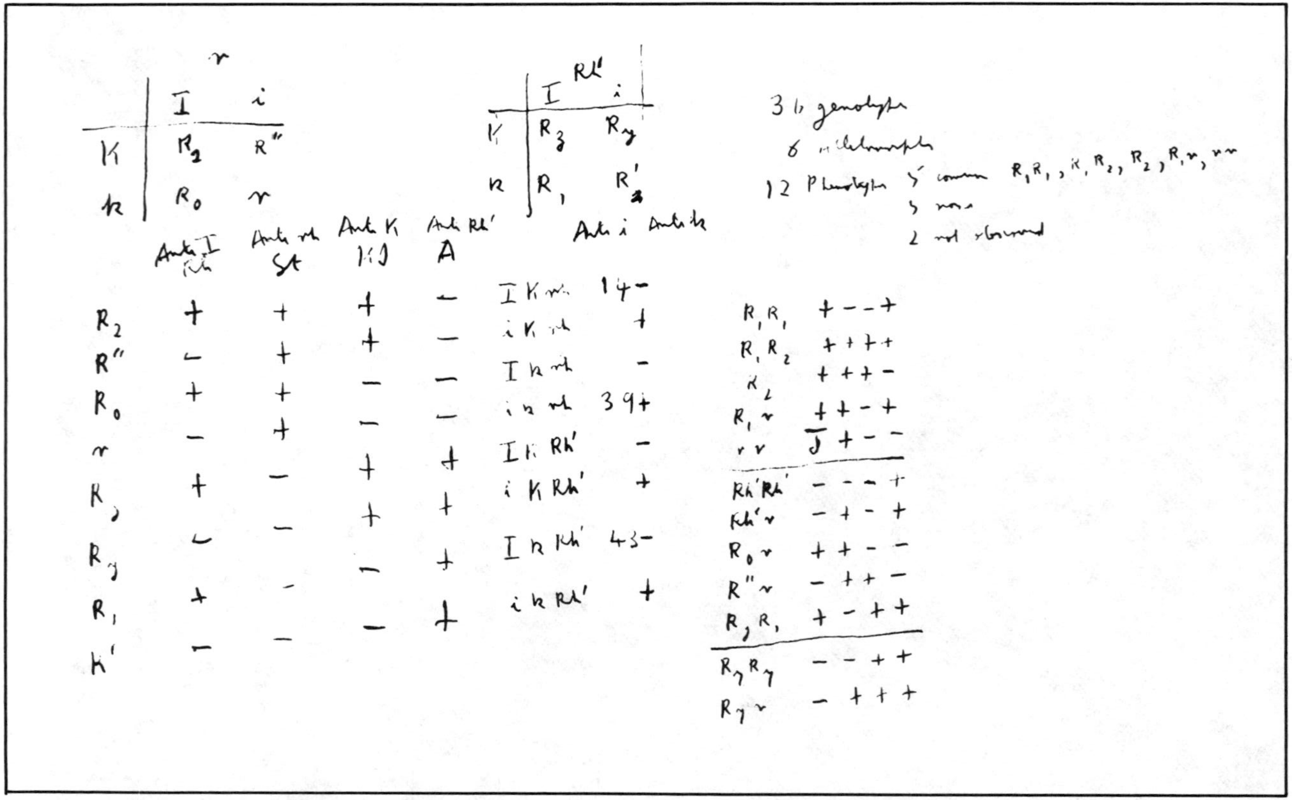

Plate 14.   *The Rh system interpreted as three closely linked genes, as first written in the Bun Shop, 1943 (by courtesy of R. R. Race).*

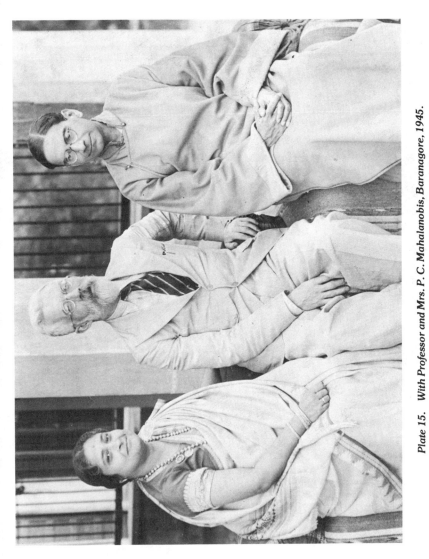

Plate 15. With Professor and Mrs. P. C. Mahalanobis, Baranagore, 1945.

Plate 16.   At Lake Junaluska Assembly, N. C., 1956. Photograph by Mrs. Horace Norton, reproduced with her permission.

Plate 17. Portrait by Leontine Camprube, painted at Lake Junaluska, N. C. 1946. Reproduced by permission of G. M. Cox.

Plate 18.   In the garden at Whittingehame Lodge at the 100th Meeting of the Genetical Society.

Plate 19. With Gertrude Cox and Frank Yates at Blue Ridge, N. C., 1952.

Plate 20.   A butterfly-hunting party at Blue Ridge, N. C., 1952.

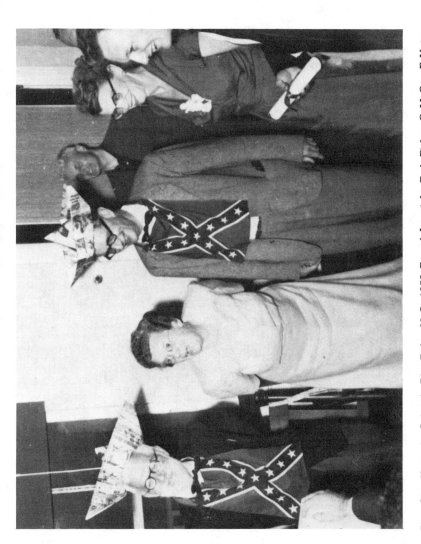

Plate 21. Kentucky Colonels, Blue Ridge, N. C., 1952. From left to right: R. A. Fisher, G. M. Cox, F. Yates, Besse Day.

Plate 22.   With Chancellor Lawrence A. Kimpton at Chicago University, 1952.

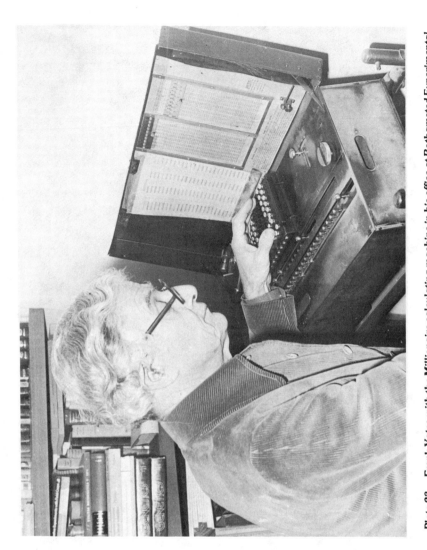

Plate 23.   Frank Yates with the Millionaire calculating machine in his office at Rothamsted Experimental Station, 1974.

Plate 24. Portrait by A. R. Middleton Todd, R. A. painted 1957. The original is hung with other presidents of Gonville and Caius College, in the college hall. Reproduced with permission.

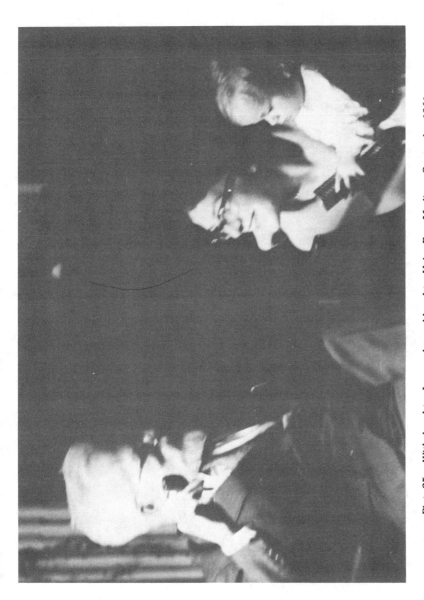

Plate 25.   With daughter Joan and granddaughter Helen Box, Madison, September 1961.

Plate 26.   Calcutta, February 1962.

10
Galton Professor of Eugenics

When, in 1929, Fisher contemplated the possibility of succeeding Karl Pearson in the department of applied statistics at University College, London, he characterized the position as "the only one for which I have qualified myself." This was strictly true. He was uniquely qualified to become both professor of eugenics and head of the biometric laboratory. He had, in fact, made himself the only possible candidate for the position held by Pearson almost since the beginning of the century. At the same time, Pearson's position was unique. Statistics was taught at no other British university, nor was there another professor of eugenics, charged with the duty and the means of research into human heredity. If Fisher were to teach at a university, it would have to be as Pearson's successor.

Early in 1933, when Karl Pearson announced that he would retire at the close of that academic year, Fisher was the obvious candidate. However, the decision of the university in respect of Pearson's successor was complicated by the fact that Pearson's son, Egon S. Pearson, had been working in the statistics department for the previous 10 years. It was felt to be impossible to pass him over and injudicious to promote him in the same department under Fisher. A compromise was arrived at, intended to satisfy the reasonable aspirations of both men but, in its nature, guaranteed to embarrass and hamper their relations. Fisher was invited to become Galton Professor in a eugenics department split off from what was to become a separate statistics department under Egon Pearson, Pearson being promoted immediately to a readership and a couple of years later to professorship of statistics.

Fisher received the invitation to stand for the Galton Professorship along with strong assurances that Egon was well disposed to his appointment. Before responding to the invitation, however, he wrote to E. S. Pearson (24 May) a very cordial letter but touching on a potential difficulty:

. . . During the last few years a number of students have come to me under the guise of voluntary workers, with a view to learning statistical methods, especially in those points in which I have differed from your father. I think that we must anticipate that these would continue and only hope we can avoid their becoming an embarrassment. I suppose the academic ideal is that they should hear both views and use their own brains but I have not enough experience to know that this would work out well and shall be glad to know how you view this particular difficulty, which I have put before you as frankly as possible.

The difficulty was not to be easily solved, for Pearson had formed a different view. With expressions no less cordial and hopeful of cooperation than Fisher's, in his answering letter, he wrote:

I would suggest that you should not start by giving any *lectures* in your department in pure statistical theory. I imagine that the training you have been giving in statistical methods need not take lecture form. Later on, when you are quite sure of the ground, I would like to think that you would be ready to give some lectures on your outlook on statistical theory which would fit into courses of the Statistics Department. At the same time, I have no doubt that some of your students will choose to come to my lectures.

If my approach to statistics were the same as my father's, I think there might be difficulty but as it is not, I anticipate there will be less cause for embarrassment than you think.

Fisher responded, asking Pearson to reconsider his position on statistical teaching. Pearson replied, reiterating that, in effect, only if each kept strictly within his own territory could they avoid conflict.

Fisher was nettled. As he wrote to Gosset on June 6:

I had hoped to get on rather cordially with Egon, but he has not made too good a start by suggesting that I should not lecture on Statistics, as though I should get any students worth their salt on any other subject, on the interesting ground that he has already adopted and teaches my methods. I feel the embarrassment of the position quite as much as he does; but I cannot expect him to say more in his lectures than he cares to say in print. . . .

It does not make it any easier for me, or for students who come to University College just because they think that my work, in some respects, has superseded

Pearson's, or because they want to avoid the inconvenience of getting at it second hand, to explain that I was appointed, rather like a ticket-of-leave man, on condition that I "didn't do so any more."

E. S. Pearson, in correspondence with the author, recalls being in his office at University College about this time, when Fisher unexpectedly knocked at his door and, after some more general exchange of friendly remarks, came out suddenly with the proposition that, despite the decision of the college and university's appointments committee, "Could we not in fact agree to make the eugenics and statistics departments into a single department?" Pearson said he would think about it, though he saw (and those whom he consulted confirmed his view) that "such unification would be impossible with on the one side an older man of well-established reputation and strong views, also a Professor, and on the other hand a younger man (only a Reader) who must build up a small department of four persons and who was still in the process of establishing his own philosophy of Statistics." In the end Fisher acceded to Pearson's stipulation that he teach no "pure statistical theory" and, instead, proposed to lecture on "The Philosophy of Experimentation."

This first difficulty, and the way each handled it, showed the basic weakness of the university plan. Pearson felt compelled to defend the integrity of his department against the outsider even before he took office, and Fisher, in the one university where he might teach, was forced into rivalry as an exponent of his own ideas. It did not help that the departments shared the same building, for differences in statistical understanding, tolerable perhaps at separate institutions or within a single department where both views were taught, became intolerable by juxtaposition. Thus although members of the statistics department introduced Fisher's methods and ideas in their lectures, the interpretation was their own, and they expressed their own reservations and criticisms. Students would sometimes raise such points with Fisher, only to discover that he regarded the lecturer's interpretation of his work as perverse or wrongheaded. It was a situation well designed to frustrate the undoubted goodwill on both sides and to enhance any differences between the departments, personal or statistical, to causes of mutual acrimony.

Though Fisher foresaw some of the difficulties of the situation, he did not hesitate to accept the invitation to stand for the professorship. Indeed, on May 28, he wrote to J. B. S. Haldane, thanking him for his help in putting the invitation in his way and saying he would accept, "as soon as I can find the Registrar's letter on the subject."

Haldane had been appointed recently as professor of genetics at University College, part-time, in addition to his part-time work at the John Innes Institute. There might have seemed a conflict of interests between him and Fisher, for the two population geneticists had not often agreed about their science; but, in fact, coming together to the same university, both were eager for friendly cooperation. Fisher proffered houseroom for some of Haldane's genetic stocks. Haldane reciprocated with an offer of some of his inbred lines of mice. They accepted complementary roles in the training of students, readily agreeing to accept and send suitable students to each other. Haldane came with his students to hear Fisher's lectures, and, meeting at the staff tea beforehand, they enjoyed, as Fisher suggested they might in his invitation, "a chance to talk over points of common interest as they occur to us from time to time."

Another possibility of awkwardness existed within the Galton Laboratory, for Fisher's staff were Karl Pearson's people and staunchly loyal to him, and the quarrel between him and Fisher was notorious. On both sides there were doubts as to how well they could work together. Fisher was careful to respect their personal loyalty to Pearson; Pearson was made an honorary member of the Tea Club, and when he joined them in the Common Room, it was observed that Fisher did him the unique honor of breaking out of conversation to step forward and greet him cordially.

The difficulty which arose was consequently not of a personal nature but concerned the work of the department. There were four staff members: Dr. Julia Bell continued the genealogical records of genetic diseases as before; Miss M. N. Karn continued computational work; but Dr. P. Stocks and Dr. G. M. Morant found the transition more difficult. Stocks was moving from the department, part-time, to the General Registry Office. Fisher tried hard to persuade him to use this liason to get cooperative studies made of the British population, but he was rebuffed and Stocks soon transferred wholly to his new job. Dr. Morant had specialized in craniometry. Fisher hoped to broaden his interest and his usefulness by interesting him in photography. He did, in fact, take a course in photography and set up a photographic room.

As lecturer in medical statistics, with a full-time assistant, Morant represented a large commitment of departmental funds, and it became evident that Fisher did not value the craniometric work at so high a rate. He showed what could be done with such data by developing the discriminant function and its application to cranial measurements (by E. S. Martin in 1935 and M. Barnard, 1935). But, in 1936, in an article on " 'The Coefficient of Racial Likeness' and the Future of Craniometry" [C P 141], he criticized the use of Pearson's coefficient of racial likeness and showed reason to believe that the fundamental problems for the ethnographic interpretation of cranial remains could not be solved by further accumulation of skull measurements, however numerous and meticulous they might be. Such data had been collected apparently with an "intuitive confidence . . . that, by amassing sufficient statistical material, all difficulties may ultimately be overcome." While attention had been concentrated on skeletal remains, however, Fisher observed, "necessary and preliminary enquiries seem largely to have been ignored." Knowledge of the heredity of head measurements was needed and, above all, an assurance of adequate numbers, known to be representative of the population sampled, and this could be obtained only when living populations were examined, with the collateral information available about them regarding sex, blood relationships and factors (like nationality, language and religion) that affect intermarriage.

Consequently, Morant felt his long labors were unappreciated in the Galton Laboratory and, in 1937, he transferred to the anthropology department, carrying his salary with him.

□ □ □

Although available funds were limited while the original staff remained on the books, the transition to a biological outlook embracing genetics was partially achieved by appointment of new staff. Initially Fisher found he had departmental funds enough to make a single senior appointment and promptly invited E. B. Ford to join him as Lecturer. Ford would have liked nothing better but, when he realized that an offer such as he could consider would seriously strain Fisher's resources, he reluctantly declined. Fisher then decided the best plan would be to make two junior appointments of one genetical and one statistical assistant. Knowing the work of K. Mather, he invited him to apply for the genetical appointment, which he did successfully. W. L. Stevens was appointed assistant in statistics.

Even before these positions were considered, however, Fisher sought a personal assistant. Here again he really wanted two for the price of one. As he had written to Haldane in May 1933, the post of assistant to the professor did not exist:

The Professor seems to have a Secretary up to £150 a year, who may not be versatile enough to feed snails and work the calculating machine when she is not typing letters. My best hope seems to lie in the allocations for the wages of the dog-man and especially for their food—I have great hopes of their food.

In the event, in March 1934, he appointed an honors graduate in botany, Sarah B. North (later Mrs. Holt) as his genetical assistant. She was without secretarial training and had never used a typewriter, but Fisher asked her to help with office work until a secretary was appointed. Having got the person he wanted for the job, however, he was at first inattentive to her official status. Consequently, she was entered in the university books and, despite his later protests, for many years remained on the secretarial staff, while the real secretary was graded as her assistant.

Fisher was not successful in this or some of his other dealings with the administration, and this added to his frustrations. In setting up his department, he applied to the university for the larger items he required. He asked for a greenhouse and had to accept a windowbox instead. He asked for a microscope for Mather's cytological work and was told that the money must come from departmental funds. In the first year he overspent his budget by some £37. After much correspondence with outraged accountants, he was finally permitted to borrow the amount from his allowance for the following year, being given to understand that such a thing must never happen again. Having obtained funds from the British Association, he offered a place in his department to Dr. F. Gross, who came to England as a German refugee in 1933. Again, he found himself in trouble with the university, for he had failed to ask permission in advance to appoint an immigrant. The petty constraints and irritations of working under a bureaucratic administration did nothing to endear its officials to him, and he did not hide his contempt for those who seemed to him to exhibit their self-importance by obstructing the real work of the university.

Fisher's difficulties in relation to College officials as well as with the Statistics Department are worth mentioning only because they continued throughout his professorship, and although they affected him unequally, they affected him in the same sense. He felt the lack of personal support in the very places where he most depended on it. On this account he was probably more impatient of the obstructions put in his way at this time than at any other period of his life. Possibly the slight friction between him and the college officials primed both for the confrontation between them in 1939. With the statistics department, he could not discuss statistical matters privately in a way that might reconcile differences, and his criticism of public statements tended to raise hostility between the departments and frustrate all hope of communication.

The limitations of financial support troubled him less, for he had some latitude in guiding current work and he was able to win sponsorship elsewhere. His department did grow in a direction he wanted when the Rockefeller Foundation funded establishment of a serological unit at the Galton Laboratory in 1934. As professor he became editor of *The Annals of Eugenics,* a quarterly started by Karl Pearson in 1926. College support for the journal was simply not enough, but the Eugenics Society responded to Fisher's appeal with a generous annual grant that enabled him to maintain regular quarterly publication well into the war period.

□ □ □

In 1934 Jerzy Neyman joined Pearson's staff. At that time there was no ill-feeling between him and Fisher; indeed, Neyman had applied to Fisher as well as Pearson when he was looking for a job in England. Although Fisher had no money to appoint him in his own department, he did him the honor of putting forward his name as a candidate for membership of the International Statistical Institute, to which Neyman was elected in 1934. The situation changed rapidly thereafter. Neyman sniped at Fisher in his lectures and blew on the unquenched sparks of misunderstanding between the departments with apparent, if undeliberate, genius for making mischief. Differences in their whole philosophy of statistics began to emerge, and open conflict began when Neyman [45] read a paper on "Statistical Problems in Agricultural Experimentation" before the Royal Statistical Society in March 1935.

At the beginning of the paper Neyman complimented Fisher and acknowledged the importance of his work to agricultural experimentation. There only remained, he said, certain details that required more attention than had yet been given them, and these he proposed to consider. These details challenged the unbiased nature of the test of significance for randomized block and Latin square designs and claimed that Latin squares had been overrated. On both issues Neyman undertook to rectify the deficiencies of Fisher's thought.

The condescending attitude would have been galling, even if the conclusions had been sound. Actually, however, when Neyman discovered bias in the z test and challenged the validity of its use for Latin squares, he was not testing the null hypothesis Fisher thought appropriate, for which the z test had been designed, but was testing a different null hypothesis arising from the novel mathematical model for field experiments which he introduced in the first part of his lecture.

Although the point was not always clearly in view during the debate which followed, the argument was really about which kind of model and which kind of null hypothesis was appropriate in practical circumstances. It is a commonplace that science proceeds by viewing natural phenomena in terms of some model, preferably a model that is extremely simple and yet adequately represents the real and complex world. Fisher's ability to conceive of such models was rooted in his thorough familiarity with the practical aspects of the problems he studied. His mathematics, though often inspired, frequently appeared almost as an afterthought and was never pursued for its own sake. In contrast, Neyman felt much more confidence in the insights to be gained through mathematical manipulation.

In the matter of agricultural experimentation, it had been clear to Fisher that explicit mathematical modeling of agricultural field trials was likely to be fraught with hazards; correlations between adjacent plots, due to fertility gradients, drainage patterns, and the like, would be complex and unknown and impossible to model adequately. For this reason, at an early stage he had introduced randomization. In his original introduction of the analysis of variance [*C P 32*, 1923] he had discussed the average values of the mean squares appearing in the analysis of variance, not on the assumption of independent observations but over all possible rearrangements. So far as the test of significance was concerned, by introducing randomization he had been able to avoid many of the pitfalls of more explicit modeling. The null hypothesis he thought appropriate was that the treatments were entirely without effect so that, under randomization, when the null hypothesis was true, the treatments could be regarded as mere labels that would have fallen differently if another randomization pattern had been adopted.

Neyman, however considered a null hypothesis which supposed that although different treatments might give different mean yields when applied to different plots, the mean yields over the whole field would be equal. To deal with this possibility, a more complicated model was required, with the disadvantage that because more assumptions were necessary, it was more vulnerable. If Neyman's assumptions were true, then it would follow, as he said, that Fisher's z test would be biased and might detect significance more often than it should, especially in the case of the Latin square design. But were his assumptions appropriate in the field? Even if they were, Neyman offered no alternative test of significance. Moreover, the actual effect on the z test was still undetermined, for Neyman's prediction was based on expected values, and the distribution of the test value appropriate to his model was unknown.

Despite these unanswered questions, the mathematics was impressive. Wishart [45], for example, congratulated the speaker on his "critical survey of the modern methods of field experimentation of the English school, accompanied by a somewhat penetrating analysis of the mathematical aspects of the question." Since others might be led to worry about trusting the z test, Fisher in his discussion of the paper [C P 128] aimed to allay any doubts Neyman might have raised as to the theoretical adequacy of his methods. He demonstrated in a very neat way that so far as his null hypothesis was concerned there was no doubt of the unbiased nature of the test for the Latin square, and he broadly hinted that Neyman did not understand the problem and had been led astray by the mathematics. Neyman responded by repeating the argument that for *his* null hypothesis and for *his* model the test would be biased. Fisher retorted:

I am not going to argue the point. It may be that the question which Dr. Neyman thinks should be answered is more important than the one I have proposed and attempted to answer. I suggest that before criticizing previous work it is always wise to give enough study to the subject to understand its purpose. . . . Neyman would like a different test to be made and I hope he will invent a test of significance, and a method of experimentation, which will be as accurate for questions he considers important as the Latin Square is for the purposes for which it was designed.

Yates, in written discussion of Neyman's paper, presented justification of the z test by reference to the randomization set, an idea then in the press in

Fisher's forthcoming book, *The Design of Experiments* (1935). Referring to a randomization experiment that he had conducted with T. Eden [31], Yates wrote:

What the experiment does show is that the randomisation process effectively generates the distribution of z, and the need for the postulation of any parent population from which the thirty-two values are to be regarded as a sample is entirely avoided.

Thus in the case of a randomized experiment he dismissed the need to specify any model such as Neyman had proposed.

A couple of months after the debate over Neyman's paper, in May 1935, the Royal Statistical Society met again to hear F. Yates [46] talk on "Complex Experimentation." The discussion that followed banked the fires of controversy between Neyman and Fisher. In a very long contribution, Neyman, again after expressing his high appreciation of the work of Fisher at Rothamsted, raised many objections to factorial designs. He seemed determined to establish his superiority by picking holes in Fisher's work. To this and other criticisms, Yates replied:

A general survey of the remarks immediately brings out one fact of considerable interest. Those speakers who are actually engaged in experimental work or are in close contact with experimental workers, appear completely satisfied with factorial designs, while those who are not so engaged have raised several objections to the method. Now if the method were as fundamentally unsound, misleading, and unreliable as the critics would have us believe, one would have expected some at least of those who had most used it to have discovered the fact, both from bitter experience and because they had at one time or another devoted a considerable amount of thought to it. But this is not the case. I am encouraged, therefore, to believe that the force of these criticisms is perhaps not as great as their volume would indicate.

Yates was right. Forty years later factorial designs continue to be some of the most useful not only in agriculture but wherever experimental work is conducted.

The conflict between Neyman and Fisher was primarily conditioned by their different approach to the whole subject: Fisher was a research scientist using mathematical skills, Neyman a mathematician applying mathematical concepts to experimentation. Now, mathematicians did not readily appreciate the central role of randomization in Fisher's statistical theory. To Neyman, Fisher had suggested that "before criticizing previous work it is always wise to give enough study to the subject to understand its purpose." In his written reply, Neyman reflected back the comment on Fisher himself, who "not only criticized my paper, but blamed me for a variety of sins of which I am not guilty." As for understanding its purpose, he continued:

Instead of guessing the desires of Professor Fisher, I was interested in the problems of agricultural experimentation as I understood them and in the adequacy of the methods of their solution in frequent use. I noticed the inconsistency of assumptions underlying the z test and the procedure of agricultural experiments. The z distribution is deduced from that of two sums of *independent* squares. The methods of Randomised Blocks and Latin Squares are based on *restricted* sampling, and *restricted* sampling means the mutual dependence of results. From the point of view of statistical theory the inconsistency in the application of the z test to the data of Randomised Blocks and Latin Squares is therefore striking and it did not occur to me that Mr. Yates and still less Professor Fisher could have overlooked it.

Mathematicians tended to formulate the argument in these terms, that is, in terms of normal theory, ignoring randomization. Nevertheless, in doing so they exhibited a fundamental misunderstanding of Fisher's work, for it happens to be false that the derivation of the z distribution depends on the assumptions Neyman criticized.

The two independent squares to which he referred are the sum of squares due to regression (or model) and the sum of squares due to error, whose relationships in the analysis of variance have been discussed in Chapter 5. Although it is true that if the distribution of the observations were normal these quantities would be independent, it is not true that the z distribution could not be valid when these quantities are not independent. The requirement is only that the vector **y** shall have an equal chance of lying in any direction in the sample space. As mentioned earlier, randomization produces a roughly symmetrical distribution, not one in which the observations are independent; the symmetry induced by randomization can produce a distribution in which the angle θ has very nearly that expected on normal theory.

Neyman was not the first mathematician to find difficulty in understanding why Fisher felt certain that his conclusions were justified by randomization. Very much the same points gave trouble in the work of S. S. Wilks, done under Wishart in 1933. Wilks had a master's degree in mathematics and had worked for some time with H. L. Rietz, a theoretically oriented mathematical statistician, at Iowa City. Having then come to England to study statistics under Pearson and Wishart, Wilks submitted a paper to the Royal Society in which, using characteristic functions, he proved the independence property of comparisons made in the z test.

Fisher was asked to referee the paper. Rather than communicating his opinion to the editor, he wrote directly to Wilks to explain why rejection might be necessary and to suggest that Wilks withdraw his paper and submit an amended version to a mathematical journal. He pointed out that Wilks's result, although produced by a novel route, was not new, being covered by Fisher's very simple and general proof in *Metron* in 1926. Wilks, defending his paper in his reply, showed that it had not been ignorance but misunderstanding of the *Metron* paper which had prompted his work. Moreover, Wilks had included in his paper a proof that the estimates given by Latin square designs were biased, and Fisher pointed out the error of this demonstration.

Unfortunately, he did not convince Wilks about the statistical points. (It was when W. G. Cochran, a fellow student at Cambridge, showed him the very simple algebraic proof of lack of bias for the Latin squares that Wilks withdrew his erroneous argument.) Instead, Wilks was hurt. He had set his heart on publication in a Royal Society journal and never quite forgave Fisher for having prevented the recognition which he felt was his due. We wonder whether this misunderstanding of Fisher's work (and of Fisher's attitude) might never have happened had Wilks been confronted earlier by an experimental philosophy and accepted it, rather than mathematical theory alone, as the basis of statistical understanding.

It was a consequential misunderstanding for, turning Wilks away from Fisher, it confirmed his reliance on a theoretical approach. On his return to the United States, Wilks set up the teaching program in mathematical statistics in the mathematics department at Princeton University, and his bias in favour of theory has affected the tradition of education in mathematical statistics in the United States from an early date. It helped prepare the way for the postwar propagation of highly theoretical mathematical statistics at the expense of statistical practice in American universities.

The affair with Wilks would not have developed as it did but for the personality of Wishart. Wishart had read but, in Fisher's opinion, probably did not understand the proof in *Metron;* and, being currently preoccupied with characteristic functions, he would tend to discount work by other methods. Anyway, even after the rejection of Wilks's paper, he managed to ignore Fisher's proof, and this led to a public confrontation on the issue in 1934 [47].

The occasion was an exposition on the analysis of variance, presented by Wishart before the new Industrial and Agricultural Section of the Royal Statistical Society. In the course of his talk, Wishart mentioned a recent paper by J. O. Irwin, written at Rothamsted, which he said proved the unbiased nature of the comparison in the test of significance, and he promised the paper by Wilks to prove the independence property. He did not mention Fisher's proof, which covered both points, nor did he observe that Irwin had actually referred to it in his paper and had acknowledged Fisher's priority. These facts Fisher pointed out in his discussion of Wishart's paper. Wishart appeared impervious to argument; even after the altercation at the reading of the paper, he expected to publish his contribution unchanged. Only after Fisher made it clear that if that were done, then the whole of Fisher's discussion of it would also be published was Wishart prepared to add a footnote reference to the *Metron* paper and to change the reference to Irwin and to Wilks.

□ □ □

In these years Fisher was defending and expounding his views on a variety of subjects in tough but generally goodhumored debate. This was the time of the most lively discussion of the eugenic policies that he endorsed. There were public differences of opinion in genetical circles on dominance theory and natural selection. Fisher's repudiation of inverse probability in 1930 was still a subject of debate in the literature; Fisher was also trying to make himself understood on other questions of estimation and inductive inference. In contrast, the debates on randomization were acrimonious. Fisher even embroiled himself in hot dispute with W. S. Gosset [48], when Gosset, speaking before the Royal Statistical Society in March 1936, advocated a systematic field arrangement, Beaven's half-drill strip method, and argued that its lack of randomization was an advantage.

Fisher revered Gosset, his oldest friend in statistics, whose early work had inspired his own. He had kept close contact with Gosset in extending the applications of Student's test, based on randomization, and he might have hoped that Gosset of all men should understand his aim. Yet he knew that Gosset had always been ambivalent about randomization. Could he not have let it pass if, 13 years after the subject had entered their correspondence, Gosset still believed a systematic experiment to be inherently more accurate than a randomized one?

Fisher could never let pass what he believed to be wrong reasoning. He was known on other occasions to argue the case, even with someone who produced a correct answer, if the reasoning on which it was based was not sound. To leave unchallenged what was irrational was to open the door to all untruth.

Of course, the matter did not appear in this light to Gosset or to his audience. Gosset had chosen a balanced sandwich design to counteract the expected fertility gradient of the field. He had chosen a systematic design for ease of cultivation. In the long strips of the barley field, adjacent half-drills were seeded with the same variety and the varieties were alternated across the field, so that an array of sandwiches resulted in unvarying form: *ABBA, ABBA. . . .* For this arrangement he had sacrificed the advantages of randomization, and there was perhaps no cause to quarrel with him on that account, if it was his deliberate choice.

The way he presented it, however, anticipated criticism of the lack of randomization by claiming that because the experiment was not randomized the results would be more precise. Referring to Fisher's introduction of randomization, he remarked, "This enabled us to obtain a certainly valid estimate of the variability of our results, though usually at the expense of increasing that variability when compared with balanced experiments." And, again, "Since the tendency of deliberate randomizing is to increase the error, a balanced arrangement like the half-drill is best." In both cases he referred to a "balanced" design, meaning systematic. A randomized design can also be balanced, but Gosset did not mention such an arrangement (except in a footnote in his later paper [49] to dismiss it as impracticable in this instance).

Fisher challenged the assumption that randomization tends to increase experimental error. From the beginning this had been a common objection against randomization, despite the fact that Fisher had shown, at its introduction, that randomization does not detract from the precision possible with properly designed experiments. The experience of a decade supported his original demonstration. Gosset, he thought, was arguing from a false premise.

Gosset went further. Although he admitted that the lack of randomization of the arrangement might be used to impugn the validity of his conclusions, he asserted that the bias introduced would tend to increase the estimated error and thus assure that chance effects would rarely appear to be significant. Indeed, in his subsequent paper [49] he claimed that "Conclusions drawn from experiments giving significant results by such [systematic] arrangements are *more* valid in the ordinary sense of the word." Fisher objected that the bias might go either way to an unknown extent. It could not rationally be claimed as a virtue in the arrangement that the estimates were contaminated by bias.

The debate grew hot. Gosset had the courage of his convictions, the same pertinacity that had fortified him to pursue his own researches through the years when they had seemed important to no one else. He stood his ground against Fisher and left him seething with rage.

After the public quarrel, Fisher clarified his position in a paper with Dr. S. Barbacki [*C P* 139], using rather abundant data from a uniformity trial with wheat. With these data he compared the actual precision with the estimates for two cases: the systematic sandwiches as compared with randomized sandwiches and the systematic pairs as against randomized pairs of observations. With these particular data (in contrast with Gosset's expectation and Tedin's findings in another case), the actual error was considerably *increased* by the systematic arrangement and the estimated error was decreased. Moreover, the estimate of error was very different according to whether one chose to consider pairs or sandwiches. Since with the systematic arrangement used there was no objective justification for choosing the one rather than the other, the results were ambiguous in a way that was impossible with a randomized arrangement, where the sandwiches, pairs, or any other chosen units of randomization would themselves dictate the appropriate estimator.

Gosset responded immediately in a letter to *Nature* (which Fisher answered, [*C P* 140]) and later in an article which appeared in *Biometrika* after his death [49]. They were still at cross-purposes and Gosset reiterated his defence of the balance rather than of the system in his arrangement.

On the basis of what was practicable in field trials, Gosset's arrangement had something to be said for it which Fisher's criticism did not touch. Nor did Gosset's response touch the correctness of Fisher's argument on behalf of randomization. It is a pity that the two great scientists and great friends should have spent the last year of Gosset's life divided by a quarrel which did credit to the understanding of neither. Their papers are sharp, if not actually unfair to each other. Fisher's initial rage was excessive, and his later actions lacked magnanimity. He had been touched on a tender point; it was a matter of right reasoning and he was adamant.

It is as well to recall here that, although this scientific disagreement was hotly pursued on both sides, it was not a negation of the friendship and mutual admiration they felt for each other but rather the reverse. They would have been less moved had they valued the scientific opinion of the other less highly. Earlier differences of opinion between them had been argued with energy, though less publicly, and their friendship had grown. Gosset had become a fellow of the Eugenics Education Society and much interested in evolutionary genetics, so that they shared an ever broader range of scientific interests. The obituary and appreciation of Student written by Fisher in 1937 [C P 165] expressed in most generous and warm terms, his genuine admiration both of the man and his work.

□ □ □

The debates on randomization theory illustrate a real problem faced by Fisher in expounding his ideas: statistics lies at the intersection between two different modes of thought, of natural science and of mathematics. Whereas it is obvious to a scientist that his understanding of the biological situation does not in itself enable him to handle statistical aspects of his work, it is less obvious to a mathematical statistician, though equally true, that his understanding of statistical theory alone does not enable him to handle statistical problems of the real world. Indeed, it is even possible to be led astray by mathematical considerations, as, it seemed to Fisher, Neyman had been in the case of the z test. Neyman himself observed that the bias he found disappeared under certain conditions, and Fisher pointed out that these were precisely the conditions that applied when the null hypothesis was true, so that although the bias existed in the mathematics, it was irrelevant in practice in testing the null hypothesis.

Fisher was aware that this aspect of statistics was disconcerting for mathematical students and that they were often reluctant to encounter the real world as a genuine statistician must. His attitude to statistical training is apparent from the way he phrased his proposal for collaboration with Haldane in May 1933:

Supposing mathematically trained lads come to me hoping to get some sort of doctorate by working in my department, knowing nothing, and not very willing to know
anything, of experimentation with living material, can I make them attend lectures in
your department on genetical theory, as, at any rate, one step towards apprehending
the kinds of reasoning used by experimenters?

For him, contact with experimental material and problems of experimentation
were a necessary complement to understanding of the theory.

A regular exercise set to students in the 1930s was working out complex
designs. Every day Fisher would put up on the blackboard a brief specification of the experimental problem, and everyone was expected to work out an
appropriate design, the factorial, the confounding scheme for balanced incomplete blocks, partially balanced incomplete blocks, as required.

Fisher's lectures on the logic of experimentation were not conceived as exposition of mathematical theory of statistics but mainly concerned the logical
connections between mathematical concepts and practical problems of
design or analysis. Students profited according to their resourcefulness, for
the lectures were meaty and Fisher's exposition was not always clear. Sometimes he would skip over connecting links of the argument, leaving students
to fill in for themselves if they were curious enough. Probably he did not
prepare the lectures in detail, for he was known occasionally to forget a relevant proof and have to work it out (or fail to) on the blackboard during class
or when students came to him for explanation.

For example, we quote Cochran's [50] account of "Fisher explaining a
proof."

In one of his lecture courses, he quoted without proof a neat result for what appeared
a complex problem. Since all my attempts to prove the result foundered in a maze of
algebra, I asked him one day if he would show me how to do the proof. He stated that
he had written out a proof, but after opening several file drawers haphazardly, all apparently full of a miscellaneous jumble of papers, he decided that it would be quicker
to develop the proof anew. We sat down and he wrote the same equation from which
I had started. "The obvious development is in this direction," he said, and wrote an
expression two lines in length. After "Then I suppose we have to expand this," he
produced a three-line equation. I nodded—I had been there too. He scrutinized this
expression with obvious distaste and began to stroke his beard. "The only course
seems to be this" led to an expression four and one half lines long. His frown was now
thunderous. There was a silence, apart from beard stroking, for about 45 seconds.
"Well," he said, "the result must come out something like this" and wrote down the
compact expression which I had asked him to prove. Class dismissed.

Fisher's students bear unanimous witness to his generosity in handing out
ideas for them to work on. D. J. Finney recalls:

He put before me ideas for a whole series of papers on genetic linkage in man. Certainly the detailed algebra was mine but the inspiration and almost all the originality was his. He did the same kind of thing for Panse in his work on quantitative inheritance. As I came to know this habit of his, I learned to look at publications of others who were at some time his students and to detect ideas that almost certainly came from him.

Once the work was in hand, Fisher was no less generous in giving help and encouragement when students brought their problems to him.

In 1934, C. I. Bliss [51] was analyzing data on the effect of insecticides by a method of assay similar to that developed by J. H. Gaddum (1933). He has recorded what happened when he started talking to Fisher about it:

We were discussing probit analysis on a Saturday in Harpenden. In computing the regression of percent kill in probits against the logdose of insecticide, the empirical probits for zero and 100 percent kill are minus and plus infinity. I proposed assuming in these groups that one half an insect died or lived so that they could be included in the regression. He countered that these points were of little value and might better be omitted. I protested that they were as real observations as any others in the series and should not be dropped just because of the limitation of the transform at these levels. He was silent for a moment and then remarked, "When a biologist believes there is information in an observation, it is up to the statistician to get it out." After lunch that day he derived the maximum likelihood solution of the dosage-mortality curve, in which the empirical probits led to working probits based upon all of the information in each quantal response, including those at zero and 100 percent.

Fisher's contribution was published in an appendix [C P 126] to Bliss's paper in 1935. It was almost his only direct contribution to the subject of assay of drugs and toxicants. Like his singular contribution to the extreme value problem, the question arose as a practical difficulty of statistical analysis and, having solved it, he went on to consider other problems, leaving it to others to exploit his discovery.

In addition to Fisher's sense of responsibility to the biologist, to get the information out for him, his characteristic response was to formulate the problem in an original manner so as to render it tractable. A few days before Christmas 1935, H. Fairfield Smith handed him a long typescript in which he had attempted to compare different varieties of wheat which had been characterized for a number of heritable qualities. On Boxing Day, a public holiday, he and Fisher were alone at the laboratory when Fisher appeared at his door and asked him out to lunch. In the restaurant, Fisher turned over the menu, remarking, "I think this may be what you want," and reformulated the problem in terms of discriminant functions. These achieved the object with elegance, within the bounds of the menu card. In this case it happened that

Fisher's paper [C P 138, 1936] on the discriminant function applied to a taxonomic problem, was already in the press, but he was quick to see how the new idea could be applied to Smith's different problem, and to a range of others. Often reformulation of a problem made it accessible to treatment by known and tried statistical techniques.

□ □ □

As the eugenics department settled down under its new professor, the Galton Laboratory appeared a perfect example of a father-image society, as R. M. W. Travers, who was a psychology student there in 1935–1937, re- marked. The "children" had their own interests, and they squabbled among themselves, while Fisher rode imperturbably over all. Fisher's administration was loose and paternal. He did not interfere with the way his assistants chose to work (if Stevens did not often arrive before midday, that was his affair, so long as the work progressed) but he did keep in touch with each of them, partly through his regular work with the mice and in the computing room, sometimes by dropping in to their rooms for a talk or lunching together, and most surely by making afternoon tea in the Common Room an occasion for talking things over.

Fisher liked to have members of his department sit down at tea time on the cushioned benches round the walls of the room, so that they could all join in the conversation. Tea was laid out on a table in the center. At Fisher's insistence, two loaves of bread were provided, white and brown, and a large pat of butter, in addition to cake or biscuits made for them by the senior cleaner, Mrs. Randall. Tea was to be enjoyed in an unhurried manner and made the opportunity for discussing ideas and questions arising day by day.

Staff and students sometimes sought him in his office or with his secretary, Miss Simpson, and the laboratory cat. This animal Fisher had named Hathor, after the Egyptian goddess of love, but to the laboratory she was always known as Chi-squared. The name expresses something of the awe and affec- tion with which they regarded Fisher.

They saw him often in the computing room. He liked his Millionaire calcu- lating machine and was disdainful of the up-to-date desk calculators which were in plentiful supply by the mid-1930s. The Millionaire stood on its own stand. It was clumsy to move and operate and it made a noise like an old- fashioned threshing machine. Fisher defended it because it did multiply instead of doing multiplication by repeated addition like other calculators. He could make it perform complicated calculations both quickly and accurately. (Others also who worked much with the machine appreciated its virtues, like Frank Yates at Rothamsted, who for this volume (Plate 23) has kindly permitted himself to be photographed at the Millionaire, still in his office at Rothamsted in 1974.) Helen Turner recalls how, coming from Australia in 1938 to be his student, she first met Fisher in the computing room. Having heard tales of the Millionaire, she quizzed him about one speculation, that he had picked it up in the Portobello Road for a mere song. He laughed heartily and said he only wished it were true; in fact he had paid quite a lot of money for it.

At that time he was making calculations of the discriminant function for some rough subjective scores for blood characteristics. Between the reactions scored as negative and positive (and given arbitrary values 0 and 1), there were three intermediate categories: dubious, weak, and probably positive reactions. To these he wished to attribute numerical values. The process required solution of a fourth-degree equation which he did by a numerical method in which rounding errors could become important. He therefore carried his calculation to more decimal places than the Millionaire could hold and had to enter his figures twice. Students were somewhat amused at the contrast between the roughness of the raw data and the precision of the calculations. (When these calculations were used to illustrate the method in later editions of *Statistical Methods*, the values of the three intermediate scores were given to six decimal places.)

Among the group at the Galton there was an affectionate solicitude for Fisher. Every evening Miss Simpson would remind him when it was time to go to catch the train home. One day, after he had left, Horace Norton was standing by the window looking out into the foggy evening and the busy traffic along Gower Street. Norton could see Fisher crossing the street below. He shook his head and said, "You know, that man's too valuable to let out alone." Cochran [50] remembers a day when he and Fisher were standing at the corner of Euston Road and Gower Street, waiting to cross the road on their way to St. Pancras Station.

Traffic was almost continuous and I was worried, because Fisher could scarcely see and I would have to steer him safely across the road. Finally there was a gap, but clearly not large enough to get us across. Before I could stop him he stepped into the stream, crying over his left shoulder, "Oh, come on Cochran. A spot of natural selection won't hurt us."

Students who lived in Harpenden often traveled on the train with him. At one time or another all must have visited his home. In part, this was a way to ensure that they saw living experimental material not only at the Galton Laboratory but at Rothamsted and in the field. When C. I. Bliss spent a year at the Galton, he, being interested in entomology, was taken out to hunt grouse locusts. One Sunday Fisher and he traveled down to the hunting grounds in Savernake Forest in Wiltshire, where their host, A. G. Lowndes, showed them round. On Monday Bliss set out on a borrowed bicycle, carrying containers for his catch, to beat the bracken on the fringes of the woods. Because of Fisher's interest in breeding grouse locusts, his friend Lowndes, science master at Marlborough College, also took groups of schoolboys to collect these insects.

On most weekends Fisher entertained guests at his home. During the morning they would walk over to the poultry experiments which continued on Rothamsted Farm until 1937. Or they might make a longer expedition to count rooks' nests. When Fisher had realized how little information existed about actual numbers of birds of different species, he had begun a continuing census of the rooks in his neighborhood. Even with his poor eyesight he could observe rooks' nests, for they are large structures and obvious in the topmost branches of tall trees. He covered an area 4 by 6 miles in extent surrounding Harpenden and recorded the position of all rookeries found, together with the numbers of nests, in each successive year from 1931 until he moved from Harpenden in 1943.

In winter, rooks of a whole neighborhood gather for the night at a central rookery. They arrive not in small groups from individual rookeries but in flocks that fill the sky like soot: for Fisher's guests it was a great spectacle, viewed against a winter sunset. Each arriving flock rouses earlier arrivals who fly up from their nests and circle about, cawing noisily. The gathering of the clans, however, takes place all through the afternoon. The rooks make a tour of rookeries, picking up local populations at each stop. Fisher began to trace out the regular itineraries the rooks followed, piecing them together gradually, for afoot he could not travel "as the crow flies." Wondering how a chosen route might have developed, he was interested to observe that sometimes an unscheduled stop at unoccupied trees during the winter was followed in the spring by the building of nests in a new rookery there.

He would take guests at a brisk pace for miles around Harpenden. He had an uncanny faculty for getting hold of important people and making them do what he wanted. Ford recalls an occasion when he unexpectedly produced the Swedish ambassador (whose home in Harpenden contained a small rookery), took him to lunch, and "made him with myself, trudge over Hertfordshire fields and countryside in the rain to look at some archeological site which, as far as we could determine, seemed represented by a gateway in an ordinary field hedge." After a strenuous afternoon, guests were glad to sit down to family tea and afterward, perhaps, have a chance for less distracted conversation with Fisher in the library.

After 1934, when the work on purple loosestrife, *Lythrum salicaria,* began, Fisher had a plot of *Lythrum* to show visitors to Harpenden, as well as plants in the wild. Occasionally he invited friends to join him on holiday at his sister's home in Argyllshire where he found several colonies growing wild. The plants grew in the wettest parts, like those "in a colony extending round two sides of a swampy field in Auchnagarron Farm at the head of Loch Riddon," and he would return from his expeditions wet to the knees.

Visitors were taken to Rothamsted, of course, where Frank Yates was head of the statistics department. Yates had been appointed assistant statistician in the department in 1931, when Wishart left for Cambridge. He quickly grasped the essentials of Fisher's approach and became enthusiastic about the new factorial designs and application of the principles of experimental design to new problems. His promotion followed naturally when Fisher moved to the Galton Laboratory in 1933. He was to remain at this post until his retirement in 1968, a key figure not only in the growth of the department and of the experimental station but throughout the world of statistics, during the period of tremendous expansion and consolidation of the subject which followed.

Fisher saw Yates frequently in these years in connection with the volume of statistical tables which they were preparing, *Statistical Tables for Biological, Medical and Agricultural Research,* which appeared in 1938. The authors took great pains to make the volume serviceable to research workers. Their pains were rewarded; as early as 1943, despite the restrictions of war, an enlarged and revised second edition was brought out. Today, the volume, in its sixth edition, is considered indispensible in almost every research laboratory or statistical department.

It was in Yates's office that Fairfield Smith, coming to England to be Fisher's student in 1935, first saw Fisher,

holding forth to two other visitors on a new idea he had just hatched for incomplete blocks. This was exciting since we had been wrestling for the last few years in Australia with how to put about 64 varieties into an 8 × 8 Latin Square. Later we went for a walk round the Farm. It was a tremendous experience to see how intimately he knew every detail of what was going on.

On his first visit to Milton Lodge with Ken Mather one Sunday, Smith was struck by the presence in the diningroom of a chicken coop and in the library by the bottles of alcohol and potash and test tubes in which transparencies of chicken bones were being prepared, among which they set their glasses of cherry brandy. On another occasion a reporter for the local newspaper, waiting for his interview with Fisher, was rather overwhelmingly entertained by innumerable children coming and going who displayed in turn a young family of experimental puppies, a litter of nonexperimental kittens, the baby chickens and the toffee tins sitting on the shelves in the drawing room, used for the breeding of snails.

Other visitors had glimpses of different aspects of Fisher's home life. Dr.

H. N. Turner recalls that on her first visit, someone had just given the professor a new set of intelligence tests to experiment with, and the first thing she was presented with on arrival was the test to go through. Paul Rider recalls, in particular, a time when Stefan Barbacki and he were guests at Milton Lodge:

Fisher took Barbacki and me to see the Rothamsted Experimental Station. He also took us on an exploratory tour of the site of the old Roman fort in the neighbourhood. He pointed out to us where the entrance to the fort had been, where certain walls and fortifications had been, where the horses had been kept, etc. There was one sort of hole or pit which he said was where bears had been kept for baiting. Barbacki, who was Polish and did not know much English, could not understand the word "bear." Quickly Fisher used the Latin word "ursus" and Barbacki understood at once.

Another thing that I remember about Harpenden was a display of mouse skins that were the result of a genetic study. Some of these were beautiful in both colour and markings and if the mouse were a larger animal they would have made some fine lady a lovely coat.

Rider also remembers the day when he with his wife and three children left their lodgings in London to join in an almost Dickensian English Christmas feast: the big family around the laden dining table, the good food, and the English cider that accompanied it.

Tradition shaped and ordered the events of the Christmas season. According to custom, on Christmas Eve Fisher led the older children out to collect holly to decorate the house; for weeks they had been scanning the neighborhood for the most richly berried trees. As usual, they marched home with the holly slung across the length of the long-handled clippers they had used to secure the boughs, singing Christmas carols as they went. In the afternoon the children organized a streamlined assembly unit to glue strips of colored paper together into chains to be strung across the ceilings of the living rooms. Special china had been brought out for the festive table, silver had been polished, and stored apples selected and rubbed to a high gloss and set in the silver basket with other fruits. The coal-burning kitchen stove for days past had been stoked steadily as Christmas cake and mince pies were prepared and a big ham boiled. On Christmas Eve ingredients for the Christmas pudding were prepared and the mixing bowl was carried out for Fisher to stir the Christmas pudding ceremoniously. All the household stirred and wished after him.

On the great day itself tradition still ruled: at 7 a.m. the children were allowed to get up. Each found his largest knee sock, hung at the foot of his bed, had been stuffed by Father Christmas in the night. (Once only the existence of Father Christmas was challenged, and one little boy stayed awake and saw for himself the figure who entered and filled the socks; he wore an unfamiliar long gown of scarlet silk, whose pink facings might have revealed to a more sophisticated watcher that Father Christmas was a doctor of science from Cambridge University; a mask with curling silvery beard hid his features.) Stockings in hand, the children went to the boys' bedroom to explore the contents together, down to the tangerine which bulked large in the toe of each. When that was eaten, the family gathered again, this time outside their parents' bedroom door, and "awakened" them with a Christmas carol. Then everyone burst into the room and into the bed for a romp.

There was a special breakfast, with fruit salad or eggs and bacon in addition to the usual porridge, and on this as on every other feast day the porridge was treated to a handful of raisins. Then the formal presentations began: first the youngest gave her gift to her father, the oldest present, and so on down to the youngest; then the second youngest took her turn; and so it continued until the father had given his last gift. When the gifts within the family had been all presented, parcels from relations and neighbors were distributed.

Then it was almost time to go to church, and the whole family trooped out to morning service. They did not stay for the sermon, however, for Mrs. Fisher and one or two of the girls would return to complete preparations for Christmas dinner, and everyone else took a walk for an hour to get an appetite for the meal. When they returned they would find the table laid, with all the splendor of bright linen, china, and silver, and the bowls of delicacies in the center of the table; the ham clothed in golden breadcrumbs and white paper ruff was set on the sideboard, to supplement the roast—a turkey or goose, or perhaps a couple of cockerels from the yard; bottles of beer and cider stood ready, and even the children were allowed to taste cider with their Christmas dinner.

When the gong in the hall was rung and the family assembled, each by his alloted place, the father said grace, the long Latin grace which was reserved for birthdays and feast days, delivering with slow solemnity the syllables he had heard read daily at dinner when he had been a student at Cambridge. It was the father's task to carve the bird, after an expert flourish of the carving knife against the steel. It was the father's task also to portion out the Christmas pudding when, in a blaze of brandy, it was brought into the room. If it were to contain silver threepenny bits, as it sometimes did, it was his responsibility not to disappoint any child of one of the hoped-for coins. Then, after the last nut or grape or syrupy piece of preserved ginger was consumed, the paper crackers were lifted from their decorative piles on the table, and everyone chose a partner and pulled; the halves tore apart with an explosive crack, releasing the "surprises" within; paper hats appeared on many heads and it was time to move from the table, the adults to drink coffee in the drawing-room.

Visitors were more or less coopted into family life, and they put up with it remarkably well. Those with cars would carry off some of the family party on expeditions to Whipsnade Zoo, while others bicycled. On one drive Fisher was pleased to direct Jack Youden to tunnel under the high hedges, down the narrow rutted lanes which, doomed to disappear after the war, still existed within a mile of Milton Lodge, thus giving a new dimension to his experience of driving. If there were visiting children, they too were coopted among the wild surge of Fisher children. Fairfield Smith recalls them "teaching my shy four-year-old to climb trees." Also, "There was one Sunday evening of hilarious charades in the drawing-room. I remember, too, their boisterous greeting of a previously unknown visitor from overseas whose name happened to coincide with a well-known limerick!" (D. Dugué may well have been somewhat disconcerted at the association of his name with "the gay dugong and the manatee.") On Sunday evenings Fisher regularly organized word games with the children. In the summer he would often organize and score their efforts at high jumping.

The children also were brought into Fisher's activities and investigations. They accompanied him on walks and visits to the chicken pens. They tried out the intelligence tests. They were as familiar with the "asp" of the serologists as members of the department, and they did not take any more kindly to it than Paul Rider did:

All of us in the Lab were subjected to bites by a particularly vicious blood-letting instrument referred to as an asp. We had sore fingers and ear lobes for days. Fisher did not spare himself. I remember one day when he went about with a bit of cotton wool stuck on to an ear lobe. The asp must have taken an especially savage bite, because the cotton did not stop the flow of blood, which had run down and soiled his white shirt collar.

Fisher boasted to E. W. Barnes, Bishop of Birmingham and godfather of his youngest daughter, that at 6 months she had already been recruited as the subject of serological research. Dr. W. J. B. Riddell visited Milton Lodge more than once to classify their eye and hair colors. Fisher brought home calipers to measure length and breadth of their skulls. The older children visited the Galton Laboratory, and, like all other visitors and students there, they went through the battery of taste tests from *a* to *z*. The samples of clear fluid were to be classified as sweet, salt, acid, or bitter. In fact, they tested discrimination of the presence of phenyl thiocarbamide, which was present in some of the samples at various different concentrations. Thus, not only were tasters and nontasters distinguished, but as data accumulated, the variability in threshhold of individual ability to taste could be assessed.

Fisher was very aware of the role he could play as host for visitors from overseas, in introducing them while they were in England to English scientists and to their experimental work. Such contacts were made at the Galton, Rothamsted, and Milton Lodge. The Natural Science Club at University College, too, was a place of pleasant meetings with fellow scientists, which Fisher enjoyed with his visitors. It was Fisher who had the idea of getting glass-bottomed pewter beer mugs for the Club and, through a former school friend associated with a brewery, he acquired the genuine article he had in mind. When he left the Galton Laboratory in 1943, it therefore seemed to his staff appropriate to present him with a glass-bottomed pewter mug; this they had engraved with a symmetrical 6×6 Latin square, using the letters of his name: the exact form of the square was determined by the letters in the first two lines:

F I S H E R

. . F R S .

He was delighted.

Naturally, visitors often came to England with plans to visit individuals and institutions elsewhere, and Fisher opened up opportunities of the same sort which had not occurred to them. He encouraged them to attend the meetings of the Genetical and Royal Statistical societies, and he invited them to go with himself to other meetings of special interest. Some he took to the Royal Society Club or to soirees at the Royal Society; many he persuaded to attend the annual meeting of the British Association. During the latter week-long gatherings of scientists, introductions were made between the visitors and a wider range of English scientists than they had previously met, new contacts explored in the informal atmosphere of shared meals and expeditions together, and a desirable chance offered to present and to discuss their own work in public. In 1938 Fisher was particularly gratified to have no less than three of his students from the United States, C. C. Craig, W. J. Youden, and H. Norton, together with W. L. Stevens, his statistical assistant at the Galton, and F. Yates of Rothamsted, all speaking at one of the meetings of the British Association Mathematics Section. Several of their papers appeared soon after in the *Annals of Eugenics.*

Fisher was grateful, during these years, that as Galton Professor he was editor of the *Annals of Eugenics.* He took his responsibility seriously, both in its obligations to the authors and to the public. The quarterly became distinguished for the quality of the articles that appeared in it on genetical, statistical, and eugenical topics. Contributors found it exceptional also for the speed and ease with which their articles passed into print. This was due to Fisher's dislike of the referee system, with its delays and often petty criticisms. His experience as an author led him to attribute dilatory and irresponsible refereeing to editors who shuffled off their responsibility on unnamed referees who could not be called to account. He held that whatever refereeing was done, the responsibility for the decision to publish or not rested with the editor alone, and, if something was worth publishing, the sooner it was done the better for all concerned. He welcomed statistically amateurish contributions if they dealt with the data of some real problem, and he would offer ideas for improving the statistical techniques.

The *Annals* contained a number of Fisher's own articles and many articles by his students and colleagues. This was natural with a journal whose scientific role had yet to be shaped, and that was shaped by the editor's definition of his subject. In this sense it was the organ of a philosophy of science deriving from Fisher himself. It reflected the breadth of his own interest in the whole gamut of subjects that might, directly or indirectly, contribute to an understanding of human heredity, the emphasis in his own work on the problems of the real world and his readiness to use unconventional and, where necessary, highly theoretical mathematical means of dealing with such problems. There is much in its pages during the decade of Fisher's editorship that is still interesting and stimulating to the modern scientist. But it is a difficult journal to place satisfactorily in a niche on the library shelves. It resembles its editor. Predominantly genetical in its subject matter, largely mathematical in its treatment, its interest medical, agricultural, sociological, statistical, it is too deep to be classified as general, too broad to be lodged with any one speciality.

□ □ □

On considering eugenic policies it had become obvious that the genetic basis for failures of the nervous system was usually unknown. Facts were wanted in general and in particular. Dr. Bell, collecting pedigrees for *The Treasury of Human Inheritance,* turned her attention to genealogies of cases of Huntington's chorea, a rare type of heritable insanity. The disease was known to be due to the presence of a dominant genetic factor whose effect usually becomes evident in middle life but which occasionally fails to exhibit itself. On appearance it results in progressive nervous disability leading to insanity. The children of a father who develops symptoms of this disease, therefore, have an equal chance of being normal or affected: either they carry the factor or else they are entirely free from the dangers either of developing the disease themselves or of passing it on to their children. But they usually do not know which is their fate until after their own children are born.

Considering the genetic possibilities of improving prognosis in such cases, Fisher decided it would be worthwhile to determine as many as possible of the linkage groups in man. With only 23 pairs of human chromosomes, he felt it should be possible in time to map increasing portions of these chromosomes. Hence he investigated among his students and, where possible, in clinical and geneological studies the many slight but distinct differences known to be hereditary. Classification was made for earlobe attachment, hair on the second joint of the fingers, blood groups, and ability to taste phenyl thiocarbamide (and in this connection of smoking habits).

Fisher argued that any one of these inherited traits, harmless in itself, might be found to be linked with some factor, like that for Huntington's chorea, which it was desirable to have "marked." If such a linkage were found, it would be possible to rate the chances of possible victims more accurately, to be in a position to say, for instance, that the child who exhibited the linked factor in the same form as his father had an 80:20 chance of carrying the disease factor and, on the other hand, the child who did not so exhibit the "marker" had a 20:80 chance. Though such statements would not be ideal, they would give some real guidance to potential victims in making the difficult decisions that faced them. Fisher knew that the search for linkage would be long and at first disappointing, but he felt that it ought to be begun and might eventually revolutionize prognosis of heritable diseases.

Soon after moving to the Galton Laboratory, he suggested that it was desirable to set up a counseling service at the eugenics department which could serve those of the public who desired eugenic advice, those, for instance, who were found to carry or feared they might carry some dangerous or debilitating genetic factor. At the time the idea was not adopted, but a generation later this sort of service was beginning to be performed by members of genetics departments in cooperation with medical organizations.

Of Mendelian factors in man, Fisher was most hopeful of the blood groups. He had reason to suppose that there would be found to be a very large number of identifiable genetic distinctions to be made through serological research, in addition to the familiar ABO series. Landsteiner and Levine in 1927 had already discovered one other group, involving the M, N, MN genotypes, which were distinguishable using immune serum from rabbits. Fisher, in initiating serological research at the Galton Laboratory, looked forward to the discovery of others, through the use of immune serum from a variety of animals.

In 1935 the Eugenics Education Society established two Leonard Darwin Research Studentships for the study of questions of race mixture and the inheritance of intelligence. Several students elected to these scholarships worked in the Galton Laboratory. First R. B. Cattell [52], between 1935 and 1937, undertook a survey in Leicester and the surrounding country districts, in which the performance in intelligence tests of school children was recorded, together with the number of siblings of each child and the type of school attended (state supported or fee paying). The analysis of these data provided a quantitative measure of the variability in "intelligence" as defined by the test, a measure of the differential birthrate for this measure of intelligence, and a measure of the association between family size and educational opportunities. The results illustrated the force (and the incidence) of the economic advantage of small family size. Similarly, R. M. W. Travers [52] between 1936 and 1938, made a study of psychometric differences and occupational selection.

Such studies at the Galton Laboratory were naturally hedged about with precautions in the statistical design and analysis. This was not always the case elsewhere, and Fisher was saddened to see investigations into birthrate and human intelligence, undertaken at this time on a scale he could not possibly afford, rendered worthless for lack of statistical forethought. One sociological investigation, subject to comment in the literature, exhibits the sort of errors in planning which so easily squandered research effort in the social sciences. This was the famous Lanarkshire milk experiment.

In this experiment, nearly 10,000 children received a supplementary ration of three-quarters of a pint of milk daily for about 4 months, and their increase in height and weight during that period were compared with that of children receiving no milk supplement. About half the children received raw milk and half milk from the same source that had been pasteurized. The recipients of the milk, however, were not similarly divided, for each school received either raw or pasteurized milk only. Consequently, as Gosset [53] pointed out in his criticism of the experiment, the equivalence of the contrasted groups of children (of like sex and age) was in doubt from the beginning. In this case Gosset entirely agreed with Fisher about the need for randomization. First, it might be that the children had not been selected at random within the school but that humane staff had been tempted to put those who most appeared to need the extra nourishment into the milk-fed group. If the children were selected strictly at random, the experiment might still be vitiated. In a poor district the increase in weight might be due to the additional nourishment irrespective of the kind of milk supplied; it might be further enhanced, in comparison with more prosperous districts, by the generally lighter weight of winter clothing worn by the poorer children at the time of the first measurements, while the weight of summer clothing at the time of the final measurement was equal. In the absence of records from the separate schools it was impossible to eliminate the doubts that the chosen method of allocating the milk had introduced.

The analysis of the data was equally misleading. The conclusion of the report was that "the effects of raw and pasteurized milk on growth in weight and height are, so far as we can judge, equal." This seemed to be wishful thinking by promoters of pasteurization, for comparison of the gross increases suggested there was a difference in favor of raw milk, which a more sensitive analysis might bring out. Fisher and S. Bartlett (C P 92, with S. Bartlett, 1931) therefore calculated from the published tables the average increases in height and weight, standardized for age, for the whole group of children observed— the control group and the two milk-fed groups. The excess due to milk feeding was then obtained by subtraction, and the relative value of raw and pasteurized milk could then be assessed. Although the increase in weight due to supplementary milk amounted to only 4 ounces, pasteurized milk gave a distinctly poorer result, the return being for boys only two-thirds of that for raw milk and for girls a smaller difference in the same direction. In respect of increase in height, again, there was little difference due to milk-feeding but that little showed the value of pasteurized milk to be for boys only half that of raw milk and for girls about 70%.

□ □ □

The study of human genetics is in its nature limited by the fact that controlled breeding is impossible with human populations. It is possible, however, to learn about human inheritance by analogy: what is found to be true of a sufficient variety of animals may be applicable also to men. For this reason breeding programs with animals and plants became essential to the elucidation of human genetics. For example, if a theory of dominance were shown to be applicable to a number of species, it might plausibly be used as a guide to the general character of the dominance relations in man. Therefore, investigations were undertaken to test whether Fisher's theory of the evolution of dominance was confirmed and supported in detail by a wide range of different organisms: the poultry experiments continued, and snails, grouse locusts and mice were bred at the Galton Laboratory.

A number of problems could be helpfully formulated in terms of population genetics, and Fisher made studies, for example, of the detection of linkage with dominant and with recessive abnormalities and looked into the conditions in which deleterious mutations might be sheltered or accumulated in the genetic complex [C P 131, 132, 133, 1935].

The common occurrence in some species of lethal factors and the discovery of systems of inheritance involving more than one lethal factor presented interesting problems as to the evolutionary formation of such groupings. H. Muller's [54] brilliant demonstration of balanced systems of lethal factors in *Drosophila* was followed by other discoveries of balanced lethality, such as O. Winge's [55] explanation of eversporting varieties of stocks, in which the plants with single flowers, when crossed among themselves, give progeny half with single and half with double flowers. According to Winge's theory, the ever-sporting singles, from which the doubles are derived generation after generation, are heterozygous for two closely linked lethals carried on homologous chromosomes. Apart from recombinations, the pollen all contains the gene for doubleness, while the ovules are of two kinds, one containing the gene for doubleness and the other the pollen lethal.

Interpreting data on ever-sporting stocks, on the basis of Winge's theory, Fisher [*C P* 105, 1933] was able to show that the slight predominance of double plants observed in the progeny was due to differential viability in their favor. He also adduced evidence that under human selection the linkage between the pollen lethal and doubleness had become much closer, presumably because those strains in which recombination occurred more freely were continually eliminated. The study made a nice illustration of the plasticity of gene arrangement on the chromosome, by which desirable genetic combinations could be formed and stabilized under natural selection.

Another case then thought to be due to two closely linked genes, both lethal as homozygotes, occurred in the purple loosestrife, *Lythrum salicaria*. The trimorphic flowers differ somewhat in other respects but are characterized by the long, mid, or short length of the style. Early studies by Charles Darwin (1877) and by his granddaughter Lady Barlow (1913, 1923) had shown that the long style was due to a homozygous recessive and that the short style was due to a different genetic factor which reduced either long or mid to the short manifestation. Mid, however, seemed anomalous. It could not be attributed to a single dominant at the same locus as long for two reasons. First, the segregation of progeny from some long × mid crosses did not give a 1:1 ratio of mid to long but something like a 5:1 ratio. Second, no homozygous mid had been found. In 1927, therefore, E. M. East [56] proposed that mid might be due to the presence of two closely linked genes, both of which were lethal when homozygous. In 1932, however, he modified his proposal, for he had found an apparently homozygous mid plant which in his experiment gave all but one mid progeny. East's publications interested Fisher in the problem, and in 1934 he calculated the selective consequences of East's (1927) interpretation and found that if the lethal genes existed, their close linkage was to be expected as a result of the type of selection found [*C P* 130, 1935].

The theory of two linked lethal factors, however, required experimental verification, and in October 1934 Fisher sent his discussion of East's theory to Haldane, asking whether it might be possible for him to set up an open-pollinated plot of *Lythrum* plants at the John Innes Horticultural Institute at Merton (where Haldane worked parttime). He wanted to obtain seedlings from his open-pollinated plot, "say 1,000 plants annually for five years," and check the theoretical equilibrium values against values actually obtained. Earlier he had suggested the idea to Lady Barlow; now he was planning to put it into practice himself. He did not believe that satisfactory counts could be obtained in wild populations, to establish the relative abundance of the phenotypes and check the theory, but he thought supplementary counts worth making; and in succeeding years he himself made counts during summer holidays with his sister. In particular, in one Argyllshire colony, he got useful counts in 1934, 1935, 1937, and 1939. The open-pollinated plot was, in fact, set up at the Chelsea Physic Gardens in 1936, and many seedlings were later grown at Merton, after Mather moved there in 1938.

In nature, fertilization generally takes place only when the pollen is derived from anthers of the same length as the style receiving it. In contrast to previous experiments, using artificial means, the open-pollinated plot would rely wholly on such "legitimate" fertilization. Although the series of experiments might therefore take longer, this technique had the advantage that the actual frequencies of phenotypes derived from each sort of plant would represent its contribution under natural conditions to the population of the next generation. Moreover, in practice, open-pollination saved the labor of hand-fertilizing, and large progenies could be relied on, since it avoided the semisterility experienced with illegitimate crosses. The possibility that stray pollen might be brought in from outside the plot proved negligible; only very rarely was there evidence of foreign pollen, even though Fisher learned of *Lythrum* plants growing only 150 meters from his plot at Harpenden. Their progeny, if they had occurred in his plants, would certainly have been distinguishable.

The plan, and the collaboration with friends, reveal themselves in Fisher's report of the experiments in 1940 [*C P* 178, with K. Mather]. "Four Short plants from seed of an open-pollinated plot grown at the Chelsea Physic Garden in 1936 were tested by open-crossing with Long at Harpenden in the following year and sufficiently large progenies were grown at Merton in 1938." One of these families contained mid plants. The short parent was then grown with a mid daughter (which therefore contained the identical factor for mid), "in isolation in Dr. F. Yates's garden at Harpenden and two progenies from the reciprocally crossed seed were grown this year [1940] at Merton." If the homozygous mid had been lethal, these progenies would have segregated with respect to mid and long in a 2:1 ratio. In fact, however, they showed frequencies which accorded well with expectations for 3:1, but incompatible with a 2:1 ratio. It was virtually certain, therefore, that the gene tested was not lethal.

Fisher then turned to consider alternative reasons for the anomalous results obtained by Barlow and East, in particular the possibility of polysomic inheritance for the mid factor with four or six homologous chromosomes instead of two as in the usual diploid. The possibility of polyploidy was in full accord with the fact that *Lythrum* had a chromosome number of 30 (as was then believed, though later investigations established the number to be 60) while the basic number of the genus is 15. To distinguish between diploid or polyploid inheritance, Fisher first calculated the expected ratios on the basis of each type of inheritance, and then performed the one cross that would give unambiguous results [*C P* 192, with K. Mather, 1942]. By crossing short or mid plants with long, it was possible to distinguish from the proportions of the progenies whether the parents had received no mid gene or one, two, or more.

The results showed that the plants were polyploid: eight families proved to have more than one mid gene; seven of them had two mid genes and gave a ratio of mid:long about 4:1, according well with expectation on the basis of hexaploidy, though not excluding the possibility of tetraploidy. Theoretical examination had already shown [*C P* 184, 1941] that results for these alternatives "are so closely similar that no simple observations will distinguish the possible cases." The eighth family appeared to be derived from a plant with three mid genes.

The record of experimental facts only tells a part of the story; it gives no idea of Fisher's enthusiasm. He positively enjoyed the field trips with his assistants. He was charmed to go to Down House with Lady Barlow to plant a *Lythrum* seedling grown from a seed that Charles Darwin had collected from his experiments. He wanted to set up an open-pollinated plot in the gardens of Down House where Darwin's research might be not only memorialized but continued. Besides, the whole program was intellectually most rewarding. Sending the full report [*C P* 196, with K. Mather, 1943] to E. L. LeClerg in May 1944, he wrote:

[*Lythrum*] presented a problem the solution of which gave me immense satisfaction, chiefly because the large amount of thought invested in planning the line of approach which now seems so obvious bore good fruit in that the experimental results have been, from first to last, "as good as gold."

Referring to results from the 1943 season, he added, "Presumably the mother plant was a triplex tetraploid. It may also be a quadruplex hexaploid . . . The final test, as I believe it will be found to be in all complex polyploid problems, lies in the double back-cross." In 1947 [*C P* 218, with V. C. Martin], he could report that further work had confirmed the tetrasomic inheritance of the factor for midstyle. Work on *Lythrum* continued to give him satisfaction of the same sort, for the combinatorial problems and possibilities had a great fascination for him. The *Lythrum* plant became a vehicle through which he explored these problems, while the genetical factors themselves for him held slighter interest.

11
Evolutionary Ideas

In 1932, Fisher took the opportunity offered by an invitation to present the Herbert Spencer Lecture in Oxford, to address a distinguished scientific audience on "The social selection of human fertility" [C P 99]. In this talk he placed his advocacy for family allowances in a wider philosophical and scientific context. His introduction was, in fact, a consideration of a subject apparently unrelated to his title, the question of philosophical determinism.

In brief, he said, the "laws of nature" exemplified preeminently by the mathematical expressions of experimental physics had seemed to imply a universe governed by exact and necessary laws of causation, as if it were a deterministic mechanism. The more exact confirmation of these laws which had been possible after their enunciation, by the use of more refined techniques of measurement, had suggested that the determinism might be complete. The regularity and predictability of the motions of heavenly bodies, indeed, had suggested the possibility of a like predestination in human affairs. By contrast, the "laws of chance," which were called in where mathematical laws had not been discovered, seemed an illusion of causelessness, a mere product of human ignorance. With adequate information, it had been felt, the social sciences might become as exact as the physical sciences.

Recently, however, from the most deterministic subject of experimental physics and from that branch of the subject dealing with the most reliable materials, the gases, had arisen the realization that the "laws of nature" had their basis in "the laws of chance." The reliability of physical material was found to flow, not necessarily from the reliability of the ultimate components, but simply from the fact that these components were very numerous and largely independent.

Humanly speaking, therefore, we are in a position to recognize only a single principle of regularity in the world—only a single reason why the behaviour of a system should be reliable—why it should, if it does, react to the circumstances in which it is placed, in a way which can be satisfactorily predicted from a knowledge of the circumstances. The only reason known to us for such reliability is that its properties are in some sort the average or total of a large number of independent items of behaviour.

From this statistical point of view, organization of the components, rendering them not independent of each other, actually reduced the reliability of the system. Thus a battalion, moving "as one man," was less predictable than a battalion of men acting independently—at least to the opposing commander who had no control over them.

Birth was an independent event in the sense that the heredity of each individual was decided independently, by chance, even within a single family. Therefore natural causation was paramount in respect of the hereditary endowment each generation received from its progenitors and, in a population of 20 million or so to a generation, the quality of the output was quite precisely determined by the circumstances governing parenthood. Similarly, insofar as the incidence of births was consistently unequal, as between groups of persons of unlike genetical constitution, so far would the genetical composition of each generation differ from that of its predecessor. Thus it appeared that the half of the contemporary British population "above average" was reproducing roughly two-fifths of its numbers, whereas the half "below average" was reproducing about four-fifths of its numbers. In the next generation, therefore, the ratio of these groups would not be $1:1$ but $2:1$ below the current average. Given the current differential birthrate, with a continuing social promotion of ability and of all congenital tendencies unfavorable to reproduction, the outcome was inevitable: an increasingly rapid fall to ruin of national abilities.

Fisher suggested that social organization and the influence of public opinion might be used to modify the effects of the operation of natural law. In particular, the economic motives, at least, for family limitation, which operated most effectively in the middle classes, might be removed by the institution of graded family allowances. Thus he concluded, with guarded optimism:

The consideration of an evil, widespread, automatic and destructive of all that cooperative man can achieve or aspire to—destructive of the very powers of achievement or aspiration—is apt to induce a mood of fatalistic dejection. I hope I have made it clear that I see no ground for fatalism, in the action of natural law . . . It is, as it seems to me, part of our business as scientists to distinguish between what is inevitable, and what is subject to control; but it is our business as citizens to see that the possible control is exercised.

While Fisher's ideas for family allowances were meeting impediments in the outside world, his thoughts about the implications of indeterminism continued to develop quietly.

□ □ □

To many physicists the discovery in 1930, of the uncertainty principle of W. K. Heisenberg, creator of quantum mechanics, presented a philosophical problem, for it removed the historical basis of their faith and was irreconcilable with the perfection of their art. To Fisher, in contrast, it came as a natural extension of statistical philosophy. The certainty of physical laws being seen to be based on the innumerable and independent events involved, man could play a creative role in the fringe of uncertainty at the borders of the realm of natural law. By making his observations as it were independent by randomization, he could get a valid measure of biological variation and hence of operative laws. On the other hand, by constraining events to dependence, he could modify the effects of the operation of natural law.

Being invited to contribute an article for the new journal *Philosophy of Science* [C P 121, 1934], Fisher chose to consider indeterminism and natural selection. From the standpoint of logic, he discussed the implications of the different systems of causation with special reference to evolution by natural selection and showed how what were grave difficulties on a deterministic view of the world became creative opportunities on an indeterministic view.

Introducing indeterminism, as he had before, by way of the new physics and as a resultant from the fact that all laws of natural causation are essentially laws of probability, he again remarked that, in consequence, the predictability of a system has the same basis in the "natural" as in the "social" sciences. Thus the *appearance* of social determinism, evinced, for example, by the history of the rise and decadence of many great civilizations, was illusory:

The course of social evolution in the history of civilised man, gives the appearance of being controlled by necessary and unintelligent causation, merely because intelligence has not in fact yet been applied to its guidance. Things which are allowed to drift will naturally appear to be at the mercy of the current.

So far, his presentation followed the same lines as in 1932, but he now considered further implications of indeterminism for evolutionary theory.

When brought into the new perspective of an indeterministic view of the world, the concept of evolution assumed majestic proportions. The existence of order and harmony remained observable facts, and causation was none the less recognizable if it turned out to be a statement of probability, but certain limitations, once regarded as rigid and necessary, appeared now as conditioned or even casual. This casual component removed the logical difficulties encountered by deterministic theory and recognized the *creative* element in evolutionary change.

The first difficulty on a deterministic view of the world was, he wrote [C P 121], that every cause must itself be an effect of an antecedent cause, which, in turn, was merely an effect. There was no end to the chain of antecedent causation, unless one postulated a First Cause extraneous to the logical system. Indeed, to the logician who perceived the logical consequences of his deterministic theory, causation was as nonexistent as freewill.

Only in an indeterministic system has the notion of causation restored to it that creative element, that sense of bringing something to pass which otherwise would not have been, which is essential to its commonplace meaning.

Closely associated with this was the explanation of man's experience of time, in that he can remember but not foretell. For the determinist this presented a second difficulty, being an arbitrary constraint, inconsistent with a deterministic system:

For the instantaneous state of such a system must be related to its subsequent states by equations identical with those which relate it to its previous states. That this is not so with the human mind is, like the power of effective choice, universally verified by subjective experience. Unlike this latter belief, however, it is also demonstrable by objective experiment.

The relation of consciousness to past being demonstrably different from its relation to future events, an indeterministic system was required, which allowed the objective reality of a creative element in consciousness, as in nature, and made sense of unidirectional experience of time.

The real existence also of purpose and choice were indicated by biological considerations. In a deterministic system the subjective experience of purpose and choice would be an illusion. Any organism that was to achieve certain purposes connected with survival and reproduction would have to have evolved so that it behaved as though its actions were purposive, but the illusion, in itself, since it could not affect action, would be of no selective advantage. The appearance of this illusion in mankind must then be postulated as a purely fortuitous accident, an accident to be expected equally in the lowest and the highest organisms. In an indeterministic world, by contrast, human qualities of aspiration, planning, and foresight were rationally possible and might be advantageous. We could recognize primitive precursors of these qualities as having a real part to play in the survival or death of the organisms that evince them.

Biologically it might be said that purposive action by the organism as a whole is the crowning stage of an evolutionary process by which relatively large masses of living matter have come to achieve that cooperation of parts and unity of structure which we call individuality. For, on a statistical view of causation, spontaneity or creative causation is at its highest when perfect unity is achieved.

A third difference in the theoretical implications of the different systems of causation was apparent in that only on a completely deterministic scheme of causation were the earlier links in the chain of events causally equivalent to those that followed them. In an indeterministic theory intermediate events could not be neglected or eliminated. In consequence, scientific research was interested not merely in the manner in which precedent events determined or influenced these consequences, but in locating in time and place the creative causation to which effects of especial importance were to be ascribed; for evolutionary effects, this was found in the interaction of organism and environment, in the myriad biographies of living things. This accorded with only one evolutionary theory, Darwin's theory of natural selection, for the _selective_ value of choice must always be in its harmony with the world around, its capacity to utilize its advantages or penetrate its undiscovered possibilities.

Thus in an indeterministic world certain happenings entailed consequences or entailed systems of probability for various consequences, other than those that could have been foreseen from antecedent happenings. Statistically speaking, the casual or creative component might be defined as the deviation of the event in question from the average to be expected. This component was thus casual in relation to preceding causes but creative in relation to the consequences it entailed. The event in itself was not creative but might be thought of as creative in the aggregate of all such casual components. Genetic mutation was of this sort, not capable in itself of propelling the organism in any particular direction but, in the aggregate, under natural selection, having consequences which could not have been foreseen in advance.

Fisher concluded that the kinds of effects as to the causes of which scientists commonly inquire were limited to those which presented this sort of statistical simplicity. We might inquire meaningfully why a pebble is round and how it came to be so; we could not hope to specify in detail how more complicated geometric configurations of masses of matter had come about. We accepted the explanation that the pebble had been rounded by the abrasive action of the surf, because we recognized that such action would be less destructive and more uniform on a round pebble than on one with edges or corners; in this we did not require any special rounding tendency to the chipping or scratching caused by each particular impact. In the aggregate of these causes, rather than in any particular element, we discerned creative evolution.

□ □ □

In 1950, presenting the 4th Arthur Stanley Eddington Memorial Lecture, Fisher chose the title "Creative Aspects of Natural Law" [*C P* 241]. Picking up the same train of thought as before on modern physics and evolution, he now expressed more particularly how it affected man in his self-conscious nature. As before, he defined the word "creative" in its logical and scientific sense as qualifying causation; as implying an indeterministic world in which, had the nature or intensity of the causal system been different, the effects that flowed from it would also be different; a world in which the causal system really could have been different.

He now introduced into the discussion the subjective strand of meaning in the word "creative," which is charged with emotion. The word seems appropriate to us only when applied to matters of importance, especially to something new in its nature and potentialities. "Work to be creative in this sense must have value, intellectual or aesthetic, moral or social value; consequences which excite wonder, or admiration."

Approaching the matter on this level, it is, therefore, almost axiomatic that the process by which living things, as we know them, have gradually come into existence, is, in the fullest sense, a creative process. It has created new things, pregnant with potentiality; it has produced among other things growth, voluntary movement, and appetite; striving and effort, joy and pain; consciousness and, in Man at least, conscious self-criticism. It would be strange if the word did not fit, seeing that for ages it has been used precisely for the coming into existence of these things, however variously the process was conceived. It is almost like saying that Creation is creative; the only new implication, and it is an important one, that the phrase now has is that for us creation is still going on, whereas in the childhood of our race it was thought to have been all finished a long while ago.

"For us creation is still going on": it is as if time stood at the sixth day of creation and man was even now coming into being in the image of the creator.

The implications of an indeterministic system of causation, which had satisfied Fisher as statistician, logician, and biologist, satisfied him also in suggesting a moral philosophy in consonance with his personal commitment and faith. An individual life could be of unique consequence, for creative causes would have scope through a man's active involvement in the world of living things and in the world of ideas; in the drama of his success, or failure, actually to accomplish good results; and in his positive acceptance of responsibility for the creative role, now largely belonging to man, in shaping the environment not only of his kind but of almost every evolving organism on earth.

Just where does the theory of natural selection place the creative causes which shape evolutionary change? In the actual life of living things; in their contacts and conflicts with their environments, with the outer world as it is to them; in their unconscious efforts to grow, or their more conscious efforts to move. Especially, in the vital drama of the success or failure of each of their enterprises. . . . Creative causation [is] a function of every organ through the entire life history, of the brain in devising, and of the hands in execution. . . .

The surface or limit separating the inner from the outer life of each living thing is also, in our experience, the true seat of our consciousness, the boundary of the objective and the subjective, where we experience, through our imperfect sense-organs, what comes to us from outside, and, with at least equal obscurity, that which rises into consciousness from within. If consciousness is, as it would seem, the symbol, or even the means, of unification of our being, this is the region to which creative activity could most fitly be traced. . . .

We come here to a close parallelism with Christian discussions on the merits of Faith and Works. Faith, in the form of right intentions and resolutions, is assuredly necessary, but there has, I believe, never been lacking through the centuries the parallel, or complementary, conviction that the service of God requires of us also effective action. If men are to see our good works, it is of course necessary that they should be good, but also and emphatically that they should work, in making the world a better place. It is not necessary that others should know by what particular agency the result has been brought about, but there must be in the result something for them to thank God for. We must face the difficult and responsible task of getting good results actually accomplished. Good intentions and pious observances are no sufficient substitute, and are noxious if accepted as substitute. . . .

For the future, so far as we can foresee it, it appears to be unquestionable that the activity of the human race will provide the major factor in the environment of almost every evolving organism. Whether they act consciously or unconsciously human initiative and human choice have become the major channels of creative activity on this planet. Inadequately prepared we unquestionably are for the new responsibilities, which with the rapid extension of human control over the productive resources of the world have been, as it were, suddenly thrust upon us. Yet there have in recent times been some signs . . . that we do not feel that ruthless exploitation is good enough.

Thus, Fisher's thoughts about indeterminism in nature extended their scope. Far from finding scientific study led to a sense of disillusionment about the nature or potentialities of man, to religious doubts or to discordance between his thoughts and his feelings about life, he drew from science itself the confirmation of the unique value of man as a product and as a participant in creative causation.

□ □ □

In the paper of 1934 referred to above [*C P* 121], Fisher spoke of scientific interest in locating in time and place the creative causation to which effects of especial importance were to be ascribed. He was referring on that occasion to evolutionary effects in nature. At the same time, he was interesting himself in such creative causation in human consciousness, through his studies in the history of evolutionary science.

He had studied in some detail the writings of Charles Darwin in order to understand the development of his thought about evolution by natural selection and the reasoning which had led him nevertheless later to accept the incompatible theory of inheritance of acquired characters (*Genetical Theory*, 1930). He had attempted to follow the reasoning of biologists who, after the rediscovery of Mendel's work, had failed to appreciate that Mendelism complemented Darwin's theory and was inconsistent with any other evolutionary theory [*C P* 98, 1932], and in "Creative aspects of natural law" [*C P* 241, 1950] referred to above, he was to reconstruct the world pictures and reasoning of two early twentieth century evolutionary philosophers who had rejected natural selection because for them it had been closely associated with determinism, which they rejected.

In 1932 he attributed the misunderstanding of Mendel's work by its rediscoverers to their lack of appreciation of its logical implications. In contrast, to him the logical and mathematical aspects of Mendel's work were most impressive. He felt that despite the popularization of ideas attributed to Mendel as the father of modern genetical knowledge, very little attention had been paid to Mendel's paper itself. It was in a missionary spirit, wishing to exhibit Mendel's genius in its own terms and not as it had been perceived, that, being invited to contribute an article to the new journal *Annals of Science* late in 1935, he chose for his title "Has Mendel's work been rediscovered?" (*C P* 144).

At the same time, he was aware that Mendel's paper presented some rather remarkable problems. In studying it, he sought afresh the answers that it could give to his questions: What did Mendel discover? How did he discover it? And what did he think he had discovered? He also considered the reception and interpretation of the paper at its first publication and after its rediscovery.

Mendel had stated, after his description of the first seven experiments, that: "In the experiments above described plants were used which differed only in one essential character." Unless he had known *in advance* of the separate inheritance of the characters he was studying, he could scarcely have used seven such pairs of varieties. Perhaps, as Bateson had suggested, the statement should not be taken literally. But, on consideration, alternative interpretations appeared unacceptable and difficult to reconcile with the order and style of Mendel's report. Thus Fisher concluded:

If Mendel is not to be taken literally, when he implies that one set of data was available when the next experiment was planned, he was taking, as *redacteur,* excessive and unnecessary liberties with the facts. Moreover, the style throughout suggests that he expected to be taken entirely literally; if his facts have suffered much manipulation the style of his report must be judged disingenuous. Consequently, unless real contradictions are encountered in reconstructing his experiments from his paper, regarded as a literal account, this view must be preferred to all alternatives, even though it implies that Mendel had a good understanding of the factorial system, and the frequency ratios which constitute his laws of inheritance, before he carried out the experiments reported in his first and chief paper.

In the following section of Fisher's paper, the reconstruction was made and the conclusion emerged that there could be no doubt whatever that Mendel's report was to be taken entirely literally and that his experiments had been carried out in just the way and much in the order that they were recounted.

The reconstruction of the details of the experimental program presented difficulties of which Fisher remarked, "Seeing how often it is taken for granted that all clouds were cleared away at the rediscovery in 1900, it is singularly difficult to ascertain exactly how Mendel's experiments were conducted and, indeed, what experiments he carried out." It was necessary to date the different parts of the experiment; to estimate the numbers of plants grown in every year and the area and the labor available in the monastery garden; to consider how and when the plants would have been scored in various characters and how the physical process might have affected the recorded results; to check the yields for plausibility on this basis; to check the actual course of Mendel's argument as he introduced it and to follow it through from one experimental trial to the next; and, finally, to analyze his results.

By his analysis, Fisher made an "abominable discovery": Mendel's data had been "cooked." He unearthed two suspicious facts: statistical tests applied to Mendel's data showed that the very close agreement between his observed and expected series was most unlikely to have arisen by chance; in particular, there was a very large discrepancy in one series of results, where the observations agreed closely with the 2:1 ratio which Mendel expected, but differed significantly from the expectations when corrected so as to allow for the small size of the test families.

Writing to Ford soon after, he described this "shocking experience":

The first thing that struck me was that in testing homozygosity in plant characters, Mendel used F_3 progenies of only 10, and did not notice that the chance of a heterozygote being classified as a homozygote is not negligible, being between 5% and 6%. Nonetheless, Mendel's data agreed with the 2:1 ratio, requiring a compensating chance deviation which would only come about once in 30 trials. And then the same thing happens again later and there is not a sign that Mendel saw the complication and allowed for it. Now, when data have been faked, I know very well how generally people underestimate the frequency of wide chance deviations, so that the tendency is always to make them agree too well with expectations. So I tested all the larger experiments and finally the whole of his recorded data and, in the aggregate, the deviations are shockingly too small for 64 degrees of freedom. I have divided the data in several different ways to try to get a further clue, e.g. by years and by the absolute sizes of the numbers, but . . . can get no clue to the method of doctoring.

Having written the paper in the Christmas vacation, he submitted it to the editor, Dr. Douglas McKie on 8 January 1936, with the comment:

I had not expected to find the strong evidence which has appeared that the data had been cooked. This makes my paper far more sensational than ever I had intended, and adds another mystery to those that have been puzzling me, some of which I think I had made some progress with. As it stands, my title is now more ironical than I had intended it to be, but I cannot help it if circumstances proceed to emphasize so strongly my main point, that Mendel's own work, in spite of the immense publicity it has received, has only been examined superficially, and that the rediscoverers of 1900 were just as incapable as the non-discoverers of 1870 of assimilating from it any idea other than those which they were already prepared to accept.

To account for the rather startling evidence that "the data of most if not all, of the experiments have been falsified so as to agree closely with Mendel's expectations," Fisher suggested that Mendel was possibly deceived by an assistant "who knew too well what was expected." Suspicion later fell on Mendel's gardener, a man reported to be of inebriate habits and untrustworthy character by H. Iltis, in his biography of Mendel (published in German in 1924). Fisher did not see the biography before his paper was published. Early in 1936, however, he was consulting J. Rasmusson, an expert on the genetics of the pea, who gave him valuable hints as to experimental details relevant to Mendel's report. Rasmusson referred him to Iltis' book, to which, consequently, a footnote reference was added in Fisher's paper.

But, if the assistant had known too well, it seemed certain that Mendel himself must have known precisely what to expect. It appeared, first, that he must have known in advance of the experiments that the seven characters he investigated were inherited separately and that he had carefully selected each of the pairs of varieties as being suitable for his purpose, because he knew them to be contrasted essentially only in the single factor under study. Then, the simplicity of his experimental plan and the adequacy of the numbers of the first crosses reported were further indications that he knew very much what he intended to do. This evidence was consistent with the observation that in a number of cases Mendel had scored only a small sample from very much more abundant available material, even where a fuller enumeration might have demonstrated facts about which he might be supposed to have been in doubt. He seemed, in fact, confidently to have assumed the answers. For example, he appeared to have assumed that each of the parents contributed equally to the inheritance of the progeny, for he took no care to prove it from his material, although he used reciprocal crosses from which he could easily have demonstrated the equivalence. Similarly, he appeared to have assumed independent assortment of the different factors, for, again, he recorded smaller numbers than he might easily have done from plants actually grown. In fact, two of the seven factors that Mendel declared to be independent have in later tests been shown to be linked!

These indications together made a strong case for the hypothesis that Mendel had already become aware of the phenomena of dominance and segregation and that the experiments of 1856–1863 were not the means of their discovery but were carefully planned as a demonstration of the factorial scheme he had conceived earlier. (In his biography, Iltis records that Mendel hybridized gray and white mice in his rooms; it is possible that this work led him to recognition of the Mendelian laws.)

That the experiments were planned as a demonstration was a new conception to Fisher, who had assumed, like everyone else, that they had been exploratory in nature and had led by a logical progression to the discovery of the principles which they illustrated. Nevertheless, it was thoroughly in accord with his earlier assessment of the particular area of Mendel's genius. As he wrote in 1935 [*C P* 144]:

In 1930, as a result of a study of the development of Darwin's ideas, I pointed out that the modern genetical system, apart from such special features as dominance and linkage, could have been inferred by any abstract thinker in the middle of the nineteenth century, if he were led to postulate that inheritance was particulate, that the germinal material was structural, and that the contributions of the two parents were equivalent. I had at that time no suspicion that Mendel had arrived at his discovery in this way. From an examination of Mendel's work it now appears not improbable that he did so and that his ready assumption of the equivalence of the gametes was a potent factor in leading him to his theory.

Fisher made no explicit remarks about Mendel's methodology in 1935, but in 1955, in preparing introductory notes on Mendel's paper to accompany a new translation,* he rectified the omission; it was his answer to the question raised in 1935: what did Mendel think he had discovered?

If we read his introduction literally we do not find him expressing the purpose of solving a great problem or reporting a resounding discovery. He represents his work rather as a contribution to the *methodology* of research into plant inheritance. He had studied the earlier writers and tells us just in what three respects he thinks their work should be improved upon. If proper care were given, he suggests, to the distinction between generations, to the identification of genotypes, and, to this end, to the frequency ratios exhibited by their progeny, when based on an adequate statistical enumeration, studies in the inheritance of other organisms would yield an understanding of the hereditary process as clear as that which he here exhibits for the varieties of the garden pea. There is no hint of a tendency to premature generalisation, but an unmistakeable emphasis on the question of method.

When it was recognized that Mendel was principally concerned to justify a method of investigation, many of the peculiarities of the paper became explicable. He was reporting a carefully planned demonstration, rather than the protocol of the first observations that led to the formation of his ideas. The assumptions, the omission of confirmation of results, the almost complete absence of repetition of tests, all conform to pattern. "He is acting as it were on principle, and without the opportunism with which research workers usually seek vigilantly for supplementary information."

Fisher's later notes emphasize Mendel's interest in the combinatorial properties of his data. In Mendel's original paper the theoretical consequences of the factorial system had been thought out carefully; Mendel had envisaged the combinatorial situations which might be important to understanding multifactorial inheritance. To Fisher, this had always seemed an important aspect of Mendelism. In 1935 he was moved to comment:

This understanding of the consequences of the factorial system contrasts sharply with many of the speculations of the early geneticists, such as that new species might be formed by the mutation of a single factor, or that mimetic groups, found among butterflies and other insects might be explained by the paucity of the genetic factors con-

*A new translation of *Experiments in Plant Hybridisation* by Gregor Mendel, with Fisher's commentary and assessment, was actually published for the Mendel Centenary in 1965 (Oliver and Boyd: Edinburgh and London).

trolling the pattern and colouration of the wings. In these respects it has taken nearly a generation to rediscover Mendel's point of view.

In 1955, he pointed out the stress which Mendel laid on combinatorial mathematics, "which has been almost constantly overlooked by commentators" and "very well deserves the attention of all who teach the subject." He remarked that young workers needed preparation in combinatorial manipulations, because the enumeration of the genotypes to be expected when several factors are segregating was an obvious first step toward exploring the possibilities of a cross.

More recently, indeed, in the exploration of the complexities to be expected in tetrasomic and hexasomic inheritance, combinatorial mathematics, of by no means an elementary character, has provided the only possible means of clarification.

In Fisher's own studies of tetrasomic inheritance in *Lythrum*, the combinatorial manipulations were not only necessary but, for him, provided the more fascinating problems of the investigation, and he enjoyed them for their own sake.

If one accepts that he was right about the negative reception of Mendel's work, that each generation found only what it expected to find and ignored what did not confirm its expectations, one may suggest, perhaps, that his own positive appreciation of Mendel's genius was augmented by the progressive discovery of his own intellectual accord with Mendel, as to the logical and mathematical means by which facts could be established.

☐ ☐ ☐

The introduction of Fisher's theory of dominance modification (1928), the publication of *The Genetical Theory of Natural Selection* (1930) and the simultaneous intensification of public interest in eugenic policies, together presented the case for natural selection in a way that could not be ignored. Moreover, Fisher took the numerous opportunities that came to him in the next few years to give the theory of natural selection its proper prominence in genetical and evolutionary discussions. In particular, he spoke repeatedly of the logical connections between theory and biological fact, whose neglect he felt had led biologists into untenable positions.

So, as we have seen, in speaking of the bearing of genetics on theories of evolution to the Royal Society of Dublin [*C P* 98, 1932], he showed Mendelian genetics to be consistent with a theory of evolution by natural selection only; speaking of the social selection of human fertility at the Herbert Spencer

Memorial Lecture at Oxford [*C P* 99, 1932], he considered logical implications of indeterminism, and he made the same argument more explicit in writing of indeterminism and natural selection 2 years later (*C P* 121, 1934). In presenting results of poultry experiments to the Linnaean Society (*C P* 116, 1934), his plea to his audience was to avoid the *assumption* of any evolutionary theory, in advance, but to let the evidence speak. He did not then argue that alternative theories of evolution were logically untenable, but this was the subject of his address to the Science Masters Association in January 1934 [*C P* 122]. Again, in relation to the reception of Mendel's work [*C P* 98, 1932 and *C P* 144, 1936], he blamed the misunderstanding of biologists of the time on their lack of appreciation of the logical situation.

By 1936, "the current state of the theory of natural selection" had become sufficiently prominent to warrant a discussion at the Royal Society. On this occasion an impressive number of scientists presented qualitative results highly suggestive of selective pressures in a variety of natural situations. To carry conviction about the existence of natural selection, however, required more than qualitative demonstration; it needed the irrefutable proof of quantitative measurement in the field. Fisher's contribution referred first to the logical situation that led, in particular, to rejection of theories of evolution which, like mutation theory, failed to explain adaptation:

The explanatory content of a theory of evolution only reaches its absolute zero with the mutation theory. Organisms evolve, that is to say, heritable changes take place in them, because they mutate — because unexplained heritable changes take place.

He then turned to the possibility of measurement of selective intensity.

He recalled a suggestion he had made in 1931 [*C P* 93]. Polymorphic species provide natural markers maintained in equilibrium by a balance of selective forces. Moreover, in these cases selective intensities can occur greater than can exist elsewhere. Consequently once the genetic basis of the polymorphism was known, it would be possible to compare the frequencies observed in the field with those to be expected from random mating in the absence of differential elimination, and so to estimate the selective intensities that maintained the stable polymorphism. He recalled that this had actually been done:

I had some correspondence with Dr. Nabours of Manhattan, Kansas, who was able to organize an extensive collecting trip in Texas and Mexico with a view to ascertaining the frequencies with which different gene combinations, recognizable on the basis of Dr. Nabour's genetic work from the colour pattern, were found among a number of species of grouse-locust. I must admit that I believe neither Dr. Nabours nor I have published anything about it. He sent me quantities of illuminating data, and I sent him very lengthy dissertations about them. I am confident that Dr. Nabours will forgive me for mentioning the matter now, for certainly every one of the species . . . for which large samples were collected, showed unmistakeable evidence of differential survival among the genotypes in their wild habitats. In most cases the differential elimination was sufficiently moderate for it to have been due either to death only in the period between the formation of the zygote and the time of capture, or in the other half of the cycle between the time of capture and the formation of the next generation, when both differential mortality and differential fertility would be effective. In one case, I remember, in *Acrydium arenosum*, no amount of differential fertility, even complete sterility, would suffice to explain the disparity in numbers observed, for crossing over alone would produce more than were observed. In fact the figures could only be explained by the differential elimination down to about 40% of one genotype . . . compared with its competitors.

After the Royal Society discussion, Fisher obtained Nabours' figures and substantiated the claims he had made both as to the measured differential elimination (estimated from field counts) and differential viability (estimated from laboratory breeding) which affected the different genotypes (*C P* 167, 1938).

Thus he showed the possibility of measurement of selective intensities in polymorphic species; that is, cases in which selective intensities would be large enough to be measured with a reasonable number of observations and when frequencies of each genotype would be high enough to provide good estimates. Even in this case, however, one could only hope to observe evolutionary change if the population size were constant, for, with fluctuating numbers, the effects of differential elimination would be confounded with changes in the size of population, due to an excess of births, deaths, immigration, or emigration. And insect populations, while they had the advantage of providing adequate numbers of specimens, were notoriously variable in size.

□ □ □

A beginning had been made with marking insects for population studies. Early in 1930 C. H. N. Jackson, engaged in tse tse fly research in Tanganyika (Tanzania), spent some months in Fisher's laboratory; he had developed a marking technique and Fisher was greatly interested in its possibilities as a means of estimating numbers. In 1935 Jackson returned to Fisher's department, with data from a regular release and recapture program. Fisher suggested a rough but easy method of estimating the population number, using the combined information from a series of recaptures. The supposition was made that the recapture ratios at given dates $t, t + 1, t + 2, . . .$ would fall off approximately in geometrical progression. When Jackson visited again in 1939, he had enough data to confirm that the geometric progression was appropriate and that the estimates made from previous and subsequent recaptures checked.

In 1935 Fisher suggested that his assistant, W. L. Stevens, should look into the method to see if its efficiency could be improved. Stevens was able to show that where the numbers caught and released on successive occasions were constant, the method of maximum likelihood yielded a simple solution for the geometrical progression but that when these numbers varied there was no very elegant method of combining the evidence. In the summer of 1938, following earlier consultation with Fisher, Ford obtained singularly thorough data of an insect whose population size was affected by irregular emergence and rapid changes of numbers. Stevens was asked to do the statistical analysis, as joint author of the paper, but when he tried the method on these data, it was clear that, as it stood, the method was quite inadequate for an organism with rapidly changing numbers.

In the early months of 1939, during Jackson's third visit, it occurred to Fisher that a backward and forward progression would supply separately the rates of birth or immigration and of death or emigration. In Jackson's data the chief importance of this step lay in the assistance it gave to the estimation of population movements. At the same time, Ford appealed to Fisher. W. H. Dowdeswell and he had completed their part of the paper, describing the location and habits of the butterfly *Polymmatus icarus,* on the Island of Tean in the Scilly Isles, and the experimental procedure whereby, as nearly as possible, 50 specimens had been captured, marked, and released each day through a period of two weeks of late summer. They needed the statistical analysis. To Stevens' relief, Fisher then took over the problem, with a view to trying out his new method of separating contributions of emergence from those of death and of weighting the unequal numbers recorded in a new, much simpler manner.

The result was, first, his representation of the release and recapture data in a triangular trellis, showing on the top margin the dates throughout the period of observation and on the two diagonals respectively the numbers released and recaptured each day. This exhibited all the relevant data simply and clearly. Then, applying his "backward and forward progression," he worked forward from the early dates to the end of the period of observation to obtain his first estimate of the population on successive days by reference to data already collected on each day. Reversing direction, he then worked backward to correct for the distortion of the estimates during the earlier period, due to a large emergence on one or two later days. It became apparent that there had been about 360 insects in the beginning and that, during the whole period from August 26 to September 7, about 450 – 500 butterflies had died in all, including 100 which had emerged during that period, so that the population of imagines decreased progressively from about 360 to almost nothing.

Thus Fisher extended the theory he had been developing for interpreting such recapture data. The new theory, the idea of making the particular calculations, and, indeed, the actual arithmetical calculation of the ratios were his own. He was pleased to have realized the possibility of eliciting the main facts of population size, births, and deaths by a procedure involving no more than simple arithmetic, and he hoped it might be a real encouragement to biologists who might be thinking of sampling studies of natural populations. So, having been interested in the subject for more than a decade, he came to contribute his first paper in capture-recapture sampling [C P 177, 1940, with W. H. Dowdeswell and E. B. Ford].

The analytical section of the paper was quickly written and sent to Oxford. The three co-authors delayed only long enough to make final corrections on the earlier manuscript, in their own hands, before submitting it to the Royal Society; Fisher's name was inserted as an author and that of Stevens covered with a strip of paper. And the referees laughed it to scorn: the illegibility of the hand-written corrections, and the piece of sticking paper, got more attention than the originality and detailed care of the experimental procedure and of the analysis, for one referee charged the authors with illiteracy: any secretary who undertook to reduce the article to correct English style, he asserted, would better deserve to have his name on the article than its originators, for Fisher's contribution was plagiarism; to this referee it seemed obvious that Fisher had stolen Stevens' work, and replaced his name by his own!

These criticisms were as extraordinary as they were virulent. One could hardly find men less guilty of verbal carelessness than Ford, who in his famous monographs had already proved himself to be meticulous in his use of the English language and a master of lucid exposition, and Fisher, who rarely failed to say precisely what he meant. The charge of plagiarism was pure venom. Fisher erred on the side of being overgenerous with his ideas so that, in fact, some confusion had already arisen about the real authorship of some papers from his department that did *not* bear his name. In one case, he had indignantly denied his responsibility for excellent work done in his department, which a correspondent mistakenly attributed to him, and had insisted that the whole of the credit belonged to its author. In another case, a friend had written to Fisher suspecting that he had written a passage that had appeared under another name. To his correspondent the passage had seemed nonsensical in its context, and Fisher had explained how, in conversation on the subject of the paper, he had written down the argument in question and handed the sheet of paper to the author who, apparently unable to understand it, had included it *verbatim* in his discussion, no doubt in blind confidence that any remark of Fisher's would prove both wise and relevant.

The referees recommended rejection of the paper and, on learning why, Fisher wrote to the editor, explaining the whole course of his interest and activity in the development of the ideas presented, and of the part which Stevens had played. Nevertheless, the paper was rejected. It was published a year later in the *Annals of Eugenics* [*C P* 177, 1940].

□ □ □

The new success in estimating the births and deaths taking place in a natural population whose size was changing from day to day suggested immediately the desirability of making similar counts to estimate the differential elimination of identifiable types within species. This work was, in fact, the basis of the great discoveries by the Oxford group in the 1950s when it was found, for example, that the different types of polymorphic snails are eliminated differentially, each type suffering heavy predation if it appeared in a microhabitat or at a season in which its particular coloration was conspicuous. Similarly, the melanic and normal forms of the peppered moth, *Biston betularia,* were tested against each other in the contrasted environments of an unpolluted Dorset woodland and a city park. Differential elimination at very high rates was found to affect the different forms according as they were visible respectively against the "peppered" appearance of lichen-covered white oaks or against the bare black trunks of city trees.

This work was also the basis of the beautiful demonstration of evolutionary change actually taking place by natural selection, which Fisher and Ford published in 1947 [*C P* 219, with E. B. Ford]. The work was actually begun in 1939, when Ford observed a gene beginning to spread in an isolated colony of a tiger moth. For some time he had been on the watch for such an indication in his field work. Fisher was no less enthusiastic than he about the news, and responded with glee, "This is a very ripe plum waiting to drop into our mouths." The game was afoot, and during the following 8 years before publication they had great fun with it. Indeed, for 18 years they continued the annual collection, together, of moths from the same colony. Starting in his fiftieth year and continuing until his retirement at the age of 67, Fisher made a point whenever possible of joining the parties which went out from Oxford; Ford, inviting him, made a point of warning him to bring suitable clothing— and change of clothes—"Remember, we shall be practically lying in a marsh all day!"

The moth itself made good sport: being easily disturbed when resting on the herbiage and flying actively in the sunshine with a characteristic undulating flight which made it rather difficult to catch; on warm days rising to a considerable height and circling the tree-tops well out of reach; on rainy days hiding under low-growing herbiage, but, the authors reported, "Even then it is not difficult to find. We have ourselves made large catches in drenching rain

by crawling through the vegetation and boxing the specimens as they sit."
The site of the colony comprised about 20 acres of fen and marsh. One part
was relatively open, bearing coarse grass in large hummocks "which are a
great barrier to rapid progress"; the other consisted of a dense jungle of reeds,
shoulder-high in July when the collections were made, and several ditches
intersected the site. Thus the hunting-ground presented hazards, especially
for a semi-blind collector; as Fisher's sight failed further in the later years, he
knew himself to be no real help, but he continued to enjoy the expeditions.

Collections made in 1939 and 1940 confirmed the interest of the situation.
A local variant, var. *medionigra* of the moth *Panaxia dominula* L., had long
existed in this particular colony at a low frequency which, judged from the
specimens collected before 1929, could not at that time have exceeded a
gene frequency of 1.2%; in 1939 the frequency had reached 9.2% and in
1940 stood at 11.1%. Laboratory breeding was begun to determine the
genetic basis of inheritance, and this established that var. *medionigra* was the
heterozygous form of the rarer var. *bimacula,* and intermediate in wing pigmentation between the two homozygotes. It seemed likely that in this colony
the heterozygote enjoyed a slight advantage over both homozygotes, thus establishing the polymorphism. Its spread must have been triggered by some
change in the immediate conditions which gave the *medionigra* form greater
selective advantage.

After the first 2 years, there were changes in the field work. The investigators were satisfied that

a situation of sufficient interest had developed to warrant estimating year by year the
absolute numbers of the population in which the *medionigra* gene was spreading. Accordingly we began this work in 1941 and have continued it each season since, using
the method of marking, release and recapture, developed by Dowdeswell, Fisher and
Ford (1940).

New calculations were devised to make the estimates and to check assumptions against a variety of indications in the data. In successive years the total
emergent population was found to vary from 1000 to 8000 or more, and the
gene frequencies of the *medionigra* gene fluctuated between 4.3 and 6.8%,
showing no tendency to return to the high value of 1939 – 1940 or to the low
one observed previous to 1929.

The fluctuations in the gene frequencies did not bear a simple direct relation to population size, nor were they small enough reasonably to be ascribed
to random sampling in the reproduction of a population of 1000 or more
each year. They were much greater than could be ascribed to random survival only and must therefore be attributed to random fluctuations in natural
selection.

This was a very nice result. The only alternative to natural selection, as an explanation of the observed fluctuations of the gene ratios, was random sampling in the process of reproduction, which would be most effective in just such a small isolated colony as was here considered. Sewall Wright had long advocated the view that under such circumstances random sampling would have important evolutionary consequences, but the hypothesis had not been tested in the field. In fact, the analysis of the data with the *medionigra* gene was the first in which the relative parts played by random sampling and selection in a wild population could be tested. The data did not support the view that chance fluctuations in gene ratios, such as might occur in very small isolated populations, could be of any evolutionary significance; instead, it forced the conclusion that, even in such extreme cases, natural selection must be held responsible for the observed evolutionary changes.

Fisher and Ford also pointed out [*C P* 239, 1950] that the new data were "completely fatal" to Wright's theory of "random drift," in that not only small isolated populations but also large populations experience fluctuations in gene ratios. Any effects that flowed from fluctuating variability in the gene ratios would therefore not be confined to such subdivided species but would be experienced also by species having continuous populations. Presumably, random fluctuations must always be present but they are accentuated in a small population while other causes, such as selective survival varying from year to year, are equally influential in large as in small populations. Yet the data showed that, even in such a small population as had been investigated, other causes must be acting with greater effect.

In Fisher's sequential exploration of quantitative aspects of evolution by natural selection, we see how logical considerations suggested what sort of observations might profitably be made, and how when these observations were made, new statistical methods were required and developed to handle complexities in the field, and so extended the possibilities of the field work, until, finally, data could be obtained that were capable of confirming the theory of evolution originally indicated by logical considerations, and of excluding alternative theories.

□ □ □

In 1940 Fisher became interested in ecological investigation of natural populations of the heterostylic primrose, *Primula vulgaris*. It appeared that a genetical situation existed whose evolutionary effects would prove well worth observation. Fisher's observations were begun in 1942 and continued for 15 years.

Commonly, primrose colonies consist of about equal numbers of pin and thrum plants. In the pin phenotype, the stigma protrudes from the mouth of the corolla, and the short anthers are hidden within; in thrum, the positions are reversed so that the anthers are clustered at the mouth of the corolla tube and the short-styled stigma is hidden about halfway down the corolla tube. In each case self-fertilization is prevented, partly by the physical separation of the sexual parts, and since pin and thrum are mutually fertile the polymorphism ensures cross-fertilization. A stable population, therefore, contains about equal numbers of pin and thrum plants. A single genetic factor is responsible for the polymorphism, pin (ss) being recessive to thrum (Ss), so that the gene ratio is normally *3s* :*1S*.

In 1940 in certain primrose colonies in Southern England, J. L. Crosby [57] observed a third form, the long homostyle, characterized physically by having stigmas and anthers at the same upper level in the mouth of the corolla. Unlike pin and thrum, the homostyle is self-fertile. Genetically, homostyle was found to be controlled by an allele of pin and thrum, dominant to pin and recessive to thrum. Crosby observed homostyles occurring in two restricted areas in Berkshire and Somerset, respectively, and in these areas they occurred in a high proportion of the population of most of the colonies.

Arguing from theory, since pollen from homostyles is effective on pin plants, the introduction of this homostyle allele into a primrose colony must introduce an imbalance in the gene frequencies of the mutually dependent pin and thrum, and since homostyle is also self-fertile, it has a great selective advantage over both its competitors. Indeed, if homostyle incurred no physiological disadvantage, the homostyle gene must gradually replace both pin and thrum and tend to 100%. To this extent the high proportion of homostyles found by Crosby were to be expected, and his predictions that homostyle would take over seemed justified.

To Fisher the situation presented a most peculiar evolutionary problem: cross-fertilization had given the species a selective advantage sufficient for a dimorphism to have evolved excluding self-fertilization; then a gene had appeared whose effect was not only to introduce self-fertilization into any colony where it became established but, apparently, to gain a selective advantage through the very mechanism by which the plant normally ensured cross-fertilization. In the end it must inevitably replace its alleles and exclude the possibility of the colony reverting to its former habit of cross-fertilization.

Had nature any defence against such perversion of its means? Some evidence suggested it might, for, although the mutant had established itself in a few colonies and apparently suffered no physiological disadvantage, the inevitable had not happened; even those colonies where it occurred contained a good number of pin and thrum plants. And in most colonies the species remained distylic. More than 120 species of *Primulaceae* were, in fact, distylic in nature, though rare homostyles were known to occur in a number of them. Presumably, the recurrent mutant homostyle had usually been suppressed. Although it had recently been able to establish itself in these two small, widely separated areas, because of some local and perhaps temporary selective advantage, still some check might exist to prevent its reaching 100%.

In April 1942 Fisher, having occasion to visit the Marine Biological Station at Plymouth, decided to break his journey in Somerset and spend a week investigating the primrose colonies there for himself. Taking lodgings in the village of Sparkford, at the center of the homostyle country, he tramped around the neighborhood seeking colonies of primrose plants and making counts of the different style lengths they contained.

He lodged at the home of a Mrs. Windsor, who proved a countrywoman of parts. After the monotony of plain wartime fare in Harpenden, Fisher found himself treated, to his delight and astonishment, to fresh cauliflower topped with local cream. Over the years, he was to become a familiar figure in the neighborhood of Sparkford, and he was to cherish the local acquaintance and seek out his old friends there during his annual pilgrimage.

In 1943 he paid a second visit to Sparkford and enticed Ford with the invitation

to join me in Somerset where I shall be trying to make sense of the peculiar primrose colony. The place is quite rough and I shall be out nearly all day, if weather allows, on long tramps locating patches of the plant. It would, as you know, be the greatest happiness to me if you found it possible to join me.

That year they stayed in private homes. The following year Ford joined him again, and although accommodation in advance was secured only for 3 days, they always managed to find somewhere to sleep, even though on one occasion they had to share a bed.

The first visits to Sparkford were a much-needed holiday and refreshment to Fisher, and in later years he anticipated his return as a happy escape from routine into a world in which he found himself very much at home, where the traditions of an older England lingered on in the family loyalties, the rural interests, and the country wisdom of the people, a way of life protected by rolling fields and pastures, orchards, woods, and streams. It was a holiday he liked to share. When his son Harry was up at Cambridge, they went to Sparkford together in the Easter recess. Usually in later years he took one or more of his children with him.

He set out on these expeditions in a holiday spirit, heading toward a delightful part of the world in the beauty of early spring, toward the physical activity and the research interest which gave a focus to his days without in the least detracting from their other charms. In the evenings when he returned through the hamlets and rich farmland, the bellringers would be practicing for competitive campanology and, on Easter Day as he walked to morning service, the air rang with birdsong and the fine peal of bells from the church at Little Camel. On the road he met the warm greeting of other foot-travellers, and in the public bars he joined in the talk of dogs and cows and cider apples. Rain or snow might drive him to the nearest inn for shelter; but soon the sun would emerge and he would go on his way.

While observations continued in the field, the population genetics problem deserved examination both in its own right and for comparison with field data. Given a colony in which homostyle established itself, therefore, Fisher asked what was the expectation on Crosby's theory as to the proportion of homostyles in succeeding generations, how fast homostyle would take over, and how the imbalance introduced with homostyly would affect the relative numbers of pin and thrum plants.

In 1949 he published the investigation of a theoretical system of selection for homostyly in *Primula* [C P 234]. Since the transformation due to one generation's breeding is nonlinear in the genotype frequencies, pretty stiff mathematics was required for his development of a method of treating the iterated nonlinear transformations as he did using terminal expansions in two variables. In the outcome, however, each terminal expansion was found to define a consistent method of enumeration of generations in a continuum, with genotypic frequencies changing during the course of 28 generations in his tabulation. He showed how the observed populations could be represented by particular generations in this continuum, in terms of the two functions he used.

Assuming, as Crosby had, the self-fertilization of homostyles and an initial invasion of 2% homostyle genes, it was found that by the eleventh generation more than 70% of all plants would be homostyles and more than half of these homozygous. In succeeding generations thrum and later pin would be progressively eliminated.

The evolutionary picture remained baffling. Plants homozygous for homostyle were apparently viable, and Fisher found no evidence that they suffered any physiological disadvantage. Nevertheless, colonies in which they occurred did not tend to 100% homostyle. In those that he observed he rarely encountered colonies of any considerable size with more than 80%. It appeared that some check of unknown origin must come into play when the proportion of homostyle was high, and perhaps in other circumstances, so that the almost irreversible condition of 100% homostyle had not occurred. Nor had high percentages of homostyle spread to other colonies over the country.

A student of Fisher at Cambridge, W. F. Bodmer, [58] carried the work further. He demonstrated (1958, 1960) that the model used by Crosby was imperfect in assuming total self-fertilization of homostyles. Bodmer found that homostyles possess one of the devices that favor outcrossing of pins and thrums, that is, the stigma accepts pollen before the anthers have dehisced and after the bud has opened sufficiently for the access of pollinating insects. The average interval before dehiscence is 2.75 days. Moreover, Bodmer showed that outcrossing of homostyles not only can but does take place to a very considerable extent, even up to 80% in certain circumstances. The opportunities are greatest when the stigma is relatively long and the anthers dehisce late. Both these variables are controlled partly by polygenes and partly by the environment. Thus it comes about that the amount of such cross-fertilization varies from year to year.

From an evolutionary point of view the cross-fertilization of homostyles, being in part genetically controlled and therefore responsive to selection, gives the colony the option either to progress by self-fertilization toward more complete homostyly, with consequent relative invariance in all plant characters or, by cross-fertilization, toward the original distyly with a concomitant increase in variability. Evidently, in the Sparkford area and in the patch of colonies containing homostyles in Berkshire, natural selection has favored the relatively invariable form, and homostyly in most of these colonies has increased fairly rapidly up to about 80%.

It is clear that selection is responsible, for it has been found that the crossover which gives rise to homostyles is not immensely rare; indeed, if large numbers of primroses be scored anywhere in the country occasional homostyle flowers will be found among them, and many of these are genetically true homostyles. Yet homostyly has not spread through the population outside the two restricted areas, and this must be because the greater variability gained through cross-fertilization is usually at a selective advantage. Moreover, even in the Sparkford area, homostyly has been observed to decrease in certain colonies as selective pressures changed: in Sparkford Wood South the reduction from 82.4 to 73.2% over the period 1941–1948 was statistically significant and so was the decrease from 34.5 to 20.1% between 1943 and 1944 in a linear colony along Laurel Copse hedgerow, the data being homogeneous before the change (1942, 1943) and after it (1944, 1948, 1949).

No obvious ecological reason has yet been deduced for the existence of the two predominantly homostyled colonies within the ordinary range of heterostyled primroses. Yet it is not surprising that a few areas should exist where inbreeding is favored; the countryside of southern England, of course, not only differs immensely from its primaeval condition but has been much modified by changing forestry practice and by the enclosures during the last two hundred years. It may be predicted also that the modern widespread use of insecticides and of selective weed killers may alter the ecology of many primrose habitats.

12
In the United States and India

In the 1920s a handful of research workers in various scientific fields were struggling with statistical ideas in North America; Fisher early made contact with them. Several he met personally in the summer of 1924 at the International Mathematical Congress in Toronto or at centers of learning in the eastern United States which he visited before his return to England.

E. B. Wilson had been a distinguished physicist. In 1922, however, he became head of the Institute of Public Health at Harvard, and being now concerned with eugenic and population problems had turned to statistical work. His interests in human heredity and statistical methods were close to Fisher's. He was quick to appreciate the value of Fisher's innovations, because he was impelled to employ these methods to deal with scientific problems in his new work. But he was a scientist first. In 1956, he wrote to Fisher:

What we need in statistics is scientists who know some statistics. It is astonishing how few real scientists know statistics and how few statisticians know science (I exclude mathematics from the sciences). When Johns Hopkins came to set up its School of Public Health they could find no statistician to make Professor of Vital Statistics . . . so that they appointed a biologist, Pearl, who had studied with Pearson but could not have read your 1922 paper. When Harvard had the same problem in 1922, the Dean told me that professional vital statisticians were hopeless and would I leave the headship of Physics at M.I.T. and come to Harvard in Vital Statistics. That is how I got into the field at the age of 43.

312

Fisher was in entire agreement about the relative merits of "scientists" and "statisticians" in statistical work. After World War II he went so far as to avoid the appellation "statistician" altogether, and referred (for example in *Statistical Methods and Scientific Inference*) to the "scientific worker," whose duty it was to "form correct scientific conclusions, to summarize them and to communicate them to his scientific colleagues."

Fisher's work met real wants felt by the scientists rather than the statisticians. T. L. Kelley was an educationalist who, following the early work of E. L. Thorndike at Columbia University, was exploring the new field of psychometry. To meet problems in this field he had published a book on *Statistical Method* in 1923. In 1924 he met Fisher and the following year he wrote enthusiastically that he had been reading Fisher's *Statistical Methods for Research Workers;* he seemed to welcome the need to revise his own volume in consequence. E. V. Huntington, mathematician and professor of mechanics at Harvard, also met Fisher in 1924 and quickly appreciated his work. Huntington had contributed to a *Handbook on Mathematical Statistics,* published 1924. The author of this *Handbook* (and in 1927 a *Monograph on Mathematical Statistics*), was H. L. Rietz, professor at the University of Iowa, at that time the leading center in the United States for the teaching of mathematical statistics. In 1931 E. S. Pearson spent the summer session teaching and consulting in this department. Pearson and his audience both found it natural to discuss the new methods in mathematical terms; nevertheless, a member of this department, E. F. Lindquist, distinguished himself a little later by his application of the new statistics to psychological and educational data.

Fisher's work applied most immediately to agricultural research. It was George W. Snedecor, working with agricultural applications, who was to act as midwife in delivering the new statistics in the United States. Soon after joining the mathematics department at Iowa State College of Agriculture (now Iowa State University) in 1913, Snedecor became interested in statistical applications and began teaching basic statistics. Then, in 1924, Henry A. Wallace, an alumnus of Ames and assistant editor of *Wallace's Farmer* of Des Moines (whose father was then Secretary of Agriculture and who in his turn was to hold that office, and later become Vice President of the United States), recognizing the importance of statistical applications in agricultural research, persuaded 20 members of the faculty at Ames to meet on Saturday afternoons to study multiple regression and other statistical methods. In consequence, Snedecor and he prepared a bulletin on *Correlation and Machine Calculation* (1925). Also, his lectures stimulated a growing demand at Iowa State for professional help with statistics, and in 1927 a Mathematics Statistical Service was instituted with Snedecor in charge, together with A. E. Brandt, geneticist and professor of mathematics. Snedecor was universally acclaimed a grand man to work with, and interest in the work of the laboratory and demands for its services grew rapidly. In 1933 it was recognized as an autonomous Statistical Computing Center directly under the college president.

This was the first statistical center of its kind in the United States. In addition to teaching basic courses, it took on the functions on the campus at Ames that the statistical department served at Rothamsted. Both were associated with agricultural experimental stations; both served the research workers on the farm and in the laboratories through consultation and computation; both accepted the responsiblity for basic research in statistical theory and statistical method; and, from the beginning, the development and use of statistics was recognized at Ames as a cooperative enterprise between experimenters and statisticians. Naturally, statistical work coming out of Rothamsted was of the first interest to the Ames Statistical Service.

During the 1930s and 1940s it was a regular custom at Ames, largely due to Dean R. E. Buchanan of the graduate school and Professor E. W. Lindstrom of the department of genetics, to invite an outstanding scientist as visiting professor for 6 weeks each summer. At Snedecor's suggestion Fisher was invited in 1931, one of the first to make the Summer Session visitors program a success. The visitor's field of specialization had to be of major interest to several departments, and in this way Fisher was ideal. His *Statistical Methods for Research Workers* offered a text on subjects keenly interesting to the departments of mathematics and genetics and to many individuals in agronomy and animal husbandry.

Plant and animal breeders were not only involved with statistics in testing varieties against each other but in planning breeding programs. Fisher's method of maximum likelihood was of particular interest in interpreting the basic distributions of genotypes in terms of unknown parameters such as gene frequencies and recombination frequencies. Indeed, it may be argued plausibly that Fisher's development of the method owed something to its usefulness in dealing with genetic problems. For geneticists, Fisher's visit had added interest because of the publication in 1930 of his *Genetical Theory of Natural Selection*.

Cereal geneticists in particular were demanding statistical help. In 1930 three of the voluntary workers who had come to Fisher at Rothamsted had been cerealists: C. H. Goulden, the officer in charge of the Cereal Breeding Laboratory at the Dominion Experimental Farm at Winnipeg, Manitoba, visited during the summer; F. R. Immer spent the year at Rothamsted before returning to his work at the University Farm, St. Paul, Minnesota; and J. W. Hopkins spent 2 years gaining a Ph.D. at Rothamsted before taking up work as biometrician with the National Research Council of Canada.

This interest was stimulated by current work with maize or Indian corn. Before World War I G. H. Shull at the Carnegie Institute and E. M. East at

Harvard had published work on the genetics of corn which showed that, in comparison with the original varieties, inbred strains, when hybridized, would produce seed giving a more uniform crop with increased yield. In the 1920s this work was applied to the practical problem of producing better commercial seedcorn. Varieties, superior in yield and in certain cultivational qualities, were inbred for a number of generations to reduce the genetic complement to near homozygosity. Then inbred strains with different desirable qualities were crossed to produce hybrid seed combining the advantages of both parent strains but suffering none of the disadvantages of inbreeding, since dele-terious recessives no longer manifested themselves in the largely het-erozygous genetic complement of the hybrid. It was discovered that further advantage could be gained by crossing the first hybrids and so introducing three or four of the inbred strains into the commercial seed. The new hybrid corn showed, even then, prodigious advantages over the varieties from which it was derived in uniformity, in yield, and in selected qualities like resistance to lodging and to disease.

Hybrid corn was not yet much grown commercially. The great industry that grew up to supply the farmers' annual demand for fresh hybrid seed was a thing of the future. As late as 1933 only 1% of the corn acreage of Iowa was devoted to hybrid corn. In less than 10 years it was to be 99%. But plant breeders were already excited by the promise of the method, and in 1931 Fisher was delighted to see for himself the work in progress and to talk to the men who were doing it, in the heart of the American Corn Belt, at Ames.

The invitation was a great opportunity for Fisher to introduce his statistical methods personally to a potentially very large audience. Ames was one of the great land-grant colleges of the Midwest which, from their institution by the Act of 1864, had been charged with special duties in teaching and research in the agricultural and mechanical arts. The need for statistical methods in agri-cultural research had been recognized at Ames; already courses in statistics were available for biologists there, and statistics was a recognized subject for a master's degree*. Thus, Fisher was invited to introduce his statistical philosophy to that nucleus of teachers, research workers and students that was most ready to appreciate what he was talking about. Moreover, the same enterprising spirit which brought about his visit assured its success. He was impressed by the responsiveness of his American audience, by their indi-vidual energy and drive, and by their initiative and success in getting things done and changes made.

□ □ □

Fisher gave three lectures each week on statistical methods, closely follow-ing the exposition in *Statistical Methods for Research Workers*. He gave three

*The first such M.A. degree was awarded to Gertrude M. Cox in 1931.

further lectures each week either on topics in *The Genetical Theory of Natural Selection* or in the theory of statistics. Of course, many of Fisher's audience had no more statistics than they had learned from Snedecor's basic course, and in those days even the experts in statistics were relatively unsophisticated newcomers to Fisher's ideas. A. E. Brandt recalls that until that time he had regarded the controversy over degrees of freedom of χ^2 as no more than highly technical nitpicking. Yet as a trained geneticist and practicing statistician he was in a good position to appreciate Fisher's work. The appreciation was mutual; Fisher learned that Brandt had developed a formula for calculating χ^2 in a particular case, a useful technique to which he immediately drew attention in a lecture at Ames and which he introduced in later editions of *Statistical Methods*.

Practical problems were given special attention. At seminars, held about twice a·week, members of the staff, or occasionally a visitor, took turns in presenting some of their own experimental results with their statistical analysis. Then Fisher and the others present would be asked to comment on any errors or ambiguities they had noticed in the procedure or in the speaker's interpretation, on whether the question which the experiment answered was really the one the experimenter had intended to ask, and what additional inferences might have been drawn. Besides these scheduled or "clinic" sessions, Fisher was available by appointment to individuals singly or by twos or threes, if they wished to inquire more fully into some topic he had mentioned in his lectures, or to ask him about some methodological problems they were encountering in their own work. J. L. Lush, [59] describing all these activities, comments that "As consequences of this summer session, the work and interest in statistics continued to expand; experiments were designed better, with the questions in sharper focus; and more instruction in the theory of statistics was given."

Fisher found the hot weather of the midcontinent physically trying. He lodged upstairs in the Kappa Sigma fraternity house, a few blocks from the laboratory, and every morning he brought his bedsheets downstairs and left them in the refrigerator until evening. He would have been glad to do the same with his mattress had there been room for it. In the office the small electric fan was inadequate to refresh the humid air. He suffered constant physical discomfort, for, apart from the weather, he was plagued with a fistula all through the summer. The condition had been diagnosed when he first arrived in the United States, and Harvard had offered to have the necessary operation performed without charge. But this would have put him out of commission for a month. He preferred to finish the job that he had come to do and to have the operation performed after his return to England. Meanwhile, the sitting position was uneasy, so he rigged up the office fan sideways on the desk and lay down on the floor beneath. More than once visitors were alarmed to find him there, apparently fallen in a faint but in reality comfortably asleep.

The indoor swimming pool in the gymnasium offered welcome relief. There was an occasion when Fisher lost his glasses in the water there. James Snedecor, then 14 years old, recalls joining with other swimmers in the search for the spectacles at the bottom of the pool. It would have gone ill with Fisher if the glasses had not been successfully fished out, for he had no spare pair with him.

Fisher's conversations with his students would often take place on the way down to the Student Union for a cool drink around the coffee tables there. In those days of prohibition, beer was nowhere available, but one could get tea, and Fisher came to enjoy fruit drinks, too, especially the fresh fruit drinks Mrs. Brandt prepared.

Obviously, he needed looking after, tactfully. Es Brandt became a self-appointed protector during the summer of 1931; he looked after his comfort and made sure that he arrived where he was expected, suitably prepared. Brandt was embarrassed by the "baggy grays" worn by his eminent visitor, embarrassed on Fisher's behalf when he appeared shabby among all the smartly pressed suits of the Americans. One day they sat drinking at a restaurant whose windows faced, across the street, an establishment bearing a large sign, "Pantalorium." Brandt had enjoyed interpreting Americanisms which struck Fisher as strange or piquant, and now he explained "Pantalorium" rather carefully, pointing out how handy the place was for visitors and students at the University who wanted their trousers pressed. Fisher heard him out and then, looking him straight in the eye, responded: "I understand you perfectly." Whether or not he ever made use of the pantalorium is not recorded.

During the summer session Fisher visited other Midwestern centers. He lectured on cumulants at Minneapolis, and stopped off at Rochester, Minnesota, to see H. L. Dunn at the Mayo Clinic. He hunted butterflies with Dunn and his sons in the lake country of Minnesota, was taken boating on Lake Mendota in Madison by the mathematician, M. H. Ingraham, and was later shown the sights of Chicago. Before coming west he had visited friends on the East Coast; he had been welcomed in New York with a dinner party by Arne Fisher, and had met E. B. Wilson in Boston, agricultural research workers at Cornell and Ithaca, and geneticists at Philadelphia. On his return eastward he attended the Third International Congress of Eugenics at New York. It was a busy, stimulating, and rewarding summer.

It was also the curtain-raiser to the dramatic spread of his influence in America during the following decade. Although he could not manage to accept the invitation he received to be visiting professor at New York State College of Agriculture at Ithaca for the first semester of the 1932 – 1933 academic year, he did make the trip to Ithaca in the latter part of August 1932 for the Sixth International Congress of Genetics. On the voyage over he met other European geneticists and especially appreciated the company of Prof. O. Winge from Denmark. Before the end of his stay, he ran out of money and borrowed $10 from F. A. E. Crew and another $8 from T.·H. Morgan. From Ithaca he made an expedition in T. L. Kelly's car to Gilmanton, New Hampshire, to see the total eclipse of the sun of August 31. After the conference, which he reported to have been "hot but delightful owing to the excellent exhibits and personal discussions," he returned to England through a magnificient storm in the Atlantic.

□ □ □

After the summer session of 1931 Brandt sought leave to join Fisher and in the autumn of 1933, and again in the year 1935–36, he brought his family to Harpenden and attached himself to Fisher's department in London. After his return, he was eager to have Fisher visit Ames again.

The opportunity arose in 1936 when Fisher was once more visiting professor at the summer session, from mid-June until near the end of July. Brandt took over the arrangements not only for the summer session itself but for the whole summer, contacting the universities, making engagements, studying train timetables, and finally filling every minute with 60 seconds worth of distance run, without intolerable discomfort to Fisher. At Ames Fisher stayed in his home and was given the abundant fruit drinks and the tactful shepherding in daily affairs he needed, especially if he were not to be diverted from the timetable or forget appointments.

In addition to lectures Fisher gave biweekly seminars, supervised research projects of several advanced students, and attended the three special weekend meetings, one for statisticians, one for geneticists, and the third for agricultural economists, which drew speakers and auditors from the surrounding states to join him in discussing matters of common interest. Finally, at the commencement ceremony at the end of the summer session, he was awarded an honorary doctorate of science from Iowa State University.

This time, Fisher did not lecture on genetics but he referred to the subject and its problems often in the "clinic" sessions. Nearly half his lectures were devoted to the design of experiments, exploring the area covered by his book with that title, that had come out the previous year. Nearly all the rest concerned topics on the theory of statistics.

The weather was oppressively hot, worse even than in 1931. That summer he began to take a pinch of salt in his cup of tea. He must have been suffering from salt deficiency through excessive sweating, for he found the slight salinity delicious. Ever afterward he continued to doctor his cup of tea in the same fashion. His demands for tea, and perhaps Brandt's remembered enjoyment of tea time at the Galton, resulted that year in the institution of tea at Sne-decor's weekly seminars, seminars that provided an opportunity for dialogue between staff and students on research problems.

There was by this time an air-conditioned restaurant at Ames, and, since prohibition had ceased in 1933, one could drink beer there. An open-air swimming pool had been opened at Boone, 15 miles away, and parties frequently drove out after lectures to refresh themselves. Fisher was an unusual sight to local swimmers, with his long-legged, old-fashioned swim-ming suit, his blanching gingery beard, and his pipe. (The pipe appears again in Plate 12, a snapshot of him paddling in the sea during a family holiday in 1938, with jacket pockets bulging as usual. On his next visit to the States in 1946, he is described as having evolved a manner of swimming almost up-right in the water, while he drew on the unlit pipe.) The novel sight spurred a cheeky Boone boy, jerking a thumb in Fisher's direction, to ask one of the party who was putting on the Carnival.

The real carnival of Fisher's stay was Independence Day. There is no holi-day in the United States like the Fourth of July whose celebration involves every age group in day-long festivities, with funfairs, costume parades with elaborate floats, and finally firework displays. Fisher joined in it all. As they drove home in Brandt's car, one in the party was exulting over the celebration of American Independence when a sudden thought struck him: "You don't celebrate the Fourth of July in England, do you?" Fisher admitted that this was so, paused and added thoughtfully, "Perhaps we should." For a few mo-ments the remark seemed innocent, even complimentary, before its ambi-guity dawned upon them.

Fisher was relaxed and in good form during these weeks; he was having fun. That is, he was surrounded by enthusiastic researchers and their prob-lems. Mrs. Mauss (then Miss Besse B. Day, assistant to F. X. Schumacher of the Forestry Service) recalls:

Our friendship developed rapidly, firstly because I had so many unsolved statistical problems and secondly because of Ames' unbearable heat. Jack Youden, Nancy Millon the young Canadian, Ron and myself used to run off to the Boone swimming pool at every opportunity and dine later in the one air-conditioned restaurant. I came with a brief case full of the hardest kind of problems of design and analysis, with lots of data in the field of forestry. I was swamped and about ready to sink in deep water! My situation was a field day for Fisher. Had not Brandt been his guardian angel and pulled him away one time, Fisher would have missed the President's reception and dinner in his honour, to help me with the analysis of a mass of tree growth data.

Probably the most gratifying part of all his activity was the contact with these advanced students at the seminars and in connection with their re-search. And, like Brandt after Fisher's visit in 1931, some of his students in 1936—Jack Youden, C. C. Craig, Horace B. Norton III—managed within the next year or two to join him at the Galton Laboratory; and Besse B. Day was on her way to spend a year there when war broke out in September 1939 and the visit was cancelled.

Among the more senior men, he was no less deeply involved. George Snedecor had just completed his book *Statistical Methods*, which appeared in 1937. C. H. Goulden, visiting Ames for the summer session, was working on *Methods of Statistical Analysis*, which was to appear in 1939. F. X. Schumacher also visited for the summer session; he was already the joint author of *Forest Mensuration* (1934) and was soon to undertake the writing with Roy Chapman of the important statistical text on *Sampling Methods in Forestry and Range Management* (1942).

Some measure of the importance attached to Fisher's statistical ideas and the rapidity of their spread may be gained by observing the statistical textbooks that appeared in the 1930s. A student of statistics during World War II might have been referred to Fisher's *Statistical Methods* and *The Design of Experiments*, to the works of Snedecor, Goulden, and Schumacher and Chapman referred to above, to L. H. C. Tippett's *Methods in Statistics* (1931), and perhaps also to D. Mainland's *Treatment of Clinical Laboratory Data* (1936), P. Rider's *Introduction to Modern Statistical Methods* (1938), E. F. Lindquist's *Statistical Analysis in Educational Research* (1940), G. Freeman's *Industrial Statistics* (1940), or K. Mather's *Statistical Analysis in Biology* (1943). All these books were, in effect, introductions to Fisher's statistical ideas and commentaries, explications and elaborations of what he had written. Several of them he saw in manuscript form and discussed with their authors; only Lindquist, from Iowa City, and Freeman, from Harvard, were not immediately associated with Fisher. The books exhibit also how men in diverse fields saw the need of texts devoted to applications in their own speciality:—to agriculture, forestry, education, psychology, medicine, biology and industry. The interest in America is apparent when one observes that, apart from Fisher's books, two of those listed originated in England, two in Canada, and five in the United States.

American interest in Fisher's work was reflected in 1936 in the eagerness with which friends seized the opportunity of his presence to invite him to visit them all across the United States and to speak in their universities. During the summer sessions he managed to slip away one weekend to visit St. Louis, to meet Prof. Paul Rider again (who had spent the year 1935 – 1936 with him at the Galton Laboratory). In St. Louis he also met Edgar Anderson, a former voluntary worker with him at Rothamsted who was then working at the Missouri Botanic Gardens. It was Anderson's data on species of *Iris* that formed Fisher's first published example of the application of the discriminant function. The article had just appeared, and Anderson had more *Iris* data to talk over.

□ □ □

After the summer session followed 5 weeks of travel: to Colorado Springs to lecture for the Cowles Commission; to Rawlins, Wyoming, to visit his aunt Laura, the long-widowed Mrs. Alfred Heath; to Chicago to talk on the logic of experimentation and visit Wright and his geneticists; to Ann Arbor, Michigan, for three more lectures, and to Ithaca for two.

Traveling everywhere by train, Fisher was impressed by the great railroad systems and by the comfort and interest of transcontinental travel by train which at that time had reached its zenith of streamlined efficiency. Americans were justly proud of their railroad system. On a broader loading gauge than British railways, travel was smoother and faster. Pullman service was almost luxurious. The glass rear cabin gave magnificent views of the moving landscape. Approaching the Rocky Mountains there was a stop long enough for passengers to stroll along the platform and watch while additional powerful engines were promptly coupled to the train to help pull it over the mountains. On the network of railroads connecting with the transcontinental lines passenger service was frequent and good. Fisher was charmed when a friend who wished to discuss something with him took a train from Madison to meet him in Minneapolis. Joining his express as it passed though early in the morning, he breakfasted with him and had a chat before dropping off again at Chicago, where he would pick up a connection home to Madison. On one journey Fisher noticed that there was a great press of people in one car, which was anomalously decorated as a Western Shack. It turned out to be the bar car, where passengers were stocking up before the bar closed at the border of a dry state; liquor sales resumed only when they reached the Illinois state line.

After a train journey by night, he arrived early one morning in Ashville, Georgia, for a part of the tour which he particularly enjoyed, with the foresters. After the lecture-seminar he was driven with Dr. Osborne and Dr. Schumacher through the forested areas of south Georgia and northern Florida. There was something very pleasant in being out in the field again, and a quiet sanity in contemplating with the foresters the growth of trees a hundred years to come. Then he moved north again to attend cancer research meetings; by the end of August he met with geneticists at Woods Hole, Mass., for a few days before joining Mrs. Fisher on her arrival in New York and becoming immersed in the tercentenary celebrations at Harvard University.

On this occasion, during the first two weeks of September 1936, Harvard University awarded honorary degrees to 62 eminent scholars from around the world. Fisher was invited with Mrs. Fisher to attend the tercentenary celebrations and to receive from Harvard University the honorary degree of doctor of science. Such an honor deserved a very special celebration. Mrs. Fisher had never before left the family for more than an evening, but now she did so in order to spend 3 weeks with her husband in the United States. It seemed natural that it should be a family friend from Rothamsted, Miss S. G. Heintze from the chemistry department, who, with the help of the resident housemaid and a temporary nursemaid, undertook to keep the home fires burning. It was a noble gesture from a spinster in her middle years to give up her summer holiday to look after an undisciplined bevy of six little girls and their two big brothers, ranging in age from 1 year to rising 17; Fisher had true friends.

Mrs. Fisher arrived safely, though very seasick, and was carried off with her three unprecedented evening gowns, made or bought for the occasion, to Boston to join in the equally unprecedented social whirl of dinners and lectures, interviews with journalists, and sightseeing expeditions associated with the tercentenary celebrations. On the third tercentenary day she attended the central ceremony when, as the official record shows, "after the singing of Handel's 'Let Their Celestial Concerts All Unite' by the Tercentenary Chorus, the significance of the Tercentenary Celebration . . . had its dramatic fulfillment. The honours of the Tercentenary were conferred." Proudly she heard her husband cited as a "student of heredity who has improved statistical methods and assisted agriculture by the application of his science."

It is characteristic that with the journalist who interviewed him for the Boston Globe's series "The World's Wise Men—Scholars Coming to Harvard's 300th Anniversary," Fisher took the opportunity to enlarge upon the question of the differential birthrate rather than to discuss his contributions to statistics and that the article carries a photograph not of Fisher but of Francis Galton whose theory explaining the dying out of the male line of many British peers, through marriage with heiresses, Fisher had quoted in his discussion. It was from other sources that the journalist culled the famous synopsis of Fisher's statistical work: "Fisher taught experimenters how to experiment" and discovered the uniqueness of the honor bestowed on Fisher for his services to agriculture. Harvard had not recognized similarly any man from the 50 agricultural colleges in the United States which for 50 years past had poured out "shelf full after shelf full" of research bulletins.

That year several learned societies chose to have their annual conferences in Cambridge, Massachusetts, at this time. With E. B. Wilson, Fisher attended meetings of the American Mathematical Society. During the American Astronomical Society's meetings Wilson introduced the young astronomer, T. E. Sterne of the Harvard Observatory, whose work on least squares Fisher had already met. A series of luncheon meetings at the Harvard Union was arranged with Sterne and other astronomers for informal discussions with Fisher of statistical methods and small samples.

On September 19 the Fishers visited Princeton, New Jersey, and on the following day reached Washington, D.C. Fisher gave lectures at both places and received the always astounding hospitality of Americans. Then he saw his wife off from New York on her return voyage to England and himself took the train to the West Coast. There he enjoyed the hospitality of Dr. and Mrs.

Horace Gray at their home in San Francisco and "anticommuted," as he put it, from the city across the river to the University of California at Berkeley, where he lectured for a month on designed experimentation. Dr. Gray, a medical doctor working at Stanford University Hospital, had spent the previous year with Fisher at the Galton Laboratory. Fisher took the chance also of visiting genetical friends on the West Coast, particularly C. H. Danforth and his mice at Stanford. Finally, he was homeward bound on a route that took him through Canada so that he could stop for a few days to see Goulden in Winnipeg and Hopkins at Ottawa, and on October 26 he sailed from New York for home.

This was not quite the end of the affair. Fisher was warned to expect a representative of the U.S. Department of Agriculture to meet him aboard ship before he sailed to make a formal offer of a professorship at Ames, to be jointly sponsored by the university and the U.S. Department of Agriculture. Arrangements had not, in fact, quite reached this point when he sailed, but the possibility of removing from England either to the Midwest or to California was definitely in the air. It was much discussed within the family during the autumn and winter. The children approved, their father promised that if they went to Ames they should have ponies like the Brandt children, with whom he had himself ridden that summer; if to California, they should have date palms growing in the garden.

Fisher was less easy to satisfy. Having endured two summers at Ames, he was not at all sure that he and the family could stand the climate there and suggested that the contract with the USDA might include an option to transfer its contribution to another American university if that proved desirable. To this the USDA agreed, but the State College, fearing to lose him, would not accept the escape clause. Moreover, the appointment would have to be renewed annually. Fisher, contemplating the expense of removal from England and the hiatus which would result in the education of the children and in his own career if the contract were not renewed, reluctantly decided that he had better remain at the Galton Laboratory. But the idea of moving was very attractive; in contrast to the halfhearted support of University College, it seemed in America he could expect that the energetic people of that great land would, in their phrase, "run with the ball," welcoming, adopting, and supporting his schemes with some alacrity.

It is tempting to speculate what differences might have occurred in the history of statistics had Fisher decided to move to the United States in 1937. One may be certain that satisfactory solutions to the problems of climate, salary, and tenure would have emerged. Possibly, even in the war, Fisher might have been able to serve his country better in the United States than he was permitted to do in England. Moreover, his contact with statistical developments in America would not have been broken during these critical years when new influences made themselves effective. J. Neyman emigrated to America in 1939 and A. Wald in 1940; in teaching, theoretical and mathematical aspects of statistics received increasing emphasis. Changes took place in the mood of the American statisticians which extinguished the enthusiasm that had so charmed Fisher in 1936. When he returned to the United States in 1946, he was welcomed by the younger statisticians as a great originator and authority certainly, but also as a foreigner whose ways were not always their ways, nor his thoughts their thoughts. The transition had begun from the reality to the myth, from acceptance of Fisher as an exciting and helpful colleague in their research to his reception as an oracle, of uncertain temper and controversial meaning.

□ □ □

In the 1930s Fisher's influence was extending elsewhere in the world through the voluntary workers who had worked with him either at Rothamsted or at the Galton. Thus the Swedish geneticist, J. Rasmusson, working with quantitative characters of the garden pea, came to Rothamsted in 1931—1932; subsequently, he arranged for Fisher to visit Astrand, Sweden, to lecture on statistics during the summer term of 1937. In 1938, he invited Fisher, with his wife and sons, to visit Sweden and stay with his family for a week so that he might see the work at Landskrona.

J. B. Hutchinson visited Rothamsted in the summer of 1928. Working from 1933 with the Empire Cotton Growing Association in Central India, he introduced the application of genetical and statistical principles to cotton breeding. In his book, *Genetics in Cotton Improvement* (1959), Sir Joseph Hutchinson records, for example,

At Indore, the use in progeny row breeding of modern experimental techniques involving replication and randomisation was demonstrated (Hutchinson and Panse, 1937), and the way was opened to the measurement of the genetic variance on which improvement by selection depends. The direct relation between genetic variance and the prospect of advance in plant breeding (Hutchinson, 1940) and the loss to the breeder from 'the misguided selection of lines with low variance' (Hutchinson, Gadkari and Ansari, 1938) became accepted as fundamental concepts in a new approach to breeding problems.

When Fisher was in India in January 1938, Hutchinson mentioned to him his assistant, V. G. Panse, and later that year Panse, in turn, came to the Galton Laboratory to study under Fisher.

Fisher's first visit to India in December 1937 and January 1938 was made at the invitation of P.C. Mahalanobis on behalf of the Indian Statistical Institute and, more generally, of statistics in India.

Mahalanobis, since 1923 a professor teaching physics and meteorology at Presidency College, Calcutta, had been gradually led away from physics into consultation, research, and teaching of agricultural statistics and to the founding of the Indian Statistical Institute. Meteorological problems had early involved him in statistical studies and led to discussions with his colleagues of statistical problems in anthropology and agriculture. His analysis of an experiment on paddy (rice) brought him to Fisher's attention in 1924, and Fisher wrote, sending his recent papers leading up to the analysis of variance and z test. When Mahalanobis visited England in 1926–1927, he introduced himself to Fisher at Rothamsted, enthusiastic about these new ideas, and during their day together he mentioned his plan to tabulate the variance ratio. His first tabulation of 5 and 1% points appeared in 1932. [Today this quantity is commonly referred to as Snedecor's F, since Snedecor, in ignorance of the Indian work, later tabulated the variance ratio under the symbol F, chosen in honor of Fisher. In *Statistical Tables for Biological Agricultural and Medical Research* (1938), Fisher avoided using F, since this symbol was not used in the tabulation of Mahalanobis, which had priority.]

Soon after his return to India in 1927, Mahalanobis was surprised to receive an invitation from the government of India to apply for a grant from the newly formed Imperial Council of Agricultural Research (ICAR); he was impressed when an official was sent from Simla to reinforce this invitation. He had not been involved to any extent with agricultural statistics and was hardly aware of the existence of the Imperial Council; but the prospect interested him and he applied for and received the grant. A little later he was asked to undertake a project sponsored by the government and, then, to take an agricultural research worker into his laboratory for 6 months to analyze data he had collected and to receive instruction in statistical methods. Mahalanobis attributed these successive invitations directly to Fisher's influence with the Indian government, and he was almost certainly right. Rothamsted was consulted in setting up the ICAR. Fisher, concerned to involve a good statistical consultant in their work would naturally have mentioned his recent visitor as a man likely to be interested and helpful to the council. When Mahalanobis accepted the first invitation of ICAR, he became the obvious man to turn to for their later statistical needs.

What Mahalanobis did with his opportunities was something else. Being asked to teach the research worker, he realized there was no institution in India where such men could learn statistics. As a professional teacher, he saw the chance to extend the educational functions of his group and establish a center capable not only of undertaking basic research and doing special projects but of providing statistical instruction for agricultural and other workers in government service. His proposal to start a teaching institution was strongly supported by the Viceroy, Lord Linlithgow, and the Indian Statistical

326
The Life of a Scientist

Institute came into existence in 1932, together with the Indian Journal of Statistics, *Sankhyā*.

Only the vision was large. The work was done at Mahalanobis's home and, apart from voluntary workers, Mahalanobis had three trained assistants supported by the ICAR. S. S. Bose was a brilliant young physicist, hired in 1929, who, studying and using Fisher's methods extensively, worked much on problems of design. Within a few years he was preparing designs for agricultural experimentation all over India and was helping supervise the actual carrying out of the experiments. R. C. Bose and S. N. Roy, both mathematicians without previous statistical experience, joined the group after the ICAR grant was increased in 1931. The Indian Statistical Institute itself received a training grant initially of R600 or about £45 a year and employed one man, its director, Mahalanobis, part time.

Although separately sponsored, the teaching and research flourished hand in hand. When Mahalanobis wrote 5 years later to invite Fisher to India, he could boast of a certain amount of useful research in the theory of sampling and other theoretical topics, a large number of applied problems in hand, and a series of short courses that had been attended by more than 70 officers or deputations from government departments, scientific institutions, and universities.

Despite this growth in the work and its evident utility, it was clear that modern statistical method was making little headway in academic and scientific circles. In the universities statistics was represented only by one or two questions asked in the masters examination in economics. When Mahalanobis suggested that the Indian Science Congress ought to set up a special section for statistics, he was told that if statistics deserved a section, they might as well set up a section for astrology too!

In response to the disdain of the committee of the Indian Science Congress, Mahalanobis determined that in 1938 the Indian Statistical Institute should hold a special Statistical Conference, independent of the Science Congress, though run within the same framework and more or less at the same time, and he wished to have Fisher as president at this conference.

He told Fisher frankly that the Indian Statistical Institute itself needed his moral support and patronage. Its financial situation was not satisfactory. It still relied much on the work of volunteers and on grants for special programs and from the ICAR—and since the ICAR was setting up its own statistical section, their support to the institute might be withdrawn. The journal *Sankhyā* was run at a deficit of £250 a year, met entirely out of private funds.

So Mahalanobis asked Fisher, first, to lend his name in support of the institute and, in token of his cooperation in their work, accept honorary fellowship of the institute; second, to come to India for 3 or 4 months (he came for 6 weeks), to see and advise on their work, preside at the Statistical Conference, travel, lecture and consult wherever visits could be arranged; third, to serve as moderator for the Statistical Diploma Part I, for which examinations had been devised by the institute and were to be set for the first time that year; and finally, to support the journal *Sankhyā* by himself contributing occasionally to its pages. This assignment Fisher accepted in all its aspects.

□ □ □

He sailed for Bombay in November 1937 in the company of the delegation from the British Association to the Indian Science Congress. The 3 weeks on board were most interesting. Lord Rutherford was there, a special guest of the Congress. Also attending were a galaxy of distinguished psychologists: Carl Jung, whom Fisher had last seen at Harvard when they were both awarded honorary degrees, and two Fellows of the Royal Society, C. S. Myers and C. E. Spearman. He was at the time on the editorial committee of the *Journal of Neurology, Neurosurgery and Psychiatry* as an authority on statistical and genetical aspects of the subject, and he welcomed the opportunity of a long sea trip to hear the views of these experts in its medical and psychological aspects.

The voyages to and from India were a refreshment Fisher would not forego for the sake of a longer time in India; he refused the offer of a passage by air. Aboard ship he had time to think and leisure to converse, and he considered the time to think was a right worth preserving at some apparent sacrifice to his work.

In Calcutta he stayed at the home of Professor and Mrs. Mahalanobis while the Congress was in session. He was told that any guests whom he cared to invite to dinner would be welcome any day, to the capacity of a dinner table seating a party of ten; so he brought in guests every day: members of the delegation from the British Association who might not otherwise have recognized the existence of the Indian Statistical Institute and statisticians he felt should meet their host at the Statistical Conference personally.

The most potent public opportunity for the expression of his views on the future of statistics in India was the presidential address [CP 159], delivered at the opening of the Statistical Conference in the presence of His Excellency the Governor of Bengal, the president of the Indian Statistical Institute and distinguished guests, many of whom held government and university posts in India.

He introduced statistics as being originally the factual basis of statecraft. Deploring the segregation in England of two traditions, the one the practical administrative tradition of state officials and the other the academic tradition of mathematical statisticians, he suggested that:

In developing her statistical services India might learn from the difficulties which England has encountered, and somehow contrive neither to allow official statisticians to be blinded by ignorance of method, nor to allow academic statisticians to be sterilized by lack of responsible experience.

In England, recent statistical developments had led, he said, to the

somewhat ludicrous spectacle of entomologists, foresters, plant physiologists and others with no trace of mathematical pretensions, applying freely and with understanding in their daily work mathematical refinements which most official mathematicians could not understand and which too many teachers of mathematical departments were unable to expound.

From this experience he drew the moral that

responsibility for teaching statistical methods in our universities must be entrusted, certainly to highly trained mathematicians, but only to such mathematicians as have had sufficiently prolonged experience of practical research, and of responsibility for drawing conclusions from actual data, upon which practical action is to be taken.

The new statistical methods showed the potentiality of statistics as an intimate union of mathematical and scientific understanding, and, having discussed these methods briefly, Fisher concluded that India had a great opportunity to utilize the newly discovered possibilities of statistical science; the Conference itself showed the will to seize this opportunity and to carry the work forward, as "the brilliant school of workers that Prof. Mahalanobis has gathered round him" was already doing.

This public statement was supplemented by personal meetings and discussions with university and government officials. Fisher saw the Viceroy several times in Calcutta and at the end of his stay was urged to write to him his further observations and recommendations.

In these ways Fisher lent the weight of his prestige to spread appreciation of the new statistics in the service of India. More particularly, he saved one project that was threatened with extinction. In 1936 Mahalanobis had begun a sample survey of the jute crop in Bengal. It was the first of its kind. The minister responsible for funding the project from year to year, newly appointed in 1937, simply could not understand how by taking a relatively small number of samples from the hundreds of millions of acres of jute, it could be possible to make reliable estimates of the whole crop. He was proposing to put a stop to the survey work when Fisher arrived in Calcutta. Though Fisher did not manage to convince the minister, his support for the sample survey had its effect. He spoke with Bengal government officials, the Governor of Bengal, and the Viceroy himself (since it concerned not only the state but also the central government). He insisted that the sample survey should continue; there was no prospect by any other means of discovering the facts of the case, for jute or for any other factor affecting life in the vast uncontrolled enterprises of the subcontinent.

Thus at a critical stage the sample survey was saved and the way cleared for what grew to be the National Sample Survey of India, which now brings into the government a continuing record of economic and agricultural statistics. It has proved indispensible to state planning in such varied aspects as the standardization of weights and measures, the distribution of food supplies, the planning of agricultural and industrial developments, and the allocation of educational and medical resources. As Fisher declared, congratulating Mrs. Mahalanobis on her husband's achievement, "Alone in the world you have a National Sample Survey and your government can have real information about what is going on."

□ □ □

In Bombay Fisher was particularly glad to meet R. S. Koshal, formerly a voluntary worker at Rothamsted, then working for the government. The work Koshal [60] had done in England had resulted in publication in 1935 of a paper which had aroused the old lion, Karl Pearson, to a last great roar in *Biometrika* [61, 1936], a violent attack against Koshal's work on application of the method of maximum likelihood.

For Koshal the attack had been deeply unsettling. Because of it he had found himself under suspicion with his superiors and, being a government employee, was unable even to get permission to write to Fisher. He had confided in Prof. J. Maclean of the University of Bombay. Maclean had written Fisher that Koshal was very distressed because he had made a slip in his paper (he had omitted to apply Sheppard's corrections) and feared Fisher's anger; he explained how it was that Koshal could not write directly; he observed that retrenchment was in the air and Koshal's job might be affected, that none of the Central Cotton Council for whom Koshal worked could understand his sophisticated statistics, so that they might well dispense with him in ignorance of his worth and, finally, that if Koshal had to go, he, Maclean, would feel that one of his mainstays had collapsed.

Koshal had been concerned to show how the practical difficulties in using the method of maximum likelihood could be overcome. As Fisher described it [*C P* 149, 1937]:

Koshal knew that the theoretical objection to the method of moments had never been answered; failing theoretical justification it had been put about that it was a "practical" method, while the equations of maximal likelihood were impossible of solution. Koshal determined to take the case at its most difficult, i.e. a heavily grouped Pearsonian curve, and to show that the direct numerical approach to the solution of maximal likelihood was not impracticable.

The likelihood equations are complex and nonlinear and insoluble by analytic methods. Koshal's method was an application of numerical analysis, applicable quite generally.

Suppose there were two unknown parameters θ_1 and θ_2. Then, as a preliminary, some appropriate method, such as the method of moments, might be employed to supply the first trial values. These values define a point in the parameter space that lies somewhere in the vicinity of the maximum of the likelihood function. Now, in such a vicinity, the likelihood may be approximated by a quadratic equation:

$$f(\theta_1 \theta_2) = A + B\theta_1 + C\theta_2 + D\theta_1^2 + E\theta_2^2 + F\theta_1\theta_2$$

Koshal chose a triangulation of six points and calculated the likelihood values at these points from the data. It was an arrangement of extreme neatness and economy, for from these six values he could easily solve the equations for the six constants required and calculate the position of the maximum of the approximating quadratic surface. If a better estimate were required the same procedure might be repeated in the vicinity of the new trial value. Convergence is usually rapid, and the extreme economy made possible by the selection of these particular six points reduces the labor of computation to a minimum. Finally the coefficients D, E, and F may be used to obtain approximate variances and covariances of the estimates.

The work stemmed immediately from Fisher's work on the method of maximum likelihood. In Pearson's view it was "clearly planned to show how the 'Method of Moments' is much inferior to the 'Method of Maximum Likelihood,' " and in his response he aimed, on the contrary, to prove the general superiority of the method of moments over the method of maximum likelihood. His quarrel was with Fisher, not Koshal. Fisher felt that he should reply, and in doing so [C P 149] he followed the argument presented by Pearson step by step, making plain where he felt that "errors of reasoning" and "errors of arithmetic" occurred, and what the consequences were of these errors and of what he characterized as "tricks of presentation" which together led to the conclusions Pearson desired.

Pearson's paper opened with the italicized and arresting query: *"Wasting time fitting frequency curves by moments, eh?"* But Pearson died in 1936. Fisher, writing shortly afterward, felt free in the last section of his paper to consider in all seriousness the question Pearson had raised. He pointed out that the considerable place occupied by "fitting curves by moments" was allotted at the expense of attention to other subjects. Students had little time left for adequate study of the theory of comprehensive computational processes, the exact treatment of small samples, the analysis of variance and covariance, the theory of estimation, and practical computation on topics of interest in contemporary research. He concluded:

So long as "fitting curves by moments" stands in the way of students obtaining proper experience of these other activities, all of which require time and practice, so long will it be judged with increasing confidence to be waste of time.

Fisher's paper appeared late in 1937. Koshal met him at the quay-side on his arrival in Bombay, and during his day or two in that city they were able to discuss the matter. Fisher showed Koshal what he had written, reassured him again of his confidence in him and considered with him what amendments were required to improve upon the earlier work. He made a point of meeting Koshal's superiors and left them in no doubt that he thought very highly of the work and the character of their young colleague.

☐ ☐ ☐

S. S. Bose, the earliest of Mahalanobis's assistants, was the representative of the Indian Statistical Institute who traveled to Bombay to meet Fisher and accompany him on the long train journey back to Calcutta. He went with him also on the pre-Congress tour in central and Northern India. Fisher found him a delightful companion—modest, interesting and accomplished. It was with real regret that he was to learn, the following summer, of the death of this young man. It was a great loss to the Institute.

In Calcutta Fisher gave a series of six lectures at the University of Calcutta, speaking on modern statistical methods. He also lectured at the Anthropological Museum on the subject of eugenics. Workers at the institute itself were excited at the prospect of talking over their problems with him. Mahalanobis had taken R. C. Bose and S. N. Roy to Simla during the previous summer recess so that they might read and discuss Fisher's papers on estimation. On his arrival began a series of seminar-type discussions, during which they raised questions that had emerged during their studies. Questions of design were also discussed and problems of their current work.

A perplexing statistical problem concerned quantitative measures of racial likeness. Mahalanobis had early been interested by his anthropological friends in the characteristics of build or appearance of the numerous races of India that might be found to distinguish or relate the different races and languages (or language groups) within the subcontinent. In one of the first articles to appear in *Sankhyā,* Mahalanobis [62] investigated the tables of anthropological data collected by W. Risley, which had seemed previously to have been discredited by K. Pearson's criticism, and, by introducing certain corrections, had been able to answer the criticism and virtually add the data to the scientific evidence. Even earlier, he had found himself dissatisfied with Karl Pearson's coefficient of racial likeness and had proposed, as an alternative, a measure of distance, which he called D^2. Considering the various measurements to be represented by a point in n-dimensional space, with two populations there would be two clusters of points in the space of the measurements; the D^2 statistic was a measure of the distance between such clusters. To use this statistic, however, the distribution of D^2 was required or, at a more elementary level, the moments of the distribution. This in turn raised general questions of multivariate statistical distributions, for, if one has a whole series of measurements, their *joint probability distribution* would need to be considered in order to obtain distributions of the relevant statistics.

As mentioned earlier, Fisher found a surprising unity for normal theory distributional problems that come from very different sources. In particular, it turned out that the distributions required when the null hypothesis is true—in ordinary correlation, multiple correlation, comparisons of means with Student's test, the χ^2 test and analysis of variance problems, problems described in terms of the correlation ratio and in terms of intraclass correlation—were all really covered by the distribution of the multiple correlation coefficient, which he had discovered in 1924 (*C P* 37). Fisher had pointed out that by suitable transformation these null distributions are essentially all derivable from one. By 1928, he had also solved the distribution problem in the nonnull case for the multiple correlation coefficient. If regression of y on a series of variables $x_1, x_2, \ldots$ is considered, it turns out that whereas in the null case it does not matter if the x variables are random variables or fixed constants such as distinguish between treatments, in the nonnull case it makes a difference. Fisher derived the distributions for the nonnull case both for the case when the x's are random variables and when the x's are fixed. Consequently, his solution provided the distributions in the non-null case of all the quantities previously mentioned, covering problems in ordinary correlation, multiple correlation, Student's test, noncentral χ^2, and noncentral F.

At the Galton Laboratory Fisher himself became interested in problems of quantitative differentiation of groups through multiple measurements, and for this purpose he had devised discriminant functions. The ingenious simplicity of the idea is well illustrated by Fisher's application [*C P* 138, 1936]. Four measurements were made on the flowers of 50 plants each, of two species of *Iris* found growing together in the same colony: the petal length, the petal width, the sepal length and sepal width. The problem was then formulated in a manner that took a long step toward solution. Fisher wrote:

We shall . . . consider the question: What linear function of the four measurements

$$X = \lambda_1 x_1 + \lambda_2 x_2 + \lambda_3 x_3 + \lambda_4 x_4$$

will maximize the ratio of the difference between the specific means to the standard deviations within species?

That linear function he called the discriminant function.

The most direct application for the discriminant function was in deciding how best to classify a new multiple observation, for example, a new skull has been obtained and we ask if it belongs to population A or to population B. An extension of this basic idea was given in Fisher's paper, using the numerical data for the two species of *Iris*. A third species of *Iris*, intermediate between the two previously discussed, was suspected of being a hybrid derived from them. On this hypothesis the theoretical value of the discriminant function of the hybrid was calculated, taking into account the fact that the presumed ancestors were, respectively, diploid and tetraploid, and the presumed hybrid was hexaploid. A test of significance was made of the difference of the theoretical from the actual value, and no significant difference was found.

Fisher showed how such tests of significance could be developed, making an ingenious application of analysis of variance applied to regression. The test criterion of the distribution implied by the preceding steps turned out to be exactly equivalent to the generalisation of Student's ratio presented by H. Hotelling [63] in 1931, Hotelling's T^2.

It was by no means clear initially that the researches of Mahalanobis, for comparison of populations on which multiple measurements had been made, were closely related to those of Hotelling and Fisher and that they could all be viewed as divergent applications of ideas of Fisher's paper on the distribution of the multiple correlation coefficient. Having introduced his measure of generalized distance more than 10 years earlier, Mahalanobis had first estimated the moments of the distribution in the null case from random sampling data; then in 1930 he had found the distribution of D^2 for the null case [64]. Only quite recently, in 1936, R. C. Bose [65] had found the exact distribution for the nonnull case in which the dispersion matrix was taken as known. When Fisher arrived at the institute they were puzzling over the solution of the case of greater practical importance, the nonnull case when the dispersion matrix was not known but was replaced by an estimate.

Discussion of the problem, indeed, began immediately on Fisher's arrival, when Mahalanobis drew him into the office to tell him about this work. There they sat talking while Mrs. Mahalanobis waited in the diningroom and the dinner got cold. A small imperial person, voluble and full of laughter and lively conversation, Mrs. Mahalanobis was also a perfect hostess. (She is shown in

Plate 15 with her husband and Fisher in 1945.) When the men joined her, she sensed their constraint and came straight to the point:

When I sat down for dinner I found Professor Fisher was paying attention to me, asking trivial questions, making conversation. After three or four minutes I told him "Will you please do me a favour?" He said, "What is it?" I said, "One thing: you must forget my presence here. I have to be here to see that the food is properly served but I do not want to disturb you. I know that my husband has been waiting very eagerly for a long time to ask you many questions when he first sees you." He thanked me, and immediately began talking shop to my husband, and answering questions. They had pen and paper brought out and I enjoyed seeing that he was completely relaxed. From that day, he never stood on ceremony with me.

Discussions at the Indian Statistical Institute played their part in bringing together the work of Hotelling (T^2), Mahalanobis (D^2), and Fisher's discriminant function. This resulted in 1938, in Fisher's publication of a paper ($C\,P$ 155) that brought these diverse researches under a common point of view and compared the results in a common notation, while adding to the theory of discriminant functions a test of significance of deviations in direction.

The distribution of D^2 in the null case was χ^2; the distribution found by R. C. Bose for the nonnull case was, of course, a noncentral χ^2 identifiable with one of Fisher's distributions in the 1928 paper [$C\,P$ 61]. This was what the excitement was about in Calcutta in 1937. The realization of this fact led on, in 1938, to the solution by R. C. Bose and S. N. Roy [66], in what Fisher described as "a very brilliant research," of the distribution of D^2 when the dispersion matrix was unknown, which Bose referred to as the Studentized general distance. The latter distribution also was derivable from one of the forms obtained by Fisher in 1928.

□ □ □

During his stay in India many government officials came to consult Fisher. Their questions were often economic; for example, a representative of the Ministry of Posts and Telegraphs asked how they should estimate the future volume of postal income. Fisher never refused a question. Mahalanobis admired his versatility and the patience with which he attended to long and sometimes incoherent recitals, seeking to understand the problems and to give helpful suggestions, which he then set down in writing. More appropriately, he was consulted by officers of the Agriculture Department and ICAR in New Delhi about setting up their own statistical service.

In Calcutta Mrs. Mahalanobis took ever-present care for his comfort. She guided him with great good humor and experience on shopping expeditions and advised him on his purchases. He was impressed by the silverware of Indian design and fascinated by the quarter of the bazaar occupied by silversmiths, and she shared his enthusiasm. He bought various articles of silk and sandalwood to take home, and she took charge of his purchases and of the gifts he received during his stay—saris, silver, and jewels—for security locking them in her own closet until his departure. There were very friendly feelings between them by the time he was due to leave.

On January 15 they planned a farewell gathering at their home, a garden party in the afternoon to be followed in the evening by their first and only meal alone together, before Fisher left the next day. Before the garden party was over, Mrs. Mahalanobis was called away to her father's bedside; he had been ill for some 6 weeks and now had taken a turn for the worse; he was dying. Without waiting to tell Fisher, she hurried to her father's house and arrived 15 minutes before he died. The next morning, very early, she returned to leave a message of apology to Fisher that she had not been present for the quiet dinner they had hoped for, and to unlock his treasures. Then she went back for the funeral.

In those days, the attitude of superiority commonly adopted by Englishmen in India did not invite confidences. Mrs. Mahalanobis expected her famous English guest to depart, neither knowing nor caring what family affairs had called her away. Fisher asked Mahalanobis what had happened, and then he asked about the funeral arrangements. He learned that the body would be taken that day to Presidency College, to lie in the prayer hall during a brief ceremony and to be shown outside the college before going on to cremation. Professor Mahalanobis would, of course, go to Presidency College, as would all the friends of the family, to pay their personal respects and their homage to a distinguished member of the Bramah Sumaj. Fisher asked whether he might come. Unexpected as it was, his wish was accepted; he also laid flowers on the bier, and, standing among the friends of the family at the Indian ceremony, he identified himself as their personal friend.

War intervened, and it was not until 1945 that Fisher was again in India, at the invitation of Mahalanobis, for teaching and consultation at the Indian Statistical Institute. In addition to R. C. Bose and S. N. Roy, he met a brilliant new recruit to the institute, C. R. Rao, who was eventually to succeed Mahalanobis as director.

Apart from the seminars, Mahalanobis was confident Fisher's presence would be helpful to him in relation to the government, since the ICAR would naturally consult him when he visited New Delhi. The institute was severely under attack from the Agriculture Department, and the methods of Mahalanobis, for example, his use of interpenetrating samples, were being challenged. Fisher knew only that relations had been strained since the ICAR withdrew its grant to the institute. When his flight reached Karachi, he was met by an official representing the ICAR with a confident invitation, almost a summons, to stop off at New Delhi to have consultations and give a lecture before proceeding to Calcutta. He answered that Mahalanobis had invited him and was expecting him that evening; he would go direct to Calcutta as planned. On arrival at the institute, Fisher noticed that the head of statistics of ICAR was not there, but he said nothing. Mahalanobis also held his peace, although he had invited the man not once but several times and had been refused. By chance Fisher did learn the facts and, when he went to Delhi at the end of his visit, he was prepared for tough negotiations. His defence of the work of Mahalanobis, which he believed to be important, saved the day.

The great difference between the 1945 visit and that of 1937–1938 was that war was going on. There was strict blackout in Calcutta for fear of Japanese bombing. There was little dining out, and few foreign visitors. They were almost a family party at the institute, with time for quiet conversation as they sat back together on the divan to smoke the hookah after dinner. Fisher at first found it a little difficult but went on to become quite a good smoker and to find it very pleasant. He brought a little hubble-bubble back with him to Cambridge, a gift from Mahalanobis, and he was disappointed to discover he could not get the appropriate smoking mixture.

Mahalanobis begged him to stay longer with them, but he would go. He became quite brusque and barked out rudely: "No, Prasanta, I have got to get back to London." So he went back to London, in time to attend the Council of the Royal Society and to support the candidacy of Mahalanobis, whom he had put up for election. As Mahalanobis commented, "It was quite characteristic: he would not tell me about it. He never told me." He was equally silent on his role in proposing Mahalanobis for the Weldon Medal.

On the eve of his departure from Calcutta, Fisher celebrated his 55th birthday. Blackout prevented an evening party, but during the afternoon some 500 guests assembled, and there were speeches and congratulations. That night Mrs. Mahalanobis had 55 oil lamps lit and set, not on the roof (because of the blackout) but inside the house, and Professor and Mrs. Mahalanobis and their guest, just the three of them, sat down to the birthday dinner in true Indian fashion. Mahalanobis had mentioned earlier that the traditional way was to sit cross-legged on low wooden stools, call *bili*, and Fisher insisted on doing so although his hosts begged him to make himself comfortable in his own way. He had a little difficulty at first in gaining the position but soon became comfortable. In later visits he often took up the cross-legged position for preference. He enjoyed Bengali food, the river fish, the spiced vegetables cooked in melted butter, the lightly curried sauce. (He suggested that the menus should be written in Bengali rather than in French, for the style of cooking was as distinctive as the language.) Mrs. Mahalanobis had prepared a wholly Indian feast for his birthday, and they ate it from marble platters with their fingertips.

□ □ □

Of the many subsequent visits Fisher paid to India, the greatest occasion for the institute was in 1957, with the celebration of its 25th anniversary. For the occasion 80 statisticians from every part of the world were invited to stay in the special guest houses, built under the supervision of Mrs. Mahalanobis beneath the *lichi* trees of the compound. Everything, from the making of the breese blocks to the weaving of bedspreads and curtains, was done by the workmen of the place. By then the institute employed nearly 2000 workers. The block of office and laboratory buildings stood five stories high. Mahalanobis was the director of a great and internationally famous institution and, in celebrating the quarter of a century of its sometimes precarious existence, his first guest in honor and affection alike was Sir Ronald Fisher.

13

Blood Groups in Man

Fisher's interest in work on human blood groups was precipitated out of his interest in genetical theory. It was, in fact, in 1930, on an occasion when Haldane and he were discussing the nature of dominance, that Haldane mentioned the work which brought him suddenly to the brink of this new field of exploration.

The work was that of Dr. Charles Todd of the Medical Research Council's research center in Hampstead, London, on the serology of poultry and oxen. Red blood cells bear antigens, which result in agglutination of the cells when exposed to an alien serum that contains the corresponding antibodies. A serum may thus be exhausted of its various antibodies by mixing it, in turn, with a series of alien red blood cells. Todd had investigated reactions of such exhausted sera and had found, in particular, that serum exhausted with the cells of both parents did not agglutinate the cells of the offspring of those parents. This finding appeared extremely important to Fisher, and he wrote Todd at once a long letter discussing the implications of his work and possible extensions of the genetical investigation.

The rule Todd had discovered suggested that the agglutinative reaction was determined by the direct products of the genes responsible so that the genes could be investigated directly. If one could obtain a serum sensitive to a particular gene, the dominance question could be settled. If dominance were a primary biochemical phenomenon, the genes of recessives always being defective, inactive, or less active in some special respect than the corresponding genes of dominants, then the liability to respond by agglutination to any

338

particular ingredient in the serum would always be completely dominant. On the other hand, if, as Fisher believed, dominance were a phenotypic reaction, then the liability of recessives so to respond would be shared by heterozygotes; that is, the two alleles, initiating characteristic but different reactions, could *both* be distinguished in the heterozygote.

Fisher suggested a test that might be made to distinguish between these hypotheses and added that if it were possible to obtain serum sensitive to a particular gene, other possibilities opened up. The magnitude of the reactions due to a single gene in comparison with those ordinarily observed would give an idea of the number of such genes in which the group of individuals tested ordinarily differed, and, if one could detect a single gene, the total mutation rate of genes having no visible effect would appear in the small proportion of exceptions to Todd's rule.

Todd was much intrigued by Fisher's suggestions for extending investigation of the genetical aspects of his work. He asked Fisher over to lunch at Hampstead, and they had a long talk. A few months passed and Fisher wrote, saying he would like to visit again and this time bring with him D. W. Cutler, assistant director at Rothamsted and a zoologist much interested in the Eugenics Society. He was by then contemplating the possibility that if individual recessive genes in man could be detected in the heterozygote through their characteristic serological reactions, blood groups would be of diagnostic importance for eugenic applications; in particular, recessive genes responsible for human abnormalities might be serologically distinguishable.

There was no doubt about Fisher's enthusiasm. He talked of Todd's work at Ames in 1931 and was pleased to think he excited a good deal of interest in it with Lindstrom and other geneticists there. Meanwhile, he pressed on the Eugenics Education Society of London the uniqueness of their opportunity to promote work that promised rich rewards for eugenics; the Society had a research committee and funds that might be spared for research, and Fisher almost suggested that it was their duty to offer Todd an assistant to encourage him in this bypath to his proper work. Simultaneously, he was urging Todd to accept the help of an assistant. Neither side seemed very keen, and the matter dropped until the beginning of 1932, when Fisher again suggested that an assistant might facilitate Todd's work and offered to propose that the Medical Research Council's newly formed Committee on Human Heredity should sponsor the work:

As you know, I am inclined to think that your serological work is going to lead to a greater advance both theoretical and practical in the problems of human genetics than can be expected from any further work along biometrical or genealogical lines . . . Could you make any use of it if I were to persuade the Committee that yours is the work best worth backing?

The committee met for the first time a few days later, before Todd had sent a reply. Nevertheless, at the risk of being premature, Fisher was tempted to speak at once about Todd's work and waxed eloquent on his behalf.

When he told Todd about it afterward, he excused his precipitancy by expressing again the need he felt for a direct demonstration in animals that genes capable of serological detection could have other genetic effects. Meanwhile, he wanted to develop a technique that could detect individual blood in man and "consequently sweep up the big aggregate of serological factors."

Fisher's earlier genetical thought predisposed him confidently to expect very numerous genetical factors; moreover, Todd's work already provided a striking confirmation of his expectation. At Fisher's instigation Todd had begun work with poultry, in which he attempted the serological detection of sex by immunizing cocks with hens' blood, exhausting the cock serum on numerous hen cells, and finally testing with cells of other hens and of cocks. The work proved unsuccessful in its first object; no sex difference was detectable. Improved technique improved the sensitivity of the test fourfold, but the results were unchanged. However the thoroughness of the attempts to exhaust sera accentuated their failure. It seemed impossible to avoid individual differences. However, many exhaustions were done, the sera were still active on the cells of a number of individuals of both sexes. Learning this, Fisher remarked:

I do not think there is any escape, unless your observational findings are revised, from the view that the whole of the reaction developed is a reaction to alien genes, or, of course, their immediate products. It is evidently possible to form antibodies to an enormous number of such alien genes, and perhaps to all.

A stunning thought, that serology might be the key to the detection of the whole of the genetic complex! But this seems to have been Fisher's hope, even though he foresaw that its realization might be complicated: "Your results do not prove that all possible reactions always take place, that is, . . . there may be conditions necessary to bring off the different kinds of reactions which are potentially available." He conjectured that it might be possible to get an immune reaction, even for the elusive sex factor in fowls, if the appropriate conditions were provided. (A sex antibody in fowls, inhibited in the females, was actually discovered by French workers in 1951.) More immediately important was the fact that the number of individual differences that were exhibiting themselves offered an immense field for further research. For the genetics of man, in particular, in whom few individual genes had been detected, serology promised access to a "big aggregate" of such genes.

Fisher had not written about human blood groups in *The Genetical Theory of Natural Selection*. When W. C. Boyd, serologist at Harvard, read the book in 1934, he wrote to Fisher, expressing surprise at the omission. Several long letters passed between them which, with a brief historical introduction, may give a notion how Fisher's genetical theory impinged on contemporary attitudes to serology.

The branch of human biology called blood group serology came into existence in 1900 when K. Landsteiner observed that the red cells of some of his colleagues were agglutinated by the serum of others. Landsteiner took samples of blood from six of his colleagues, separated the serum, and prepared saline suspensions of the red cells. Each serum was then mixed with each cell suspension, in some mixtures the cells were agglutinated, in others they were not. On the basis of the reactions Landsteiner was able to divide human beings into three distinct groups; the fourth and rarest group was discovered by his pupils in 1902. Only two antigens in the red cells were required to explain the four groups of people: those with one (A), those with the other (B), those with both (AB), and those with neither (O). The serum was found to contain the reciprocal set of antibodies, A serum containing anti-B, B serum anti-A, AB serum neither, and O serum both anti-A and anti-B. The pattern of spontaneous agglutination quickly became important clinically for purposes of blood transfusion, the O group being called the "universal donor," because, having no antigens, the cells were not agglutinated by the blood of any recipient.

In 1908 it was suggested that the ABO blood groups were inherited, in 1910 it was proved, and in 1924 the method of inheritance was determined by F. Bernstein. The four groups were determined by three allelic genes, the dominants A and B and the recessive O. Subgroups of A were observed soon, and in 1930 V. Friedenreich established that two identifiable A alleles existed, A_1 and A_2. Other forms of A were observed during the 1930s.

The group O appeared to be entirely recessive, and no antibodies reacting with O could be discovered. It could not be distinguished in the presence of A or B. This caused no difficulty on current genetical theory, either for those who accepted the O group as a genetic absence or as a (complete) deficiency of biochemical action. It did, however, pose a problem on Fisher's theory of dominance, and he was optimistic that antibodies reacting with O would be discovered. In a letter to W. C. Boyd in 1934 he expressed the view that:

At first the ABO series seem to show some analogy with what is found in several polymorphic species, namely a relatively common and widespread recessive with a number of common allelomorphic variants. The evidence for dominance in blood group work is, however, rather exceptional, and I think it would at present be premature to conclude that no antibody reacting with O can be produced in immune sera. If this were done the heterozygote could be detected.

In response, Boyd referred to his own lack of success, and that of many others, in seeking antibodies to O, which had led him to believe it was completely recessive "or, more likely a complete absence."

Group O was, in fact, to prove generally recessive, but the picture was far from clear, and for many years occasional cases of what appeared to be anti-O sera continued to sustain hope that reliably anti-O sera would be secured. Eventually, another genetic factor was found to be responsible for some recorded "anti-O" sera, and today O is believed to be an amorph.

Apart from the ABO series which produced spontaneous agglutinins, few other blood groups were then known. K. Landsteiner and P. Levine had announced a second discovery in 1927, the more brilliant because in this case the antibodies were not found but had to be made. Seeking new antigenic differences, the authors had injected rabbits with different samples of human red cells and adsorbed the resulting rabbit immune serum with other red cells until they found what they were looking for, namely, antibodies that would distinguish between blood in a way that was cutting across the known ABO distinctions. A year later they were able to discuss the manner of inheritance of the MN blood groups and to put forward the two-allele theory, now universally accepted. Each of the alleles M and N determines the presence of the corresponding antigen on the red cells. There is no recessive, and the three genotypes MM, MN, and NN give rise to three corresponding phenotypes M, MN, and N.

In the same series of experiments in 1927 another agglutinogenic reaction occurred, unrelated to MN or ABO reactions. The two types of the new blood group were called P+ and P−. Soon after, the anti-P antibody was found occurring naturally in the serum of a human being and in ordinary nonimmune sera from horses, rabbits, pigs, and cattle. It was shown that the P antigen was probably inherited as a Mendelian dominant character, and later work supported this view. The early anti-P sera, however, produced only a weak reaction to some blood samples, and the frequency of the two groups could not be determined. The P system was not important in work in the 1930s. It was not until after 1951 that the P group was recognized as a system as strong and complicated as the ABO system. In that year Levine and his collaborators announced the discovery of a new and powerful antigen, which Ruth Sanger (1955) recognized as part of the P system, an antigen of the system shared by P+ and P− people. Thus, like P+ (renamed P_1), P− was recognized as a dominant (P_2), and rare individuals were found without the new antigen, the recessives (p).

This sort of development has shown the terms "dominant" and "recessive" to be somewhat misleading in a manner which, interestingly enough, justifies both Haldane and Fisher in their original views. Haldane was right in that there are in many blood systems the ultimate recessives, now more frequently referred to as amorphs, which are defined by the fact that they appear to have no effect on the biochemical system. Fisher was right that there are "recessives" like P− that can produce a characteristic antigen. When good antisera are available both alleles can be distinguished in the heterozygote.

One other genetic factor was known which was associated with the blood groups. Antigens were found not to be confined to red cells but widely distributed throughout the body. In 1926 antigens A and B were discovered in saliva. Then in 1930 exceptions were found to the rule of antigen presence in the body secretions and a dimorphic genetic factor identified, secretor (Se) being a Mendelian dominant and nonsecretor (se) being recessive. The genes responsible were found not to be linked with the ABO genes. The strength of the antigens in saliva could be variable, and the distinction of secretors from nonsecretors was sometimes difficult. Blood tests were paralleled by tests for the secretor factor.

The three series, A_1A_2BO, MN, and P+ P−, were the only human blood groups known at the time Fisher wrote to Boyd of the possibility that antibodies could be formed to an enormous number of alien genes. In answer Boyd pointed to the current evidence:

Todd's work with poultry . . . did show a large number of factors and . . . his work with cattle showed an even larger number, but similar work with other animals has either produced only a few antibodies or none at all. I enclose a reprint of some work I did with guinea pigs and mice in which I found it quite impossible to demonstrate any individual blood differences. Castle and Keeler found only three differences in rabbits and, though they made many attempts, could not demonstrate more. I also made an attempt to find other blood grouping factors in rabbits but could not. Work in press. So I wonder if you are justified in supposing that many, or most, genes can stimulate the formation of antibodies.

Today, a generation or two later, when grouping work on the red cells has discovered a score of different blood group systems and elucidated many of their genetical complexities, when the work has extended into the fields of the genetics of the haemoglobins, serum groups, enzyme groups, and the so splendidly complicated antigens of the white cells, it requires a conscious effort to imagine that the field could ever have seemed to afford a paucity of genes for investigation. If Fisher's vision had a simplicity that was not to be realized literally in the detection of most genes through serological research, it had also a scope and grandeur that was to be realized beyond the reasonable hopes of serologists at that time.

Fisher's claim that the stable polymorphisms in respect of blood groups must have evolved and must be maintained by a balance of selective forces has had a more checkered history. In 1934 it was usually assumed that blood groups were completely without selective effects. Thus although Boyd agreed that the ethnographic distribution of blood groups was difficult to explain without selection, he suggested (following Sewall Wright), as a partial explanation, that tribal isolation might have resulted in chance differences in the gene frequencies now found in different races. He could not entirely accept Fisher's explanation that natural selection alone would be effective. Fisher had written:

A gene would not be found disseminated among many millions of people without the positive aid of selection if it had arisen within 10,000 generations or so in only a single mutation, as, I think, the first speculations about the ethnographic distribution of the blood groups were inclined to assume. If, moreover, not a single mutation but a definite rate of mutations is postulated, the question arises why the mutation rate should be different in different races. Consequently, I cannot see any escape from the view that the frequencies have been determined by more or less favourable selection in the different regions, governed not improbably by the varying incidence of different endemic diseases in which the reaction of the blood may well be of slight but appreciable importance.

The logic seemed inescapable. In his serological work, therefore, Fisher was eager to demonstrate the fact. His plans for the first investigation of the rhesus factor in Great Britain in 1942 were influenced by this hope, to the embarrassment of his serologists. He was enthusiastic when in the 1950s the first large-scale investigations were made, and associations found, between blood groups and a variety of diseases. Later work did not sustain the early findings of direct associations. In the few instances established by serological studies, the associations are indirect: effects of the secretor factor, or of the form of the haemoglobin or immunoglobulin molecules. The selective effects, necessary on Fisher's reasoning, proved more subtle than he had anticipated; many would say today that they do not exist.

☐ ☐ ☐

In the autumn of 1934 Dr. D. P. O'Brien came to England as the representative of the Rockefeller Foundation to consult with the Medical Research Council how best the foundation might sponsor research into human genetics and, in particular, into the genetics of mental disease. Among the members of the MRC's Committee on Human Heredity he met Fisher; in consequence, an interview took place soon after between Dr. A. Mawer, Provost of University College, Dr. O'Brien, and Professor Fisher. The interview was followed up immediately by a long letter from Fisher to O'Brien in which, after discussing various research proposals, he set out in Section 7 the proposal, which he had made personally during the interview, for research in serological genetics at the Galton Laboratory.

Having outlined the present state of knowledge of human blood groups, he pointed out the most obvious applications that he foresaw of work from the new research unit if it were approved, concluding, "The first step would seem to be to form a small but efficient research unit devoted to serological studies of accessible pedigrees of medical interest." Turning to practical details, he specified the staff he proposed for the project, the first 5 years of the program he planned, the equipment he required, and the cost he anticipated. He must have worked fast to have been ready not only to persuade Dr. O'Brien to consider serological research rather than more direct investigation of mental disease but also at once to present concrete proposals for consideration.

Six months were to elapse before he knew whether his research proposal would be approved, but immediately he began to look for a pathologist capable of taking charge of the serological work and willing to start from scratch at the Galton. Dr. Todd, being on the point of retirement, was not a candidate, but Fisher sought his advice about the medical appointment and urged him to look in at the Galton to see the new work and to use their facilities for any blood group project he wished to pursue in his retirement.

Fisher's inquiries led rather unhesitatingly to Dr. George L. Taylor, who had given up general practice 6 years earlier because of a heart condition, and since then had been doing research in the pathology department at Cambridge University. His record showed him to have made a brilliant academic career, a body of painstaking and meticulous research, and a loving circle of friends; acquaintance showed a comfortable, cautious, gentle man, inspiring complete trust. Fisher's proposition interested him and, when the project was funded in April 1935, he was ready to accept the post of medical officer in charge of the new serological unit.

It was a happy choice for both men. For both it was a new enterprise in which they constantly discovered cause for mutual respect. Taylor's technical skill, care, and precaution provided a springboard of facts for Fisher's imaginative flights, and Fisher knew he was a man worth persuading to the theory. For his part, Taylor found his new chief gave him wholehearted support with the authorities and was keenly interested in the work and in its implications. Taylor's personality, his loyalty, kindness, and good humor, earned a warm and equal response, and they worked well together.

Taylor was a perfectionist: this was the basis of his great contribution to serology, not only in fundamental research but, during 5 years of most demanding war work, simultaneously in its applications. Given his temperament and his unremitting devotion to his task, there was every reason to fear that he would become the victim of heart attack. Unhappily, the end of the war was to come just too late to ease his burden, and he died in March 1945, almost exactly 10 years after he had joined the Galton Laboratory.

With Taylor's letter of acceptance, Fisher on May 20, 1935 sent to the Medical Research Council an enquiry whether he was authorized immediately to order the equipment for the new laboratory: the refrigerator, centrifuge, hot-air oven, tables and chairs, and electric fittings. He was so authorized, and he began also to prepare suitable housing for the laboratory and its animals. Although it was made explicitly clear that at no time would University College itself be prepared to spend a penny on the serological work, no objection was raised to the new use of space in the existent eugenics department. The new space, a former museum, belonging jointly to the statistics and eugenics departments, was therefore divided into two sections and that part allocated to eugenics was given to the serologists. This area required improved supplies of electricity, gas, and water. In the animal house concrete dog pens had to be demolished in one room in preparation for the rabbits. In July Fisher suggested that "such alterations in my experience are liable to be expensive" and asked permission to postpone appointment of a serological assistant and to use the saved cash for a few months to pay for construction costs. Such requests were invariably treated sympathetically by the Rockefeller Foundation and MRC, and the work went forward financially unimpeded.

The museum was, at this time, still occupied by Karl Pearson's private collections. Fisher therefore wrote Pearson on May 7:

I understand from Dr. Morant that the skeletons and skins of experimental dogs now housed in the museum of this department are your property and that you sent me through him an offer to arrange for their removal whenever it is desired. I should be glad if, without putting yourself to any inconvenience, you could make arrangements during this spring or summer for housing this material elsewhere as it is certain that the museum will be increasingly needed for other purposes.

This missive produced no immediate action.

Meanwhile work on the museum resulted in objects being shuffled about and after Pearson dropped in on June 13, he was moved to write:

In our correspondence about clearing out my belongings from the museum you stated that it would be quite convenient if I removed them in the course of the summer. This I fully intended to do in July. I was therefore much surprised to find today that you had already emptied some of the cases and put some of the skins on the top of the central standing cases where, in the course of a week or two, they will be full of moth. I am sending trunks down to Dr. Elderton and she will remove the remainder of the skins. The skeletons have gone today. But please do not expose them to danger of moth by turning them out in the open again before they are removed, or at least warning Dr. Pearson or Dr. Elderton when you have done so. There are certain belongings of mine in the long central standing cases. Do you wish these removed before July 15th when I return to London and will remove them? Please let me know. I had thought when you said any time during the summer would do that July would suit your convenience. I am getting the swinging cases cleared of my photographs of albinoes etc. I believe the Royal College of Surgeons will appreciate their value and they will thus be preserved.

So, in July, the last of the photographs, the skins, skeletons, and other objects were taken out, and equipment for the serological unit could replace them. Dr. Taylor began to settle in and he hired a technical assistant, Miss A. M. Prior.

At the end of the first year's work, Fisher reported exultantly to the Medical Research Council that "the most sanguine expectations that could at first have been formed are likely to be realized." Dr. Taylor had found abundant evidence that the unknown but recognizable reactions in human blood corpuscles were extremely numerous;

Indeed, he has scarcely examined any animal serum without finding constant reactions distinguishing different human individuals not referable to the spontaneous OAB series or to Landsteiner's M and N allelomorphs. . . . It has not been possible to follow up all the promising indications of the sort which have appeared, though we think that one factor found in pig serum can now be used with the same confidence as any of the standard reactions. It is in the pig serum also that Dr. Taylor has detected and repeatedly confirmed a very remarkable series of reactions of blood of patients in institutions for the mentally deficient. . . . Much work will have to be done both on sera from other sources and on conditions of preparation and preservation of cells needed to give the most reliable results before any definite genetical interpretation should be attached to them. Dr. Taylor cannot do both the exploration and the exploitation of this find.

It looked as though the serological research might already have discovered something important on mental disease. Fisher therefore suggested that a second technical assistant should be added to the permanent staff in 1936, instead of 1937 as planned, as well as a new junior boy, and that they should begin to look for a young medical assistant. The MRC agreed. Miss A. M. Prior's appointment was made permanent in July 1936, Miss E. W. Ikin was engaged as technical assistant, and a little more than a year later, with the appointment of Dr. R. R. Race, the quartet, which was to make itself famous in serology in a series of papers by Ikin, Prior, Race and Taylor was complete.

In September 1936 Taylor began a year's work under V. Friedenreich at the fine new buildings of the Retsmedicinkske Institut Universitets, Copenhagen. The lavish, up-to-date equipment filled him with admiration. It was a wonderful place to work and to acquire expertise in the most modern and exact techniques of serology. After his return he was able to pass on the fruit of his experience not only to his permanent assistants but to several temporary assistants who came to train in the Galton Unit in the summer of 1939; this had happy consequences during the war.

On Taylor's return in September 1937 Fisher was ready to appoint the new medical assistant, as soon as Taylor could approve the choice. Dr. Race was a diffident young man with a fiery head of hair and a capacity for enthusiasm equal to Fisher's own. In 1968 he was to open the 2nd Fisher Memorial lecture [67] with the words: "Fisher's enthusiasm for blood groups was extremely infectious and it continued to the end of his life." Race caught the infection immediately:

The first time I met him he was at the Galton Laboratory when he saw me about a job I had applied for: he said I could have it, and I walked down Gower Street on air for I had not met anyone of his quality before. Though my salary was to fall from £350 a year to £250 I well knew it was promotion and, of course, so it was: the turning point of my life.

For a married man like Race, with a medical degree which could have assured him a better income, it was worldly folly to sacrifice £100 a year. His attitude recalls Fisher's dictum in 1913: "We must be ready to sacrifice social success at the call of nobler instincts. And, even as regards happiness, has any better way of life been found than to combine high endeavour with good fellowship?" That was the way of life in the serological unit.

The enlarged team was harmonious and increasingly skilled in their researches, with Fisher a very interested spectator and, where possible, participant. In 1937, foreseeing the loom of groups just over the horizon, he suggested the cross-injection of a few cubic centimeters of blood between members of the Galton Laboratory, or at least those members who might prove as enthusiastic about the idea as he was. The medical men, Taylor and Race [67],

vaguely remembering their Hippocratic oaths, thought this would never do; and so, [Race later reflected], the discovery of Rh was probably postponed for two years and left to the Americans. Had Fisher's plan been carried out, anti-Rh, and perhaps anti-Kell, would very likely have been made by someone. Anti-Kell because Fisher was later found to be Kell positive. Had the injections been confined to males they would probably have done little harm: later knowledge showed that the experiment could have been disastrous to the offspring of some of the females who might have been involved, for there was no effective treatment for haemolytic disease of the newborn in those days.

For better or for worse, that particular experiment was not made.

The serological unit did, as Fisher had originally proposed in 1934, collect serological data relevant to accessible pedigrees of medical interest, and some interesting results were obtained. Also as proposed, they maintained close contact with mental insititutions through Dr. L. S. Penrose in Colchester and Dr. J. Fraser Roberts at Stoke Park, who held research posts under trusts administered, among others, by the Galton Professor. Nothing so straightforward eventuated from this work as the first appearances had seemed to indicate (the antibody in the immunized pig turned out to be the anti-P of Landsteiner and Levine) but the work provided exceedingly valuable experience.

The Galton serological unit was unique in Britain. Landsteiner and Levine in the United States had famous serological laboratories, and Friedenreich's Institute in Denmark enjoyed an equal renown during the 1930s. F. Schiff in Germany, Boyd at Harvard, and A. S. Wiener at Brooklyn Hospital in New York were doing notable work, but in Britain blood groups had been little studied, except for the needs of blood transfusion; even for that purpose few records had been kept, blood being usually donated by a relative of the patient, cross-matched for the immediate transfusion, and omitted from permanent hospital records. Taylor began to collect data in the London area and by 1938 had collected and typed the blood of some thousands of individuals.

With such data, it was desirable to have a method of deriving the actual gene frequencies of the population that gave rise to the observed proportions of the four ABO phenotypes. This is a valuable first step in comparing the blood group content of different populations; it is useful in checking theories of the inheritance of new blood groups, and, what Race and Sanger [68] later asserted in general was true in 1939 of the ABO system, "Most usefully of all we can use the gene frequencies to show that a sample contains a reasonable distribution of the groups, and we can then take confidence in our technique." Wartime experience was to make this abundantly clear.

Bernstein (1930) had provided a method of calculating the gene frequencies when anti-A and anti-B only had been used for the blood tests. In 1938, W. L. Stevens, applying Fisher's method of maximum likelihood to the problem, devised a refinement of Bernstein's method. This was employed when, in 1939, Ikin, Prior, Race, and Taylor published the results for 3459 persons in southern England giving the gene frequencies for O, A_1, A_2, and B. By August 1939 they had accumulated data for the O, A, and B groups for over 58,000 persons, and this they put on record at the International Genetical Congress at Edinburgh. The Edinburgh meeting, however, scarcely completed its business, because delegates anticipated the outbreak of war.

In May the serological unit had received instructions that immediately on outbreak of war it was to become a part of the blood transfusion service, move to Cambridge, and take over the task of providing grouping serum for the whole country. On Monday, August 29 the order came to bring the unit to Cambridge "by Tuesday." On the following day, therefore, Taylor, Race, Ikin, and Prior put their things together at the Galton and drove to their new laboratory in the pathology department at Cambridge, where Taylor had formerly worked, and began to set out their sera to be accessible there. Late on Tuesday, Dr. and Mrs. Race and Miss Prior hurried back to London to pick up clothes and other personal belongings for their stay. It was all rather hectic. Bacteriologists and other Medical Research Council people had arrived in Cambridge the day before, and the influx of wartime residents produced confusion as well as congestion.

The serologists found University College deserted. Most people were on holiday. Fisher and his scientific assistants were still at the Genetical Congress. It was understood that the college was to be closed within a week of the outbreak of war. The serologists, compelled to leave their precious animals in charge only of Munday, the laboratory assistant, impressed upon him his duty "to look after the animals at all costs." Fisher, too, was worried when Miss North, having returned to London, telephoned on September 4 that the university had ordered her to destroy his stocks. He encouraged her to resist the order, and telegraphed Mrs. Fisher to "Save the animals." Mrs. Fisher was busy packing off three daughters to stay with their aunt in Scotland, refusing evacuees (4000 of whom arrived in Harpenden on September 4 from London) in the hope that she could provide a home for some of Fisher's staff, and attempting to answer callers from the Galton who were without instructions. Despite his own busy life, Taylor managed to call at Milton Lodge, giving recent news and reassurances about the Galton Laboratory and taking charge of the three children, putting them on the train for Glasgow; never was his constant kindness and understanding more welcome to Mrs. Fisher than at that moment.

Within a few days Fisher was back at the Galton, and on September 9, the eve of the university's threatened closure, Taylor and Race arrived to carry off the most valuable of their animals and to slaughter the rest. They offered to find housing for some of Fisher's mice. Taylor reckoned "that 400 survivors may be enough to keep going the factors we want." In fact, some mice had already been sent to Cambridge before Fisher's return. But Fisher did not choose to be separated from his mice; he did not believe the university could mean to carry through a scheme of such panic precipitancy.

□ □ □

In Cambridge the Galton Serological Unit began a new phase of its existence, but there was never any question of their parting company from Fisher. Visits to London and to Cambridge supplemented their correspondence, and their work together grew.

On October 30 a letter appeared in the *British Medical Journal (BMJ)* in which E. Billing stressed the importance of the genetical and ethnological data of the OAB system of blood groups, which should be obtainable from the extensive activities of the blood transfusion service. Fisher and Taylor [*C P* 169] followed this up by a letter in the same journal a month later, appealing

to all empanelling centres now active to cooperate by sending in from time to time numbers classified in eight classes [the four OAB groups in each sex]. By doing so they will not only swell the totals and so throw light on points which require very large numbers for their elucidation but will open up the field, at present wholly unexplored, of the homogeneity or heterogeneity in respect to blood groups of the population of these islands.

Through the chance of war, for the first time an opportunity was offered to obtain numerous, countrywide data on the frequencies of blood groups in Britain, and the Galton Unit seized upon it. It was arranged that Fisher should receive, record, and return the forms to the various sections of the blood transfusion service. Already Taylor was in touch with officers at the large storage depots: Oliver at Sutton (whose assistant, Miss Dodd had trained at the Galton), Janet Vaughan* at Slough (who had worked with the Galton before the war, collecting cases of acholuric jaundice for blood grouping), and Maizels at Maidstone. By the time their letter appeared in the *BMJ,* the first 32,000 returns had come in from these centers.

With these returns, the first anomalies also appeared: it was noted that the number of males grouped as AB was deficient. As other results continued to accumulate, it became certain that systematic errors, not all of which were understood, were undoubtedly affecting the recorded frequencies of AB to such an extent that it was considered safer for the time being to ignore this small group in calculating the gene frequencies.

*Later Dame Janet Vaughan, Mistress of Somerville College, Oxford.

The letter to the *BMJ* brought a good response. In November forms began to come in—from Preston and Glasgow, Cambridge, Slough, Dorking, Hull, Dundee, Cardiff, Newcastle—in the hundreds and thousands. It became obvious that the efficiency of blood grouping varied greatly in different batches of material. In January 1940 a meeting of the Society of Clinical Pathology at Cambridge was devoted to blood grouping, with exhibition of data gathered and discussion of technique. Later that month Dr. Vaughan sent 3000 cases, regrouped, and she persuaded the other storage depots to adopt the tube technique used at the Galton Laboratory in preference to the slide technique generally used. Their regrouped cases began to come in in February. It was found that the slide technique had failed to detect about one half of the A_2 genes. Gradually, it was abandoned. As Fisher commented in a letter to Taylor,

> The misgroupings have not erred on the safe side from the point of view of transfusion, as one might possibly have expected if workers were concerned to test large numbers as rapidly as possible with a view to finding a number of reliable O donors. The errors have been predominantly in the opposite direction, in favour of false negative readings.

In addition to the special difficulties with AB grouping and the technical care needed in diagnosing A_2, there was a more general trouble. Receiving data in March, which he described as "wet," Taylor was moved to protest to Fisher:

> Before I experienced the results sent in by all sorts of workers, I should have been inclined to agree that most hospital pathologists could be entrusted to do a simple ABO grouping. They can do no such thing. They would be saved from all sorts of mischief if every time they used them, they would see that their reagents are what they believe them to be, by using controls. We find it necessary to do so.

Well might Taylor claim they were the only "professsionals" in the country: working with him, the "amateurs" recognized the unprecedented standards set by his meticulous care and began to learn from it.

Looking around for a good source of grouping serum, Taylor and Fisher first thought of using the army. Then, in the spring of 1940, Taylor approached the Air Force authorities and, after a long delay, a bargain was struck: the Cambridge Unit agreed to group all the young men of the flying training wing, that is, personnel training to be pilots and observers; in return they were to ask for volunteers for pints of blood from among those they found suitable. The men were then at Bexhill, within reasonable distance from Cambridge, but before the Cambridge Unit could begin work, they had been transferred to Torquay in the far west. This meant an hour's train journey from Cambridge to London, transit from Liverpool St. to Paddington Station, and then several hours on the train to Torquay. Taylor and Race began to visit Torquay every 2 weeks, taking the new samples back to Cambridge for grouping and returning to take pints from suitable volunteers 2 weeks later. It was slogging hard work, and as Taylor confessed on August 16,

The work entailed in dealing with this RAF business and with the difficulties raised from time to time by the senior officials of the Air Force have been terrific and we have had a really harassing time. I think, however, that the scheme is well under weigh and hope it continues to work satisfactorily because it provides a splendid source of grouping serum and we think besides that our testing the groups of what at the present time is certainly the most important group of young men in the country is a thing worth doing for its own sake. At last it has been agreed that the blood groups should be stamped on the reverse side of the identity discs.

The latter was a significant practical achievement. It was gratifying to know that the blood group, tested by themselves and therefore reliable, would appear on the identity disk of each flyer, who might at any time or place need emergency blood transfusion.

One of the young men who trooped in to be grouped at Torquay was George Fisher, who had abandoned his medical studies and was training as a pilot. Next time round, Taylor and Race took him out to dinner and swapped news. They were old friends from University College days. Even earlier, Dr. Taylor had visited Milton Lodge, of course, on blood-grouping expeditions and knew all the family. (On the outbreak of war they had all received emergency donor cards showing their long-established blood groups.) He took a special interest in George and followed the news of his movements in the Air Force with fatherly concern.

The Air Force visits continued—with the long drag across country in trains sometimes unheated, delayed, or crawling through air raids, with the arduous collection at Torquay and the painstaking laboratory work at Cambridge. On December 12, 1940 Fisher received 10,000 Air Force forms. "Thank you", he wrote, "for sending me the first 10,000 not only of the Air Force but, very probably, the first 10,000 yet accurately grouped by anyone." That was the sort of thanks Taylor and Race appreciated.

□ □ □

Vaughan's prompt return of large numbers recorded at Slough aroused ethnographic interest, for recent industrial expansion in the district had attracted considerable immigration of workers from Wales. To find out whether the Welsh had blood groups homogeneous with the general sample from Slough, a number of characteristically Welsh surnames were chosen and their contribution to the different blood groups calculated. They were 5.2% of the whole sample and were found to contribute a significantly smaller percentage in group A than in group O and in AB than in B. The surnames did indeed, indicate a source of local heterogeneity. Fisher and Vaughan [C P, 171] published a note on their findings as early as December 1939.

By the new year a more general pattern of the distribution of the ABO blood groups in Great Britain was becoming apparent, and in January this was confirmed by new data from Inverness, Aberdeen, and St. Andrews. In southern England the A group had a gene frequency of 26.7%, but the proportion fell in northern England to 24.5% and in Scotland to 20.8% in a consistent gradient.

From the outbreak of war Fisher and Taylor had been trying to get hold of records of blood groups on the Continent—Thomas's data from Stockholm, Friedenreich's from Denmark, and compilations such as those made by A. S. Wiener and by W. C. Boyd in the United States. From these it appeared that the proportion of A on the Continent was everywhere higher than in Britain and that it was lowest and closest to the British figures just across the Channel. Values for the phenotypic ratio A/(A + O) were in France 50.1%, in Belgium 46.6%, in Holland 48.6%, in Southern England 48.8%, in Northern England 45.3%, in Scotland 39.7%, and in Iceland 36.6%. Now, Iceland had been colonized a thousand years before by Norsemen who had at the same period settled in the northeast of England, in Scotland, and northern Ireland. The blood of contemporary Icelanders, in the sample of 800 recorded, provided just the blood group constitution needed to explain the distribution of A found in Great Britain. The intermixture of blood of their ancestral stock with that of local populations would have resulted in just the sort of gradient that was observed.

The discovery of the early Scandinavian influence on the British blood groups reflected also startling historical changes in Europe. Contemporary Scandinavians had not lower but higher frequencies of A than any country bordering the North Sea A/(A + O) ratio in Denmark 50%, Norway 58%, Sweden 59.6%). It could be inferred that the repeated contacts of Scandinavian countries with their eastern neighbors must in the intervening millenium have greatly raised the frequencies of A.

Fisher and Taylor were delighted with these discoveries. They presented their results at a meeting in January 1940 of the Society of Clinical Pathology at Cambridge and immediately prepared a preliminary note for publication [C P 179, 1940]. Meanwhile data poured in—from the army recruiting center near Bristol, from Wolverhampton, Brighton, Leeds, Bolton, Peterboro', Wakefield, Preston, Plymouth. As the map filled up, the gene frequencies recorded in each district fitted into the expected gradient so perfectly that when a large batch of new groupings arrived in November from Preston, Lancashire, Fisher remarked comfortably that his calculated value of 45% "ought to make a Lancastrian feel at home."

Fraser Roberts, in Bristol, was looking among his Welsh donors to see if he could confirm in Wales itself the low A frequencies found among the Welshmen at Slough. At first he was unsuccessful, using the more-or-less local donors with Welsh names and therefore mainly from South Wales. When data were brought in from North Wales, however, he noticed a difference. Separating names characteristic, respectively, of the north and the south of Wales he discovered that the gene frequencies in the south bordered those of their English neighbors but changed abruptly within Wales, so that on going perhaps 40 miles north from South Wales, as great a change occurred as one might expect to find on going 200 miles northward in England. The contrast should have pleased the Welsh, for every Welshman recognizes the difference in kind between the races of the North and the South and is suspicious and disdainful of his Welsh neighbors on the other side of the line.

In February 1941 Fisher, receiving a new batch of data from Cambridge, remarked that the men and women differed in the ratio of O to A. Taylor explained that the men were mainly students and visitors to Cambridge, whereas the women were largely of local stock; they were not comparable populations. Later on, however, the same phenomenon cropped up again in massive returns from the West of England and, less strikingly, in figures from Yorkshire; Fraser Roberts and Fisher investigated [C P 197, with J. A. Fraser Roberts]. Then, again, as the testing improved in respect of detection of A_2, differences in the proportions of the two A alleles were discerned in different parts of the country. This sort of watch for clues that might either mean the detection of technical failures or of genetical and ethnological heterogeneity, relieved the boredom of the recording and calculating of the thousands of new blood groupings which continued to come in during the years that followed.

To the serological unit the calculation of gene frequencies made a valuable check on the accuracy of grouping, by allowing comparison of the relative frequencies of phenotypes observed with those expected from the calculated gene frequencies. Their own technique was found to do very well; Taylor sent groupings from Cambridge in February 1941 that came so close to expectation that Fisher admiringly warned: "Do not forget the sad story of Gregor Mendel"; Taylor responded that he "had thought the 15,997 total figures were very good. We shall however have to put up with such perfection." The question of reliable technique continued to be a real concern, and the Galton Laboratory Unit, preparing an article on the subject, wrote from wide experience when they introduced their paper with the cautionary words:

Complete accuracy in the diagnosis of the ABO blood groups is most difficult to achieve and it is impossible unless the following essential requirements are adequately met: (1) a good technique, (2) an experienced worker and (3) sera of sufficient strength and reliability.

□ □ □

Taylor was much concerned about possible causes of the adverse reactions that sometimes occurred in patients who had received repeated transfusions. Early in the war he was diagnosing large numbers of M and N groups, in case antibodies of this series might be responsible. Then, in 1940, an article by Wiener and Peters [69] demonstrated that a newly discovered antibody could be found in the serum of certain people who had incompatible transfusion reactions following transfusion of the blood of the correct ABO group.

The new human antibody was indistinguishable from one which Landsteiner and Wiener [70] had just succeeded in making by immunization of rabbits and guinea pigs with the blood of the monkey *Macacus rhesus*. On testing the immune serum, they had made the very surprising discovery that the resulting antibodies agglutinated not only the monkey red cells but also the cells of about 85% of white people in New York. The 85% whose red cells were agglutinated by the rabbit anti-rhesus serum the authors called Rh positive, the remaining 15% Rh negative. The clinical importance of the anti-rhesus reaction quickly became evident when apparently the same antibodies were found in the sera of patients showing incompatible transfusion reactions.

Many years later it came to be realized that the rabbit antirhesus and the human anti-Rh antibodies are not the same. By that time, the vast literature which had accumulated on the subject made it impossible to change the name of the human antibody.

Actually, the discovery of the new human blood factor had been made in all but name in 1939, when Levine and Stetson [71] published their historic paper describing how the mother of a stillborn fetus suffered a severe hemolytic reaction to the transfusion of her husband's blood. Then it was found that the mother's serum agglutinated the cells of her husband and those of 80 out of 104 ABO compatible donors. The antigen responsible was shown to be independent of the ABO, MN, and P groups, and an unsuccessful attempt was made to immunize rabbits against it. Levine and Stetson's interpretation of the case in 1939 is now known to have been entirely correct. They said that the mother who lacked the new antigen had become immunized by her fetus, which possessed this antigen, having inherited it from the father. When the husband's blood was transfused the maternal antibody reacted with this same antigen on the red cells of the father.

In 1941 Levine and his associates [72] showed that hemolytic disease of the newborn was the result of Rh group incompatibility between mother and fetus. In the same year Landsteiner and Wiener [73] published a full account of their work on the Rh antigen.

As soon as possible, Taylor began investigation of the new blood group. Fisher offered to get hold of rhesus monkeys through his zoological friends. The monkeys were delivered in January 1942 and the preparation of the rhesus antibody began in Cambridge, soon followed by testing on human blood.

To gain adequate numerical data it was desirable to investigate the blood groups of mother and child in all cases of hemolytic disease diagnosed. In May 1942, therefore, Taylor and Mollison issued an appeal in the *BMJ* for sera from the mothers and children suffering from hemolytic disease and, if possible, the fathers also. All the advances in knowledge of Rh in Britain came from this appeal. The response was good; many practitioners and pathologists in Great Britain and Northern Ireland were cooperative, and in March 1943 the data for 56 families had been tabulated. Race, Taylor, Ikin, and Prior [75] explained:

The material is far from complete, for the great difficulties always present in the pursuit and investigation of human families are in wartime enormously increased. For instance, men and women are away from home in the forces; travelling is difficult; and, with doctors so fully occupied, personal visiting is practically impossible.

This is a reminder of the restrictions surrounding all the rhesus research undertaken by the British serologists during the war, in addition to their appointed work.

□ □ □

Fisher had been much excited by the discovery made by Levine and Stetson. In the summer of 1939 he had been consulted by a Leonard Darwin student as to the research she might undertake, and he had at once suggested that this would be an extremely important and fascinating topic for study. (She had liked the idea but, with the outbreak of war, had found the situation unfavorable for starting such work.) For Fisher, investigation of the hemolytic reaction promised the sort of confirmation he expected of selective effects of blood group genes. He anticipated that the balance of selective advantage within any blood group system (and probably between different systems) would be complicated. In the case of hemolytic disease one complication was soon evident; incompatability did not always result in the production of maternal antibodies. In fact, when the Rh gene frequencies were established, the number of families in which hemolytic children were to be expected was found to be about 20 times the number in which they occurred. Fisher wanted to investigate all the antibodies produced during pregnancy in the hope of throwing light on the situation.

Once Rh research was begun at Cambridge, the opportunity for Fisher's investigation seemed ripe and in March 1942 E. B. Ford visited Cambridge with the intention of discussing with the serologists a plan he and Fisher had hatched: Ford volunteered to collect samples of blood and saliva from mothers and infants in maternity hospital in Oxford to be tested for Rh and other antibodies at Cambridge. Taylor was sceptical about selective effects in serology and less than enthusiastic about the plan. Later the same month, however, he learned that evidence had been secured of enhanced antibody content in certain cases of pregnancy. This evidence was not, in fact, sustained by later work but, coming at this juncture, the report gave support to Fisher's expectations, and Taylor agreed to collaborate with the work. He knew that for years Fisher had wanted to secure just such evidence, that Fisher was confident that blood group factors were of selective value and was hopeful of demonstrating the fact—even though he recognized that with a source of danger which had been in equilibrium for perhaps millions of years, all sorts of complications were likely to have arisen to puzzle the investigator.

Fisher did not hide his eagerness. The letter he sent to Taylor with the plan of action reveals the impatience with which he had awaited his opportunity:

You will recognize it as essentially what I have proposed many times before though I am afraid you may have thought it a rather fantastic scheme. I should have much rather the reaction of antibody concentration in pregnancy had been a find to the credit of the Galton Laboratory than to the Scandinavian you mentioned. However, with such important facts as he has established, all pointing towards a selective equilibrium of blood groups, it seems foolish not to look further, or at least to be deterred by the quite thoughtless dogmatism of the many writers who have asserted that blood groups are unconnected with disease or survival.

Inevitably there were delays and difficulties initially in setting up the arrangement at Oxford. Perhaps this was just as well for Taylor, who was laboring in heavy waters. In July 1942 he wrote that the numbers being taken into the Air Force were enormous. They were completely occupied in dealing with them; he could not hope to attend the meeting of the Genetics Society, although he was a member of the committee, nor even manage to see Fisher as he had wished! By October, when Ford's persistence had won out and the first samples began to arrive, conditions in Cambridge were slightly easier.

In Oxford Ford was excited by the prospect of starting the work. Doctors at the Radcliffe Infirmary, and later also from an evacuee maternity hospital in Oxford, drew the samples of blood from mothers and umbilical blood from the infants, while Ford personally collected their saliva, swabbing out the mouths of the infants with little pledgets of cotton wool. At the bedside he recorded the child's sex and date of birth, the mother's name, age, previous children, and miscarriages and whether the marriage was between cousins, and to this he added remarks as required, noting for instance, the occasional death of an infant in hospital. Carrying his collection home, he immersed the tubes containing saliva in boiling water; these had been labeled by a writing diamond, since paper labels would have lifted off the tubes. He packed the tubes of blood inside wooden postal packages provided from Cambridge and enfolded them, two per envelope, in the addressed envelopes also provided. Then he posted the lot to Dr. Taylor.

Only on certain days could the Cambridge unit do what was necessary; thus samples were sent just twice a week. The batch sent on Saturday was 2 or 3 days in the post, and some of the first batch were found to be decomposed and stinking on arrival; later samples however usually seemed fit to analyze.

Though he had won official permission, Ford found he had a further task in winning cooperation of the doctors and nurses to whom his visits were a nuisance. They were understaffed and very busy. Ford called for support, and in January 1943 Fisher spent a day at the Radcliffe Infirmary discussing the scientific and clinical interest of the hemolytic reaction, suggesting the importance in preventive medicine of measuring the maternal antibodies several times during pregnancy to detect mother-fetal incompatibility as early as possible, and emphasizing the importance he attached to the current serological investigations. He was duly impressive and proved to be just the influence needed to interest the medical staff and win their sincere cooperation. The difficulties of Ford's task eased, and changes were even made in the hospital routine for his sake, a remarkable gesture at any time, and under wartime conditions most remarkable.

Nothing came of the Oxford work. Probably the state of preservation of the samples was unsatisfactory; certainly the frequencies of the groups recorded were very disturbed. In any case, very few of the Oxford families can have had hemolytic disease and it was to be through investigation of hundreds of cases of the disease that the facts were established, and an explanation provided for some of the cases in which parental Rh incompatibility did not result in diseased children.

In 1943 Levine [75] reported an interaction between the Rh and ABO systems which explained, in part, why mother-fetal incompatibility led to hemolytic disease in only a small proportion of cases. ABO incompatibility between the parents seems to protect the infant from Rh-induced hemolytic disease. Most strikingly in cases of Rh incompatibility in which the father is AB and the mother O (so that every embryo has an antigen for which the mother has a corresponding antibody), hemolytic disease was not found to occur. Probably any fetal cells that leak through the placenta are destroyed by the anti-A or anti-B antibodies before they can persist long enough to stimulate the formation of anti-Rh antibodies.

Another factor found to affect the development of maternal anti-Rh antibodies was the genotype of the father. It occurred to the British workers that an Rh-negative mother would be more liable to be immunized if every pregnancy was Rh-positive and provided the antigenic stimulus, as it does with a homozygous husband, than if he were heterozygous. In 1943, the serum from an Rh-positive mother of a hemolytic Rh-positive infant was discovered, which made it possible to test the hypothesis. Because the donor's name was Stedman, the serum was originally called St. It was found to react with all Rh-negatives (rh rh), all heterozygotes (Rh rh), and about half the homozygous Rh-positives (RhRh). Thus every St-negative blood was homozygous Rh-positive. It thus became possible for the British discoverers [77] to distinguish about half of the Rh-positive homozygotes and to test whether the proportion was the same among the fathers of hemolytic children as it was in the general population. A preponderance of the Rh-positive fathers of hemolytic infants were found to be homozygous. Indeed, the homozygote was revealed to be four or five times more dangerous than the heterozygote, and prognosis in the case of an Rh-negative wife and homozygous Rh-positive husband was described as "entirely unfavourable."

□ □ □

In America and England further research was revealing the Rh factor to be more and more complex. In the Spring of 1943 Wiener defined six different alleles and had three antisera; the British had defined five of what turned out to be the same alleles. By August the Cambridge group had defined seven alleles and had found a fourth antiserum. The excitement of discovery sharpened by rivalry was matched only by the bewilderment caused by the obscure relationships of the new factors. The symbolism was chaotic, with capital and lower case Rh, Hr, primes, and subscripts. The genes of Weiner appeared as composite reacting molecules instead of having each its own specific antigen and antibody so that, for example, anti-Rh_1 serum recognized antigens for both Rh_1 and Rh_2 and anti-Hr recognized both Rh_2 and rh. Despite changes made by Wiener, the terminology was hopelessly confused.

Fisher has described how order was brought out of this chaos in a remarkable article on "The Rhesus factor, a study in scientific method," which was first presented during his visit to the United States in 1946 and later published in *The American Scientist* [C P 214]. In it he introduced to American scientists the story, still unfamiliar to them, as it had unfolded itself to English workers during the war years.

The first phase follows from the American discoveries up to 1940 of a serum (Δ in Table 2) that distinguished two phenotypes in known proportions, 85% Rhesus positive and 15% Rhesus negative. Since the Rhesus negative phenotype lacked a hypothetical dominant gene R, it must therefore consist wholly of the recessive genotype rr; the frequency of r genes must, therefore, be $\sqrt{.15}$ or 39% and the frequency of R genes $1 - 0.39$ or 61%. The frequencies of the genotypes RR and Rr could now be calculated, RR being $(.61)^2$ or 37% and Rr, the heterozygous remainder after subtracting the homozygotes, being 48%. These inferences controlled deductions from the next group of experimental facts discovered.

The second phase opened with the discovery, among sera from mothers of hemolytic children collected by Taylor, of the St serum giving strikingly different results from the standard anti-Rh serum and evidently containing a different antibody (γ in Table 2). Of English blood donors 66% reacted positively to both sera. When the same series of donors was classified simultaneously with both sera, it appeared that all Rh-negative donors reacted with this antibody. It was discovered on studying the families that this was true not only of the donors but also of their parents and their children. This could only be interpreted as proving that the 48% designated by the genotype Rr were included in the 66% reacting positively with both sera, whereas the 37% designated by RR were divided, some having positive and some negative reactions to the new antibody. There must be at least two genes for R, say, R_1 and R_2, which were distinguishable through their opposite reaction with γ.

Table 2 Reactions of Four Antibodies with
Three Common Alleles (from *C P* 214)

	Δ	γ	H	Γ
R_1	+	−	−	+
R_2	+	+	+	−
r	−	+	−	−

There must, therefore, be at least six genotypes: R_1R_1, R_1R_2, R_1r, R_2R_2, R_2r, and rr whose frequencies were calculable as above; all were fairly common except R_2R_2.

The first effect of the conception of these genotypes was the recognition of two new kinds of antibody when they turned up in 1943, the one reacting only with R_2 and the other only with R_1 (H and Γ in Table 2). In the light of the genotypes already postulated, it was easy to recognize what these new sera were doing. The finding of the new sera thus strongly confirmed the view that six main genotypes were present and, indeed, as may be seen from Table 2, enabled all to be recognized by objective tests, with the exception of the rare genotype R_2R_2 which was still indistinguishable from the more common R_2r.

□ □ □

In the autumn of 1943 Fisher moved to Cambridge and came into constant contact with the serologists. After the day's work they used to meet at the Bun Shop, a little public house handy to the pathology department and much frequented by university men. Almost every evening Fisher and Race might be found drinking beer there and talking over the tangle of emerging Rh results.

The third phase was initiated by the discovery of the new sera. In Fisher's words:

The new sera not only confirmed the position won so far, but in conjunction with those previously in use provided a network of tests sufficiently fine to catch and identify certain rare types of donor blood which did not fit into the five big classes already distinguishable. About 4% of those donors tested with all four sera gave results in one way or another anomalous, and with increased experience in the reliability of the different sera available it was possible to recognize four new genes, all rare, constituting together only about 5% of the total gene frequency, yet with reactions as distinctive as the three more abundant genes first recognized. These are shown in the enclosed portion of the following table.

Fisher's table is reproduced in Table 3, with the addition of the symbols CDE appropriate to each of the eight alleles defined. Brackets showed the reactions of R_z not yet confirmed when the scheme was born. Everything outside the enclosed portion was predicted from theory, and there was no objective evidence for its existence at the time.

What had happened was that Fisher, studying the results of the British work with four antisera, had noticed that the reactions of two of them were antithetical, and he supposed that the antigens and the genes recognized by these two antibodies were "allelic"; he called them C and c. The reactions of the

Table 3 Reactions of Seven Alleles to Four Known Antibodies, with Extensions Suggested (From *C P* 214)

		Δ	γ	H	Γ	δ	η
CDe	R_1	+	−	−	+	−	+
cDE	R_2	+	+	+	−	−	−
cde	r	−	+	−	−	+	+
cDe	R_0	+	+	−	−	−	+
cdE	R''	−	+	+	−	+	−
Cde	R'	−	−	−	+	+	+
CDE	R_z	+	(−)	(+)	+	−	−
CdE	R_y	−	−	+	+	+	−

remaining two sera were not antithetical: the antigens they were recognizing Fisher called *D* and *E* and supposed that they also had allelic forms, *d* and *e*, which would be capable in favorable circumstances of stimulating their own antibodies, anti-*d* and anti-*e*. All the observed reactions appeared explicable on the basis of the three closely linked genes *C*, *D*, and *E* each with alleles.

Race [67] has recorded the occasion at the Bun Shop when the idea was first put forward:

Fisher saw the quite simple but at first to me at any rate almost unbelievable pattern behind the bewildering reactions of the various Rh antisera: he saw Rh as controlled by a system of three very closely linked loci, each with alleles. And this is more or less how most people still think of Rh. He must have spent very little time on the scheme for he produced it the day after being given an important piece of the puzzle, perhaps two hours after dinner were enough. But tens of thousands of blood group workers and students were to thank him for making Rh intelligible. The slide shows his first writing down of the synthesis. The symbols were parochial, in a few days he changed them to the now familiar *C, D* and *E*. . . .

The three loci were assumed to be so close together that crossing-over between them could be only an extremely rare event: if crossing-over could occur at all freely the frequencies of the various combinations of *C* or *c*, *D* or *d*, *E* or *e* would have differed greatly from those observed.

Plate 14 is a full size photograph of the piece of paper shown on Race's slide. In it one may observe Fisher's tiny handwriting, so often referred to as "squiggles" and the typically open top of his figure 8.

Characteristically, Fisher conceived the situation and represented it graphically in the form of 2^3 factorial as shown in Figure 12 [*C P* 214], which Fisher explained thus:

We may represent the eight heritable antigen complexes geometrically as the corners of a cube, while the six elementary antigens are represented by the faces; each allelo-

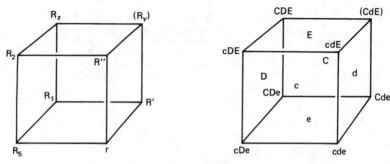

Figure 12. The 2 ³factorial of Rh genes. (From C P 214.)

morphic pair of antigens is then a pair of opposite faces, and the three faces meeting in any point specify the antigens in each complex.

The genotype of each individual is determined by the pair of complexes supplied by the two parents. If these are wholly alike the person in question is triply homozygous; alternatively, he may be singly, doubly, or triply heterozygous.

The eight triple homozygotes obviously correspond with the eight corners of the cube. The 12 single heterozygotes correspond with the 12 edges of the cube and, like the first eight, given anti-*d* and anti-*e*, would be all recognizable by serological tests only. Each face of the cube has two diagonals giving six pairs of genotypes, all doubly heterozygous. Members of the same pair contain the same five antigens and give the same reactions; there are, therefore, 12 genotypes and only six phenotypes, and family evidence would be required to distinguish between pairs. Finally, four genotypes, triply heterozygous, constitute a single phenotype which should react with all six antibodies, the commonest of these being *CDe/cdE* in 0.8% of the population.

In the Fisher Memorial Lecture Race [67] continued: "Another very beautiful idea that Fisher wrote down, again in the Bun Shop, was that very rare crossing-over could account for the puzzling difference in frequencies of the various CDE combinations." Displaying the piece of paper, he added "on the lower part of the page are to be seen some faded beer stains: Fisher was writing on a table with a beaten copper top and with his very short sight couldn't see that some of the valleys were in flood."

It was a pretty consequence of the postulated three closely linked genetic loci for Rh that the relative frequencies of different combinations of the three factors might be supposed to be maintained by occasional crossing-over from the more common combinations. As shown in Table 4, there are three common heterozyotes making together 54% of the population of Britain, *CDe/cde* (33%), *CDe/cDE* (11%), and *cDE/cde* (10%). All are heterozygous at two of the three loci, and all would produce *cDe* by crossing-over.

Table 4 Crossover Products of Common Rh Heterozygotes
in Great Britain (from *C P* 214)

		2.62%	1.0%	0.2%	1.2%	
CDe/cde	33%	*cDe*	*Cde*			By crossover *c/d*
CDe/cDE	11%	*cDe*		*CDE*		By crossover *c/e*
cDE/cde	10%	*cDe*			*cdE*	By crossover *d/e*

In addition they would produce, respectively, *Cde, CDE,* and *cdE*. If these four rarer gene combinations in the British population were maintained by such occasional crossovers, this would explain the fact that *cDe* was about as common as the rarer three put together and that the only remaining combination, *CdE*, which could not be produced by crossovers of these common genotypes, was exceedingly rare. It had not yet been found.

On this view, the frequency ratio representing a crossover between *d* and *e*, being considerably larger than those representing crossovers either between *c* and *d* or between *c* and *e,* the order of the genes within the chromosome could be inferred to be such that *c* lay between *d* and *e,* as illustrated in Figure 13.

D   C   E

Figure 13.   Relations of Rh loci on the chromosome. (From C P 214.)

□ □ □

Fisher found the solution to the intellectual problem most beautiful and satisfying, but, as Race says, the scheme was "almost unbelievable." In the following months, Fisher discussed it repeatedly with Taylor and Race, and with Dr. D. F. Cappell, their serological colleague in Dundee, when he visited in January 1944. Fisher's solution did seem just possible on current evidence but the situation had never before arisen in blood group work that more than one genetic locus was involved in the same blood group; to suggest three such loci, very closely linked, yet distinct, was the sort of bold innovation that takes time to grow on the imagination and to carry its own conviction. Fisher waited; later he explained,

Of course I could have published the thing myself but I have been always through the last thirty years entirely averse from butting in on subjects on which I am not working practically. I have never regretted my constant policy of waiting until those who are practically concerned with the science in question are sufficiently interested to see eye to eye with me and make use of any ideas I can put forward.

The conception of three linked Rh loci suggested to Fisher that a study of the properties of mixtures of different types of human anti-Rh sera might throw light on the problem of antibody absorption. Race [78] accepted this suggestion, and the result of his study convinced him that Fisher was right. In June 1944 reporting the discovery in consequence of this work of an "incomplete" Rh antibody, he boldly introduced Fisher's tabulation and his new symbols into the literature as a means of distinguishing the three categories: the antigens, the genes or allelomorphs, and the antibodies "for which provision must be made in a satisfactory notation."

Writing to Dr. Cappell in September 1944, Fisher invited his avowal of support:

I am most heartily glad to see that you are inclined to accept the theory of Rhesus allelomorphs which I drafted after some discussion with you and of course a great deal of discussion with Taylor and Race. Somehow I got the impression that you were not satisfied that the theory fulfilled the primary condition of fitting the genetic facts so far ascertained. But it is a fact that for my own part I should have liked to publish the scheme earlier if I had had the encouragement of your support at the time. I imagine that on thinking it over the thing has grown on you, as it has on Race, and as a good theory should, but I do not think you ever brought yourself to write giving your full concurrence. Please do not think that such support would not now be as welcome as ever, as Taylor still has his doubts and Wiener will not readily admit, I am sure, that any considerable clarification has been due to work on this side. He still ignores the Stedman antibody on which, indeed the whole scheme was based. Even Race's publication of the scheme was quite non-commital. Indeed no striking confirmation is to be expected unless immune sera are produced containing δ or η, for R_y is certainly very rare and could only be detected with certainty in a parent and child R_1R_y, rR_y or the other way about.

Confirmation was closer than Fisher had anticipated. First, a family was found in which the seventh gene (R_z) was exhibited combined with R_1 in one member and with r in another, and on testing, reactions that had been predicted for R_z were realized. Even more exciting, within a year, A.E. Mourant [79] a newcomer to the Galton Unit, found one of the new antibodies predicted, anti-e, and all the anticipated interactions were confirmed at the time or, in the case of R_y, later. The first conclusive example of R_y was demonstrated in 1948 by Dr. Clara van den Bosch [80] in Holland.

The final prediction, the occurrence of the second new antibody anti-d, was announced in 1946; so that that summer Fisher [C P 214] could say of his proposed scheme:

At least one writer felt this proposal to be so conjectural that he denied stoutly the possibility of these hypothetical antibodies, yet within the year (1944) in which Race first published the proposal, Mourant had discovered anti-*e* in a transfusion reaction, and very promptly recognized its conformity with the reactions predicted. Furthermore, before the end of the following year Diamond of the Harvard Medical School had confirmed the occurrence of anti-*d* in a case of haemolytic pregnancy reaction, following earlier experience of transfusion.

The discovery of Diamond and later cases presented by Habermann and Hill, thought at first to be anti-*d*, have not stood up to the criticism of the years. It now looks as if the postulated anti-*d* does not exist. At the time the thrill of the repeated confirmations, following quickly after the publication of the scheme, was the sort of experience few scientists can have enjoyed, and none could be indifferent to.

Unhappily, Wiener had declared a private war against his British rivals. He now accepted a single tripartite gene at the Rh locus, though the *CDE* scheme was anathema to him. As author of the scheme, Fisher was involved in the subsequent unpleasantness, and in 1946 he shared with Race the balm of his reflection:

So long as things hang together in inheritance, it is a matter of indifference whether they are conceived as closely linked adjacent genes or tripartite but single genes. As a Hebrew Wiener finds perhaps trinity and unity less compatible than they really are and will enjoy the acrimony of a monophysite controversy. The only point which interests me is whether, on an adequate number of observations, recombinations are or are not found to occur.

Recombinations, which would in any case be extremely rare, have not been recorded with assurance. The case rests.

In view of this early and continuing opposition, there has been a good deal of debate as to whether the Rh genes ought to be considered to be one or three. The fundamental genetic relationships have been upheld, and, without further confirmation, Fisher's scheme has the great advantage of convenience; the simplicity and clarity it introduces into the description and communication of precise relationships in an increasingly complex subject has been an inestimable boon to students of the subject at all levels of sophistication.

At the end of the war, one of the first needs of serologists was to confer together about the nomenclature and symbolism to be employed in their subject. The guidelines for genetical nomenclature, set up by the International Congress of Genetics in 1939, could not meet postwar needs for serology.

The wartime developments, particularly with regard to the Rh factor, had brought chaos: the need to specify gene, antigen, antibody, and antiserum by distinct symbols had been recognized and met by Fisher's scheme, but there were more than one set of symbols concurrently in use. It was felt that some consistent notation would have to be agreed internationally as soon as possible.

Fisher, Ford, Mather, and Race formed an unofficial "National Committee on Blood Group Nomenclature" and, in December 1945, formulated proposals for a consistent symbolism. The chairman of the United States National Committee, L. H. Snyder, commented that "This report deserves careful consideration. If we adopt it, it will resolve most of our difficulties." The Danish National Committee reported that they had tried hard to follow the 1939 guidelines but found it impossible and subscribed with relief to the British proposals. The question of nomenclature was in the background as serologists the world over renewed contacts after the war, and gathered at special meetings in 1946 and 1947.

During the summer of 1946 Race traveled extensively to Basle, Milan, Paris, Texas, Mexico City, and Boston. In Europe he was "treated royally and alcoholically on the strength of the " 'Théorie de Fisshaire' which," he told Fisher, "just shows that it is of more than abstract value." In Paris he received a cartoon depicting a postman delivering a substantial envelope addressed *CDE*: it was "Le Facteur Rh." A duplicate that Race sent to Fisher with compliments from Paris long stood on his mantelpiece. Fisher also had the pleasure that summer of speaking of the British work in the United States and, in particular, on the occasion when he presented the Woodward Lecture at Yale, before Levine and Wiener.

By this time Race was director of the Blood Group Unit of the Medical Research Council at the Lister Institute in London. After his return to England, Race made an early opportunity to visit Cambridge, bringing with him his new assistant, Ruth Sanger, to meet Fisher. Before many days both men were enchanted by her uninhibited manners, her warm heart, and her lively intelligence. Race stood in awe of Fisher and was astonished that anyone would be so bold as to tease the great man, as she did; Fisher found it delightful to be treated frankly and without fear. Her spontaneity and joy in life brought a new lightness and warmth to the continuing collaboration between Fisher and serologists at the Lister Institute.

It was not until 1947 that any agreement was reached about serological nomenclature, at meetings held in Washington and Geneva. The outcome was somewhat surprising in view of the generally favorable reception of the British report. In effect, it was a personal victory for A. S. Wiener in respect of the rhesus factor, but a Pyrrhic victory, for, though his nomenclature was given official recognition and priority, which was sustained under renewed criticism in 1956, still serologists have continued to this day to discuss the Rh factor in whatever terms seem appropriate to them, and when complexities of the Rh system are involved they are almost forced to use the *CDE* scheme to express themselves unambiguously.

Neither Fisher nor Race was willing to be drawn into acrimonious debate with Wiener. Race and his colleagues at the Lister Institute for years suffered the brunt of the assault, in sarcastic verse, diatribe, public misrepresentation, and private invective. Such things merit no scientific acknowledgement but they are personally trying; in appreciation of Race's endurance under fire, Fisher commented in the foreword of Race and Sanger's *Blood Groups in Man* [68, 1950], that

On matters once controversial (and still occasionally so) the senior author has exercised a commendable restraint. No one would judge from the text how often his personal contributions have been ignored, and when verified have been published without acknowledgement of his priority. Our present understanding of the complex Rh situation owes much to the good temper with which, in spite of irritants, he steadily pushed on with his own problems.

The next step in the exploration of the Rh factor was to estimate the frequencies of the Rh gene complexes [*C P* 209 with R. R. Race, 1946, *C P* 215, 1947]. Race had collected a series of 2000 samples by 1947, and Fisher devised a maximum likelihood method for estimating the frequencies and did the calculations. Of the 12 gene complexes combining *C, C^w, c, D, d, E,* and *e,* only eight were common enough to provide estimates, *CdE* and some of the *C^w* alignments being very rare. (The frequency of *CdE* has since been estimated in Switzerland to be some 140 per million, roughly the same order of magnitude as the figure of 50 per million in England derived by Fisher and Race on the basis of the crossover theory.)

Since the 12 gene complexes can be paired in 78 different ways, 78 different genotypes exist, even without the inclusion of other rare variants. Yet the five common genotypes alone accounted for 88% of the population and, together with six other genotypes with frequencies between 1/2% and 3%, accounted for 98% in England and Western Europe. A striking contrast was early observed between white Americans and American Indians and soon between other races also. Fisher's method of calculating the estimated frequencies was soon to prove its utility in assessing such racial differences.

The spade work on Rh was completed in the 1940s. It was to be followed by a fairly luxurious flowering of variant phenotypes and new antibodies. But, as the curtain falls on this act of the drama, there are echoes of the refrain at its rising. On May 17, 1950 Fisher also wrote in the foreword for Race and Sanger's book, *Blood Groups in Man*:

Research people are usually so conscious of how much remains to be done that they sometimes underrate the extent of what has been already accomplished. In particular this seems to be the case with respect to the future use of blood-grouping as the principal tool of a comprehensive study of the human germ-plasm.

Much had, indeed, been discovered since, less than 20 years earlier, Fisher had first begun to think of blood grouping as the principal tool of such studies and had expressed the view that it would lead to "a greater advance in the problems of human genetics than could be expected from further work along biometrical or genealogical lines."

Three years earlier, on March 22, 1947, he had written to his old friend, Dr. C. Todd, with whom he had first canvassed the comprehensive study of human genetics through serology:

As it was largely your work which led me about twelve years ago to set up the Serum Unit at the Galton Laboratory, which has been in recent years so immensely successful in elucidating the Rhesus factor and finding others of similar effect, I think you might like to support R. R. Race whom his genetical friends feel ought now to be put up for the Royal.

The modest beginnings at the Galton Laboratory had grown to a full-fledged research institution specializing in blood groups in their genetical aspect; the diffident young man who had joined the unit in 1937 ten years later had grown to distinction at the head of the new speciality. It was time to celebrate his achievements.

14
Losses of War

Like most of his countrymen, Fisher was not convinced of the necessity of war until after the Munich crisis in August 1938; then he knew it was only a matter of time. After the whole of Czechoslovakia was overrun by Nazi troops in March 1939, plans were made which began to reveal how the onset of war would affect University College and the eugenics department. As we saw in the previous chapter in April the Medical Research Council engaged the whole of the serological unit to hold themselves in readiness, immediately on outbreak of war, to take up duties in Cambridge as the central Serum Grouping Unit of the Blood Transfusion Service.

At the end of August 1939 the International Genetical Congress met in Edinburgh. It was an uneasy session, from which, if they attended at all, some delegations and individuals withdrew in anticipation of the outbreak of war. Fisher stayed to the end, improving the shining hour. In the course of discussions about the possibility that blood group frequencies found in man were determined by a balance of selective influences, it had occurred to Ford that the evidence on the parallel possibility for the taste-testing factor in man could be obtained by testing the anthropoid apes. Fisher promptly persuaded his good friend W. J. B. Riddell (soon to be the first professor of ophthalmology at Glasgow University) to bring to Edinburgh the solutions of phenyl thiocarbamide required in the experiment. F. A. E. Crew gained permission for a party from the Genetical Congress to visit Edinburgh Zoo to taste-test their apes. Ford could not attend the Edinburgh meetings, but Crewe, Riddell, Fisher, and a number of others abandoned the conference one afternoon and made their way in high spirits to the cage of the chimpanzees.

The eight chimpanzees were given mugs containing first a small quantity of slightly sweetened water (2% sucrose), a pleasant mixture for which they were expected to return for more. They drank it up and came back. This time they were given a sugared mixture containing 6¼ parts per million of phenyl thiocarbamide; there were one or two dubious expressions but they returned again, and received a mixture with 50 parts per million of phenyl thiocarbamide. This time there was no doubt about their reaction: one chimpanzee turned his back emphatically on his visitors; another poured his drink on the ground with disgust and went to the opposite side of the cage; one sat gazing reproachfully at his audience while the fluid dribbled from a lower lip fully extended, a picture of misery; a more spirited beast spat it out in a strong jet straight at his persecutors. The geneticists were much diverted by these exhibitions. Obviously, there were tasters among the chimpanzees. There were also nontasters: two of the eight animals tested returned to receive and to drink a mixture containing 400 parts per million and still returned for more.

This result was astonishing. The small sample was, of course, inconclusive as to the exact proportion of tasters, but it suggested not only that chimpanzees were polymorphic for the taste factor in the same way as men but in approximately the same proportions. (In human races there was some variation between 25% and 30% and in the Edinburgh chimpanzees 25% nontasters.) This meant that in all the million or more generations of separate evolution of the anthropoid and hominid stocks, the selective balance in favor of the heterozygote must have been maintained practically unchanged in the lineage of both. What was the selective reason for the polymorphism could not be guessed, but that the selection existed was amply demonstrated by the amusing exhibition at Edinburgh Zoo.

In the south again, Fisher and Ford, together with Julian Huxley, their host at the Regent's Park and Whipsnade Zoos, and two members of the Galton tested all the apes they could find. Two gorillas at Regent's Park gave clear signs that they were tasters; the orangutan at Edinburgh was a nontaster but the two at Whipsnade were tasters; gibbons, though less obviously responsive, were judged to be of both sorts; and the chimpanzees continued to respond with uninhibited expressiveness. Seven of twenty-seven were found to be nontasters, that is 26%; a twenty-eighth, being the daughter of two nontasters, added nothing to the information on gene frequencies, but she was a nontaster, as predicted on the basis of the genetics of the human taste factor. Everything pointed to the existence of the same polymorphism as existed in man among all the anthropoid apes.

The Genetical Congress ended and Fisher went on to the British Association meeting in Dundee. He was there when war was declared on September 3, 1939. University College was put on a war footing: as planned, the teaching departments were evacuated to Wales, away from the danger of bombing, which some feared would strike London as soon as war was declared. In expectation of closing the whole of the university buildings within a week, the college assumed that the research work would cease at once and the workers disband. Notification was sent to all departmental staff, which read like immediate dismissal. Miss North, having returned from Edinburgh with Stevens a day or two before, was told by a college official on September 4 that all the animals in the animal house must be killed immediately. She resisted this order, saying nothing was to be done until Fisher returned. After several attempts, she managed to reach Fisher by telephone and tell him what was happening; she sent off some mice that Taylor offered to keep in Cambridge; and she succeeded in fighting off the university officials until Fisher could get back.

On his return, he refused to have a single animal killed. He challenged the right of the college to betray the objects of the Galton Trust by dissolving the department, and he continued, with his assistants, to occupy the premises of the Galton Laboratory until such time as the college should acknowledge the right of the department to exist by providing funds for its transfer to other quarters and its maintenance there. The rapid dissolution proposed by the college seemed to him a measure of panic. Evacuation might be desirable, although he doubted there was any urgent need for it. Animal stock might have to be sacrificed, but nothing yet justified the indiscriminate destruction of material of unique genetical value, and he was sure that the departmental staff, while they would, of course, find war work, could make a special contribution to the war effort if they held together as a team. Six members, including himself, had volunteered for national service as a computing unit and were entered as such on the Royal Society's register of scientific personnel.

On September 14 the superintendent of buildings at University College called on Fisher and stated that it was his intention to turn out the eugenics department. Fisher challenged his authority. Finally, he said he thought he would lock and bar the department up. Fisher "could only advise him strongly to do nothing of the kind," as he reported to the provost afterwards. But the provost upheld the authority of the superintendent and informed Fisher next day: "You must now remove such personal effects as you may desire in order to carry on any personal work which you may wish elsewhere, and cease to attend the College." Fisher arranged to see the provost personally. Meanwhile, he appealed to Sir John Russell, Director at Rothamsted Experimental Station, and was granted the loan of a couple of rooms to accommodate the eugenics department for the duration. When he met the provost he asked for funds for the evacuation of the department to Rothamsted and for its maintenance there, and the provost acceded to this request.

A week later it was all to do again. The provost wrote, promising transfer and maintenance costs for Fisher alone. Fisher protested again, and the Provost made a slight concession: "I realize that to leave you transferred alone there might make it very difficult for you to carry on any necessary correspondence etc. and I am therefore prepared to arrange that Miss Karn should join you but beyond that the College is unable to go." Simultaneously, letters were sent from the provost's office to all Fisher's staff, requiring them at once to cease work at the Galton and to devote their whole attention to the task of seeking such other work, national or personal, as might be open to them. No admission to the college would be given to them after September 30.

On September 29 Fisher wrote a letter to *The Times,* representing the plight of the Laboratory. The letter concluded:

During the last war the administrators learned, though perhaps with some reluctance, that men trained in research were essential for the success of the national effort. The remaining nucleus of my department, if I may speak in its praise, constitutes a unit for heavy mathematical computations as efficient, both in machines and men, as the country can command. Obviously no work of first-class national importance can be found for such a unit at a few days' notice. I submit that in certain contingencies its continued existence might be of the greatest value, so long as the machines and the expert knowledge had been kept together. Cannot a little patience be exercised before completing its demolition?

On the morning of October 3 Fisher, together with Miss North and Mrs. Tysser, made their way to the college, which was locked against all but Fisher. Fisher entered by the main gate and went round the corner to help his assistants climbing over the high spiked railings. Miss North, the first to climb, was nearly over when the porter ran up and pushed her. Fisher intervened— "Would you lay hands on a lady, sir?"—and tangled with the porter. There was a scuffle which made headlines in the press next day, and Fisher and the ladies entered the building. J. B. S. Haldane, envying Fisher's opportunity, reported Fisher was the only professor he knew who had knocked out an ex-prize fighter. But the porter was not knocked out. Fisher escaped with a scratched face, bleeding profusely, and the women had minor injuries. W. L. Stevens, arriving half an hour later, climbed in through a back window Fisher opened for him. Later he talked to reporters, explaining that the college had so far refused to let the staff go on working together, that they had offered their services as a unit, and that it was as a unit they could be most useful.

While waiting the college's response to his letter to *The Times*, Fisher and most of his department worked as best they could in a borrowed office in the London School of Hygiene.

The chairman of the college committee, Lord Sankey, now took an active role in settling the difficulties between the provost and the professor. The three men met on October 7, and it was agreed that the college would pay for the removal of the department to Rothamsted, to take place within 2 weeks; that the departmental staff should be free to work with Fisher at Rothamsted until they found war work; and that Fisher would be entitled to retain one member of his present staff and receive the Galton Trust funds, that is, in addition to his own salary, £300 a year to pay for one salary and for running expenses and occupancy at Rothamsted. Fisher had won the breathing space he asked for his department to find war work together, if they could.

At Rothamsted they crowded into their rooms. Fisher's desk and calculating machine faced those of Miss Simpson, and the Millionaire stood beside them, in the general office in the plant pathology building that they shared with other Galton personnel. In a cubicle or half-room further along the corridor, Miss North shared space with the genetical assistant, Dr. A. C. Fabergé, until he was called up for military service. Most of her time, however, was spent in the mouse house on the upper floor of a wooden hut in the grounds outside. As the months passed without summons to war work, the staff began to disperse. Early in 1940 Miss Kam found mathematical work at Woolwich; Norton returned to America; early in 1941 Stevens and Fabergé left and Munday, the laboratory assistant, was called up. Of the original staff only Miss Simpson and Miss North (now Mrs. Holt) remained.

Fisher was not given any work of national importance during the war, and he was hurt by his inability to serve, especially when he knew E. S. Pearson and J. Wishart and J. B. S. Haldane were doing special war work. Unfortunately he appeared as a senior biologist; in consequence, perhaps, his special abilities in application of statistical and scientific reasoning to problem solving were not widely appreciated; indeed, before the war, ideas now associated with operational research were hardly envisaged and the need for them was not felt. That operational research was developed in the British armed services during the war was largely the personal achievement of P. M. S. Blackett, a student of Rutherford and one of the most brilliant of the younger scientists advising the government. Subsequently, the American armed services attached statisticians, including A. E. Brandt and W. J. Youden, to active units as operational research workers. At the same time the utility of a statistical approach was being exploited in a new range of industrial applications by a brilliant group of young men brought together by Womersley to do research for the Ministry of Supply. These men proved so valuable in advising on problems of war production and logistics that by the end of the war, when operational research workers began to be demobilised, they were welcomed into industry.

□ □ □

The Battle of Britain in 1940 was fought just over the southern horizon, 25 miles away from Harpenden. There were daily air-raid warnings, and occasionally an enemy plane strayed near. Fisher was walking across Harpenden Common one afternoon when a string of high-explosive bombs fell right across the Common, just missing the surrounding buildings. That winter, walking home in the blackout under the wavering beams of searchlights, he could sometimes see the ruddy glow to the south, where London burned, but "the worst evil that the war has brought to this locality," he claimed, "is a nasty outbreak of petty officialdom."

Billetting officers, food officers, employment officers, the ARP Wardens, and the police were busy explaining and requiring compliance with wartime regulations. Someone seemed always there to remind anyone who failed to carry his gas mask, or to take cover during an air raid, or who showed a light in the blackout. Fisher quarrelled with the butcher whose paddock lay between the water pump and an allotment of his and who would not let him carry water across. He was much annoyed by the representations of fellow allotment-holders who objected to the barking of the dogs he kept on his allotment. The dogs were part of a breeding experiment set up under Genetical Society sponsorship in 1935. In the war, many people who had agreed to take breeding bitches felt they could not continue. Fisher kept his dogs, but others were killed, including some that were essential to the experiment and that proved irreplaceable.

Fisher joined what soon became the Home Guard, along with his younger son, Harry. What emerged from this service was an interest in the effects of vitamins on nightblindness. A number of the older men were much handicapped in the blackout by poor night vision. Fisher found his way by holding his walking stick vertically against the curb; this worked well only if he could see the curbside obstacles, pillar box, or bus stop sign. He became convinced that vitamin A capsules, which he took as a preventative against the common cold, assisted his night vision, an observation which later investigations appear to confirm.

He took an allotment in the grounds of Rothamsted, close to the mouse house. Trenching the area, he declared that the second spit must have been laid as a Roman road, it was so compactly paved. He was very proud of his produce: the line of leeks, each dressed in a paper collar to blanch its neck; the huge beets, cylinders 2 or 3 feet in length, some of them too long to be cooked whole in the large ham boiler at Milton Lodge, yet sweet and tender. He took a second allotment for the children to work with Mrs. Fisher's help. This was broken out of meadowland in Rothamsted Park, and its first yield was a cartload of flints which the local authorities carried away for road making.

Rabbits were introduced at Milton Lodge, for food, and Fisher designed runs for them, cylinders of chicken wire which the rabbits themselves could roll across the grass and so save human labor in moving hutches. In the summer he took family parties with sickles to cut clover hay from Common or wayside for the chickens and rabbits. The children were expert in discovering wild raspberries on the Common and raiding the best briar patches around the neighborhood, and quantities of fruit were collected for eating fresh and for perserving or bottling against winter. At the time rose-hip juice was recommended as a source of vitamin C, and so hips were also collected.

Es Brandt wrote from America offering on behalf of himself, Harold Hotelling, Besse Day, and Walter Shewhart together, to care for the six Fisher girls during the war. Fisher accepted this extraordinarily generous offer and, in discussing the future of what he affectionately referred to as "the whole monkey house," revealed a depth of love and concern for their happiness which might have surprised them; in daily life he gave little sign of sensitivity to their feelings, nor was he involved in practical details of their lives. Mrs. Fisher bought the six small trunks and got together and packed six sets of clothes. She spent a long day with all the girls at the American Embassy in London, getting the necessary visas. In March 1940 she visited the embassy again to obtain an adult visa for Margaret, who had just turned 15. There she was told that a transport was available, and, although Margaret must await issue of her visa, the five younger children could sail at once without her. The decision had to be made there and then. Mrs. Fisher said she would put all her eggs in one basket; the children should all travel together on a later transport. The ship sailed without them, and was sunk. This was probably the last tragedy of its kind, for thereafter evacuee ships were not run, and consequently the Fisher children remained in England throughout the war.

The education of the younger girls had been interrupted by evacuation to Scotland for some months and by the possibility of evacuation to America, and Mrs. Fisher decided not to return them to the local convent but place them in a high school providing better teachers and facilities in the higher forms. This was made possible by a fortunate circumstance: it happened that the head mistress of St. Albans High School for Girls and her second in command, Miss M. A. Finch, were both members of the Baconian Society in St. Albans and had met Fisher there. By the time Mrs. Fisher approached them they had already decided between themselves that they would welcome his daughters in their school. Showing Mrs. Fisher round the school, Miss Finch overcame all her hesitation, even about costs, and arranged extremely generous terms. From that moment she adopted not only the younger girls but the whole family under her wise, generous, and far-reaching kindliness.

In the summer of 1940 George Fisher joined the Royal Air Force. His father was pleased, despite himself, that George, who as a medical student was reserved from national service, had volunteered (keeping quiet about medicine) and had been accepted as a "student." He made him promise to sit the examination at the end of the year toward his medical baccalaureate, though they probably both knew this might be impracticable; in fact, when the examination took place, George was in Canada, training to be a pilot.

So the first year of war passed without a great deal to show for it so far as Fisher's work was concerned, and with a good deal of energy spent in accommodation to the contingencies of war, but he hoped better of the second year, for he had been consulted about some veterinary work of the Agricultural Research Council and hoped this work would expand. As he put it in February 1942:

I am not in such a state of utter stagnation as during the first year of the war when all affairs of national importance were in the hands of officials of such superlative competence and foresight that it was a positive insult to suggest that any extra help would be useful. In fact, we are now allowed to do a bit towards improving current experiments in animal husbandry with a suspicion that successful work, if acted upon, may improve the food supply or free a certain amount of tonnage.

He was expecting, too, to have some interesting people in his department that year. He persuaded Haldane to ask for an office at Rothamsted, with his assistant, Helen Spurway. Dr. Dora Lucka, an Austrian refugee, replaced one of the departed computing assistants. (Later in the war she began the translation of *Statistical Methods for Research Workers* into the German language, and this was eventually published in 1956.) Two other refugees were with the department that year, Dr. V. Myslivec, a statistician from Czechoslovakia, and Dr. Olbricht, veterinary surgeon from Poland. Best of all, E. B. Ford came as visiting professor in his department from October to December 1940.

Ford and Fisher had a lot of work to discuss. The blood group results were continuing to come in from all over Great Britain, and one evening in November Dr. Taylor joined them and talked to the Baconian Society in St. Albans. The Dowdeswell, Fisher, and Ford paper had just appeared, on the estimation of births and deaths in the fluctuating population of tiger moths, and the results which had been collected in July 1940 awaited analysis. A year or so before, Fisher had become interested in selection experiments with the short tailed *Sd* mutant mouse and was now ready to start the breeding program.

□ □ □

Talking with Ford about this tail mutant, Fisher thought it would be interesting to find the frequencies of wild mice on the Rothamsted estate, and together they began to trap wild mice there, and later in a spot near Oxford, to observe what variation in tail length occurred in nature. After Ford left, whenever Fisher became restless and wanted a walk, he would lead Mrs. Holt on a tour of inspection and baiting of the mousetraps. (He seemed unaware that in consequence she would have to stay late to finish her work and that she was needed at home; it occurred to neither of them that she would ever leave mouse work undone.) At one of the traps Fisher was amused to find that a mouse that had been caught there the previous day and duly marked had returned; released again, the mouse continued to be caught daily, enjoy his meal and, in due time, his release. Fisher was pleased to think that it made this trap its club, for he had designed the trap to seem natural and attractive to his shy guests, with a tiny grassy tunnel into the ground floor where there was food and access to a compartment above which had the makings of a nest.

One of the wild species, the short-tailed *Microtus herbensis*, showed a curiously high proportion of defective or anomalous developments of the tail, not unlike those produced in the house mouse by mutants such as *Sd*. Indeed, Fisher thought it possible that the short tail of the *Microtus* had been produced by one of these factors, the deleterious effects of which had been largely repaired in the wild stock. In order to carry further the analysis of the short-tail factor and its modifiers, he sought other species of *Microtus* from abroad to hybridize with the English population. To A.B.D. Fortuyn he wrote in January 1941 to discover if he had some species of this genus wild in Peking. In England he began trapping local *Microtus* in the hope of domesticating them for breeding.

Microtus proved shy of capture and domestication. They were nervous and difficult to handle, and in captivity they tended to pine and die. The attempt to catch and keep them was too time-consuming to attempt after the laboratory assistant, Munday, went into the army. Even before Fortuyn replied in April the trapping had to cease.

Laboratory stocks of the short-tailed *Sd* mice had been received from L. C. Dunn of Columbia University shortly before the war. Fisher encouraged his staff to use his mice in their original research. Looking after the stock, Sarah North noticed that the tails varied in length, and without Fisher's knowledge, she began to select and breed from those with the longest and the shortest tails. She had thought of making a special instrument to measure tail length accurately, and one day, having occasion to show Finney the mice, she told him of her design: a box to contain the mouse, with a slit cut out at one end through which the tail would protrude, and with a millimetre scale attached on the outside, against which the tail could be held taut for measuring. Finney being enthusiastic, they went out at once, obtained a box and together built the instrument. When he learned about it in the summer of 1939 Fisher was charmed with the tail measurer, which he thought most ingenious. The instrument was used for all their tail measuring and was taken to Cambridge in 1943. He was most encouraging about the experiment, too, and suggested the program would be improved by the introduction of modifiers from other stocks; they could make it a joint project.

After evacuation from London, they prepared independently the plans for introducing modifiers into the *Sd* line; both were delighted to find, on comparison, that the plans were identical. During the winter at Rothamsted mouse stocks suffered through the incursions of rats into the mouse house, and through cold and disease. The first matings were not put up until September 1940. Despite the depletion of stocks by the same causes in the following winters, the selection lines made good progress [*C P* 199, 1944].

In the plus line, the medial tail length remained practically stationary from the matings made in September 1940 to those of June 1941, although short tails became somewhat scarcer and long tails more frequent.

The medial tail length seemed not easily to exceed 40 mm. At this stage it was observable that many tails were externally normal for about two-thirds of the normal length, but failed abruptly, and seemed incapable of forming the delicately graded series of vertebrae which occupies the distal third of a normal tail.

In the matings made after June, however, the medial length stabilized at about 50 mm, gradually increasing during the following year to 53 mm. At the same time the tails became much more uniform in length and a graduated series of small vertebrae were formed with increasing perfection until, in 1943, mice in the plus line were approximating normality, and the heterozygotes could be distinguished only by transparencies which showed the tail bones lying a little closer to each other than in a normal tail.

In the same period the minus line quickly diverged from the plus line. The tail length diminished very rapidly, so that the second group of matings set up in February 1941 was already producing mice with an average tail length of no more than 12 mm. Throughout the next 18 months of selective breeding, however, the average tail length remained the same so far as could be judged from the small numbers bred. Evidently, natural selection had been constantly at work to mitigate the severity of the reaction to the mutant gene and so, incidentally, to make the development of the tail more normal; this selection was sufficiently strong to counteract the rather mild intentional selection used in the experiment. The breeding program in a neat way made it possible actually to measure the rate of selection.

The progress of the plus line suggested that some not uncommon modifying factors, or combinations of modifying factors, sufficed to eliminate the sharp bends in the proximal half of the tail so as to give a straight butt, and that a further set of substitutions were required to permit normal development in the distal region. Though Fisher's lines were relatively inbred, they were evidently highly heterogeneous in genes which, in the presence of the *Sd* mutation, were capable of influencing the development of the tail. Indeed, the experiment demonstrated that before the introduction of the *Sd* factor, there had existed in his stocks both genes that tended to make it recessive and genes that tended to make it dominant. It was an attractive evolutionary theory that the short tail which was normal in *Microtus* might have originated by the occurrence of a short-tailed mutant, without lethal effects, which had been suitably modified under natural selection.

The *Sd* gene in the house mouse was lethal when homozygous, and in Dunn's experiments homozygotes never survived birth by more than 24 hours. The mutant also had deleterious effects when heterozygous; mortality among Fisher's heterozygotes was systematically higher than that of normal homozygotes, except in the plus line. Thus it appeared that the same modifying factors that increased the length of the tail also improved viability. Consequently, when one of the tail-less homozygotes born in October 1941 in a plus-line litter was alive after more than a week and appeared normal, Fisher wrote exultantly to Ford: "Of course I ought to insure its life for thousands for it can scarcely be anything other than a nearly or quite viable homozygote." The mouse proved to be a genuine homozygote. At 21 days, together with the rest of the litter, it was separated from their mother. "The next morning it was found dying. Warmth failed to revive it and it refused milk administered in different ways. It died half an hour later, having lived just over 22 days."

In analyzing the record of the survival of all homozygotes born alive in his lines, Fisher [C P 199 with S. B. Holt] confirmed the association between increased tail length and improved viability. He found of his homozygotes born alive, 46% survived over 24 hours, in contrast to Dunn's record of zero survival; 13 survived over 48 hours; and the majority of these came from the plus lines.

No awkwardness arose from having Fisher with his department under the same roof with the statistics department at Rothamsted—quite the reverse. Fisher kept himself separate from the statistics department at an official level and left D. J. Finney and O. Kempthorne alone to deal with the statistical work of the station. Unofficially, of course, he was always available for discussion with the statisticians and with the other members of the Rothamsted staff. Soon he was working closely with C. B. Williams at the beginning of the series of papers relating to the logarithmic distribution, and he was working again with his old friend Gerard Thornton.

He continued to do a little teaching. In 1941 he was very happy to have M. K. Maung doing special reading under him, a Burman student, whose work seemed to Fisher exceedingly promising; V. J. Panse was also at Rothamsted, a former student at the Galton, continuing work for his Ph.D. under the supervison of Yates, who was away on war work. By Finney's initiative, Fisher gave a series of general lectures for the staff at Rothamsted. The purpose of these lectures was not to give high-level seminars on new statistical theory, but to provide some basic statistical instruction for biologists. Understandably, there was a good audience.

The first couple of lectures were quite brilliant in their presentation. Fisher began by talking about the principles of the analysis of variance and, by taking exceedingly simple numerical examples, perhaps only of four observations in a two-by-two configuration, he illustrated in nonmathematical but clear arithmetical and logical terms the subdivision or, rather, the various possible subdivisions of the sum of squares of deviations, the different fields of application, and the different interpretations to be placed on such partitions. It was beautifully done, taken slowly, and a model of clarity.

By about the third or fourth lecture, however, there was a change. The content and the exposition remained sparkling but the pace increased tremendously. Finney recalls:

I soon found myself giving additional tutorial periods to those who had managed to scribble down a few words of wisdom from the blackboard but were totally lost in the details. For the professional statisticians the lectures remained exceedingly valuable but I am afraid the latter part of the course was quite largely lost on those for whom it had originally been planned.

I have noticed this many another time in Fisher's lectures and in his writings. He could always put over an elementary point for the non-statistician and non-mathematician with remarkable clarity and make the truth come alive. For the professional mathematician and statistician his writings and lectures on more advanced matters were often obscure but always interesting and repaid careful study. In an intermediate zone where the biologist without mathematical training needed to be led along carefully if he were to understand, Fisher went far too rapidly and failed to appreciate the points over which difficulties would arise.

The difficulty could be overcome only if the scientist raised his difficulties in conversation, for then Fisher could be persuaded to take the hurdles alongside the scientist.

One had first to convince Fisher that a problem existed, and this was not easy, as Dr. S. B. Holt's experiences with him over the years show. On her first day at the Galton Laboratory in the middle of the Easter Term 1934, Fisher asked her to attend his statistical lecture. She had done no statistics, but he assured her that all she would need was knowledge of some elementary algebra. Afterward, when he had asked how she had got on, and she had told him she had understood nothing of what he said, he simply refused to believe her. By 1940 she had learned some statistics but still had many difficulties. One Saturday morning at Rothamsted she asked him to explain the derivation of a formula. He brushed the request aside. She persisted, but before he would credit that she could have any difficulty with something that was to him so simple, she had to plead with him that she was an "utter fool" and *could* not understand. Then he explained it to her in detail, and all was clear.

Some years later, when Fisher began teaching genetics to first year students at Cambridge, he confided to Mrs. Holt that he found it difficult to make his talks suitable for students who had perhaps never before heard of a gene. He asked her as a favor to attend his lecture and afterward charged her to tell him frankly what she thought of the lecture and its suitability for undergraduates. She said it was a wonderful lecture and she had learned much from it, but that it was far above the heads of students. Fisher said this was what he had feared, adding that he had made a real effort to make it intelligible.

Some months later Mrs. Holt again attended one of his lectures as critic, and was happy to report that this time she felt he had attained his object so far as the first half of the lecture was concerned. He was delighted. She was forced to add that the latter part of the lecture, an exposition on the genetics of *Lythrum salicaria*, was more suitable for his own graduate students.

Fisher's optimism in the autumn of 1940 was, in a sense, justified: he enjoyed his contacts at Rothamsted; Ford's stay was a most happy and productive event; work with the A. R. C. developed, and he became involved in the design and analysis of a series of experiments on phenothiazine as an anthelmintic applied to lambs. That year he was elected president of the Genetical Society. The preservation of irreplaceable genetical stocks was by then becoming a serious problem for members of the society. It was a difficult time to serve, but, he felt, important for that very reason. As for his own research, in 1941 conclusive results came in from his *Lythrum* experiment, and although he was very concerned about physical conditions in the mouse house and the losses to his experimental lines, still he was getting some interesting results. Nevertheless, the year was one of unprecedented trials.

Miss G. M. Cox, then director of the new statistical department at North Carolina State University, Raleigh, had invited him to attend and lecture at the summer session there in 1941. Fisher was looking forward to the visit. W. G. Cochran was professor in the department, and other old friends were planning to attend the session or to meet him in other cities. Then, late in March, his application for an exit permit was refused by the British authorities. He immediately cabled Miss Cox: "Please get USDA or Vice President Wallace to press Foreign Office to grant my exit visa." At first, he wondered if the government of the United States was responsible for his rejection. Brandt and Miss Cox inquired. President Graham of North Carolina State worked directly with the Vice-President of the United States, H. A. Wallace (their old friend from Iowa), with Harry Hopkins, Lord Halifax, and the Secretary of State to try and secure permission for Fisher to make the visit, without avail. American officials apparently knew nothing of the affair and could only guess at the motives of their British counterparts. For two months the series of anxious messages continued across the Atlantic before Fisher gave up.

In England, Fisher's inquiries met with a blank wall; he could discover nothing, nor could the president of the Royal Society to whom he appealed. Evidently he was politically suspect, his name infamously smeared, through some bureaucratic accident probably, the folly of some nameless minor official. Was it because of his having harbored communists in his department, or others who expressed themselves freely against the government? Was it because of his scuffle with the porter at University College? Could it be that his eugenic ideas had been mistaken as somehow endorsing Nazi "eugenic" practices or that his friendship for the German geneticist Von Verschuer of The Institution for Biology of Inheritance of Berlin had raised doubts about his loyalty to England? Whatever the suspicion, it was unnamed and unanswerable, and Fisher writhed under the imputation of dishonor to his name. To a cousin in the diplomatic service he expressed his bitterness 6 months later when he wrote of his son, George, in the Royal Air Force: "I hate to think that the swine, whoever he may be who has dished my own official reputation may have stabbed him in the back also."

George had returned from Canada with his wings, received a pilot officer's commission, and was stationed in Scotland flying the nightfighters called Beaufighters. He adopted a pleasing puppy from one of the experimental litters at Milton Lodge, and soon "Beau" was the mascot of his squadron. In September Fisher stayed a fortnight in the neighborhood of George's operational aerodrome in order to get a few walks and chats with him. Seeing the similarity of build and gait of the two walking figures one would suspect they must be father and son. To Fisher George's seemed "a thoroughly enviable life while it lasts."

In March, with the refusal of the exit permit for Fisher's visit to the United States, came an attack from another quarter which made it desirable for him to be in England. The remaining employees of the college, Fisher's secretary and his research assistant, received notification to the effect that they had failed in their duty to their country, and in compliance with the college instructions, inasmuch as they had not of their own accord already secured alternative employment, and that, in consequence, the college would suspend their employment from the close of the current year, September 30, 1941.

In fact, Miss Simpson had offered on the outbreak of war to do censorship work, as she had done in the World War I. (Eventually, she was called up to do this.) By 1941 Mrs. Holt had a baby daughter and so was officially exempt from war work. Fisher knew it would be useless for himself alone to appeal to the provost, and although he wrote to him, protesting that the dismissal was unjust to both of his assistants and unjustifiable on any grounds of national utility, he also wrote to the head of the Medical Research Council, inviting him to write back expressing, in the strongest terms possible, how highly he appreciated the scientific value of Fisher's work with the Serum Grouping Unit and how much he regretted that Fisher should be compelled for lack of a secretary to let drop the unique chance of saving and analyzing these data. Then he laid the letter from Landsborough Thomson before the chairman of the college committee, Lord Sankey, who had previously been so helpful. A reprieve was won for Mrs. Holt when Fisher pointed out that she was his one officially designated assistant, according to the agreement made in 1939, but he was told Miss Simpson must go.

Already Fisher and his remaining assistants had taken over the laboratory and office work of other members of the department and uncomplainingly worked long hours to do what was necessary. In particular, Miss Simpson was responsible for all the office work involved in continuing regular quarterly issue of *The Annals of Eugenics* and for organizing the heavy flow of blood group data. She shared the task of counting and recording with Fisher. By this time some 12,000 donor cards had been received, recorded, counted, and posted back to Cambridge, and more came in every week.

For immediate help in retaining Miss Simpson Fisher turned to the ARC, whom he had so far served without pay. They offered to pay her salary until the end of the year and to make arrangements to pay the department for all services subsequently received. In fact, the arrangement enabled Fisher to hire an additional assistant a year later, to deal with the increasing computations for the ARC.

Simultaneously, he wrote to the Royal Society, applying for a grant in support of the *Annals of Eugenics,* and in August appealed to O'Brien of the Rockefeller Foundation.

in the hope you may find some means of helping my department in the very great difficulties in which it is now placed. . . . If I can find no remedy, this change practically puts an end to my own scientific work, for I have not been without a personal secretary in my department for twenty years and my eyesight is so bad that eyestrain would certainly force me to relinquish most of what I can now do.

Having described the *Annals of Eugenics* and the blood group work and mentioned that the ARC, like the MRC, would endorse the value to them of the work of his department, Fisher concluded:

[I] would be willing to give my whole time to appropriate war work. So far, however, I have been left to do my former scientific work under difficulties which up to the present have not proved insuperable and it may be that the continuance of my department is the sort of scientific salvage work in which your committee is interested.

To this personal and scientific appeal, Dr. O'Brien responded sympathetically, and before the end of the year he requested a more formal exposition to present to his colleagues. This Fisher sent on December 23, 1941.

The MRC arranged a grant to cover the salary of Miss Simpson for 1942, to be renewed for the following year. In March 1942 the Royal Society approved a grant towards the *Annals of Eugenics,* from a gift of the Rockefeller Foundation, and this grant was repeated in the following March. Fisher was no longer in danger of losing the hands and eyes which his assistants lent to his work, nor the personal support which they gave him through all these vicissitudes.

But he had a new anxiety. In December a telegram arrived informing him that his son, George, had been seriously injured. Fisher took the next train to Scotland, and Mrs. Fisher followed on the next day. His life, they discovered, was not in danger. He was conscious, though heavily drugged, the injuries to face and head covered by bandages. Whether there would be permanent damage was uncertain. The orbits at least seemed unharmed, but the jaw and possibly the skull had been cracked. He had been lucky. Secretly he had been giving flying lessons to a young Observer. With the Observer in the pilot's seat, flying in cloud among the Scottish hills, the plane had dropped too low and had crashed into the hillside. The Observer had been killed on impact, and George had been rushed unconscious to hospital.

After facial surgery, George was kept in hospital longer than predicted, and Fisher wondered if the doctor had been too optimistic in his early assurances that he would recover fully. After a month he was removed to Bangor Hospital, the prognosis still uncertain. The doctors had different theories about the cause of the diplopia resulting from the accident. Happily, after some months the diplopia disappeared, and George was judged again fit for active service.

First, however, he faced court martial for the loss of his Observer and his plane. Afterward he reported that the trial had been merely a form: a trained pilot was an investment the Air Force could not afford to sacrifice on account of a technical illegality, such as was in those days practiced by a number of the keen flyers. Witnesses perjured themselves on his behalf, and he was acquitted in law. But he could not acquit himself of the moral responsibility for the death of his friend. He did not regret his decision when the boy begged him to teach him to fly, but he regretted deeply his lack of judgment later, when he had known for a moment that they were flying dangerously low and had not spoken. He had to live with the fatal consequences of his mistake, and it was a comfort to him to confide in his father in this moral crisis.

Not only had George escaped with life, and with his eyesight unimpaired, but he had fallen in love with and had been accepted by one of the nurses at hospital. Fisher met the girl and found himself ready to love his chosen daughter-in-law. He hoped George would take the present chance of happiness for himself and his girl and risk the future, that he would marry at once and assure an heir. The girl felt differently. During the war she was committed to the task of nursing, and she did not feel free to marry until it was over.

George was ordered abroad the following summer and, on embarkation leave, he and his father worked out a code by which he could inform them at home whereabouts in the world he was working. Not long afterward they learned in this fashion that he was stationed in North Africa, in the neighborhood of Bône. His observations of natural and human life around him were acute and amusing and obviously he was well and happy.

□ □ □

In January 1942 Mrs. Holt was delighted to discover three polydactylous mice in one of Fisher's lines, and a fourth in a closely related line. With Fisher's approval she initiated a study of the genetics of polydactyly in his

he Life of a Scientist

mice. As an isolated inherited entity polydactyly in mice had been reported only three times before, and within each of the polydactylous lines the degree of manifestation varied widely. Moreover, the character failed to give regular Mendelian ratios. A.B.D. Fortuyn, observing that in his stock normal as well as abnormal sons and daughters were born from normal and abnormal parents in all combinations, had rejected any mode of inheritance as an explanation of the condition and concluded that in some way the environment was responsible.

There was, however, some evidence that polydactyly was genetically controlled, and no evidence that could not be explained by genetical theory. In Fortuyn's lines the inheritance was obscure, but the proportion of polydactyls born remained fairly steady, which suggested that the variable manifestation was due to the effect of different modifiers in the different stocks.

The fact that in each case the condition had appeared in inbred lines confirmed the hypothesis that each worker had been investigating a single mutation. Since 1936 Holt had found several polydactylous mice, including five offspring of one mating in Fisher's inbred lines. None of these polydactyls bred, but the new polydactyls came from the same ancestry, though not from the same line, and probably arose from the same original mutation. If so, the mutant was not only completely recessive but was partially suppressed even in the homozygote. The extent of variability in the degree of expression observed in different lines and in different individuals within the same line suggested considerable heterogeneity in respect of the modifying factors concerned in the suppression of polydactyly.

These inferences were consistent also with the evolutionary picture. Polydactyly had been observed both in man and in poultry, and in these cases was known to be genetically controlled. In man it is a Mendelian dominant. In domestic poultry it is semidominant and Fisher's investigation had shown it to be recessive against the genetic background of the Indian jungle fowl. In mice it appeared that the evolution had been carried a step further toward the suppression of the character in the homozygote. Assuming that the mutant suffered a comparable selective disadvantage in all these groups, the modification of dominance had taken place at very different rates; rapidly in mice, less rapidly in birds, and hardly at all in man; of course, this is precisely what might have been expected, considering that a generation in mice is perhaps 10 times shorter than in birds and a generation in birds shorter than in man by a similar factor.

As Mrs. Holt realized, polydactyly in mice interested Fisher as another probable instance in which his theory of the evolutionary modification of dominance explained the facts so far known and invited experimental verification. It was not possible to determine the mode of inheritance of polydactyly until the manifestation of the character had been greatly increased. An inbred polydactyly line was, therefore, established, initially showing 52% manifestation of the abnormality. By selection after outcrossing, a line was finally obtained with 90% manifestation of polydactyly. It was then possible to show that polydactyly was indeed due to a recessive gene with incomplete penetrance.

The speed of modification indicated that the number of modifying factors involved in bringing up manifestation so far must be small. One such modifying factor had been brought in by a female from Dr. H. Gruneberg's *fidget* stock, which had been introduced as being possibly polydactylous (which she proved not to be). The F_2 generation from her showed complete penetrance. Since at this point in time the polydactylous line was in danger of extinction through the incursions of rats and other causes which had hampered the whole breeding program, the numerous mice of her line were used in setting up the selection line. Another modifier, distinct from that in the *fidget*, was derived from the laboratory stocks. Their effect was cumulative: with both, penetrance was nearly complete, with neither it was low and with one or the other manifestation was intermediate.

From these experiments it was established with certainty that the polydactyly which had seemed genetically unaccountable to previous workers was, in fact, due to a single genetic factor, for which Mrs. Holt introduced the symbol *py* in 1945 [81]. Furthermore, they established that two or possibly three modifying factors were commonly present which suppressed manifestation. Mrs. Holt's account of her work on this experimental programme won her a Ph.D. in 1945.

Fisher followed these experiments with great interest from their beginning in 1942, as he followed the *Sd* experiments. It was always his custom to spend a good deal of time in the mouse room, but at this time he worked there longer, for he had no laboratory assistant. For a short time a schoolboy had been engaged to help with the animal work. One day he forgot to feed and give water to some of the mice, including the polydactyls, and they nearly all died. After that Holt and Fisher together did everything for the animals themselves, feeding, cleaning, and recording the mice and walking the dogs at least once daily. The dogs went with him when he made counts of local rooks' nests, as he had done for a dozen years past. Plate 13 shows him mending mouse cages outside the mouse house, with his favorite, Amy, beside him.

When the weather was good, Mrs. Holt often brought her baby Susan to work and left the perambulator outside the hut while she went up the steep steps to the mouse room. Fisher encouraged her to do so: Susan was his godchild, and he was fond of babies. One day, some time after Fisher had left the mouse room, Mrs. Holt went down to see that all was well with the baby. She found Fisher standing beside the perambulator watching Susan, who was trying to bite the large, half-peeled onion he was holding to her mouth; it was rather too large for her to manage. He looked up, pleased, and said, "All babies love onions. Mine did. I have just got this out of the allotment for Susie!"

He was busy collaborating with ARC and MRC in 1942, and he was trying to break new ground. In serology he was working with Ford to get observations made on the antibodies of pregnancy, hoping to detect selective effects in the new Rh factor. He began counts of primroses in the Sparkford area that April, exploring the peculiar evolutionary situation of homostyly in local colonies. He put in a lot of heavy computational work in calculating the table for the Behrens' test in 1941 [C P 181] and published several papers on combinatorial problems of experimental design in 1942 [C P 186, 187, 189, 190]. Yet, busy as he was, the opportunity to do worthwhile war work was withheld and, with it, both the service and the recognition he craved.

Under the shadow of his more obvious tribulations a greater tragedy grew to its fruition. His marriage became a bitterness to him, tainted by every frustration he experienced, and when he escaped from the Galton Laboratory in Harpenden to work in Cambridge, he left behind him the domestic situation as well and separated from his wife.

Fundamental causes of the rift were always latent in the marriage. Both partners had exalted expectations of each other and were bound to be to some extent disappointed. The life they chose was never intended to be easy. Fisher set the pace and carried his burden largely oblivious of how his companion fared under hers. Eileen was disappointed of the sympathy and consideration that could have made allowance for her difficulties and thus eased them. Fisher was disappointed of an equal partner in his endeavors; having married a very young wife who was molded by her life with him, he missed the spontaneity of a strong and independent spirit. These disappointments might perhaps never have become unendurable if it had not been for external stresses. Considering these stresses, it seems remarkable that the marriage endured for more than 25 years and that, after its break-up, each still felt his own love for the other to be real and was passionately grieved by the separation.

The stresses were always there.

For Mrs. Fisher the physical demands of life were often overwhelming. As a housekeeper, she attempted to make ends meet, always on a tight budget and without modern conveniences. Cooking was done on a Primus stove or on the coal-burning kitchen range, which also provided the hot water supply. Clothes were washed by hand in galvanized tubs and dried with the aid of a hand-mangle in a shed outside and a clothesline in the garden. The house was heated by coal fires, and there were grates in every room. Shopping was usually wheeled home in a perambulator. Scrubbing and polishing in the big old house, with its numerous inhabitants, never ceased. The care of the poultry and, in part, the garden was her responsibility. Even with some help in the house, this was sufficient to employ her energies.

As a mother, she was usually either bearing or nursing an infant, tending toddlers, and, until 1933, teaching young children school lessons. She kept track of school clothes and school work and inspired many of the activities of the schoolchildren; she gave them their meals and baths and the songs and stories of bedtime; she tended them in their sicknesses. She made their clothes on an old treadle sewing machine and knitted so many kneesocks that she could turn a heel while reading aloud to her husband.

As a wife, she had another and no less demanding life. Before breakfast, every day after Fisher returned from America in 1931, she played "medicine ball" with him. She read him *The Times* over a leisurely breakfast and saw that his boots were ready polished on the footstool and that he did not forget his stick or briefcase when he left for work. After he went to the Galton he no longer came home for lunch, but he entertained guests at Milton Lodge almost every Saturday and Sunday. She shared in the recording and cleaning and feeding of the mice, and she made the snail cages until 1933; she helped in preparing the skins of mice and (during the poultry experiments) the transparencies of chicken legs; and she assisted with bee keeping. In the evenings she read and listened and discussed with Fisher until a late hour, continuing even in the bathroom while he took a bath; inevitably, she shared his sleepless nights when he suffered with headaches.

It goes without saying that she was not a perfect housekeeper or mother or wife: no one can simultaneously play medicine ball, feed the infant, dress the youngsters, and prepare the breakfast porridge on the Primus stove. The first difficulties in the marriage rose from the conflicting demands on her strength. An incident of the early 1920s illustrates the direction subsequent developments were to take. One day, after lunch, when Fisher was just settling into the exposition of a train of thought that was currently fascinating him, Mrs. Fisher hesitantly excused herself, pleading exhaustion. She was under doctor's orders to take a nap in the afternoon and she needed it. Fisher was extremely upset: "You are not interested," he said, "you do not love me." At the time, Mrs. Fisher accepted her husband's judgement and knew that she had failed in showing love. But the shock of the accusation remained. Many years later, when she could no longer deny to herself that her husband was sometimes unjust, the incident recurred to her consciousness, with the absurd pain it had caused them both. By then such incidents had become a recurrent nightmare.

With the egotism of a child Fisher assumed that he came first with his wife in all the circumstances of life. Mrs. Fisher assumed it, too, but she was in collusion with an essentially false expectation, that he had a right to feel hurt if ever he came second to her needs or those of the children or the demands of housekeeping. Obviously, he had no realization of how much he was demanding of her, directly or indirectly; what he did realize, in situations like that described above, was the pang of personal rejection.

In the early days painful episodes passed off quickly, but there was a bad period in 1926 – 1927 when Mrs. Fisher showed the strain. During pregnancy she developed symptoms of an incipient gastric ulcer, and after the baby was born she had little milk. The baby was puny and fretful; Mrs. Fisher felt guilty because of the anxiety which had produced her own poor health, her poor milk supply, and a baby so frail that it nearly died of measles a few months later. Her husband came first; she could not think of interrupting the evening with him to give the infant her full attention; at night she was anxious that its cries should not disturb him.

The stress on the marriage fluctuated with the stress under which Fisher was living. The level of his intellectual activity was a fairly constant factor, and the irritability of his temper appears to have been a concomitant of his mental irritability. Quite small reverses could throw him into a towering rage which, if not vented elsewhere, would bury his wife under its volcanic debris. "Don't be frightened," he once roared, "I am not angry with *you*." But, even when it was not directed at her, his anger was terrifying.

At Rothamsted, while Fisher accepted the need for patience, in one sense, he was nevertheless fretted in temper by delay and sometimes frustration of acceptance of his ideas. He was upset when his work, for example on randomization and the analysis of variance, was misunderstood and misrepresented in the 1930s. At the Galton Laboratory, he was required to deal through petty officials with a bureaucratic machine. It was a time of economic depression; the attitude of the college toward his department was not enthusiastic, and there seemed little appreciation of what might be made of the department and little wish to support his efforts and supply his needs. There was inherent tension and sometimes active dissention in his relationship with the statistics department.

At work Fisher had no outlet for the soreness of spirit caused by the abrasions of individually petty but numerous irritants arising from his professional situation. It came out at home, where the still more petty irritants could be dealt with directly. If there was no bacon for breakfast, or the coffee was vile, he let it be known he was hurt and angry. Once his wife said she did not wish to go for a walk when he suggested it, pleading to stay at home with a daughter who was in bed with pneumonia, but he had suggested the walk especially to give her refreshment away from the sickroom and was not to be refused. He began to exhort her on such occasions to examine her conscience, to recognize that in her lack of love for him she was motivated to forget to buy the bacon or to refuse his suggestion of a walk together. Such actions, he felt, were unkindly meant.

With the outbreak of war the stresses multiplied on both husband and wife. Fisher was fighting to keep his department together, to continue the enumeration of massive blood group counts, to salvage some part of the genetical work, to produce the *Annals of Eugenics*. He was frustrated in his desire to do useful war work by means of his computing group; he was shocked by the refusal of an exit permit which might have allowed him to do something useful in the United States; and he was harried by petty officialdom.

Mrs. Fisher, for the first time for some years had no living-in maid and had difficulty in finding daily help. Housekeeping became progressively more difficult as shortages of food and material became more stringent. She had extra jobs, like putting into practice ideas Fisher suggested: to make and use a haybox to cook the porridge overnight, when quick-cooking oats were no longer available; to breed rabbits and prepare and sew their skins; to organize hay-cutting expeditions. She found the heavy work she was now doing in the garden and allotment a drain on her depleted energies.

Working in Harpenden, Fisher was much at home and in a highly irritable state. Mrs. Fisher and Miss Simpson were hard hit by his splenetic attacks. Both were made nervous by the anger repeatedly aroused by some error or misapprehension of his wishes. Their very fear caused mistakes and roused Fisher to attack them. Miss Simpson confided her trouble to Mrs. Fisher, and they wept together. They had no other answer to the situation. They were both near breaking point.

□ □ □

From blaming his wife for her lack of love, Fisher came to believe that the neglects that hurt him so much were the fruits of deliberate malice. While Mrs. Fisher busied herself trying to carry on and felt her strength failing, she was further disheartened as her husband accused her of betraying and spiting him. By 1942 she was nervously exhausted and physically in ill health, her tears as commonplace as her husband's anger.

No doubt she was reacting to her husband's paranoia with resentment and anger, and it may be supposed that he experienced her resulting passive aggression and found it extremely painful. But, if it was so, she did not know it, nor did her children perceive it in her behavior. She appeared only too anxious to support, help, and please her husband. To all appearances his accusations were wildly unfair and his anger a cruel and sadistic persecution.

Occasionally one or another of the children would speak up in support of her mother and Fisher would turn on them and literally slap them down. It seemed to him then that his wife was not only betraying him but that she had corrupted the children. George alone was sympathetic with both his parents, and they both turned to him with trust and respect when he came home on leave.

Mrs. Fisher consulted a psychologist and was advised to take a holiday and rest. At first this seemed impossible advice to take. Eventually, however, she wrote to her youngest brother, the Rev. Desmond Guinness, and told him she was unwell and had been told to take a holiday. He was a bachelor and at the time rector of a small country parish enjoying a large Victorian rectory, and he had a housekeeper, an elderly lady who, like himself, had earlier been a missionary with the China Inland Mission. He found it in his heart to invite the whole Fisher family to stay with him for a few weeks in the summer of 1942, and promised that his sister would have no domestic responsibilities and could rest undisturbed. Mrs. Fisher accepted his offer gladly for herself and the children, while Fisher decided to stay home.

The two missionaries prepared a village mission for the especial benefit of the Fisher family during their stay. Children of the parish were invited to join in field games run by the rector every afternoon, followed by an open-air meeting, with singing of hymns and choruses, a short address and closing prayer. The Fishers formed a considerable proportion of the children in the tiny village. The message preached was the simplest evangelism, the good news that Jesus saves.

It was the message Mrs. Fisher needed: a message of acceptance and for-giveness instead of guilt, of strengthening and guiding instead of condemna-tion. She had idolized her husband and, despite all her efforts, he found her wanting, but the promise of God was an acceptance that would not fail. Her strength had broken. She was lost and frightened and wanted to be guided like a child, and the promise was of an unfailing way. Yet the message was foolishness to her. She was asking the same question she had asked as a teenager, for proof of the existence of God. George joined them, on leave, and responding to her, he spoke of the proofs of science that are appropriate to the physical world being inapplicable to the question she asked; one could not apprehend God by human reasoning but must accept Him by faith. Put that way, it made sense to her to put her trust in the God whose divine under-standing could transcend her need to understand. So she came to the faith of her fathers, having come to the end of herself. In their different ways all the children responded to the missionary appeal and all, like Mrs. Fisher, went to their uncle to profess their conversion and to discover each other of the same mind.

On their return home, the household was mobilized under the strong rule of Margaret, then 17, while Mrs. Fisher stayed a little longer with her brother. The daily Bible-reading and prayer in which they all participated was the first sign their father had of the revolution taking place under his roof. The fact that Margaret was in charge and was not cowed by his disapproval made another change. At the table, she presented her burnt offerings without apology, including the results of his own suggestions. He suggested that mineral oil would serve as an innocuous replacement for vegetable oil in salad dressing and mentioned the word "petroleum"; so she took Vaseline (crude petroleum jelly) from the medicine cupboard and served the new salad dressing. It was not a success. He suggested adding bicarbonate of soda to acid fruits, so as to cut down on the sugar required in the dessert; she added it to a mixture of blackberries and apples which turned a murky gray and lost all flavor.

The household she ruled strictly, her staff of office a stout 2-foot ruler with which to enforce obedience. After her mother's return she continued to be a strong support and, having left school, relieved her mother of much of the load of housekeeping until she joined the Women's Royal Naval Service the following spring. Harry was also in service by this time, with the Royal Air Force.

Although Mrs. Fisher was physically refreshed by her holiday and much steadied in spirit by her new faith, the effect of her holiday in Derbyshire was entirely adverse to Fisher. Her conversion represented a declaration of independence. The whole frame of reference of her life was changed. Fisher's accusations ceased to represent the ultimate judgement on her acts; she dared sometimes to think him mistaken or unduly upset, rather than to think herself guilty. He felt that he had been dethroned and replaced by some mirage his wife called God.

The conversion that might be regarded merely as youthful folly in his children seemed consummate hypocrisy in his wife. She had exposed them to her evangelical brother and had led them on by her own reversion to the religious teachings of her childhood. Through religion, he felt, she was effectively cutting him off from them. It was, of course, her Christian duty, and theirs, to honor and love their earthly father. Yet under the banner of Christianity, they were united in a circle from which he excluded himself. He felt a subtle rejection in their very tolerance, as when his wife received his anger without the habitual tears but with acceptance and a forgiving spirit. What for her was a first step toward selfcontrol and perhaps reconciliation was to him the last straw. He had endured all he could of her malignity.

It was the beginning of 1943 when Fisher's candidacy for the post of Arthur Balfour Professor of Genetics at Cambridge was seriously canvassed. The professorial residence went with the chair, but there was no departmental building. If Fisher accepted the chair he could use the residence for his family and have no laboratory, or he could use the residence for departmental offices and genetical stocks and leave his family at home. Since the girls were settled, quite satisfactorily, at St. Albans High School, the latter plan was not unreasonable. The department would naturally expand, according to hopes in Cambridge University, and after the war, a separate genetics department building would doubtless release the residence to be his home. Meanwhile, he decided to move with his department and his mice and dogs and to leave his family in Harpenden. He hoped the girls would visit him often and that they might grow out of the ideas of the silly season in Derbyshire.

He was not optimistic about a future with his wife. It was as an adversary that he negotiated with her the financial agreement, which was all that was to remain between them. He was anxious about the expense of maintaining two homes, and he required her to draw up a list of the expenses she must incur, based on her estimate of absolutely minimal expenditure at current prices. This amount, and no more, he promised to pay thereafter to support the establishment at Milton Lodge. Once in Cambridge, he used to forget to send the checks, and after waiting a while Mrs. Fisher, ever more pinched by rising costs, used to send letters begging for the allowance: the last link between them became a very sore point of contact. Any hope of reconciliation died.

□ □ □

The last considerable stay Fisher made at Milton Lodge, after moving to Cambridge in October 1943, was in the following Christmas vacation. He returned at once when, on December 23, the telegram reached him at Caius which told him that George had been killed on active service.

The parents had no comfort for each other; each suffered separately the realization how much of their lives was desolated by their loss. George, their firstborn, was dead—the son on whom they had set their hopes, of whom they were so proud, whom they had loved quite above any of their other children, whose very existence gave a point to their lives, a value to their achievements, which nothing could replace. Mrs. Fisher busied herself by day, tearlessly, trying to make Christmas all it should be for the children, and by night she wept. Fisher went through the motions numbed to everything but his grief.

They learned that George's plane had crashed into a mountainside by night. Photographs showed his funeral and the cross over his grave among a field of other white crosses. As if to make himself realize the physical reality and finality of his loss, Fisher looked at the photographs and wondered aloud what charred fragments from the flaming wreck they had managed to pick up to place in the coffin. He collected together George's letters home and had them bound. George's personal belongings were returned, his diary, the binoculars his father had given him, the writing case from his aunt, some articles of clothing, all touching reminders of happy experiences shared, and of the boy who had entered into all his experiences in life so heartily for just over 24 years, and would no more.

One slight hope remained to be explored. Fisher thought it possible that perhaps among the girls George had known—and there were girls who loved him not only in Scotland but in Canada and Africa—George might have left a son. He would have welcomed his son's indiscretion for so happy an outcome. He would have been grateful to know his grandson and to care for the mother as his own daughter in George's name. But his inquiries only convinced him that George's line was dead.

Condolences reached him from many old friends, and he answered them all in some such terms as he answered George Snedecor on May 4, 1944, five months after the event:

> Although I knew perfectly well how dangerous an occupation George's was, I found myself pretty badly prepared for the shock. It made me realize how proud of him I had become. Harry is somewhere in Palestine and I hear from him from time to time. He is four or five years younger and very much less mature. . . . Thank your wife for her kind message. One goes about rather automatically at first. With respect to anything I used to enjoy, I find that I still have my doubts. However, I think that I am pretty tough so surely I can take it.

The enjoyment had gone out of his life, but he would tough it out. He had seen things he had given his life to broken and he stooped and began to build again. In those days one might have observed him walking daily between the genetics department and Gonville and Caius College, from his new work to his new home, stick in hand, a solitary, erect figure under a battered hat, wearing a raincoat of Air Force blue. It fitted him perfectly, for it had been made for George. He carried it like a soldier.

15

Arthur Balfour
Professor of Genetics

In 1938 R. C. Punnett resigned from the Arthur Balfour Chair of Genetics at Cambridge, which he had filled since 1910. Fisher who, as Galton Professor, was an elector to the Cambridge Chair, immediately resigned his electorship, making himself eligible as a candidate for election. Dr. Taylor was eager to return to Cambridge and excited at the prospect of continuing serological work with Fisher there. He visited Cambridge to learn, if possible, whether the university intended to appoint a successor, and reported back that an official residence went with the professorship. The residence, Whittingehame Lodge, was, in fact, on Storey's Way, 2 miles out of Cambridge, surrounded by several acres of departmental land. To the east lay a plot owned privately by Punnett to accommodate his tennis court and orchard. There was no genetics department building; the laboratory consisted of a single room at the west end of the house.

After Punnett's retirement, the residue of the poultry work was taken over by Michael Pease at the Poultry Breeding Institute next door. The university did not at once appoint a successor and, though informal inquiries were made in 1940 whether Fisher would be interested in appointment to the Chair, they were not pursued. With the coming of war, the house was let to tenants and the land ploughed up by the War Agricultural Committee. The department ceased to exist.

In January 1943 the inquiries were renewed. Professor J. (later Sir James) Gray, professor of zoology at Cambridge, buttonholed Fisher at the Royal Society, and Dr. Taylor wrote from Cambridge, both asking if he was still interested. It appeared that Cambridge University would be prepared to advertise the Chair if Fisher would stand. He answered, as before, that he would be prepared to put in a lot of work to establish an adequate department of genetics *if* he could rely on the steady support of the university, but he was not looking for a job to retire to "in whatever affluence and dignity." As a first step, he stressed the importance of attaching to the genetics department, as soon as might be, the Galton Laboratory serological unit, so gaining for Cambridge the advantage of an organization and expertise built up over 10 years in this important branch of human genetics.

In a preliminary memorandum of February 1943 he was explicit in particular about the serological unit and, in general, wrote of the task of building up the department after the war:

The central condition upon which it must be based is that the professor shall be capably equipped for undertaking from time to time any special research project in animal, plant, or human genetics which he may regard as of special importance for the advancement of the subject and for the instruction of advanced students.

Cambridge University was eager to have him and promised all he asked. He agreed to stand, and in April the Chair was advertised (Taylor sent him a clipping of the advertisement to make sure he applied promptly). So, when the formalities were completed, Fisher became Arthur Balfour Professor of Genetics.

He was pleased to receive the recognition of his own university and, with it, the honor and opportunities the Chair might afford, and he was pleased by the pleasure of his friends who had heartily canvassed his election. He was also amused by one circumstance, remembering how, when his first biometrical paper was submitted to the Royal Society in 1916, the referees were "rumored to have been Karl Pearson and Reginal Punnett"*:

The Society's action was impeccable: they were two leading lights in statistics and genetics respectively, with the additional advantage, when two referees are appointed, that they were not very likely to agree. In fact, I suspect that the rejection of my paper was the one point in two long lives on which they were ever heartily at one.

*Although as a Fellow of The Royal Society, Fisher could have (and probably had) seen the signed referees' reports, information from this source was the property of the Royal Society, not his to disclose. Even in a private letter to an old friend, Fisher very properly referred to the "rumored" identity of the referees.

And now the author of the paper had been chosen to succeed each pundit in turn.

During the war, of course, he could do no more than lay the groundwork for postwar development of the department of genetics, using the facilities that were available at Whittingehame Lodge, the house and land, and his existent genetic stock: *Lythrum* plants, about 200 cages of mice, and two bitches from the dog-breeding program. Temporarily he asked for half a dozen kennels and dog pens to be constructed on the departmental land for the dog breeding, and he proposed to turn the residence over to use as the genetical laboratory. The University did not provide permanent accommodation for the dogs, since Pease and local residents objected that dogs might become a nuisance to the neighborhood. The two bitches lived out their lives at Whittingehame Lodge, but no breeding was undertaken with them.

Fisher provided some initial staff to the department. Mr. K. Williams, a computing assistant, moved to Cambridge as his secretary, supported by the Rockefeller Foundation, and Mrs. Rose Tysser (later Cashen) as computing assistant, supported by the Agricultural Research Council. ARC funds also contributed to the cost of laboratory assistants hired in Cambridge. And it was to the ARC that Fisher looked to supply another lack in his new department when, in February 1943, he wrote to Professor W. W. C. Topley:

If something isn't done, we shall be absolutely sunk for lack of calculating machines. . . . Obviously the Department of Genetics at Cambridge never even possessed a slide rule. . . . I suppose that one thing that would be legitimate would be for the ARC to requisition three from my department, through the Ministry of Supply or whoever is the official highwayman, and so secure something for the use of Miss Burton, Mr. Williams and myself if we go to Cambridge.

Fisher got his machines, though apparently not from University College.

Fisher did his best to persuade Mrs. Holt to come to Cambridge with him and was quite indignant when she told him it was impossible; for family reasons she must remain in Harpenden. She was still working on the polydactyly factor when the mouse stocks were moved to Cambridge. For the final part of her program she was obliged to let others do the breeding on her instructions, but she visited Cambridge as often as possible, and, like other visitors, slept in one of the two little attic rooms at Whittingehame Lodge which Fisher had reserved for his private use as guest rooms.

Fisher assumed, after the appointment of L. S. Penrose (on his recommendation) as Galton Professor, that Mrs. Holt would be able to resume her work on mouse genetics. He knew that she had plans for several experiments, including some on maternal age effects, a subject in which Penrose was greatly interested. He also had in mind that she could do linkage experiments to aid the work on chromosome mapping, and he was anxious to supply her with the mice she would need. It was, therefore, a great shock to him to find that Penrose proposed to exclude all but human genetical work at the Galton. He tackled Penrose on the subject and pointed out that Holt had done much work on mouse genetics and should be allowed to continue. Penrose had made up his mind, however, and had already gone so far as to let Gruneberg have the use of the animal house for *his* mice. The subject was closed; and, sadly, after this quarrel, Fisher's friendliness to Penrose ceased.

□ □ □

University College was glad to accept his offer to remain editor of the *Annals of Eugenics* until the election of the new Galton Professor. Mrs. Holt agreed to act as his assistant editor, and he was pleased with the arrangement. Indeed, he offered to take over the *Annals,* with a suitable change of name, permanently in the Cambridge department. He had slight hope that University College would be prepared to part with it, though it was published at a loss; but he had fostered it, had given it genetical as well as statistical interest, had attracted new subscribers who appreciated the character he had given it, and now he wished, if possible, to continue it himself. University College, however, decided to keep the journal and, after L. S. Penrose was elected Galton Professor, reclaimed their own. *The Annals* came to reflect the character of its new editor, being wholly devoted to human genetics.

In anticipating the work at Cambridge, Fisher probably hoped to become editor of the *Journal of Genetics. The Journal* had its origins at the same time as the Cambridge Chair of Genetics and had always been closely associated with it. From the first volume in 1911, it was edited by Bateson and Punnett and, after Bateson's death, by Punnett alone, right up to and including Volume 45 in 1943. When Fisher was President of the Genetical Society from 1941 to 1943, the future of the *Journal* was presented for their consideration. In 1941 Haldane was commissioned to approach Punnett with a view to acquiring the journal for the society. The following March, Haldane reported that Punnett was unwilling to have the *Journal* associated with the Genetical Society. With Fisher in the Chair of Genetics in 1943, the offer was about to be repeated when Haldane forestalled them; going directly to Punnett, he managed to buy the *Journal* for himself. This was a very bitter disappointment to Fisher. Both the *Annals* and the *Journal* thereafter were controlled by men at University College and, with Haldane as editor, the *Journal* appeared irregularly with some very long gaps. For the first time in many years Fisher had no editorial functions.

Well before the end of the war the need for a broader base for genetical journals in Britain must have forced itself on Fisher as it did on C. D. Darlington, director of the John Innes Institute. Together they developed the idea of starting a new journal, and together they founded *Heredity,* which first appeared in 1947. They personally provided financial guarantees for the new venture, depositing a considerable loan to the journal until it should become self-supporting. As joint editors for the first 10 years, they did all the work involved in producing a publication of such quality that it established a high reputation among geneticists. By 1957 they were ready to bring in other editors and eventually to hand over to them the reins of a successful enterprise. More recently *Heredity* has been acquired by the Genetical Society of Great Britain.

□ □ □

From 1943 to the end of the war, plans for postwar work were going forward with the Medical and Agricultural Research Councils, which incidentally promised a greatly extended range of usefulness for the genetics department at Cambridge, both in teaching and research. In this connection, we see Fisher's initial optimism that at Cambridge he would be influential in helping to develop more adequate educational opportunities for students of genetics and of statistics. We see also the directions he wished these developments to take, so as to offer what he considered important and neglected educational experiences to students and young research workers in these fields.

First, he counted on augmenting his department as soon as the war was over by the accession to it of the serological unit as a going concern. In his preliminary memorandum to Cambridge University he pointed out that the unit was still in existence, though working under the Medical Research Council for the duration, and that it had been cordially cooperating with the genetical work of the Galton Laboratory. It was actually located in Cambridge, and he suggested that even though it could not be freed for purely genetical work until after the war, a formal association between the university and the unit might be made immediately by appointing Taylor to a university post. It was not only for research he wanted the unit in his department; he suggested it might be asked at once to assist with teaching human genetics. He felt, as he said later, that "a large part of the point of teaching genetics in the University falls away unless students can be brought personally in contact with what is known as human genetics."

The university did not offer Taylor an appointment at that time but agreed to create a readership in serology and human genetics for him after the war. In June 1943, therefore, Fisher wrote cheerfully to O'Brien at the Rockefeller

Foundation, inviting him to have a look at the situation in Cambridge "as affected by recent developments," with a view to renewing Rockefeller Foundation support for the serological work after the war. O'Brien answered that he was confident support would be renewed to the unit as part of the Cambridge department.

In 1943, Fisher was brought into consultation with the ARC in planning to meet national needs for animal improvement after the war. Gray's initial inquiry for the views of academic geneticists on this subject was met by a universal howl for substantial recognition of their subject, at last, so that it might be taught in its integrity at the universities and receive institutional support for research on an adequate scale. Fisher, for instance, commented that the amount of institutional space, time, and expenditure given to genetics in Britain in the past had been so meager that it was somewhat embarrassing to be asked for a program touching applications of research to breeding of animals of economic importance: "It is like being asked for a programme for the development of the chemical industry in a country where the majority of compounds of elements are unknown and laboratories do not exist in which they can be recognized or prepared." Thereafter Fisher was involved in preparing and discussing detailed proposals for building up a competent livestock improvement program.

Before one could set up adequate research centers to undertake the field work, it was necessary to train the geneticists who would staff them. Plans envisaged a score of livestock selection centers; yet the only genetics taught in the School of Agriculture at Cambridge, for example, was a course by Pease on "Stocks and Crops," mainly concerned with chickens and rabbits; it contained nothing on large animals or pedigree breeding and nothing on plant genetics. Fisher offered to take ARC students into his department to study both the genetical and statistical aspects of animal breeding. The first of these, D. S. Falconer, came to him in 1945 and went on to ARC work under C. H. Waddington in 1947. In 1968 he succeeded Waddington as Buchanan Professor and Director of the Institute of Animal Genetics in Edinburgh.

To back such applied work, research and teaching of genetical theory was essential. In the whole country there were only three centers broadly interested in the development both of genetical theory and applications: the Department of Genetics at Cambridge, the John Innes Horticultural Institute, and the Institute of Animal Genetics at Edinburgh. If the research effort was small, the teaching of genetics was pitifully inadequate. At most universities various departments taught their own elementary courses in genetics, and students received incoherent and inconsistent accounts of the subject. The only department of genetics was at Cambridge, and even there it was not represented separately for examination purposes at an undergraduate level. D. G. Catcheside taught the subject in the botany department, and Fisher gave a course that was allotted two questions in the zoology examination for Part I of the Natural Sciences Tripos.

In 1945, therefore, with Professor Gray of zoology, Fisher attempted to get genetics and cytology recognized as a half-subject for examination in Part I of the Natural Sciences Tripos. In fact, he achieved no reorganization of Part I during his professorship, although a Part II course, for third year students, was introduced in 1952. The Faculty Board required that Genetics Part II students should do an ancillary paper in a so-called "experimental" subject and, since statistics somehow qualified as experimental, several of Fisher's students did the paper in statistics. Still, the development of an adequate teaching program was inhibited by this extra requirement.

Changes were achieved only after J. M. Thoday succeeded to the Chair in 1959. One of the first changes made was abolition of the ancillary paper in Part II. In 1964 rationalization of the undergraduate teaching was begun and provided, instead of the traditional courses on botany and zoology in the first year, two courses (on the biology of cells and the biology of organisms) in which genetics is integrated. In 1972–1973 similar changes were made in second-year biology courses.

Fisher had his own ideas about the education of the research geneticist. He considered the physical handling of living material to be a very important part of the training, and he expected his students to spend considerable time in routine examination of mice and *Lythrum*. When, in 1948, Dr. A. R. G. Owen was appointed lecturer to the genetics department, Fisher sought the collaboration of the zoology department in the hope that a variety of genetical material might be developed in that department for demonstration purposes and facilities provided for special projects in zoological genetics, not on the much-studied *Drosophila* only but on protozoa and, if students wished, on such groups as isopods, snails, and butterflies.

In 1952, in discussion with Von Verschuer, he listed three ingredients in the education of the quantitative geneticist. First:

I am beginning to feel sure that personal experience with the care of animals is of much greater importance than it has been thought to be and that a postgraduate year or two on some comparatively simple research problem in pure or classical or formal genetics is of very great educative value.

The second ingredient mentioned by Fisher qualified the meaning of the "quantitative geneticist," whose education he was considering. In contrast to geneticists whose orientation made the physical basis of inheritance the focal point of their researches, through cytology or, later, molecular biology, Fisher was a Mendelian, and his researches were guided primarily by the numerical relationships found in his genetical material. He wrote:

For young men or women of any mathematical ability, there is a great deal to be said for giving at least a year to training in the applied field of statistics. This applied field may, of course, be one of the aspects of genetics but the aim here should be not lectures in remotely abstract mathematical statistics but the study of the quantitative methods used in some chosen branch of scientific literature where statistical methods of some sort, often quite inadequate, have been employed.

The third requirement, physical experience of the techniques of a chosen genetic speciality, was no less practical than the first. Von Verschuer was doing serology and Fisher wrote:

Of course, in serological research work in human genetics, the students cannot spend too much time in the serological laboratory or visiting the families of subjects whose blood is required on a research project.

Feeling that the mathematical needs of students of biology had been neglected, Fisher was enthusiastic when Gray suggested, as soon as he came to Cambridge, that they should offer a course in applied statistics for examination in the mathematics paper of the Natural Sciences Tripos. He made detailed suggestions what should be included in the curriculum for a course in elementary statistics and combinatorial theory and sent it to Gray, with the comment:

Such a course will need a bit of framing. I am sure it should not be elementary statistics only but that a number of ideas of pure mathematics, for example, partitions, groups and matrices should be made familiar to mathematically-minded students taking biology, if only we can persuade the mathematicians to treat them lightly enough. The main point is gained if the student is put in a position not to be paralysed by the mere mention of such things but . . . feels that they are inherently rational and manageable and that if he encounters them he will be in a position to find out, at need, what to do with them.

Despite Gray's enthusiastic support for the course, the proposal came to nothing.

By improving the university courses, Fisher hoped, of course, to give the country research scientists better equipped for their task in life. Such projects as the animal improvement program of the ARC could only be developed as competent men became available, and he thought it would take 15 years to build up the staff the animal breeding program would need. At the same time, he foresaw that the effort might be wasted for lack of adequate statistical controls. He suggested, therefore, that a statistical unit, independent of the livestock centers, should undertake the statistical aspects of the work, providing objective analysis of the results of improvement programs, developing the statistical techniques required, and receiving from Livestock Centers data relevant to their own inquiries as well as acting as statistical consultants in collaboration with work of the Centres.

He was already working for the ARC in connection with veterinary research and knew there were difficult problems for statistical research in such subjects as biological assay of accessory food factors, virulence, and vaccine-potency. He suggested that the computing assistants he already had could form a nucleus for development of an ARC statistical unit equipped to deal with further work in biological assay and with the animal improvement program. He hoped the unit might also undertake statistical researches in the field of agricultural meteorology. He mentioned as a fourth possibility an area peculiarly his own, statistical work in ecology and population genetics.

It was clear that Fisher would not have time to direct the project himself, and he wanted to secure a director from an applied field. After consultation with Sir Gilbert Walker, the famous meteorologist, he approached Sir Charles Normand, who, like Sir Gilbert, was a former Director General of the Observatories in India and not uninterested in agricultural meteorology. Fisher succeeded in interesting him in becoming director of the unit *pro tem* in his retirement.

The proposed ARC statistical unit was Fisher's brainchild and although he was not officially to be associated with it, he offered to help in any way he could as an unofficial adviser, and he wanted to keep in close contact with the work, especially in its early days. In 1945 he suggested he could find room for it temporarily at Whittingehame Lodge and that accommodation would soon become available on the Rockefeller Field, a few minutes walk from Whittingehame Lodge. Eventually, he hoped, the university would build on the plot of Punnett's land next door to Whittingehame Lodge, which Punnett had offered the university at the extremely modest price he had paid for it before the World War I. The university had not leaped at this offer as Fisher wished, but he was still hopeful that a new genetics department might be established there, properly equipped for its work; there also the ARC statistical unit could be housed, and there they could share common interests and problems at the same tea table.

Naturally, Fisher envisaged the statistical unit as a valuable complement of academic training in mathematical statistics at Cambridge, where teaching could be integrated with project work. He had in mind a staff eventually of seven or eight first-class statisticians, experts in their different fields but enjoying stimulating opportunities for mutual discussion and assistance. Moreover, he proposed that at any one time one-third of the workers would be made available to work temporarily at field stations. Thus it would afford the types of experience and intellectual contacts needed to mature the abilities of those who had so far only academic training; it would develop their versatility and resourcefulness in handling statistical problems and it would build a competent acquaintance with the scientific problems of veterinary or genetic or meteorological research.

After the diploma in mathematical statistics at Cambridge was offered, to be awarded in recognition of 1 year's work in statistics for graduates in mathematics, Fisher expressed the opinion in 1949 that the regulations would only work well for those students who had already taken mathematical statistics to Part III and who gave their year's work principally to understanding an *applied* field. It seemed to him the attempt by means of a single-year course organized by the faculty of mathematics must necessarily fail to supply the real demand for statistical training:

There is no wide or urgent demand for people who will define methods of proof in set theory in the name of improving mathematical statistics. There is a widespread and urgent demand for mathematicians who understand that branch of mathematics known as theoretical statistics but who are capable also of recognizing situations in the real world to which such mathematics is applicable.

Some years later, being consulted about a proposal to set up a statistics center serving various research institutes in Scotland, in coordination with a university teaching department, he recalled the plans of 1944–1945 and wrote:

I was approached nearly eight years ago by the Agricultural Research Council on a project to establish, in various centres, departments capable of both teaching and advisory work in statistics. Naturally I am very strongly in favour of such a move though it certainly is full of difficulties. . . . I have no hesitation in advising that such a centre as you have under discussion should plan to integrate teaching closely with project work in which practical experience can be gained by those who are capable of learning from it, in contradistinction to the ruinous process of segregating the keener minds into a completely sterile atmosphere.

For various reasons, all the wartime plans were abandoned or modified out of recognition. The university never intended to build a genetics department in Storey's Way, and the department continued to be housed in the professor's residence. The plans for Livestock Selection Centres dwindled to the foundation in 1945 of a single ARC unit for Animal Breeding Research under C. H. Waddington at the Institute of Animal Genetics in Edinburgh, with its own statistician. Fisher's plans for a statistical unit had evolved in close collaboration with W. W. C. Topley, and built on his support. Topley died in January 1944. The unit was not formed, and ARC workers with Fisher were taken on to his staff or dispersed in the later 1940s. Fisher himself was consulted about ARC animal breeding work again after Lord Rothschild became head of ARC in 1948. He was also much interested in the work of the small statistical unit formed in 1956 to serve ARC workers in the Cambridge area. As for academic statistics, students for the diploma in mathematical statistics were required to do a little applied work but, in practice, they spent the year almost wholly in preparation for the Part III examination.

The greatest blow to Fisher was his failure to regain the serological unit after the war. Plans with the university had to be renegotiated after the death of Dr. Taylor in March 1945. At the same time, the MRC had little interest in maintaining the unit in order to hand it back to its proper work, but rather wanted to take over its personnel piecemeal, for various purposes, as soon as the transfusion work ceased. Fisher immediately wrote to propose that a medically qualified assistant should be appointed to enable Race to carry on the work without a break; he suggested that Rockefeller Foundation funds allotted to the serological unit before the war might be used to pay the salary, thus recognizing the foundation's continuing interest in the research unit. The MRC responded favorably to the first part of the request, and Dr. A. E. Mourant, from the blood transfusion service at Luton, was appointed medical officer in the Galton Laboratory serum unit.

While the transfusion work was drawing to a close, Fisher was trying to get the university to offer Race the position of assistant director of research in his department and Mourant that of assistant in research. Despite their commitment to Fisher, the university were inhibited by the fact that the MRC was in a position to shut off financial support for the unit in order to force acceptance of their own plans; the university temporized and never formally made an offer to Race. Meanwhile, the MRC strongly presented its alternative plans to Race and offered him much more, both immediately and in the long run, than could be hoped of the university. In January 1946 Race accepted the post of director of a new MRC blood group unit at the Lister Institute in Chelsea. Six months later, Mourant accepted a post at the Lister as director of the new MRC blood group reference library.

□ □ □

Disappointed of his hopes for serology, Fisher was the more eager to broaden the range of research in his department in other directions. A new line of genetical research was opened up at this time by the discoveries of the mid-1940s which mark the beginning of modern bacterial genetics. By early 1948 Fisher was planning to introduce research in bacterial genetics at Cambridge.

After pure bacterial cultures were successfully prepared in the later nineteenth century it had been recognized that bacteria share the attribute of heredity with higher organisms. Nevertheless, bacterial cultures grown from a single cell were notoriously changeable; when exposed to adverse conditions, pure cultures quickly gave rise to genetically stable strains adapted to the new conditions, and this apparent plasticity had been generally attributed to the effects of the environment, not to mutation. In 1943, however, Luria and Delbruck [82] published results that demonstrated the occurrence of spontaneous mutation, thus revealing that particulate genes in bacteria were entirely analogous with those of higher organisms. In 1946 Lederberg and Tatum [83] presented their discovery of the transfer of chromosomal material from one bacterial cell to another by direct cell-to-cell contact: some form of conjugation took place which gave rise to new combinations of the genes. Nothing was then known about the sexuality of bacteria, the shape of the bacterial chromosome, the manner of transference of chromosomal material, or its incorporation in the receiving cell, but, these problems aside, one could envisage the investigation of genetic linkages and chromosome mapping of bacteria by recombination studies and breeding programs similar in nature, though not in technique, to those with higher organisms.

Fisher knew nothing of the techniques of bacterial culture, and, as with the blood group work in 1934, his first step was to seek an expert capable of handling current techniques and developing methods suitable to the investigation of this particular genetic material. He considered the possibility of training a bacteriologist for the purpose, letting him spend a year with Lederberg, as Taylor had spent a year with Friedenreich in the 1930s, and to see something also of the pure culture work of Thornton at Rothamsted and of Winge in Denmark. As it happened, however, a man with the very qualifications he wanted, Dr. L. L. Cavalli (later Cavalli Sforza) was already making his way toward Fisher.

Fisher had met his name already in 1946, when Race wrote how impressed he was by the charming young aristocrat whom he met in Milan. Cavalli had introduced himself as a bacteriologist keenly interested in genetics and had mentioned that he was then grappling with the study of Fisher's statistical books. He was hoping soon to come to England, and Race was sure Fisher would want to meet him there. Fisher had also seen something of Cavalli's work when, as one of the editors of *Heredity,* he received an article by L. L. Cavalli and G. Magni [84]; the paper was published in the first issue of *Heredity* in 1947. In it the authors made application of simple statistical methods appropriate to the bacteriological material, so as to increase the precision of results and at the same time gain additional genetical information from them. It was just the sort of work Fisher appreciated most.

Cavalli realized his plan to come to England in March 1948, when he joined Mather in the genetics department at the John Innes Institute for a period of 6 months. In July he attended the 8th International Congress of Genetics at Stockholm. There he met Fisher for the first time and recalls that "to my great surprise, after five or ten minutes he offered me to take a job in his department at Cambridge. This was a wonderful surprise and I accepted immediately." In August he visited Cambridge, looking for a place to live with his wife and small son. In October he was officially appointed assistant in research in the department of genetics, Fisher having failed to get him the more senior position he had expected to be able to offer.

As it had been with the blood group work, so it was with bacteriology; the laboratory had to be set up and the first fumbling approaches made to research as the complexities and difficulties of genetic interpretation revealed themselves. Understanding would come years later, as it had come with blood group work. Whittingehame Lodge was, of course, quite unsuitable for making a bacteriological laboratory, but the basic equipment was bought, and Cavalli made a place for himself in part of the original laboratory room. A pure strain of *E. coli* was obtained, and 5 or 6 months after his arrival, Cavalli could start work. One of Fisher's activities in the United States in the summer of 1949 was the acquisition of the costly chemicals required, which could not at that time be obtained in England.

To Cavalli, it seemed remarkable how interested Fisher was in work of which he knew nothing from a technical point of view and how much insight he had into its genetic aspects. The bacterial chromosome appeared rodlike in current cytological studies; yet, rather early, in discussing the shape suggested by genetical reasoning, Fisher illustrated it as a large loop or inversion of most of the chromosome strand, with short tails. It was not until 1957 that Jacob and Wollman established that the bacterial chromosome is actually a ring.

The first clue about the nature of conjugation was the discovery in 1952 that chromosomal transfer requires the presence in the donor cell of an autonomous genetic element, the sex factor F (for fertility). The discovery of F made possible the recognition of two mating types, F^+ harboring F and acting as genetic donors and F^- lacking F and behaving as genetic recipients. When a population of F^+ cells is mixed with F^- cells, however, only about one in 10^4 donors actually transfers chromosomal material to recipients. For this reason the discovery of the mating process resulted only after the discovery of certain strains with a high frequency recombination (Hfr strains) which produce cells in which up to 100% of the F^+ cells are donors.

The latter discovery occurred in Cambridge in 1949. Using nitrogen mustard to produce mutations in his culture, Cavalli was thrilled by the appearance of a mutant strain with a much higher frequency of recombination than was ever observed before, which he called the Hfr strain. (A similar Hfr strain appeared spontaneously in the culture of W. Hayes 2 or 3 years later, and it is in this strain mainly that the mode of bacterial conjugation has been studied.) Fisher's reaction to the discovery was to suggest that a virus-like particle might determine sex in bacteria. At the time no evidence existed to support the hypothesis, but, Cavalli recalls, "Later, after I left Cambridge, it happened that three people, Josh Lederberg, William Hayes and myself, practically at the same time did discover indeed that there was a virus which is called F which determines sex in bacteria. In the strain I had found, the Hfr, that virus is anchored to the bacterial chromosome or, in fact, you can say becomes part of it."

If the F particle is free in the cell, it is only potentially a donor. When incorporated in the chromosome, however, it was found to be responsible for breaking the ring of the donor chromosome at its point of attachment, so that the chromosome becomes transferable. The orderly mode of transfer was discovered in the mid-1950s particularly by E. L. Wollman and F. Jacob [85] at the Institut Pasteur in Paris and W. Hayes in London. Once the theoretical basis of conjugation was clarified, many of the early difficulties of interpretation fell away.

These discoveries were years away when Cavalli was working in Cambridge, a one-man unit of bacterial genetics in a primitive genetical laboratory. Despite the excitement of his experimental results, Cavalli felt his isolation acutely. Not only was he in a genetical rather than a bacteriological laboratory, but he found it impossible even to talk with bacteriologists in England; because bacterial genetics was so new a subject, he was considered a crank by bacteriologists, as well as by most geneticists. He could talk to Fisher, but Fisher was no bacteriologist and could not share his technical problems and interests. Now that bacterial genetics has developed so far and become so successful, it is easy to forget the isolation of the pioneer in this or any other really new line of investigation.

Fisher did his best to make the Cavalli family welcome. From time to time he took them out to the best restaurants in Cambridge and, when Cavalli dined with him in College, was grateful to call him into his rooms afterward for a glass of port and an hour of conversation, and to regale him with new thoughts on geology, magnetism, and astronomy. Geomagnetism, like bacterial genetics, was a new subject and not yet academically respectable. Fisher was excited by the possibilities of both and very willing to give his moral support, with his interest, to the research workers in both.

In 1949 G. Maccacaro joined the bacteriological genetics unit, supported by a Rockefeller Foundation grant, and in January 1950 Cynthia Knight (later Booth) came on an international scholarship of the Rockefeller Foundation. Cavalli's appointment was with the University, however, and in the summer of 1950 Fisher learned that this appointment would not be renewed in the next academic year. His arguments, calling on the university to honor its promises to his department, rehearsing the special qualifications of Cavalli, expanding on their great good fortune in having attracted such a man and the luster his work in the new and very important branch of genetics must in later years cast on the university, these went for nothing. He returned from committee meetings red in the face, unable to give adequate expression to his feelings of outrage. It had been bad enough to receive Cavalli with such slight recognition in 1948 and to have failed to promote him; but to throw him out when the work was well begun was insupportable: insulting to Cavalli, humiliating to himself (who was made twice a liar thereby), and disastrous for the bacteriological work.

Fortunately, perhaps, Cavalli was offered an attractive position in Italy at this time. Jobs were scarce in Italy and, wishing to settle there sometime and eager to talk with his bacteriological colleagues again in Milan and Pavia, he took this opportunity and in October 1950 departed on unpaid leave from the University of Cambridge. Crippled by lack of university support, the bacteriological research in the department of genetics continued while Maccacaro remained at Cambridge.

In 1951 F. H. C. Crick and James D. Watson started working together at the Cavendish Laboratory in Cambridge. In 1953 their famous paper appeared showing the structure of the DNA molecule, the basic genetic material, in the form of a double helix. Meanwhile S. Brenner joined the Cavendish group and several times wrote to Cavalli Sforza inviting him to join them. It was too late. Regretfully, Cavalli realized that if he had stayed only a year or two longer in Cambridge, he would have found himself close to the fellow-scientists he needed, as Cambridge became the center of expansion of molecular biology. Instead, he began cooperative work with J. Lederberg, with such notable early results as their discovery of the sex factor in bacteria (1952) and their isolation of preadaptive mutants by sib-selection, a method of sampling subject to precise quantitative formulation (1955).

□ □ □

By 1950, seven years after his appointment to the Chair. Fisher knew that he could hope for little university support for any new or ambitious enterprise in his department. The department would continue to be housed in Whittingehame Lodge, and his work on mice and on *Lythrum* would continue to be the primary research in the department. He was probably glad to remain in Storey's Way, for he loved the garden, and having found a gardener who would bring his ideas for the garden to fruition, he found constant delight and stimulus in his perambulations of the area.

When he first moved into Whittingehame Lodge in May 1944, the departmental garden was derelict, the paths almost ungraveled, and the bricks edging them crumbled away, the hedges either needing replacement of many dead plants or else wildly overgrown (the hedge of thuya was about 10 feet thick and it completely obliterated pathways alongside). The ploughed areas were infested with convolvulus. Despite its state of neglect what remained of the original plantings near the house was still lovely in its spring array. Cherries bloomed within view of the window of his office, formerly the drawing room. Ornamental crab apples ran southward, whose deep pink flowers glowed against the dark red of their early foliage, a striking contrast against the dusty gray-green beneath of a bank of huge rosemary bushes, heavy with fragrance. Beyond these, eastward, Punnett's orchard opened up, the old trees spangled with pale blossoms, and at their foot a broad wave of pale mauve iris flowers broke at the orchard's edge and surged uncontrolled across the unmarked boundary between Punnett's property and the departmental land. From the first Fisher liked to stroll round the garden or sit in the orchard to talk with his visitors.

During the war, with the aid of an old and rather feeble individual, hired as part-time gardener, the reclamation was slow. Fisher started work on one particular bed, and, removing two spits of earth to the other side of the field, he dug into the third, the untouched white clay subsoil of Cambridge gault. He felt double trenching was necessary to get rid of the convolvulus, for the further he penetrated, the thicker grew the food-storage roots of the weed. Wrestling with the resistant gault, he dragged out great knuckled roots, triumphing that the convolvulus could never recover. In this struggle, he broke the tines of three garden forks, the heaviest forks he could obtain, before succumbing himself to a nasty bout of sciatica. He never completed the job and the professional gardener he was fortunate enough to find in 1947, Mr. G. Harding (head gardener during the rest of his professorship), was grateful that the work had gone no further, for the intermixture of gault made the topsoil intractable for years afterwards.

Fisher was an amateur, but like any true gardener, he nourished his whole garden by his daily observation and care for it, his enjoyment of its beauties, his interest in its progress, and his plans for its future. It was to be years before it was brought under control, but with Harding's knowledgeable and sympathetic help, Fisher's ideas were gradually realized. He wanted adequate compost heaps, and Harding saw that adequate compost heaps were properly prepared and maintained. He wanted a herringbone pattern for a bricked garden path, and Harding saw that the path was laid, even to turning through a right-angle while keeping the pattern intact, exactly as Fisher planned it. He admired blue Siberian poppies in his sister's garden in Argyllshire and was sent some seed. Mr. Harding removed the whole topsoil alongside the library building, and replaced it with peaty soil, acid enough to produce a brilliant display of the poppies. He wanted the field he had first dug to be laid out in small plots for experimental plants and it was done according to his specifications; the result was quite lovely and entirely practical.

From the office he could walk directly into the garden. A broad strip of lawn now ran between the herbaceous borders, through the center of what had been a continuous hedge of cotoneaster to the experimental layout. Just beyond the hedge a broad swathe of lawn was laid under the line of cherry trees, and there garden seats were placed in the shade, from which he would contemplate the experimental beds and the espalier of fruit trees beyond. Broad grass paths surrounded the area and divided it in one direction into five long, narrow bedding areas. However, the parallel arrangement was not apparent to the eye; for esthetic as well as practical reasons Fisher had some of the beds divided across their length, into either two or four sections, and the grass paths thus made an irregular little maze around plots of unequal size. Centrally, on a graceful pedeistal, stood the rain gauge, with Greek capitals carved around the pediment proclaiming "Water is best." (Fisher felt that there should be a sundial to challenge this claim with the assertion, "I count only the sunny hours," but he did not set up a sundial.)

He cherished the trees in the garden: the fine *Robinia pseudacacia* on the lawn, in whose shade he liked to have tea parties and play bowls in the summer; the ornamental cherries and apples; the gnarled old mulberry in the corner; the graceful almond at the back gate. He entangled himself in a long argument with the university after he discovered the foundations of the library building had been laid too close to a pear tree of which he was fond. (Having just discovered the mistake, he was in an extremely excited temper at the moment when the distinguished American geneticist E. B. Lewis was shown in; his visitor got a very poor reception.) He insisted that the foundations should be relaid elsewhere. He got his way, but it took an interminable time to get the library finished, and a good deal of rancorous negotiation.

In the course of time he planted many fruit trees: his favorite apple (the Cox's Orange Pippin), greengage and Victoria plum, apricot and early peach, and a Comice pear trained against the wall of the new library building. Leaning back on his stick to admire the young tree, he would quote: "He who plants pears plants for his heirs." In all this he acted as if the future of the garden at Whittingehame Lodge was assured, although it was always in doubt. In fact, within a few years of his retirement, the garden was to be swept away to make room for staff apartments of the new Churchill College. Churchill itself was being built "over the back fence," on adjoining property of St. John's farm.

While Fisher was professor, he enjoyed the garden in all its aspects, and it was a particular pleasure to him when his staff and his visitors shared his enjoyment. He was a happy host on the occasion of the 100th Meeting of the Genetical Society of Great Britain, on the afternoon when members were invited to Whittingehame Lodge and toured the summer garden in relaxed mood, surveying the open pollination plot of *Lythrum* plants, inspecting the exhibition plot of garden peas segregating in four plant characters and two seed characters used by Mendel, strolling among the experimental plots (Plate 18) to observe the three style types of *Oxalis,* visiting the mouse rooms and bacterial work indoors and the chickens of Pease next door, and afterward taking tea together in the garden. The 100th meeting coincided with the 30th anniversary of the founding of the Genetical Society by William Bateson in 1919. Punnett came up from retirement, and the past and present professors of genetics celebrated together at the official dinner of the society, held at their common college.

This meeting was of the type Fisher always thought the most profitable. There were papers read during the mornings (on one morning six of the Cambridge workers spoke briefly of their current work: N. T. J. Bailey, L. L. Cavalli, R. A. Fisher, H. Heslot, A. R. G. Owen, and M. E. Wallace), but the emphasis was on the extensive demonstrations and exhibits offered during the three afternoons, including the one at Whittingehame Lodge. One recalls Fisher's comment on the International Genetical Conf ress at Ithaca in 1932, "hot but delightful due to the excellent exhibits and pe sonal discussions." At the many conferences he attended, he used the time less to listen to formal lectures than to enter into discussion with participants. From time to time he urged the British Association to give more time at their meetings to exhibits, demonstrations, and discussion, feeling that papers could be read in the literature, whereas meetings provided a unique opportunity to see and to talk over the actual material of current research.

It was not all the same thing to read about coat colors in mice as to see skins such as Mather and North had exhibited at the Edinburgh Genetical Congress in 1939, showing the effect of the *umbrous* modifier on *agouti* mice. In the case of manifestation of the *yellow* gene in the *agouti* series, Fisher had found marked discrepancies between his observations on his own stocks and the accounts he read in the literature. The most obvious explanation appeared to be that in some laboratories unidentified modifying genes disturbed the expression of the gene of interest. A uniform genetic background was required against which to exhibit the single gene without disturbance due to others. Before he moved to Cambridge, therefore, he began to breed *yellow* mice with a view to making a museum exhibit of their manifestation on a uniform genetic background. He identified one darkening gene common in the wild, which could explain some of the current misleading descriptions, and a different darkening gene in his own stocks. Eventually he built up an exhibit showing fine gradations from pale clear yellow to a dirty brown in a two-way table of *yellow* mouse skins modified by these factors.

This study, like others involving modifying genes, pointed to the importance of getting a uniform genetic background against which to display factors of interest. When he accepted the Cambridge Chair, therefore, Fisher decided to put into practice what he had long felt needed doing, namely "the creation of permanent inbred lines in all, or as near as makes no matter, of the genes recognizable in mice." By doing it on a comprehensive scale, he hoped that the difficulties might be adequately explored. In November 1943, having just received from George D. Snell at Bar Harbor two of the new mutations he wished to introduce in his inbred lines, he argued the advantages of having such segregating inbred lines:

If one has single-factor manifestations without disturbance due to others such as ruin the value of many specimens used for demonstration or museum exhibition, they can be used to illustrate all points of interest such as factor interaction or linkages; they supply permanent standard material for qualitative study and the means of improved standard genotypes in mice used in human and veterinary medicine.

In succeeding years he continued to collect new genes, wherever possible, and gradually he developed more than 20 inbred lines, each segregating in five or six genes.

The work on chromosome mapping at Bar Harbor paralleled mapping work at Cambridge. The departments swopped news and mice bearing new genes. Indeed, in 1948 Snell was troubled by the amount of duplication of work and wondered how it might be reduced. Fisher responded that he regarded it as an advantage to their research:

What administrative officials call overlapping is in scientific research very like the replication which all who wish for well established conclusions must conscientiously strive for. What I suggest therefore is that when you have a possible linkage in view, either through thinking that the linkage may exist or needing to assess the strength of the evidence against its existence, you should let me know and I will see if my department has yet accumulated any data sufficiently helpful.

Fisher was to do the same. This loose form of collaboration continued between them and proved quite helpful in a period of intensive work in identifying new genes in the house mouse and establishing various linkage groups.

When in 1951 Fisher again discounted the need to prevent overlapping of work in his department and Snell's, he expressed awareness of the need for cooperation in a new way. For the time being he felt that laboratories with mice would be overloaded with linkage testing, which was approaching its peak load. Rather more than half of the linkage groups were now marked, but there was a great deal to be done before the mouse geneticists would have three or four available markers in each group so that the placing of new mutations would be tolerably easy. If such lines as he had built up were maintained, segregating cleanly in a number of factors, the task of filling in the chromosome maps would be achieved without the really extravagant waste of repeating in every laboratory the laborious preparatory development of elite stock. But he was seriously concerned that steps should be taken to preserve such elite stocks as existed:

Compulsory retirement in three or four years time will jeopardise the existence of the stocks I have built up, since there is no probability that my successor will even wish to keep mice.* We badly need international cooperation on this point, though, of course, nations and individual universities should also do what they can to minimize the wastage.

Work on inbred lines led to consideration of the theory of inbreeding. Fisher's lectures and further thoughts on the subject were incorporated in a book on *The Theory of Inbreeding,* published in 1949, which is remarkable for being peculiarly his own formulation without reference to, or comparison with, what others had published on the subject.

*The work on mice was continued by Dr. M. E. Wallace, Fisher's first genetic assistant at Cambridge in 1945. She became director of research under Fisher's successor, J. M. Thoday.

16
The Biometrical Movement

Peace brought honors to Fisher. In March 1946 he was offered an honorary LL.D. by Glasgow University. "I cannot say why, of course, but suspect friend Riddell of being at the bottom of the business." When the degree was presented in June 1947, he had the pleasure of staying at the home of Professor W. J. B. Riddell and of meeting his old Rothamsted colleague, Hugh Nicol, again. In April 1946 he accepted the offer of an honorary D.Sc. from London University, to be presented on Founder's Day in November that year. In April he was also informed that the Guy Medal in gold had been awarded him by the Royal Statistical Society, to be presented on November 27. In June he was elected foreign member of the Royal Swedish Academy of Agriculture; in July he was made an honorary member of the Swedish Seed Association of Svalof at their 60th Anniversary celebration and, crowning these Swedish honors, in October he was matriculated among the Foreign Members of the Royal Swedish Academy of Sciences.

In the same year he learned that Sir Charles Sherrington wished to propose his name for a vacancy in the Pontifical Academy of the Vatican. The proposal to include his name for candidacy was a flattering one. The Pontifical Academy consists of a maximum of 70 members, some of the most celebrated scientists of the world. (Sir Charles Sherrington and Lord Rutherford had been founding English members at its reorganisation in

418

1935.) When vacancies occur, they are not filled automatically. When Fisher was actually elected in 1961, there were only about 50 members.

During a visit to India in 1947–1948, Fisher received an honorary LL.D from the University of Calcutta. In 1948, he was awarded the Darwin Medal of the Royal Society in recognition of his work on the genetical theory of natural selection, and he was elected Foreign Associate of the U. S. Academy of Sciences. In 1950, he was elected Foreign Member of the Royal Danish Academy of Sciences, Honorary Member of the International Statistical Institute, and Honorary Member of the International Society of Haematology.

He had been honorary member of the American Statistical Association (A.S.A.) since 1930, and in 1950 the A.S.A., meeting in Chicago, celebrated the 25th anniversary of the publication of *Statistical Methods for Research Workers* at a special session with speeches and articles for the anniversary volume by F. Yates, J. Youden, H. Hotelling, and W. G. Cochran. It was hoped that Fisher would be there for that occasion and to receive an honorary D.Sc. from the University of Chicago, but he could not arrange it, and the degree was presented when he next visited Chicago in 1952 (Plate 22). Publication of his *Contributions to Mathematical Statistics* was planned to coincide with these celebrations, and the book was actually published in 1951.

Fisher accepted honors as they came, with a full sense of the honor conferred. He liked to be appreciated, to know that his work had made a difference and was felt to deserve honor, and he was openly and sincerely pleased by the awards lavished on him by the scientific fraternity. In fact, he kindled even to flattery, and he knew it. After saying good-bye to an extravagantly appreciative visitor, he once remarked, "He overdoes it grossly but still I feel warmed by his flattery." He liked to reflect upon the giver the honor he received by wearing it openly. As he dressed for Evensong, which he regularly attended in the college chapel, he would select the hood, perhaps, of the University of Chicago to wear over his gown, and thus honor that great university. Similarly, to him it seemed ungrateful to seclude his medals in a bank, and he wondered how he could acknowledge them suitably. Eventually he had a glass-fronted box made to house them securely for display in his rooms. The large, gold medals spoke for themselves—the Royal Medal, the Guy Medal, and, finally, the Copley Medal, the highest award of the Royal Society. Fisher expressed more interest in the Darwin Medal, which has a fine head of Darwin on its face, or the smaller bronze Weldon Medal, and he picked out the little Neeld Medal he had won at Harrow, with affectionate pride.

It was good to be honored among scientists but especially good when his country recognized his services, and in 1952, in the first Honours List of the reign of Queen Elizabeth II, his name appeared and he was made Knight

Bachelor. He was pleased as a scientist, for although many civil servants and politicians receive such honors, scientists are selected more rarely. He was romantically touched too: a chivalrous enthusiasm revealed itself when he spoke of the youth and beauty of the girl who was his sovereign, a protectiveness both boyish and paternal. When the day came and Fisher advanced— looking extremely smart and distinguished, his hair snowy white against the black of his morning suit, his figure more noticeably erect than ever, head up and beard jutting forward—to kneel before the queen, he as surely pledged his manhood and wisdom personally to her service and defence as any medieval youth could have dedicated his sword at once to his monarch and his lady. The gratifying emotions of Fisher on that occasion were those of the same youthful enthusiast who had sought the honor of patriotic service in 1914 and, in fact, ever after, and was now made proud by the acceptance of his services.

□ □ □

With the peace came the renewal of contact with European scientists and international exchange of scientific news between colleagues long isolated from each other. For the next few years the remaking of international ties, and the formation and reformation of international organizations was an important part of the work of the scientific community and especially for those who, like Fisher, were senior members of their profession and involved in work like his, which had developed much and taken new directions in the interim.

As soon as the war was over in Europe, Fisher wrote to discover what had become of F. von Verscheur in Germany and C. Gini in Italy in order to renew their cordial prewar relations. He was able to befriend von Verscheur when, in the aftermath of war, he was unemployed and almost destitute. The Genetical Society of Great Britain was prompt to arrange a reunion meeting in London in 1945. Ten distinguished geneticists from allied nations on the Continent were invited to speak at these meetings though the Russians did not arrive. Darlington and Haldane summarized British work and H. J. Muller, doyen of American geneticists, presented the Pilgrim Trust Lecture of the Royal Society. Visits were arranged for the guests to centers of genetical study in Great Britain and Fisher was glad to see A. L. Hagedoorn from Holland at Cambridge, to show his chinchilla mice for comparison with those of Hagedoorn, and to arrange the exchange of genetical stock. It was a great pleasure to see O. Winge from Denmark and to show him the blood group work and to discuss with him its problems and potentialities.

Wartime discoveries in serology were, indeed, a focus of excitement and controversy. The controversy centered on the nature and thus the terminology to be adopted to describe the genetical nature of the Rh system; leading geneticists were involved in the discussion of the various national committees on genetical nomenclature. More generally, the importance attached to serological discoveries is reflected in the number of new journals produced soon after the war: the British journal *Blood* (1945), French *Revue Hematologique* (1945), Danish *Blood Group News* (1948), Swiss *Acta Haemologica* (1949) and *American Journal of Human Genetics* (1949).

The implications of the serological discoveries were not only revolutionary clinically or semantically; Fisher already saw their implications as affecting the whole study of man. He expressed his vision at the 1st Congress of the International Society of Human Genetics, in 1956 [*C P 266*]:

> Whenever genuine progress, in the great sense, is achieved, fields of study hitherto disparate and unconnected are suddenly seen to be inseparably linked, and to fall into place as aspects, distinguishable only by differences of training and technique of a single grand science, or field of study.
>
> At the present time, indeed, the Sciences of Demography and Vital Statistics, of classical Biometry, and of traditional Ethnography in its physical aspects are pursued as if they were independent disciplines. They are, none the less, irresistably destined to be merged in a Science so far nameless under the catalytic action of the new knowledge of the blood groups.

The serologists could offer independent scientific evidence of a new kind on the questions he had interested himself in as a eugenist for so many years. Using only the ABO groups early in the war they had found it possible to characterize the race mixture of Great Britain. By 1945, using six or more blood group polymorphisms, they could discriminate between human populations with some refinement by means of the relative frequencies of the alleles. A serological profile of various races would illuminate the history of racial migration, conquest, and intermarriage and might indicate selective effects due perhaps to local geography or indigenous disease.

A. E. Mourant, who as director of the Blood Group Reference Library was charged with the task of recording blood group data obtained from all parts of the world, quickly took the initiative in exploring the ethnographic significance of blood groups. Fisher was fascinated by his findings among the Basque people of the Western Pyrennees. These are an ancient people, never Romanized, with a distinct culture and a language that does not belong to the Indo-European family; they might be the last of the ancient Europeans perhaps, driven to the mountains of the western extremity of the Continent by the waves of invasion from Asia. In their blood groups the Basques were revealed to be hyper-European in a variety of different respects. In particular, frequencies of group B, high in Eastern Europe, with a gradient falling to low scores in Western Europe, found a new dramatic low in the Basque samples. This was associated with very high frequencies of Rh-negative (*dd*) individuals. These findings led Mourant to suggest that the European population should be composed of a proto-Basque Rh-negative component with little or no B and an Eastern Rh-positive component with much B, a hypothesis that later data have confirmed.

Developments in serology were exceedingly rapid in the first decade after the end of the war. Discoveries of blood group frequencies were being explored throughout the world; discoveries of new blood groups and new complexities in the known blood group systems continued to excite the serological world. Fisher continued to follow these discoveries with great interest and to develop new means of calculation for application to serological data. The International Society of Hematology, an outgrowth of a conference on Rh and hematology held in Dallas and Mexico City in 1946, was formally organized at an International Congress at Buffalo, in 1948. The Third International Congress took place at Cambridge, England, in 1950. The choice of location was an acknowledgement of the significance of recent British contributions to serology, and the election of Fisher at this meeting to honorary membership of the International Society of Hematology recognized the primary role he had played in serological research. It was a most enjoyable meeting, with old friends and introductions to their young colleagues and with exhibits and new results presented, among others, by Mourant, and by Race with each of his assistants, Ruth Sanger and Sylvia Lawler. At one session discussion centered on the medicolegal problems in the use of blood groups.

Quite early in the war Fisher had been concerned about the legal utilization of blood group information in cases of disputed paternity. In 1944, he had personally assisted blood group investigation on behalf of a mother who appealed to him, believing her child to have been interchanged with another infant at the maternity hospital. In that case, the father's blood had been flown from India for typing in ABO, MN, and Rh groups; in these groups the blood of neither parent proved inconsistent with that of the infant. Fisher's proposal, with Taylor, that medicolegal aspects of blood group identification should be looked into, was considered inopportune during the war. By 1950, however, it was receiving some attention.

With distinguishable genotypes not only in the ABO, MN, and Rh systems but with the newly discovered blood groups, Duffy and Kell and a new antibody in the MN system, anti-S, Fisher felt that the chance of failure by the use of blood groups to determine cases of doubtful identity deserved careful statistical scrutiny. He exemplified the calculations [C P 243, 1951] by considering the chance of distinguishing monozygotic from dizygotic twins serologically, using the MNS antibodies, and he showed how the power of the test would be enhanced when the anti-*s* antibody should also be available. Anti-*s* had not yet been discovered, but it was discovered before the paper was published.

Despite the rapid expansion of the field, the first edition of Race and Sanger's *Blood Groups in Man* was published in 1950. Struggling to overtake results that were showing a tendency to gain on him, Mourant completed *The Distribution of the Human Blood Groups* (1954). At the same time, the implications of serology were being felt outside the serological laboratory. Mourant was invited to organize a meeting of the Anthropological Institute in March 1951, the first to be devoted to blood groups and, of course, to bring in Race, Fraser Roberts and Fisher with himself as speakers. Soon afterward, Fraser Roberts (and through him Fisher, at second hand) was drawn into the investigation of the association of blood groups with disease: Professor I. Aird, in testing an unrelated hypothesis, had collected data on patients with cancer of the stomach and had observed an excess in the *A:O* ratio. With Fraser Roberts, therefore, he went on to make very large counts, such as might reveal associations between blood groups and several common disease entities. Thus the question of the evolutionary significance of blood groups, through the selective effects of associated diseases, was seriously investigated for the first time.

These wider implications of serological genetics received prominence at the First Congress of the International Society of Human Heredity, held in Copenhagen in 1956. This was the culmination of the phase of the investigation of human heredity that Fisher had initiated by his blood group research. Through the blood groups, all his early interests as a eugenist—in the anthropological and cultural history of man, in genetical marking of the human chromosomes, and in the influence of natural selection on his evolution—were seen in their integrity as aspects of a single science. And in each aspect the promise seemed ripe for harvest.

Fisher was especially pleased by the evidence of selective effects that had been gathered at that date. Mourant [86], speaking of anthropology and natural selection, expressed the view that the observed ethnic variations in blood group frequencies could not be explained on any theory of random fluctuations but must be attributed to natural selection. He hoped more associations of disease with the regional and racial distribution of the blood groups would be sought and that these studies would gradually dovetail with others to reveal the mechanisms by which natural selection might operate. He exemplified his hope by reference to a case of balanced selective advantage already found by A. C. Allison in 1954, which rested on the simple genetic basis of the hemoglobins: the sickle-cell hemoglobin normally suffered a selective disadvantage, but in West Africa heterozygotes for normal hemoglobin and sickle-cell hemoglobin were found to be especially resistant to malignant tertian malaria, and the distribution of the gene for the sickle-cell hemoglobin was found to be associated with that of the malarial parasite through the action of natural selection.

Fraser Roberts [87], opening his talk on associations between blood groups and disease, addressed Fisher, in the chair, recalling, "how, twenty years ago you, Mr. Chairman and Dr. Ford, urged, indeed implored, those who had the facilities to look for associations between blood groups and adult diseases." Referring to the origin of the recent investigations, he said that I. Aird had not found what he sought of his data but "the result was magnificent. . . . what turned up was something different and even more interesting: in fact, just what you, Mr. Chairman and Dr. Ford had waited for, for so long." He went on to present the current evidence emerging from the first really large counts made on a number of diseases. This indicated the existence of several associations of the type sought and the probable nonexistence of several others. In the light of the data it seemed that a major phenomenon had been discovered. Further investigations were projected on these and other diseases not yet studied. In the thrilling period of new discoveries, in 1955, Fisher wrote to Fraser Roberts: "Once the hunt is up it seems the game is plentiful. It will be worth bearing in mind the confident inertia on this topic of the interwar period. In my experience everybody *knew* that blood groups were without medical effects." The first discoveries were not upheld, but at least what "everybody knew" was at last verified.

Fisher and Ford continued the long task of demonstrating evolution by natural selection in nature, Fisher on the primrose path in Somerset and Ford at Oxford where, with the support of the Nuffield Foundation after the war, he gathered a small group of brilliant younger men and extended his researches in the field from the Scilly Isles to the Shetlands. From time to time Fisher visited Oxford and he took great pleasure in the company at the genetics laboratory and in their investigations in ecological genetics.

In addition to the sheer fun of the work, he felt that it was important quite beyond the limits of the particular results obtained, for he perceived that, as he put it in the Eddington Lecture of the Royal Society in 1950 [C P 241], "for the future, so far as we can foresee it, it appears to be unquestionable that the activity of the human race will provide the major factor in the environment of almost every evolving organism." It was important to discover the ecological facts, the processes going on, and balances being maintained in an ecological habitat, before introducing changes that might prove disastrous and irreversible. He closed the lecture with the words:

Inadequately prepared we unquestionably are for the new responsibilities, which with the rapid extension of human control over the productive resources of the world have been, as it were, suddenly thrust upon us. Yet there have in recent times been some signs of a responsible attitude. . . . These are signs that we do not feel that ruthless exploitation is good enough. Our knowledge it is true is still in the highest degree inadequate; yet a beginning has been made with ecological studies, and what has been called population genetics, at least to explore the methods by which more effective knowledge can be obtained.

As the fascinating account of *Ecological Genetics* (1962) by E. B. Ford records, the genetical laboratory at Oxford was pioneering such methodology by its ecological investigations.

A. J. Cain and P. M. Sheppard, working on the color and banding of the snail *Cepea nemoralis* in the early 1950s, established by laboratory breeding the genetic basis for the polymorphisms responsible for color and pattern. In the field, they discovered that this genetic system was ideally suited to accommodate the sort of variation of the microhabitats of the snail colonies, to which their coloring approximated. The colonies were found to be relatively small, and the color and pattern found in each colony was closely associated with the prevailing background.

This suggested very strongly that the frequencies of color and banding classes must be determined by predators hunting by sight. Birds, which can distinguish colors, would be the chief enemies of these snails, and the species of bird which seemed to take the largest toll of them in England was the song thrush. The song thrush picks up all but the small specimens and carries them to convenient stones, called "thrush anvils," to break open the shells. The investigators took advantage of this fact, collecting the fragments accumulated at the anvils every few days, so as to decide whether the birds were collecting a random sample of their prey. There followed extensive studies by Sheppard and others on various types of background, which established the fact that the thrushes capture snails selectively, destroying an unduly large proportion of those whose color and pattern match their habitat least well, and that this predation exerts powerful selective pressure.

These studies were an exercise in utilizing the methods of ecological genetics. The limits of each colony having been established by observation, the existent proportions of the different types were established by sampling the living snails within colonies; the living population was compared with that indicated by shells collected at the thrush anvils; then, by releasing marked snails within the area under study (or in neighboring areas that might be affected) it was established that the snails brought to the anvil were, in fact, predated wholly from the local population and not carried in from outside the area. Finally, the total snail population was estimated from records of marking, release, and recapture, to establish that thrushes were destroying a considerable proportion of snails, thereby exerting a powerful selective pressure. The statistical analysis of the data was thus critically important to interpretation of results throughout the investigations.

H. B. D. Kettlewell was responsible for advances no less exciting in connection with the spread of industrial melanism in various species of moth in Great Britain. Since the middle of the last century over 80 species of moth in Britain alone, inhabiting industrial regions blackened by smoke, had been affected by the spread through their populations of a previously rare gene responsible for excess melanin production, blackening the moth. Ford had observed that these various species have one thing in common, namely, that they rest fully exposed upon treetrunks or rocks and derive their protection from a cryptic resemblance to their background. As the lichens, which are particularly sensitive to pollution, died, and the treetrunks and branches became blackened by soot, the cryptic patterns would become ineffective, whereas the melanic would derive increased protection from its resemblance to the darkened industrial background. In order to account for industrial melanism, therefore, Ford assumed heavy elimination by predators, those insects that matched their background least well being the most vulnerable. There was only one difficulty with this theory: whatever their differences, entomologists and ornithologists were fully agreed on one matter, that birds hunting by sight do not destroy resting moths selectively or indeed to any appreciable extent at all. Butterflies and moths had been more extensively collected and birds more widely observed than any other animals, and "everybody *knew*" the answer before Kettlewell carried out the first carefully planned investigation of the subject.

Again, the methods of exploration had to be considered in detail. Again, they involved a knowledge of the genetical situation in the species chosen for study and estimation of the local populations and careful design and analysis of experiments involving the release of marked insects and estimation of the rate of elimination of the different types. Kettlewell's methods were a model for such field work in the future. Even so, his finding of very heavy selective predation by birds of the moths whose coloring least resembled their background, appeared so unlikely that many naturalists did not believe the data, and were only convinced by photographic evidence. So quickly is the insect taken that the process, common as it is, can only be detected by careful observation. Kettlewell used observers as well as counts of the insects. When observers expected and looked for predation, it was seen in all its variety. Thrushes scan the treetrunks from the ground and dart up to take resting moths which they have marked down. Hedge sparrows and robins make their observations from twigs and bracken, yellow-hammers hover in front of the tree searching the surface from the air, and other species have other techniques.

From this initial inquiry flowed a series of others. The field scientist is fortunate in that every investigation he makes raises more questions for his investigation. Fisher was happy to be a party to discussions about the collection, analysis, and interpretation of evidence on such topics as recent evolution of the melanic imagine and adaptations of melanic larvae; wind patterns, the measurement of pollution and the spread of melanism in rural districts; and reasons for the dominance of the melanic forms and investigation of situations in which the melanic might have enjoyed a selective advantage long before the coming of industrialization. And in such investigations quantitative measurement and analysis—biometric procedures—were primary.

□ □ □

The science of biometry that had grown up within the household of statistics came of age in the postwar world. In 1945, the Royal Statistical Society's Research Section was formed in Great Britain, with separate publication in *Series B* of the *Journal,* and in the United States, the Biometric Section of the American Statistical Association launched its separate publication, the *Biometrics Bulletin.*

Fisher spent the summer of 1946 in the United States and was for 6 weeks visiting professor at the summer session of the Institute of Statistics of the University of North Carolina. The institute, formed in 1944 by the initiative of Gertrude Cox who became its director, consisted of two departments, one for applied statistics under W. G. Cochran and one of mathematical statistics under H. Hotelling. Fisher worked hard through these weeks, giving a course on mathematical statistics, participating in seminars, consulting, and answering questions. *The Leaky Gasket,* internal organ of the Institute of Statistics, recorded: "It is an everyday occurrence to see Annabelle and Helen chasing down the hall after Dr. Fisher with a bottle or two of warm beer. These two girls seem to be vieing for his favor."

It was hot in Raleigh and he was ready to enjoy the change of pace and location following the summer sessions, when the institute had arranged two statistical meetings to take place at the Lake Junaluska Assembly (Methodist Church center) in the Smoky Mountains of North Carolina. It is a spot of mountain airs and dramatically beautiful scenery. At these meetings the numbers were limited by invitation; 27 were present during the first week, when interest focused on applications of statistical methods to biological problems, and 18 during the second week, when discussions were directed toward mathematical statistics and general topics regarding the development of modern statistics as a science. A snapshot (Plate 16) taken there by Mrs. Horace Norton caught Fisher in relaxed mood.

Scientific meetings took place in the mornings and evenings. Fisher agreed to sit in the same place each day so that Leontine Camprube (Mrs. Gerhard Tintner*) could paint his picture. She set her easel at the right of the speaker's desk. (The oil painting (Plate 17) was purchased by Miss Cox for the Institute of Statistics, where it hangs in the conference room of Cox Hall at North Carolina State University. An oil-painted copy is hung in the R. A. Fisher Laboratories, Genetics Department of Adelaide University.) After the evening sessions, participants congregated with a copious supply of beer around Fisher's rocking chair on the verandah. At these meetings the plans were formed for the creation of a new international society devoted to biometrics.

The first conference of the Biometric Society was held at the beginning of September 1947, when 89 delegates (biologists, statisticians, and mathematicians) gathered at the Marine Biological Laboratory at Woods Hole, Mass., the traditional summer resort of geneticists. Scientific papers were presented (Fisher speaking on "A Quantitative Theory of Genetic Recombination," *C P* 225, 1948), a draft constitution for the Biometric Society was prepared and approved, and Fisher was elected the first president. The society decided to publish its proceedings in *Biometric Bulletin* of the American Statistical Association, renamed *Biometrics,* and in 1950 the Society acquired the journal as its official organ, with Miss Cox as editor.

Many of the delegates at Woods Hole went directly from there to Washington for the International Congress of Statistics, to promote representation of biometrics in the International Statistical Institute. That year their interests were covered in a section on statistical methodology, of which Fisher was chairman. A section on biometrics was later an option the institute sometimes accepted in cooperation with the Biometric Society. In December 1948 the Biometric Society was formally affiliated with the International Statistical Institute, and subsequent meetings have been planned to precede or follow meetings of the International Statistical Institute or the International Congresses of Genetics.

Fisher was a faithful attendant at these conferences and used to point out that they took him to delightful parts of the world. He was not very conscientious in attending lectures but enjoyed sharing the local scene and the company of scientists together informally. At Bellagio in 1953, he was regularly "at home" at teatime, and one or two, or a half a dozen visitors might join him daily at this time. At lunchtime he would pick up old acquaintances or new and, with a bottle of wine, a loaf, cheese, and olives, walk over the hill or down to the shores of Lake Como for a picnic meal. During the meeting in Brussels he was persuaded to visit the World Fair, and he spent most of the time sitting in the British pub, enjoying beer and good talk. His predilection for both was familiar to conference-goers. Thus Finney's [88]

---

*Coming from Vienna in 1931, Tintner spent some months with Fisher at Rothamsted. After the war Fisher visited him more than once at his home at Ames, Iowa.

memory of him, "shortsightedly manipulating complicated algebraic expressions in his microscopic handwriting cannot be dissociated from an equally characteristic picture of him, relaxed and with a tankard of beer poised, declaiming on some absorbingly unimportant question to a circle of colleagues at a conference." He treated the conferences as gatherings of his extended family in science. He took a paternal interest in the careers and the ideas of a number of the younger men and women and in developments within the society as a whole.

In 1952, when he spent the summer in North Carolina, he did his best to administer an antidote to certain poisons infecting the subject. The conference at Blue Ridge was conducted as an integrated series of fourteen conferences between June 6 and July 25. Only the special visiting statisticians, Sir Ronald Fisher, Frank Yates, Leopold Martin (from Belgium), and Bernard Colombo (from Italy) remained throughout the conference. In all, 196 persons attended some part of the conference, together with twice as many wives and children. Fisher and Yates (shown in Plate 19 with Gertrude Cox) were in a holiday spirit and having a good time as they exposed the folly of a purely theoretical approach to mathematical statistics; they incited each other like a couple of schoolboys to scorn the misguided ideas of pundits who had never done an experiment in their lives but were accepted as the modern authorities on statistics. On one occasion, after Yates presented his paper, Fisher disagreed, and there followed a heated discussion on how to analyze long-term rotation experiments. All this was mostly intended to get others to think. Fisher was convinced that if the young biometricians were to be saved from accepting concepts that might be fundamentally wrong, merely because they had been taught with authority, they must be led to use their own powers of reasoning and to trust them.

Fisher discovered some young enthusiasts were making expeditions before breakfast to swim in a little stream-fed swimming pool a little way up the mountain, and he asked to be wakened early to join them. He belied the evidence of his white hairs (he was 62) by plunging into the icy water with the best of them. In the afternoons he used to take children on hikes and teach them to catch and mount butterflies. Mothers dashed to buy netting when Fisher said no more children could go on these hikes without a butterfly net. A happy snapshot of Fisher with a butterfly hunting party (Plate 20), and another with a boy on his knee, record these activities. Also a statement in the local *Ashville Citizen Times* of July 13 reads: "One day between sessions [Sir Ronald] climbed the more than two miles of precipitous trail to High Top. He didn't simply walk it, he dog-trotted most of the way. A few summers ago, the same climb left a squad of husky highschool football players liquid-limbed and gasping."

Another weekend expedition was less successful. H. Fairfield Smith drove him to Pisgah National Park to climb Mount Pisgah. But Fisher was off-colour and Smith anxious; there was no beer for their picnic lunch; thunderclouds threatened as they climbed the trail, and as they descended rain began to fall. Fisher was in a decidedly bad humor on the return journey, while Smith fretted, imagining his feelings if Fisher should get pneumonia. On their arrival, Smith recalls:

Leopold Martin came to the rescue. He took us to dinner in the best available restaurant (though nothing to boast about) between our headquarters and Asheville. After dinner we went upstairs for some beer in the distant-country-cousin semblance of a night club. A little girl was singing "Frankie and Johnny." After her song she came and sat at our table. Sir Ronald started ribbing her, ever so mildly. Leopold Martin turned to me and said, "Why, Fisher's human after all!"

Human he certainly was, even in liking a pretty girl. The session closed with a hilarious party at which Fisher and Yates appeared as Kentucky Colonels (Plate 21).

He seemed inhuman sometimes in his lack of consideration for the feelings of others. In controversy of the sort he engaged in with Yates, he was capable of rough-handling those who opposed him with ready-made arguments that he treated with contempt; he was sometimes arbitrary and disagreeable; and he was recalcitrant to any form of coercion.

An instance of the latter sort arose when in May 1953 he submitted a paper to *Biometrics* on the analysis of variance with various binomial transformations [C P 256, 1954]. In this paper he was more or less critical of the contributions of four well-known statisticians: M. S. Bartlett, F. Anscombe, W. G. Cochran, and J. Berkson. The referees, victims of the criticism, were unwilling to see the paper published as it stood, and Fisher was unwilling to withdraw his remarks, though he did, in fact, modify their acerbity. He intended to shock readers into reconsideration of long-held opinions, and he maintained that if the article were considered worth publishing the editor should accept it on her own responsibility, without consulting the feelings of referees or subscribers. In the ordinary way, if readers wished to comment on a paper, their notes would have been considered for inclusion in a later issue of the journal, and this was what Fisher expected to happen. But, in the ordinary way, an author was expected to take notice of the editor's wishes in view of referees' comments, and, as we have seen, Fisher did not like the refereeing system. He did not wish before its publication to answer any criticisms that might be made of the paper, and he was annoyed by the delay. Miss Cox decided to accept the paper and to publish in the same issue invited comments on it by the four statisticians concerned. Fisher was sharply critical. To him it seemed "a means of stirring up unnecessary ill-feeling and controversy by instigating all who might feel aggrieved," and he asked:

Would you have so much preferred to go on letting erroneous methods be taught? You certainly make the duty of correcting them unnecessarily hard. Did you expect me to wait another fifteen years before calling attention to some fairly obvious misapprehensions? . . . Think for yourself whether I rushed in with criticisms or waited, hoping that others might put the thing right, until a correction could scarcely hurt anyone's reputation.

So, within the Biometrics Society, Fisher acted variously, as founding president, enthusiast, tonic, and father of careers in the making, and he sometimes fitted a description he applied on occasion to others, which he had heard first in his boyhood from his father's gardener, who declared his wife to be "that 'og-'eaded and hindependent."

Of course, Fisher promoted biometrics in other organizations. In 1951 – 1952, when he was chairman of the relevant coordinating committee of the British Association (BA), he attempted to persuade the BA to set up a section for biometry and genetics. It seemed high time to start such a section, for, as he wrote to Besse Day,

about 115 years ago or more, a meeting of the Section of Economics and Statistics of the British Association, featuring at that time both Babbage and Malthus, were responsible for founding what is now the Royal Statistical Society in this country and later, largely aided by Quetelet from Belgium, the International Institute of Statistics, and now here am I in 1952 plotting and planning to inject a spot of intelligent interest in Statistics into the same British Association for the Advancement of Science.

As president of the Royal Statistical Society in 1953 – 1954, he devoted his presidential address almost entirely to consideration of the scientific (as against the mathematical) importance of contributions of several early statisticians. In 1953, although he had to miss the inaugural meeting of the Adolf Quetelet (Biometric) Society of Brussels, he took a great interest in it and in planning the new multilingual journal, *Acta Biometrica Europa,* which Leopold Martin was initiating in Brussels. In 1952 the Biometric Society became the Biometric Section of the International Union of Biological Sciences (IUBS). At the IUBS meeting at Nice in 1953 Fisher helped gain the section IUBS sponsorship for courses in statistical methods. In the following year he was advising that such a course, organized by Cavalli Sforza in Milan, should spend half its time dealing with numerical examples; a year later he taught the course for two weeks at Varenna.

Meanwhile, from the end of the war onward, he was exerting his influence to turn the International Statistical Institute in the direction of the new research statistics. By this time considerable pressure had built up in this direction, but the matter was complicated by questions as to what should be the total numbers and geographical distribution of the membership. In 1945 the Institute was predominantly European, whereas the majority of statisticians in the world were in North America. Moreover, feeling that Europeans were hostile before the war, the Americans had formed the Inter-American Statistical Institute in 1940 and were inclined to write off the parent body; they doubted the Europeans would accept their leadership. And, in the postwar world, the United States alone had the money and the organization needed to begin restoration of the Institute.

An American government statistician, Stuart A. Rice, took the initiative on behalf of the Institute, acting as representative for W. F. Wilcox, its vice president in the United States. In an epic journey through war-torn Holland and Belgium in 1945, he succeeded in seeing and bringing together a quorum of the bureau, and he accepted their commission to undertake the Institute's reorganization, as rapporteur of a special committee to be set up for the purpose, and to plan an international meeting of the Institute in the summer of 1947, when the U. N. World Congress of Statistics would meet in Washington, D. C. In England and in the United States he rallied flagging support for the International Statistical Institute, begging patience of the disillusioned A. L. Bowley and E. S. Pearson in London and winning restoration of the annual subvention to the Institute of the United States government in Washington. The Institute was saved.

Reorientation away from governmental statistics was then the first consideration in many minds. In 1946 at the Lake Junaluska sessions in North Carolina a resolution was prepared, addressed to the Bureau of the Institute at the Hague and signed by sixteen leading research statisticians of the United States in which, recognizing that former social and governmental services of the Institute might now be yielded to the Statistical Commission of the Social and Economic Council of the United Nations Organization, they called upon the Institute to take steps:

wisely and reasonably to extend its activities to other fields . . . essential for the continuation and acceleration of the important advances to which statistical method is leading in such fields as public health and epidemiology, the measurement of public opinion, attitudes and wants, econometrics, psychometrics, industrial quality control and standardisation, industrial management, personnel selection, crop forecasting, plant and animal husbandry and the design of experiments and investigations in all fields of engineering and research.

To implement this resolution, it was necessary not only to recognize the paramount importance of experimental statistics in the modern world, but also to bring about a change in the membership that would reflect the fact.

The reorientation of the Institute interests, in fact, depended on the election of new members in such a way as to make the Institute membership more representative. At the meeting in London in 1934, there had been 29 professors, 5 statisticians, and 43 public servants, and relative numbers had not changed since. Moreover, the public servants had a further advantage in sponsorship of their travel to meetings by their national governments. Fisher, therefore, visiting Washington in 1946, tried hard to persuade Stuart Rice to ask for funds of the American government, as host in 1947, to bring research statisticians to the Washington Congress. Apart from the special financial needs of Europeans at the time, he hoped the American gesture would establish a precedent for other host governments, which would help to advance the scientific interests of the Institute. He did not succeed in this attempt.

At the same time he was trying to make sure that American research statisticians would be at Washington in good numbers, to represent their cause. As he wrote to C. I. Bliss, chairman of the Biometrics committee:

I gravely doubt if we shall get beyond highest intentions unless the body of research statisticians which your committee represents shows its vigilance and concern during the Washington meetings and manages to rejuvenate the American membership of the Institute by a considerable accession of new members.

And, in another letter to Bliss before these meetings:

Leading American non-official statisticians should get together with determination to get at least ten or a dozen of their number elected into the Institute. This should give quite a new complexion to the prospect of the Institute being of service to the world in the future.

Afterward, the election of nonofficial candidates continued one of Fisher's chief concerns for the Institute. He actively promoted the candidacy of a succession of research statisticians (not without opposition in the United States where more than one such candidate was asked to withdraw his candidacy to leave room for government candidates). Even so, in 1956 he felt:

We really have a terrifically long way to go in making the Institute as useful as it could be, since I think the great majority of our foreign membership quite take it for granted that it is primarily an assembly of officials concerned with national statistics, vital and economic, and of their more academic economic advisers. These people cannot deny the importance of mathematical statistics . . . and if we put in undeniably good mathematicians who insist on talking of the natural sciences and in terms of scientific research and holding sessions relevant to the applications of mathematical statistics to scientific research, we have done a pretty good generation's work.

The Washington meetings fell into two parts: first the conclusion of the 24th session which had adjourned prematurely in Prague in 1938; then, after an interlude of three days, the 25th Session itself. An Editing Committee was

434 *The Life of a Scientist*

formed at 24th Session to review the Report of the Advisory Committee on the Revision of the Statutes, and the interlude was busy with meetings both formal (the Editing Committee held three meetings lasting until midnight) and informal (Stuart Rice recalls vigorous argument in Fisher's hotel room, with Fisher half-reclined on the bed and a group debating views advanced, forcefully as always, by Maurice G. Kendall). Debate continued at the General Assembly of the Institute. At one point, after Fisher and Rice had objected to a proposal of Gini, the meeting had to be adjourned for 5 minutes to cool off. So the future shape of the International Institute was hammered out.

In 1947 the Institute was listed as containing 150 ordinary and 10 honorary members, and many of these had died since the last meeting. It was desired to increase the membership, and Fisher was made Chairman and Rapporteur of the Committee on Electoral Procedure, to be set up to consider how the increase should be regulated, to what total number, and what national quotas and rules of election should be adopted to preserve the international character of the Institute and satisfy the legitimate aspirations of individual states. When the Committee on Electoral Procedure met at Berne in 1949 Fisher presented two proposals. The first was to increase membership to a newly authorized but flexible total according to the formula, "The number of candidates declared successful at any election shall be the integer nearest to the formula $52 - \frac{1}{9} N$ where N is the number of members eligible to vote at that election." The second proposal was to ensure the international reputation of candidates elected by requiring that at least one-third of the votes cast should be for candidates of nationality other than that of the voter. The committee approved adoption of the two rules, and in March 1950 the Rapporteur completed his task by sending a formal report to the president, then Stuart Rice. When the Institute met in 1951, the rules were adopted after much discussion, as was also a quota rule limiting the membership of any nation to one-eighth of the total membership.

Some years later it was evident that these rules had not counteracted the tendency to fill up quotas of the largest national states first. In fact, the United States had not only filled the national quota of 35 members by election but had overfilled it by immigration of other nationals. The irregularities were pointed out at the Brazil meetings in 1955. A new Committee on Electoral Procedure was set up, with Fisher again chairman. It was a sensitive matter, for, as Rice expressed their feeling, the United States had more statisticians than any other country and probably more eligibles for Institute membership. He was sharply rebuked by Fisher from the chair, to whom the assumption of American superiority appeared both arrogant and unjustified. Moreover, even were some countries like Great Britain or the United States to be underrepresented, he felt it would be worth while in order to assure representation of smaller and statistically less advanced nations in the International Institute.

Fisher resisted the idea of raising the national quota from one-eighth to one-sixth of the total membership, and it was, in the end, a suggestion which he had made informally in Brazil that was finally adopted, namely, that members of advanced age (he suggested 75, the age adopted was 70) might be omitted from the national quota in respect of the formula governing the number of candidates to be elected. The 70+ rule relieved the pressure to increase United States representation for a decade. The pressure was renewed in 1967, and new committees were set up, but no further changes were agreed on until 1973, a fact which is sufficient tribute to Fisher's wisdom and tact in leading the committee to acceptable proposals in 1950 and 1956.

□ □ □

Increasingly through the 1950s, Fisher was concerned by a tendency in statistical education, and consequently in statistical practice, especially in the United States, to treat statistics as a theoretical subject, a branch of mathematics. He became highly critical of "mathematicians"; thus, although, later, L. J. Savage [89] was to point out that Fisher was himself "a very good one with an extraordinary command of special functions, combinatorics and truly n-dimensional geometry," earlier, Savage admitted, "I had been misled by his own attitude towards mathematics, especially his lack of comprehension of, and contempt for, modern abstract tendencies in mathematics." Fisher's objection was not to mathematics in itself but to the assumption that mathematical reasoning alone sufficed in statistical work, or that the subject could be apprehended by a purely deductive approach.

Through the years he had made the same point in various ways. In 1929 he had written that it would be deadly to the subject to be isolated as a self-contained study, to create chairs and lectureships to train statisticians to occupy chairs and lectureships, and added: "What we need is a fairly intensive mathematical training, together with *very wide* scientific interests, not so much in established knowledge as in the means of establishing it." In India, in 1938, he had said that the responsibility for teaching statistical methods must be entrusted to highly trained mathematicians certainly but "only to such mathematicians as have had sufficiently prolonged experience of practical research, and of responsibility for drawing conclusions from actual data, upon which practical action is to be taken." In 1952, at Blue Ridge, he and Yates took the offensive against the thoughtless adoption of the views of theorists rather than practitioners of mathematical statistics. In the same year he wrote a letter, already quoted in part, while advocating that a proposed statistical center serving research stations should be coordinated with a university teaching department. In giving his reasons for this recommendation he expressed rather more fully than elsewhere both his apprehension of what statistical education was and what it should be.

I think one must regard the problem as analogous to that of the technological students engaged, let us say, by M.I.T., in linking advanced physical investigations with entirely practical engineering problems. The experience of such places seems to be impressive and unanimously in favour of the view that the teacher should have constant experience of the technological problems arising and that his experience should refer to current problems and not to older problems which have ceased to be of interest. A Department of Mathematical Statistics, without any imagination beyond the realm of the classical theory of probability (and statistical teaching is certainly undertaken by a good many University Departments knowing no more than this) would certainly do more harm than good. Studies under such names as 'the design of experiments,' 'decision functions,' 'acceptance tests,' 'quality control,'. 'sequential analysis,' 'sample survey,' are now an integral part of the study of mathematical statistics and have obviously spilt over into commerce and industry on a large scale. With all this in view, I have no hesitation in advising that such a centre as you have under discussion should plan to integrate teaching closely with project work in which practical experience can be gained by those who are capable of learning from it, in contradistinction to the ruinous process of segregating the keener minds into a completely sterile atmosphere.

The "mathematicians" he objected to were not pure mathematicians, but nor were they research workers: they did not design experiments (make surveys, take decisions) in the real world and answer for the consequences of their advice, but they taught how optimal experiments might be designed under some set of hypothetical conditions that might never be realized in practice, or how decisions might be arrived at, given some hypothetical cost function incapable of being determined in the real world.

Teaching mathematical statistics in this way not only produced answers that were often irrelevant or misleading when applied to real situations but also led students to think not of the reality but of the mathematics as holding the key to statistical understanding. In Fisher's terms, it taught them not to think. Despite the magnificent educational effort in statistics which had been made in the United States, (as Fisher wrote to E. B. Wilson after the Brazil meetings of the International Statistical Institute),

Many of them seem to have no experience of the valuable process known as "stopping to think." I wish that what Stuart Rice assumes were really true for it would be very good to think that so fundamentally important a realm of thought as that which is nowadays associated with the word statistics was really receiving in one great country the attention which is its due. But, though there are able and highly educated men in abundance, I should guess that for the next twenty years at least progress in statistical understanding in the U. S. will depend very largely on what is allowed to trickle in from the rest of the world.

The men he wanted as members of the International Statistical Institute were, as he wrote to Bliss about this time: "undeniably good mathematicians who insist on talking of the natural sciences and in terms of scientific research and holding sessions relevant to applications of mathematical statistics to scientific work."

For Fisher, a purely deductive approach to statistics violated the inherent nature of the subject, which was based on an inductive mode of reasoning. At the inaugural meeting of the British Region of the Biometric Society, in his presidential address [C P 224, 1948], he compared the rise of biometry in the twentieth century to that of geometry in the third century before Christ, each marking out one of the great ages or critical periods in the advance of the human understanding. He spoke of the "enchanting clarity" of the concepts and processes of geometry, the "liberation of spirit experienced" and the "veneration" felt for the subject "by which the human spirit came to handle abstractions, of their nature timeless and perfect, and to handle them with confidence because they were well-defined." Moreover, "with well defined concepts the intellect found itself capable of acting with unprecedented efficiency. Men learnt to reason deductively, from well defined abstract concepts, to cogent and irrefragable conclusions."

His own feelings about the charms of biometry paralleled those he ascribed to the Greek geometers about their subject; but the nature of the subject was different:

Now, I suppose circumstances might have conspired to give to surveying, or to astronomy, or to any other subject sufficiently rich in observational detail, the honour of compassing the second great stage of intellectual liberation, by making known the principles of that second and scarcely explored mode of logic, which we know as induction; of clarifying the principles of reasoning from the particular to the general, from the observations to the hypotheses, in ways necessarily inaccessible to purely deductive logic, or to any mathematics which can properly be regarded as derivable wholly from deductive logic, of making men free to recognize with certainty the consequences not of axioms or dogmas, but of carefully ascertained fact. But, as it has happened, it has been reserved for Biometry, the active pursuit of biological knowledge by quantitative methods, to take this great step.

We note that the primary requisite for this new development was a subject sufficiently rich in observational detail. Moreover, Fisher insisted that experience in dealing with the phenomena of nature must at every stage be the inspiration and guide to a better understanding of inductive reasoning.

The primitive function of the biometric movement, characteristic of the present century, is therefore to conserve by constant use, and incidentally improve and refine, the thought forms, which make possible an understanding of variable phenomena.

These phenomena come to our knowledge by observation of the real world, and it is no small part of our task to understand, design and execute the forms of observation, surveys or experiments, which shall be competent to supply the knowledge needed. The observational material requires interpretation and analysis, and no progress is to be expected without constant experience in analysing and interpreting observational data of the most diverse types. Only so, as I have suggested, can a genuine and comprehensive formulation of the processes of inductive reasoning come into existence.

□ □ □

Variable phenomena were not restricted to biological sciences; nor were Fisher's interests. At Cambridge, living in college, he enjoyed the company of Fellows in many disciplines, and, because he was always keen to encourage young scientists and curious about new lines of work, he was particularly drawn to the new young fellows like V. M. Clark and S. K. Runcorn, the first to be elected Fellows of Caius when Sir James Chadwick became Master after the war. Clark was doing research in organic chemistry while Runcorn, taking up an appointment as assistant director of research in the department of geodesy and geophysics, was beginning to develop the new field of paleomagnetism. Fisher was especially fascinated by the latter work.

The geophysical laboratory was in the grounds of the Cambridge Observatory, a few minutes walk from Whittingehame Lodge. Fisher used to take a footpath from his back gate to the back gate of the observatory to visit Runcorn, and he soon came to know Runcorn's research students also. He was quite indignant later when a new policy decreed that the back gate of the Observatory should be kept locked and he was denied easy entry.

Fisher's attitude to the new generation of scientists is well described by E. Irving, a graduate student of geophysics at Cambridge in the early 1950s:

Formal position meant nothing to him. The help and courtesy that he extended to people seemed to depend inversely on their formal position; the higher up the ladder the greater his indifference. To a research student this was marvellous stuff, and it flattered our vanity. His main motivation in life seems to have been scientific inquisitiveness, and his attentions to the young research students and staff (where the action often is) was of a piece with this.

In fact, Fisher seemed eager to learn from his younger friends what they thought, so that he might share their understanding and interest in the new work. He was a good listener.

The geophysicists started from the hypothesis that most rocks of the earth's crust containing iron oxides have been magnetized at the time of their formation, igneous rocks taking the earth's magnetization at the time and place in which they cool through the Curie point and sedimentary magnetic particles aligning themselves with the earth's magnetic field as they settle on the sea, river, or lake bed. Although subsequent history might partially break down this remanent magnetization, many of the suitable formations retained remanent magnetism the directions of which (given by the angles of declination and inclination) could be measured with fair accuracy by the new methods. It was thus possible to determine the relation that had existed between the land mass and the magnetic pole at the time when the rock had been formed.

The simplest geomagnetic hypothesis was to suppose the core of the earth to act as a dipolar magnet. On this hypothesis, however, the magnetic pole must be supposed to have wandered relative to the direction of rotation of the earth, though where and how were unanswered questions. One of the early aims of geomagnetic work was to determine the position of the magnetic pole at known times and to build up a succession of polar determinations from rocks of different ages, so as to reconstruct the course of polar wandering. But the observations varied, either by reason of heterogeneity of conditions or composition at the time of formation, of changes induced *in situ* since that time, or of disturbances due to the treatment of the specimens. A further complication arose when, with a small astatic magnetometer, Jan Hospers, a student of Runcorn, collected the first measurements of the directions of magnetization of Icelandic lava flows. Runcorn was at once struck by the fact that these lavas grouped themselves into those along the present field and those opposite, in a manner strongly suggesting that the direction of magnetization of the earth had been reversed at some periods of its history.

Runcorn explained these results to Fisher and asked him how the particular statistical problems of testing homogeneity should be solved. Fisher, pulling out some old notes he had made before 1930 in connection with his paper on tests of significance in harmonic analysis [C P 75], set to work to apply the method he had then devised to the data now in hand and to develop a test of significance suitable to a worker whose knowledge of precision lay entirely in the internal evidence of the sample. He applied his method to the data of Hospers in a paper on "Dispersion on a sphere" [C P 249] in 1953, and Hospers quickly used the method in three very important papers which laid the basis for the paleomagnetic work in the Tertiary Age [90]. It has been basic to statistical method in geomagnetic investigations since and is easily the most frequently quoted reference in papers on paleomagnetism.

One of the interests of paleomagnetism was that it could provide decisive evidence about continental drift. As early as 1912 Alfred Wegener had put forward the hypothesis that land masses have moved over the surface of the earth; in 1915 he had published a book on it, and the idea had been hotly debated in the 1920s. However, despite the publication of du Toit's persuasive volume, *Our Wandering Continents,* in 1937, the hypothesis had been abandoned by most English and American geologists in view of the strong arguments against it on the part of geophysicists. Runcorn was initially sceptical of the theory, and Fisher once said to him, "You will end up by proving continental drift," and that, of course, was just what the early work on paleomagnetism did.

Fisher regularly attended the colloquia of the department of geodesy and geophysics, where he was made welcome by the director, B. C. Browne. He also belonged to the Kapitza Club and the ∇²V Club, of university physicists and mathematicians. Runcorn recalls:

In the autumn of 1952 we persuaded Fisher to give an evening talk on continental drift to the ∇²V Club. Blackett*, who was already becoming very interested, came from Manchester University to hear it. Fisher gave a brilliant lecture, reflecting on the fate of this imaginative idea at the hands of orthodoxy, but Blackett was disappointed. He said to me afterwards: "You or I could have made a better case." In fact, Blackett was not inclined to be humorous about a serious subject (and physics was, of course, that!), whereas Fisher could not avoid making witty comments about the failure of geophysicists and geologists to take Wegener's ideas seriously. He was not good at presenting subjects to an audience who knew little about the subject, and this was an example. He always assumed that facts were known and preferred to comment on them, which he did penetratingly but only those at home with the subject really benefitted.

In February 1952 Blackett had stayed for a week in Caius College with Runcorn to plan a program of paleomagnetic work to extend Hosper's work to the more weakly magnetized sediments, using a magnetometer Blackett had developed to test his theory of cosmical magnetism. Fisher had not met Blackett, but Runcorn had been for 3½ years on his staff, admired him and had become a friend. At that time Blackett's views on atomic weapons were receiving a good deal of adverse publicity. Fisher was deeply distrustful of left-wing views such as Blackett espoused, and although Runcorn defended Blackett's views, Fisher was unconvinced. However, the three had dinner at High Table and when Blackett had gone, Fisher said to Runcorn, "I have revised my views on Blackett. He has an infectious almost boyish enthusiasm for science which greatly attracts me."

*P. M. S. Blackett, the physicist, mentioned earlier in connection with his development of operational research during the war, later Baron Blackett.

Later, Fisher and Blackett got on very well together. Indeed, it seems fitting, in view of their mutual appreciation, that, as president of the Royal Society in 1966, Blackett should have presided at the first Fisher Memorial Lecture in London. Known political differences simply never arose in Fisher's conversation with Blackett or with others whom he liked personally; another example was Joseph Needham, the biochemist and great historian of Chinese science who was then Fellow of Caius College and has since been a successful and distinguished Master, retiring in 1976. Needham and Fisher were, in fact, allies in bringing about internal reform of their college after the war.

The initiative for change came from the new Fellows, Runcorn and Clark, who saw the need for change and proposed to try and elect two new members of the College Council. After sounding out other fellows and finding Fisher and Needham enthusiastic, they called a meeting of those who were interested, which was held in Fisher's rooms. M. Swann (now chairman of the British Broadcasting Corporation) and P. T. Bauer (now professor of economics at the London School of Economics) were elected—the latter by one vote, and that vote was Fisher's. In the excitement of the second round of voting Fisher failed to vote. The votes had been counted and the Master announced that the result was a dead heat, when Fisher said, "Master, have we voted?" The Master called for a new vote, and Bauer was elected. Consequently, though many other senior Fellows supported reform, Fisher was blamed for the shake-up of the college.

In shaking up established scientific opinion, Fisher's wit was provocative and often took the form of a neat dismissal, slightly outrageous in its obvious prejudice, and likely to be remembered and repeated. Once, when asked why his 1918 paper [C P 9] had been rejected by the Royal Society, he had answered that the paper had been reviewed "by a biologist who knew no statistics and a statistician who knew no biology." Similarly in conversation Fisher dismissed Harold Jeffreys' book *Theory of Probability* with the words: "He makes a logical mistake on the first page which invalidates all the 395 formulae in his book." Jeffreys' "mistake" was to adopt Bayes' postulate.

Neither Fisher nor Jeffreys had changed his fundamental position since the controversy about inverse inference 20 years earlier. They appear, however, to have agreed to disagree and Runcorn, having occasion quite often through Fisher's interest in geophysics to bring him and Jeffreys together, recalls:

Fisher was genial and Jeffreys friendly. I remember saying something about their controversy and, in his charming way, Fisher said he agreed with Jeffreys' approach more than the current school of Neyman, and Jeffreys very emphatically said, "Yes, we are closer in our approach."

Sir Edward Bullard* was, of course, a party to many of these geophysical discussions and recalls on one occasion hearing the story that the feud between Jeffreys and Fisher had ended on the day they both went to hear Arthur Eddington talk on the nature of scientific inference, and were so horrified that they shook hands and promised not to write any more rude things about each other.

Fisher's remark about Jeffreys' *Probability* was made with great good humour, with a twinkle in his eye and a hearty laugh. Knowing that Jeffreys had strongly opposed continental drift, Fisher now said, "He was wrong about probability and that makes me sure he is wrong about continental drift." He enjoyed his naughtiness and clearly expected others to enjoy it too, while making due allowance for the prejudice.

Geomagnetic investigation of continental drift began with the realization that remanent magnetism might provide decisive evidence as to the validity of the theory. If, as Wegener had supposed, the continents of South America, India, Australia, and Africa had once clustered with Antarctica about the South Pole, forming a single land mass (which he had called Gondwanaland) in the comparatively recent history of the earth, then remanent magnetism of rocks formed at that stage of terrestrial history would exhibit a markedly different orientation to the pole than the more recent formations, one inconsistent with the modern field of magnetization.

In particular, it occurred to E. Irving that the modern latitude of India, actually north of the equator, was markedly different from its supposed latitude in Gondwanaland, so that the change in the direction of its magnetization might show up decisively, despite the uncertainties of polar wandering. Fisher, visiting India in 1951–1952, gained collaboration there and samples from the Deccan Traps were secured and delivered to Irving in Cambridge late in 1952.

Irving found the data from the Deccan very suggestive, and on the basis of these results and an extensive study of the palaeomagnetism of the Torridonian sandstone of N. W. Scotland was prepared to submit a Ph.D. thesis to the University of Cambridge in 1954, in which he discussed the notion of continental drift as explanation of his findings. The seven samples from the

---

*E. C. Bullard returned to Cambridge in 1945–1948 as reader in experimental physics, but Fisher really got to know him only in 1949 when both were visiting professors at Toronto University and, with Runcorn at Caius, Fisher was becoming involved with geophysics. They continued to meet in 1950–1956 when Bullard was director of the National Physical Laboratory, and when he returned to Cambridge University in 1956, Sir Edward was made a fellow of Caius College.

Deccan, with inclination 56°, put that part of the world in a south latitude about equivalent to modern Victoria, Australia. In Cambridge, it put the cat among the pigeons. Irving recalls:

I failed my Ph.D. There were many reasons for this (most of them my own fault as I did not put my case very well and was in too much of a hurry to go to Australia) but one was that the examiners just did not see the implications of the work, the basis for which had just been established. Fisher saw it all too clearly and was furious. He need not have been, because the failure was very good for me—I had to destroy the record, which was a nice challenge.

So Irving went off to Australia to destroy the record, gradually building up a new record for that continent, as the remanent magentism of its rocks told him the story of its dancing progression, as it had circled around the South Pole during its long history.

One of the theoretical difficulties that prevented many geophysicists from considering continental drift seriously in these early days was the problem of conceiving a mechanism by which continental movements might have occurred. Traditionally the mantle of the earth had been regarded as a classical elastic solid, and this did not lend itself to motions of the earth's crust. The possibility that convection currents existed in the mantle had, however, been proposed by D. T. Griggs [91] in 1939 in order to provide an adequate mechanism for tectonic activity. In 1949, E. C. Bullard [92] had shown that rapid convection in the core was a necessary condition for the generation of the electromagnetic field and that this rapid convection required convection in the mantle, because the heat conducted out of the core was found to be more than could be transported by conduction alone. When the mantle was considered not as an elastic solid but as a viscous Newtonian fluid, with slow convection currents passing through it, a theoretical model eventually suggested itself [93].

First, however, considerable research was required into turbulent convective motion. Raymond Hide, in Runcorn's group at Cambridge, undertook researches between 1948 and 1953 aimed at elucidating possible movements of the earth's liquid core. Fisher was delighted, when he dropped in on the geophysicists, to observe Hide's "worm" as he called it, where within a rotating cylinder one could see the colored particles of red and blue weaving in and out, each following its convoluted course around the cylinder, between the rapid convective currents of the "core" and the reflected slower currents in the "mantle." This research uncovered hydrodynamical phenomena of great interest, even apart from their possible geophysical applications, as Fisher testified in 1953 when Hide applied for a Fulbright travel grant.

In a letter of November 1954 Fisher summed up the current status of Continental Drift:

Within a few years I expect very important evidence on this subject will accrue from the study of rock magnetism. At the present time it has gone so far as to convince the geophysical world of the reality of the large movements of the earth's pole which the proponents of Continental Drift have always postulated. It is however a logically tenable hypothesis that, though the pole has migrated about the crust, yet the relative positions of the main crustal features have remained unchanged. It will need the accumulation of paleomagnetic data from all over the world before it is clear whether this is a tenable view or whether large movements of crustal blocks are also demonstrated.

At present I find the rather equivocal attitude that although horizontal movements up to 300 miles are admitted, as in the Red Sea, movements of the order of 2,000 miles are strenuously denied. Since the magnitude of the horizontal forces required must stay the same in the two cases, I think this may be regarded as a last ditch defensive position appropriate to those who do not wish to admit that their old grounds of disbelief were untenable.

In the 1950s, Runcorn was traveling during the summers in the United States, on one occasion boasting on his return that he had climbed six times up and down the Grand Canyon collecting the rock samples of the different ages represented on this natural cutting, 1 mile deep, through the geological strata. As the data built up of the remanent magnetism of rocks of the American continent in comparison with those of Europe, the divergence of the Western from the Eastern Hemisphere became apparent. The position of the poles deduced from American strata of five geological ages, ranging from the old Precambrian to the Triassic, were found to diverge by the same amount from positions deduced from British rocks of the corresponding geological age; the conclusion seemed inescapable that there had since been continental displacement of Great Britain from North America, with the opening of the Atlantic Ocean. The evidence of geomagnetism was beginning to be difficult to explain without invoking Continental Drift.

It was, however, only beginning to be so. Hearing of the frustration of Irving's plans for travel abroad in 1956, Fisher predicted that the kind of opposition to his ideas which this represented "will surely not be the last example of very prejudiced opposition." Fisher suggested that a parallel existed between this and the earlier scientific controversy about organic evolution.

I think there is a parallelism in the nature of scientific controversy between continental drift in the last 80 years or so, and organic evolution about 100 years earlier. Each idea as it originated was necessarily speculative, and not accompanied certainly by sufficiently cogent evidence to carry final conviction. There were, however, many suggestive pointers. In consequence of this natural situation both questions have been argued with imperfect facts, incorrect theories, and often incompetent reasoning, over a long period during which many people have committed themselves to impossible positions, and many more, fearing to burn their fingers have enclosed themselves in towers not of ivory, but of very solid wood. In the period 1800 – 1850, although geological specimens were being collected and described, experiments in plant hybridisation carried out, the classification of animals and plants greatly improved, and embryological studies at least had attracted attention, yet so unwilling are ordinary men to run the risk of contemptuous ridicule, no one of consequence attempted to revive what had been left as speculative and almost poetical, ideas by Buffon, Erasmus Darwin and Lamarck. Darwin worked on the problem almost secretly from 1838, and only published in 1859 because he was forced to. After that the ice came down like Niagara, but of course the new idea was still ill-understood and ill-expounded for at least the next 50 years, during which a few subordinate causes of error had been removed by special research.

I think a lot of geologists must be timidly peering out of their holes on hearing the strange news that geophysicists are talking about continental drift, and I have often wondered how many scientific discoveries of importance have been left unmade for lack of the quality called moral courage.

By this time, a number of geophysicists were entering upon extensions of research into remanent magnetism and the evidence was accumulating by which geomagnetism was eventually to validate the theory of Continental Drift. The special causes invoked to explain particular results could not forever contain the flood of conviction carried by the single coherent theory capable of accommodating the accumulation of new data.

Fisher continued to the last to be fascinated by the geophysical explorations and to encourage and befriend the explorers. Runcorn moved to the University of Durham at Newcastle in January 1956 as professor of physics, and Hide joined him the following year. Fisher often visited the university to give lectures at which he charmed a new group of research students. As retirement approached he was very attracted by the idea of keeping close to his friends in geomagnetism by moving to Newcastle.

Runcorn had heard committees of the university discussing at length the idea of starting genetics at Newcastle, and thought the ideal way to do this would be by starting genetic research, setting up a research institute where Fisher could continue his research. Able young geneticists would then be attracted to Newcastle and a teaching department would develop. The vice-chancellor, Dr. C. I. C. Bosanquet, was very interested in furthering this plan but the university eventually withdrew. Instead of moving to Newcastle, Fisher went to Australia and, on arrival in Canberra, eagerly sought out Irving (rather rudely neglecting one senior man at least in consequence, who had hoped to entertain him).

During Irving's years in Australia, Fisher took the trouble to keep in touch by correspondence and visits and always gave great encouragement. It was a difficult time for Irving, and it meant a lot to him that Fisher "who could so easily have forgotten all about us" continued to be interested and to have faith in his abilities. Irving was hesitant about speaking about statistical aspects of his work, and Fisher responded:

I do not think at all that you should hesitate to speak on statistical matters in public, for statistics, as much or more than other branches of applied mathematics, needs scientists with an intelligent grasp of what they are talking about to keep mathematicians, who have their own uses, on the rails; and nothing is more obvious than that the judgment of the relative importance of different lines of work is very ill-developed in many mathematical departments.

In fact, the special difficulties experienced with unstable Australian samples resulted in the greater refinement of Irving's methods of analysis. As always, progress in statistical science came through that "constant experience in analysing and interpreting observational data," without which, as Fisher had said in 1947, "no progress is to be expected."

By 1962 the developments in rock magnetism had given rise to a lively interest in continental drift, which the geophysicists had once led the way in rejecting. The publication of *Continental Drift* (Keith Runcorn, Ed. 1962) [93] witnesses the extent to which geophysicists in many fields had already been led to reconsider the theory in relation to their own studies. Fisher died before the revolution of thought had fully taken place. His participation was during the critical years of conception and exploration. Perhaps he would not have been surprised that in 1971 R. Hide was elected Fellow of the Royal Society, or that Professor Runcorn, F.R.S., broadcast on the BBC a television program that began with polar wandering and continental drift, as established facts, and went on to build up a world picture quite transformed by the ideas of continental plates and plate tectonics resulting from the geophysical researches of the previous twenty years. But, if Fisher had been alive, he would probably by then have been involved in some other venture in its pioneering stage.

17
Scientific Inference

Fisher's scientific interests continually confronted him with the basic problem of statistics: how to make inferences from the particular to the general. From his first paper in 1912 to his last 50 years later, he was exercised by the need to express with precision what might justly be concluded from variable data about uncertain events. Earlier, while the problem had been perceived only vaguely, the deep philosophical difficulties of scientific inference had been unrealized. Fisher perceived the problem; his researches progressively revealed its logical consequences and forced recognition of the issues.

Fisher naturally faced opposition of two sorts, from those who were reluctant to admit the existence of a real problem and from those who disagreed with his approach to its solution. Because of continuing disagreements, in the mid-1950s he wrote *Statistical Methods and Scientific Inference* (1956) in which he brought together the various threads of inferential argument and restated his position.

This is not the place to enter into detailed discussion of the book. It seems fitting, however to sketch the place held in Fisher's thought by different sorts of scientific inference toward the end of his life and to indicate where divergence of opinion led to controversy.

Fisher's ideas had developed and matured over the years. He found different procedures to be applicable to different situations, the level of the appropriate inferential procedure depending on the sort of information available in the particular situation under study. Much of his early work had been devoted to what he came to regard as the lowest level of scientific inference—to tests of significance which make a dichotomy between hypotheses that are discredited by the data and those that are not. At the highest level, under certain conditions the whole probability distribution of the parameter of interest could be derived from the data using Bayes' theorem. However, this was only possible if the probability distribution of the parameter was known a priori. In 1930 Fisher proposed that in the absence of knowledge a priori the probability distribution of a parameter could sometimes be inferred by the fiducial argument. Realizing that the fiducial argument depended on exhaustive estimation, he was soon led to establish the conditions for the existence of sufficient statistics, drawing heavily on his earlier work on the method of maximum likelihood. In the absence of exhaustive estimation he felt that the likelihood function could be used to indicate the relative plausibility of different parameter values.

Fisher did not offer a universal solution to the problems of inference. Instead, looking back over a lifetime of pondering these problems, he considered various inferential techniques, each appropriate to certain kinds of problems but none appropriate to all. This was itself a cause for dissatisfaction, because a statistician trained only in mathematics is likely to expect a problem to have a single solution.

□ □ □

Everyone assumes he may argue from experience and makes inferences every day of his life. The scientist argues in no other way: he does a few experiments or makes some observations and then concludes that what he has seen or something like it will occur in similar circumstances in the future. In 1936 Fisher said the title of his talk on "the logic of scientific inference" might just as well have been "Making sense of Figures".

For everyone who does habitually attempt the difficult task of making sense of figures is, in fact, assaying a logical process of the kind we call induction, in that he is attempting to draw inferences from the particular to the general; or, as we more usually say in statistics, from the sample to the population.

At the beginning of the twentieth century, although a beginning had been made in quantitative biology, the conditions governing the validity of the inferences drawn had not been explored.

Over 200 years ago Thomas Bayes appreciated the difference in kind between the rigorously exact statements of mathematics—derived by deductive argument from the general to the particular—and the uncertain inferences possible by inductive argument from the particular to the general. He was also the first to realize that inductive statements might be made mathematically rigorous by specifying the nature and extent of the uncertainty involved. His argument follows directly from classical probability.

To illustrate, suppose it is known that *as the result of a toss of a fair penny* one of two urns A and B containing, respectively, 10% and 90% black balls, has been chosen and that a random sample of ten balls drawn from it contains seven black balls. Using Bayes' theorem, a mathematically exact statement may be made about the relative probability that the sample has come from the one urn or the other. The numerical odds are:

$$\frac{(9/10)^7(1/10)^3}{(1/10)^7(9/10)^3} = \frac{6561}{1} \text{ for } B \text{ as against } A.$$

Equivalently, since there are only two possibilities, the probability that the sample was drawn from urn A is 1/6562 and the probability that it was drawn from urn B is 6561/6562. In general, knowing the relative probabilities *a priori* (1 : 1 in this case) and given the *data* (7/10 black balls), it is possible to state the relative probabilities *a posteriori* (6561 : 1).

In this example for which the prior odds are known exactly, Bayes' method provides exact inferences about which of the two urns was sampled since it provides mathematically exact statements about the probabilities. However, had the ratio of the prior odds been known to be $x:1$, the posterior odds would have been $6561:x$; and had the prior odds been unknown, the posterior odds would have been correspondingly unknown. In scientific experimentation the research worker has usually no exact knowledge of the probabilities prior to making his experiment. To draw inferences in the latter case, Bayes introduced a postulate that later became a center of controversy among statisticians. He proposed that if the prior probabilities of mutually exclusive events (like sample drawn from urn A, sample drawn from urn B) were unknown they should be taken to be equal.

Fisher accepted Bayes' theorem and honored its originator as the first to use the theory of probability as an instrument of inductive reasoning. But, from the beginning, he rejected Bayes' postulate. He argued that not knowing the chance of mutually exclusive events and knowing the chance to be equal are two quite different states of knowledge. Thus he accepted and used the Bayesian method only in those cases in which the prior probabilities were known, as they sometimes are, for example, in genetical work.

□ □ □

Another inferential argument, also very old, employs a significance test. The following example contains the essence of the argument. A physicist has produced a theory about matter which makes it possible to *calculate a theoretical value* for the velocity of light $\theta = 183.4$ thousand miles per second. A method of measurement is available that has an observational error e (so that a measured value $y = \theta + e$) and e is known to be approximately normally distributed with mean zero and standard deviation unity. A single observation is made and yields the value $y = 186.0$. Does this support or discredit the hypothesis that the theoretical value is 183.4? If the physicist's hypothesis is true,

$$\frac{e}{\sigma} = \frac{y - \theta}{\sigma} = \frac{186 - 183.4}{1} = \frac{2.6}{1}$$

and the observed discrepancy is 2.6 standard deviations. By consulting a table of the normal curve, we find that deviations greater than 2.6 standard deviations occur by chance less than once in a hundred times. We may infer, therefore, that the physicist's hypothesis is discredited by the experiment and the result is said to be statistically significant at the 1% level.

The parameters of interest might be more complicated, like the coefficient of correlation. Suppose that a sample of 89 pairs of observations gives an observed correlation $r = 0.21$, and suppose we wish to test the hypothesis that the two quantities x and y are not correlated (that the correlation ρ in the population is zero). Does this observation discredit the hypothesis that $\rho = 0$ or is a deviation of this magnitude easily explained by chance? At the turn of the century the standard way of dealing with such problems was to attempt to deduce mathematically the root-mean-square error σ_r of the statistic r; then, assuming r to be approximately normally distributed about ρ, to refer $(r - \rho)/\sigma_r$ to the normal table, as above. This method worked reasonably well for large samples, but serious difficulties occurred for small samples, since the distribution of quantities like r could be highly nonnormal, with both the degree of nonnormality and also the value of σ_r depending on the unknown value of ρ.

Similar difficulties occurred in comparing with its standard error the deviation of sample average $\bar{y}$ from a hypothetical population mean μ, in which the quantity of interest is $t = (\bar{y} - \mu)/(s/\sqrt{n})$. For large samples the sample standard deviation s might reasonably be treated as a constant equal to the population standard deviation, but for moderate and small samples the approximation breaks down. In 1908, Student derived his distribution, which allowed for the fact that with small samples both $\bar{y}$ in the numerator of the ratio *and s in its denominator* are subject to error. It was clear that to make tests of significance in moderate and small samples it was necessary to know the exact distribution of the statistics of interest.

□ □ □

Fisher saw the importance of Student's work and he proved and vastly extended his results mathematically. References to Student's test are practically nonexistent before Fisher advertised it. (We recall that as late as 1922 Gosset, sending his tables to Fisher, wrote: "you are the only man that's ever likely to use them.") Beginning with his work on the correlation coefficient [*C P* 4, 1915], much of Fisher's early work aimed at finding the appropriate distribution functions that would provide exact tests of significance for a variety of comparisons that experimental scientists wished to make. He did this by deriving appropriate distribution theory, introducing the concept of degrees of freedom in connection with the χ^2 test, extending applications of Student's test, retabulating Student's *t,* and developing the analysis of variance and the *F* test.

Along the way confusion that had existed between hypothetical population quantities and statistics calculated from the sample of observations was cleared up. Fisher was first to use the word "parameters" to designate the population values, and he referred to the sample quantities as estimates of the parameters.

E. S. Pearson was an early admirer of Fisher's work, as was J. Neyman. (Gosset, introducing Neyman to Fisher in 1927, wrote that he "is the only person except yourself that I have heard talk about maximum likelyhood [*sic*] as if he enjoyed it.") Working together in the 1930s, Neyman and Pearson produced the formalization of the tests of statistical hypotheses which has come to bear their name. Although originally built in part on Fisher's work, the theory soon diverged from it. It came to be very generally accepted and widely taught, especially in the United States. It was not welcomed by Fisher, probably for two reasons. First, Fisher felt that in the situations considered by Neyman and Pearson, in which not only the hypothesis to be tested was well defined but also the alternatives, the question at issue could be discussed more profitably in terms of the theory of estimation than of significance testing. Second, he felt that the developments of this theory by its originators and especially by their disciples tended more and more to mathematical abstraction, with little consideration for the needs and objectives of scientists for whom the statistical methods were presumably intended. He believed this abstraction sometimes led to philosophical mistakes. For example, probability refers to the proportion of times an event occurs in some population of events. Fisher believed that in some instances Neyman and Pearson were misled into choosing the wrong population of events.

□ □ □

As soon as one starts to think of estimates of parameters, it becomes clear that there could be bad estimates as well as good ones, and this leads to the question of how parameters ought to be estimated. The example used earlier about the velocity of light has simplifying features: there is only one observation; and the observation is made on the velocity of light itself. In practice there would more likely be a number of measurements, subject to error, made not on the velocity itself but on some other quantity on which it depends indirectly. In such cases there could be any number of different ways in which an estimate might be calculated from error-infected data, but some would be much more precise than others.

Fisher asked, is there a method of calculation that is best and can be arrived at from the mathematical and probabilistic structure of the problem itself? To answer this question in 1912 he devised the method of maximum likelihood. Suppose the structure of a problem is known so that it is possible to write down a mathematical expression for the probability density $p(y, \theta)$ of the observed data y given the particular θ. Before the data have been obtained, this expression allows the calculation of the probability density for any set of data y for any known value of the parameter θ. After the experiment has been performed, of course, the data y which has been obtained is fixed, and the value that θ should take is open to speculation as an unknown variable. Fisher proposed calling the function $p(y, \theta)$ *with y fixed and θ variable* the "likelihood" and suggested that the parameter(s) should be chosen so that the likelihood is maximized. Roughly speaking, he proposed taking that estimate of the parameter that makes the observed data most probable.

The method not only provides a method to calculate unique numerical estimates for any problem that can be precisely stated, but also indicates *what mathematical function* of the observations ought to be used to estimate the parameter. Fisher was able to show that for large samples this particular estimate has smallest possible variance.

At this point Fisher was led on to devise a related measure determining the amount of information about a parameter that is theoretically obtainable in large samples. He showed that the method of maximum likelihood maximizes that information and, using this concept, calculated the loss of information incurred in using some other estimate. The ratio of the amount of information obtained using another estimate to that obtained using the maximum likelihood estimate he called the efficiency. He put the case for efficiency vividly, saying that if instead of a maximum likelihood estimate an alternative with 50% efficiency is used, this is equivalent to throwing away half the observations.

The method of maximum likelihood has come to have enormous practical importance. Fisher found many uses for it, particularly in estimating genetical characteristics of populations from observed frequencies. Over the years it has found applications in engineering, psychology, chemistry, economics, and wherever data are associated with models containing unknown parameters. Scarcely an issue of a statistical journal has appeared in the last several decades without reference to this method.

The desirable properties of maximum likelihood estimates were demonstrated mathematically for large samples. However, in investigating measures of spread [*C P* 12, 1920] Fisher discovered a property of certain maximum likelihood estimates that applies for any sample size. In this paper Fisher discussed the relative merits of two estimates of the variance σ^2 of a normal distribution and showed the superiority of the estimate s^2 which is based on the mean square deviation of the observations. He went on to demonstrate that if from the same observations *any* other estimate v of the variance is calculated, the distribution of v conditional on s^2 (that is the distribution of v for data sets yielding a particular fixed value of s^2) *does not contain* the parameter σ^2 of interest. It follows that once s^2 is known, no other estimate gives any further information about σ^2. Later, Fisher referred to s^2 in this instance as *sufficient* for σ^2 or s^2 as a *sufficient statistic*.

The property of sufficiency is as surprising as it is important. For example, the sample average, the sample median and the sample mid-range (half the sum of the largest and smallest observations) may all be used as estimates of the mean μ of a sampled normal distribution. It might be thought, therefore, that some combination of these three could provide a better estimate of μ than any one of them alone, but this is not so. The sample average happens to be sufficient for μ, and any combination of the three estimates can only provide a worse estimate of μ than the sample average.

If sufficient statistics always existed, the estimation problem would be completely solved. But sufficient statistics do not always exist. Sometimes estimation may be made exhaustive by using, with the maximum likelihood, ancillary statistics of the likelihood function. Fisher first suggested the use of ancillary statistics in 1925; in the early 1930s, with the realization that sufficient statistics would not always exist, such means of gaining the additional information from the configuration of the sample took a more prominent place in his thoughts; as he said [*C P* 124, 1935]:

All statisticians know that data are falsified if only a selected part is used. Inductive reasoning cannot aim at a truth that is less than the whole truth. Our conclusions must be warranted by the whole of the data, since less than the whole may be to any degree misleading.

Finally, the likelihood function is always available. It denotes the probability density associated with the fixed data that have been obtained for varying values of the parameters, and as Fisher pointed out, it is always sufficient for the parameters, containing as it does all the relevant information about them coming from the probability model and the data. He pointed out that credible zones for the parameter could be obtained, which included those values giving likelihoods exceeding some suitable fraction of its maximum.

The concept of looking at the whole likelihood function has been enthusiastically pursued by G. A. Barnard and others and is a most valuable tool for the practicing statistician. Problems can occur, for example, where, for certain data and models, several distinct values of a paramater are plausible. A multipeaked likelihood function points this out to the investigator. The electronic computer can be made to plot contour maps of the likelihood function when there is more than one parameter, and the pictures that are plotted can be extremely informative. For example, if for two parameters θ_1 and θ_2 the contour map around the maximum likelihood indicates not a symmetrical peak but a sharp oblique ridge, this could show that the experiment, though very informative about the *difference* between θ_1 and θ_2, provides almost no information about them individually. Seeing the map, the investigator might then be led to devise an experiment to supply the necessary information (such as information about the sum of the parameters) to make individual estimation possible.

Before the need for a deeper understanding of scientific inference was realized, an investigator would typically be interested in some population quantity θ which was to be represented by some function T of the data. This might or might not be an efficient estimate of θ. In any case an attempt was often made to calculate the standard error, or equivalently, the probable error of T, and a statement of the kind $T = 186.0 \pm 1.2$ would summarize the results from the experiment. This statement presumably had something to do with how close T lay to θ, but its interpretation was ambiguous. It might mean that θ had a probability distribution, perhaps roughly normal, centered at 186.0 with standard deviation 1.2. To this there are obvious objections. Some would argue that since θ is a fixed quantity it cannot have a distribution. If, on the other hand, one adopted the view that probability was associated with degree of belief and further was prepared to adopt Bayes' postulate and to assume, for example, that a priori the distribution of the unknown parameter was uniform, then a posteriori a distribution of θ of the required kind would be produced, although not necessarily that referred to above. But many would not accept this view of probability, nor Bayes' postulate, and Fisher was certainly among them.

In our example, the problem of the velocity of light, we supposed it known that to a sufficient approximation the measurements y were distributed about some unknown true value θ with a standard deviation one and that a single observation $y = 186.0$ was obtained. What, then, could be said about θ? Fisher argued (in 1930 and subsequently) as follows: with $y = 186.0$, the knowledge that $y - \theta$ is distributed normally with standard deviation unity induces a normal distribution *on* θ centered at 186.0 with standard deviation unity. He called this induced distribution a fiducial probability distribution. If there were 25 observations instead of 1, then $\bar{y} - \theta$ would be normally distributed with standard deviation $\sigma/\sqrt{25} = 0.2$ and θ would have a fiducial distribution centered at $\bar{y}$ with standard deviation 0.2. However, a fiducial distribution would *not* be produced if the median m were used to estimate the mean of the normal distribution. This is because m is not a sufficient statistic for that mean and some information about θ would have been lost in using this estimate. While $m - \theta$ would have some distribution, which in principle could be derived, the induced distribution from the median could not be a distribution of θ derived from the whole sample of observations, for it would not tell us all we could know about θ.

Fiducial inference thus required sufficient statistics and pivotal quantities in which parameter and statistic are in a special kind of reciprocal relationship. This realization led Fisher [*C P* 24, 1934] to study the conditions on which depended the existence of sufficient statistics or by which estimates might be made exhaustive. In the preface to this paper written for *Contributions to Mathematical Statistics* (1951), he commented:

From a logician's point of view one of the most surprising results obtained by the theory of estimation is that not only the mathematical form of the inferences which can be rigorously drawn concerning the unknown parameters of the populations sampled, from the frequencies observed in a random sample, depends on the particular mathematical specification of this population, but that the logical nature of these inferences depends on this also.

Although the fiducial argument could only be used when sufficient or exhaustive estimation was possible, a number of important problems satisfy the requirement, and Fisher soon extended the argument to these instances.

At this time, in the 1930s, a different approach to the problem of estimation, closely related to significance tests, was being developed, primarily by Neyman. To illustrate his argument we refer again to the problem of the velocity of light. Suppose an observed value $y = 186.0$ has been obtained, and is known to be subject to a normally distributed error having standard deviation unity. The hypothesis that the true value of θ is 180.0 would be discarded at any reasonable level of significance, since

$$\frac{y - \theta}{\sigma} = \frac{186.0 - 180.0}{1} = 6.0$$

and a discrepancy as great as or greater than six standard deviations is an extremely rare event. However, we can imagine postulating other values for θ, some of which would be discredited at a particular fixed significance level and some not. If, for example, we decided to reject all values of θ that gave a significant result at the 5% level, then all suggested values of θ within the interval from 184.04 to 187.96 would be accepted, and all values outside this interval would be rejected. This is because for the normal distribution a deviation of ± 1.96 standard deviations is exceeded in exactly 5% of cases (184.04 = 186 − 1.96σ, 187.96 = 186 + 1.96σ). Neyman called the interval 184.04 to 187.96 a confidence interval.

He showed that the confidence interval has the property that if repeated samples are drawn from the same population, 95% of the intervals cover the supposedly fixed value of θ and 5% do not. If there was more than one observation, a similar argument would be made with y replaced by an estimate of θ, such as the average $\bar{y}$. Such confidence intervals may be readily obtained, whether or not the estimate is efficient, longer intervals being produced by inefficient estimates. The theory was later developed so that the shortest intervals might be obtained in some cases. Helpful though it may be on occasion, as Fisher recognized, the theory has never managed wholly to exclude absurdities such as attaching only 90% confidence to statements that are certainly true, except at the expense of allowing what Fisher saw as the theory's principal defect—allowing a multiplicity of mutually conflicting statements on the basis of one and the same set of data.

The sampling property of confidence intervals, stated above, says nothing at all about the probability of θ given the result of the one particular sample of data actually obtained. Rather, it makes probability statements about the interval calculated from hypothetical samples that have not been obtained. On these grounds Fisher questioned the relevance of confidence intervals for making statements about θ. He felt that stronger inferences could be drawn via likelihood or, where the fiducial argument could be used, an actual probability statement about θ could be made for the *particular* data observed.

For elementary examples, like that of obtaining an interval for the normal mean, the confidence interval when based on sufficient statistics corresponds exactly with the fiducial interval that includes 95% of the fiducial distribution. At first, therefore, confidence interval theory was thought to be a restatement of the same idea. However, for more complex cases like the Behrens' problem discussed below, it soon became clear that this was not so.

Many of the controversies about the fiducial argument came to center on the solution Fisher provided for the problem of estimating the difference between the means of two normally distributed populations with unknown variances that are not assumed to be equal, for which Fisher provided a fiducial distribution in 1935 [*C P* 125]. Tables of this quantity were calculated by

P. V. Sukhatme [94] as early as 1938. Fisher's tabulation followed in 1941 [*C P* 181], and the numerical values gave objective matter for disagreement. Criticisms of the Behrens' test continued to the end of Fisher's life, and even later. In his travels in the United States in 1957–1958 and in Australia in 1959, he made the Behrens' test the subject of a number of his lectures to mathematical audiences, attempting, by introducing the concept of the relevant reference set or subset, to clarify the reasoning on which it was based.

It happens in the special case of Student's t that the test gives the correct level of significance under repeated sampling, *whatever the true variance* σ^2 of the population from which it is drawn. It does not follow that all tests of significance should have this property in respect of such nuisance parameters as σ^2. Nevertheless, the notion was put forward as a criterion that tests of significance generally should satisfy. For the Behrens' test the nuisance parameter is the ratio σ_1^2/σ_2^2 of the unknown variances, and the significance level is not independent of this ratio. In 1947 Welch, working from the Neyman–Pearson viewpoint, developed a formula for significance levels of a criterion that at least approximately yielded significance levels which were independent of σ_1^2/σ_2^2. The function was subsequently tabulated, and the tables were included in *Biometrika Tables* (edited by Pearson and Hartley, 1954). They gave different results from the tables for Behrens' test.

The case against the sampling theory approach may be put very simply. If, say, the ratio of the sampling variance $s_1^2/s_2^2 = 5$, we can be almost certain that the variance ratio σ_1^2/σ_2^2 is neither 50,000 nor 0.0005. However, Welch sought to make the test appropriate for all values of σ_1^2/σ_2^2 whether or not they could reasonably be expected to occur. Fisher argued that, being forced to have this property, the test must assuredly lose sensitivity and that for the Behrens' problem the infinite population sampled should not include all possible samples but only that relevant subset for which s_1^2/s_2^2 takes the value actually observed. He further argued that in giving rise to s_1^2/s_2^2 not all values of σ_1^2/σ_2^2 were equally relevant. In fact, the existence of the sample ratio s_1^2/s_2^2 induced a fiducial distribution on σ_1^2/σ_2^2, and when the Behren's criterion was sampled over this distribution of the nuisance parameter the exact Behrens' significance level was obtained.

□ □ □

In 1935 Fisher was already aware of difficulties in extending consideration to more complex problems and he pointed out the next steps, which were to prove critical for extension of this approach. He suggested [*C P* 124]:

To those who wish to explore for themselves how far the ideas so far developed on this subject will carry us, two types of problems may be suggested. First, how to utilize the whole of the information available in the likelihood function. Only two classes of cases have yet been solved. (a) Sufficient statistics, where the whole course of the function is determined by the value which maximizes it, and where consequently all the available information is contained in the maximum likelihood estimate, without the need of ancillary statistics. (b) In a second case, also of common occurrence, where there is no sufficient estimate, the whole of the ancillary information may be recognized in a set of simple relationships among the sample values, which I have called the configuration of the sample. With these two special cases as guides, the treatment of the general problem might be judged, as far as one can judge of these things, to be ripe for solution.

Problems of the second class concern simultaneous estimation, and seem to me to turn on how we should classify and recognize the various special relationships which may exist among parameters estimated simultaneously. For example, it is easy to show that two parameters may be capable of sufficient estimation jointly, but not severally, because each estimate contributes the ancillary information necessary to complete the other.

Not only were these the critical problems facing the subject at that time, they were to continue recalcitrant to Fisher's as to other approaches. Fisher had advanced the subject as far, apparently, as it could go (at least without some radically new intuition into its nature) and he was reaching out to apprehend the form of the general problem. In 1936, presenting the problem of the Nile, quoted earlier, he sought, at the least, indications of the existence or nonexistence of solutions or, in general, the conditions of solubility of the general problem.

He continued to worry at these problems but to the end of his days was not satisfied with the further solutions he could arrive at. Referring to a problem of simultaneous estimation in 1962 when he was considering whether to have immediate surgery, he said, "If I have another year of life, I feel I might get to the bottom of the thing." Even then, he felt a new intuition on which he wished to work, stimulated by an observation made to him by Graham Wilkinson, about a constraint implicit in the structure of the problem. But he had less than a month to live. He did not unravel the puzzle.

□ □ □

Today there are two main reasons why the fiducial argument is rejected by nearly all statisticians. First, some never accepted the idea that a distribution could be induced in the parameter via the pivotal relationship. Whether one can take the key step depends on whether, when one comes to know the data and to insert the estimates, the new information affects the probability statements one can make about the pivotal: only if one believes these statements remain unchanged is the fiducial argument legitimate.

Another reason for some who have seriously considered fiducial probability having rejected it is the nonuniqueness of the pivotals, especially in cases of multivariate estimation. There are cases in which one may take different pivotals for the same problem and these provide different probability statements. In particular, with more than one parameter, one may choose one pair of pivotals giving one set of probability statements and another distinct pair of pivotals that give a nonequivalent set of probability statements. This was the central difficulty shown up by a paper presented by Monica A. Creasy in 1954 [95] which Fisher denounced. This is a case in which Fisher was clearly unfair; he gave no criterion for his rejection of one of the sets of pivotals Creasy used. He appeared to regard pivotals alternative to those he used as artificial, and in *Statistical Methods and Scientific Inference* he attempted to show how, although his argument is not complete.

To many statisticians fiducial inference has come to seem an unpromising line for further inquiry, and schools of thought employing sampling theory and various modified versions of Bayes' theorem and postulate have come to dominate the world of statistics. Some Bayesians, however, like Harold Jeffreys, feel they are far closer to Fisher than either party is to the sampling theory approach, for both regard the parameter as variable and both make full use of the likelihood function. In contrast, the school of Neyman and Pearson regard the parameter value as fixed and the estimate as varying about this true value. Hence they can say the true value of the parameter lies within their "confidence interval" with a certain probability but not that different values are more or less probable.

Many modern statisticians would probably agree with the assessment of L. J. Savage, a Bayesian who described the fiducial argument as a gallant but unsuccessful attempt to make the Bayesian omelette without breaking the Bayesian eggs. But with the Bayesian approach no less than with Fisher's, very serious obstacles arise. Recently, D. A. S. Fraser [96], Graham Wilkinson [97], and G. A. Barnard [98], working from distinct though related points of view, all going back to the ideas Fisher put forward in 1934 and 1935, have suggested ways in which the difficulties of the fiducial argument arising from the multiplicity of pivotals can be avoided. It is clear that the debate on foundations has yet a long time to run, and that the fiducial idea has a continuing contribution to make to that debate.

18
Retirement

It is said that the only measure of the general board of Cambridge University on which Sir Ronald Fisher voted with the majority was that which proposed to increase the age for compulsory retirement. Fisher was due to retire in 1955 when the measure was approved by senate vote that permitted a professor to postpone his retirement for 2 years. Whatever his disagreements with the university, Fisher had no wish to retire in 1955; in 1957, when he was retired, he had made no plans to go elsewhere. In this, one consideration was the continuation of the elite lines of mice and of the genetic garden he had established at Whittingehame Lodge.

As early as 1953, he began to try to interest the ARC (as he wrote to Besse B. Day) in establishing a genetic garden,

I pointed out that in the eighteenth century as a result of Linnaeus' system of botanical classification it was thought proper to establish Botanic Gardens in various parts of the world, where representatives of the vegetation of the planet could be assembled and studied. I suggested that a rather similar need existed in this century for collections of plant material of which something is already known genetically, and which have appeared therefore in the genetic literature, but are not for the most part available as teaching material, or conserved for their research potentialities.

Like the garden at Whittingehame Lodge, the proposed genetic garden would contain living examples of the genetic factors in garden peas that had been explored in Mendel's classical experiments; exhibition plots where students could see the crosses Mendel had made, and themselves make the counts and verify Mendel's results. It would contain a permanent open-pollinated plot of *Lythrum,* again a classic of genetical experimentation, whose research potentialities were still being explored. Examples of other tristylic plants, like his *Oxalis* and *Primula,* and of plants of serological interest (discovered in the 1950s), would be assembled for use both as research and teaching material. The ARC had not established such a garden when Fisher retired, but he hoped they might be persuaded still to do so.

In 1957 Fisher was elected president of Gonville and Caius College. In normal times the functions of the president are mainly social. He presides at High Table and is responsible for the reception and entertainment of visitors and guests of the college, and Fisher enjoyed this social role immensely. He was not only a gracious host with a taste for good wine but an attentive listener with a taste for good conversation. The president also presides over the Fellows in the election of a new Master, and in 1958 the election of a new Master was the first responsibility of the Fellows. The lengthy series of meetings, both formal and informal, involved tactful handling of differences of opinion, and Fisher, with the aim simply of assuring that the new Master should be acceptable to all, was able to enjoy the process and the personalities revealed in the course of their confabulations. In the end he was more than content, he was pleased with the election of Neville Mott* as Master; the job had been well done, the college well served.

The official portrait of Fisher, as president, (Plate 25) was painted by A. R. Middleton Todd, R. A., in the spring of 1958 and is hung in the college hall.

Meanwhile, the appointment of a new professor of genetics was delayed. Fisher often visited Whittingehame Lodge, where his office and calculating machine were reserved for his use. J. R. Morton, starting graduate research in the department of genetics "during the strange interregnum period, when Fisher, to his disgust and fury, had been compulsorily retired on grounds of age," recalls:

Fisher continued to inhabit the Prof's room at Whittingehame Lodge and greatly to influence the Dept., while the appropriate committee tried vainly to appoint his successor. Of course, the fact that Fisher showed no signs of moving out and that the rest of Whittingehame Lodge, supposedly the Prof. of Genetics' residence, was almost entirely filled with mice, did not make the appointment committee's task any easier. But the point is that the Department of Genetics at that time remained highly 'Fisherian.'

This situation continued throughout 1958 and until Fisher departed for a trip to Australia in March 1959.

*Nobel Laureate in Physics, 1977

Of course, he did not spend all his time in Cambridge. During the first semester of the 1957–1958 academic year he was visiting professor at Michigan State University, and while he was in the United States he visited friends and lectured at New York, Princeton, Washington, Chicago, Ames, Minneapolis, Detroit, Milwaukee, Bloomington, and Urbana. During the Christmas recess he spent a week with C. I. Bliss at New Haven, visited E. B. Wilson in Boston, spent a few days with Besse Day and saw Jack Youden and other statisticians in Washington, and went on to Chapel Hill to visit Gertrude Cox. He was ready to entertain the idea of settling in the United States and might have done so if he had received a reasonable offer. Instead, he returned "shocked" by the developments in mathematical statistics in the United States. Although he had earlier expressed the view that the teaching of ideas "fantastically remote from the natural sciences" had left the country in the position of "a rewarding field for missionary enterprise," the situation appeared even worse then he had supposed; anyway, he was not called to this missionary field.

On his return to England in 1958, he pursued possible opportunities of work in Great Britain, in particular at the University of Durham at Newcastle, where Keith Runcorn was professor of physics, and at the University of Aberdeen, where David Finney was Reader in statistics. These friends hoped their universities might establish a genetics department for him. In fact, Runcorn had suggested the idea as soon as he moved to Newcastle in 1956, and in the latter part of 1958 Fisher paid a series of visits to the university, meeting the faculty and lecturing in the physics and biology departments. He expressed a wish to keep a few mice if he came to Newcastle and to establish a genetical garden. This would have meant finding some land, animal house and assistance for him. The university applied for a grant from the ARC to establish the garden, but they failed to raise the money and the project to bring Fisher to Newcastle was abandoned.

Fisher had visited Australia for 3 months in 1953, at the invitation of E. A. Cornish, head of the Division of Mathematical Statistics of the Commonwealth Scientific and Industrial Research Organisation (CSIRO), and, ever since, Cornish had looked forward to the opportunity for him to spend an extended period in Australia. As soon after his retirement as it could be arranged, Cornish invited him for 6 months, from April to September 1959, giving him a chance to travel to all the capital cities and many CSIRO centers in Australia and to spend more time with the workers at his headquarters in Adelaide. The University of Adelaide took the opportunity offered by his visit in 1959 to award him its honorary degree of doctor of science.

This was so happy a visit that at the end of July, Fisher decided to make his home in Adelaide. He wrote laconically to Cornish that his "few remaining years would be as well spent in Australia, in particular in Adelaide, as elsewhere," and Cornish at once arranged to get his appointment as senior research fellow at CSIRO extended indefinitely.

□ □ □

For Fisher one attraction of the city of Adelaide and the continent of Australia, was the welcome he received, first and foremost from his old friends and former students and then from newer and younger scientists. Alf Cornish, professor of statistics at Adelaide, had been his student at the Galton Laboratory in 1937−1938, and cared for him with solicitude, understanding, and tolerance; honored him; and treated him with comfortable familiarity. Fisher had stayed with the Cornish family in 1953 and was very much at home with them and fond of the children. J. Henry Bennett, professor of genetics at Adelaide, had been his student and later assistant in research in Cambridge in the 1950s. He and his wife welcomed Fisher warmly, and Fisher was charmed with their little children, especially Belinda, the toddler. The baby Nicola became his goddaughter at this time. Graham Wilkinson, in Cornish's department, and his wife and two little boys brought him into their home and shared something of their family life and their interests with him.

In Sydney, Helen Turner, his student in 1938−1939, was always ready to welcome him, enjoy his company, and share her problems and interests. In Brisbane he was welcomed by Mildred Prentice (nee Barnard), a fine redhead at Rothamsted and University College in the mid 1930s. He was now introduced to her family of four redheads. In Canberra Ted Irving, the geophysicist, carried him off on a picnic with his charming new wife and baby and they visited the Observatory at Mount Stromlo. In Melbourne Rupert Leslie took him home to his family of four children and, as they talked, music rang out from all parts of the house as each youngster practiced on his own instrument. Leslie was not Fisher's student. They had met in Melbourne in 1953, and when Leslie came up to Cambridge that autumn for graduate work in psychometry, Fisher interested himself in Leslie's welfare and was able to lend him space to work at Whittingehame Lodge in a time of need. In these, as in new relationships formed in Australia, it was obvious that Fisher found great satisfaction in being brought into the family atmosphere. He needed family, needed the presence of children, and took delight in them. Like a visiting grandfather, he enjoyed not only the attentions and the conversations of the parents but the expeditions and activities of the whole family.

This was, of course, nothing new, but it was seen afresh now that Fisher was no longer the host in his own home, college, or department. He was a family man and liked to have family around him. He had always involved his own children with him in any of his activities that at any given age, they could share. When they were quite little they helped him picking the white rambler roses daily when they were in bloom about his bedroom windows. After the age of 4, they accompanied him on his walks, looking for snails, grouse locusts, or rooks' nests or visiting the experimental chickens on Rothamsted Farm; they made expeditions with him to the Zoological Gardens at Regent's Park and Whipsnade and to archeological sites around Harpenden. During World War II they often went with him to Rothamsted at the weekends to walk the dogs and help with the allotment and the mice. (Mrs. Holt recalls more than one occasion when she was called on to cover up for them, for example, when two small girls decided to let a wild mouse run free for a bit, and nearly lost it.) As adults, his children accompanied him on expeditions to Somerset to count the primrose colonies and quite frequently traveled with him to scientific conferences in Europe. His daughter Joan accompanied him to India in 1956 and to Australia in 1959.

When he moved to Cambridge and was without his own family, he liked to gather other young families around him. Margaret Wright (later Wallace) was his first genetical assistant in Cambridge, and he followed her professional and family careers with appreciation. She left the department only to fulfill the residency requirements at Girton College for her bachelor's degree and, having married, continued work, was promoted to demonstrator, and later won her Ph.D. degree while adding four children to her household. Fisher attended the baptism of at least one of the children. He was enchanted by Mrs. Cavalli and her tiny son when they were in Cambridge, and he met the two Cavalli Sforza boys at a later age with real interest, remembering them with a gift or two of scientific books which he thought they might enjoy; he did the same with Leopold Martin's two sons in Belgium. At Whittingehame Lodge he liked to have families to tea on weekends, especially in the summer, when tea was served in the garden with strawberries and cream as a treat for the children.

His charm and kindness were most happily expressed by the respect with which he treated the interests and questions of the children. Sir Edward Bullard recalls Fisher in this aspect:

One day he was in my house in Cambridge and was talking about how he wanted to work on the genetics of a particular kind of snail, but could not get enough of them. My daughter Emily, a quiet child aged, I suppose, about 6, said nothing at the time but next day came in with a pail full of snails and said "Daddy, you know Prof. Fisher said he wanted some of those snails, there are lots under the cabbage leaves in the field." We at once went to look for Sir Ronald in his lab. I said, "Look, you wanted some snails, Emily has found you some, there they are." He reacted wonderfully and at once shouted for a lab assistant: "Tom, snail houses, quick." Tom brought what were clearly mouse houses. Then Sir Ronald called for snail food and a large tin labelled "SNAIL FOOD" was produced. Emily was fascinated that anyone should keep a stock of snail houses and snail food and asked, "Prof. Fisher, what's in the food?" He replied, "The basis is bran, but it has 17 ingredients. It contains everything a snail needs to keep it well and happy."

What he did with the snails I don't know, maybe they really were the ones he wanted, but he certainly had given great pleasure to Emily. She had liked him and done something for him; then he had taken the trouble to make her feel appreciated and had shown the understanding, humour and showmanship which he could produce so well when he wanted to.

When Fisher settled in Australia he had no grandchildren of his own, which might have induced him to stay in England; in fact, in his lifetime, he had no grandchildren living in England. His son was unmarried. He had attended the wedding of his youngest daughter in 1956, but the young couple had not yet started a family. He saw something of them on his way to Australia, however, for they were living in Calcutta, and he stayed there for 2 weeks. His daughter Elizabeth was married in Kenya in 1958. He had met the groom, Peter Cox, a year or two before, at the time of his engagement to Elizabeth, when he was a newly graduated doctor with a call to the mission field and Elizabeth was completing training as a missionary teacher. Their eldest child and Sir Ronald's first grandchild, Stephen Ronald Cox, was born in September 1959. That autumn Rose was married to William Newsom in Cambridge and Joan to George Box in the United States. Finally, in the summer of 1962, Phyllis was married to John Hodson in Kenya, and within 2 years they emigrated to Australia. Eventually there were 15 grandchildren; Fisher met three of them and planned to see the fourth, who was born in his lifetime—in May 1962—in the autumn of that year. They were all very young.

His best companions among children were about 8 eight years old. When he visited Finney in Aberdeen for a few days, he walked out every day with Finney's boy, who at that age was entrusted by his parents to bring Sir Ronald safe home. A real attachment developed along the way, with serious, happy conversations between them. In Australia the same thing happened.

When he decided to stay in Australia, he expressed a wish to live with a family instead of at St. Mark's College, and in the autumn of 1959 he was lucky enough to be offered lodging with Mrs. Eric Mayo. Mrs. Mayo was a war widow with two college-age sons, John and Oliver, and a younger unmarried sister, Audrey Simpson, living with her; she had four brothers in a big group of family concerns in Adelaide. Fisher remarked, "The house is at least 100 years old in an enormous wilderness garden. Naturally it is the haunt of uncles, aunts and cousins"; and he added, "It is heavenly to be in civilised surroundings at home." In this lively company he enjoyed conversation and sometimes hot argument. Oliver Mayo was already planning to be a geneticist, and there were stimulating discussions with him. Later, when Fisher, the Mayo boys, and a Simpson cousin were all resident at St. Mark's

College, Fisher would sometimes leave the High Table where he presided (he was president of the Senior Combination Room) to talk to the students.

There were also younger cousins, Matthew and Katy Simpson, living near the old house, and Matthew, 7 or 8 years old, became a special friend of Fisher. They walked together in the enormous wilderness garden and about the neighborhood and had happy and long conversations, for Fisher had a lot to learn about Australia, and Matthew was full of curiosity and wonder about the things of nature. Fisher brought back rocks for his collection when he went abroad. It was a red-letter day when, being abroad, he received news of Matthew and Katy. One day a neighbor, answering a knock at her door, came face to face with a white-haired stranger with a little band of children at his heels, for whom he begged a drink of water. It was hot that day, and they had walked a long way. When they had drunk, she watched Fisher hoist Katy on his shoulders and, with Matthew beside him and the other children about him, walk down the hill. They were singing.

Even more picturesque was the way he gathered children around him on his visits to the Indian Statistical Institute in 1961 and 1962. He seemed a veritable Pied Piper, strolling in the gardens of the institute with a little throng of children of various ages, the children of bearers or drivers who were residents of the compound, and when he sat down, a little one would climb on to his lap or lean against his kness. In this case, the fascination he held for them was clearly not only in his words, which they could not understand, but in himself. Charmed by this recurrent scene, Mrs. Mahalanobis had a photograph of Fisher with the children taken in 1962 (Plate 26).

□ □ □

There was a resonance between the freshness of perception of the child and Fisher's perception, his curiosity and wonder at what he saw. There was also a sense of sympathy with the freedom of thought and imagination the child enjoyed. Fisher was always a rebel against official constraints and was irked by the need to conform in everyday affairs. In Australia, he was happy to discharge the responsibility for making and keeping his engagements largely on his hosts, and he did not always make it easy for them. For example, he was already in India when he decided to take his daughter Joan along with him to Australia, and his telegram announcing the fact reached Cornish in Adelaide only the day before he was due to arrive in Sydney. Thrown off balance by the many sudden changes of plan entailed, Cornish relayed this wholly unexpected news to Helen Turner in Sydney in a voice that rose an octave as he spoke. Miss Turner reassured him she would arrange accommodations in Sydney; calming down, Cornish suggested getting an oyster supper for their arrival. Next day he arrived in Sydney, and Miss Turner met him at the airport (with a large quantity of oysters), and they went together to the BOAC reception area, having half an hour to spare before the Fishers should arrive. On the announcement board, no time of arrival was listed against the flight from India, and on inquiry, they learned that the flight had been grounded in Singapore and would be delayed 36 hours, awaiting an engine. Cornish had to return to Adelaide before the Fishers arrived, but he and Helen Turner enjoyed the supper of their favorite oysters and wondered what other surprises Fisher might spring on them.

On one of their visits to Sydney, Helen took them to visit Kuringai Chase, a big national park north of Sydney, famous for its Hawkesbury sandstone flora. Many flowers were in bloom and, though picking is prohibited, they gathered a few so that Fisher could hold them close to his eyes to see them. Some she could not identify (to answer all the questions Fisher might ask on a picnic, she said, one would need to have a botanist, a geologist, a zoologist, and so forth in the party); so she made up a tiny bunch to take away for identification. Fisher carried it back to the car and, as they got in, Helen said, "Give me the loot, Ron. I'll hide it in the glove box till we're past the gate." Voice from behind: "It's too late for that, madam," and there was the ranger. "What have you got there?" They handed over their flowers, Helen pleading that she was showing a distinguished scientific visitor around. "You should know, madam, that in such circumstances it is possible to get a special permit to pick flowers," he said, "And what is more, from the remark I overheard as I came up, it is clear you knew you were breaking the law." Names and addresses were duly taken. Helen recalls: "Sir Ronald seemed most distressed on my account—though the tale he told my laboratory colleagues over a beer in the pub next afternoon lost nothing in the telling, and he obviously regarded it as a great joke, as I did myself."

He did not like to feel that he was being organized, or even protected by others. Joan lived at St. Ann's College, just across the green from St. Mark's College where Fisher was, and at the weekends she would call for him at St. Mark's and they would go out together. One Sunday morning he was not there, nor did he reappear during the morning, nor telephone. About noon, she was sufficiently anxious to telephone Cornish and ask if he knew anything about Fisher's whereabouts. She learned that he had turned up at Cornish's door a few minutes earlier, thirsty from his walk, and had cheerfully demanded beer, which he was enjoying at that moment. Joan was asked to join the family for lunch, and afterward they drove out to the beach and swam. His shepherds sighed with relief that the adventure was well over, for the Cornishes lived on the other side of the city of Adelaide, in the suburbs at least 4 miles away. It was dangerous for him to go alone, picking his way blindly along the curb with the aid of his walking stick and across the city traffic. But they could not say this to him, for their solicitude might provoke more dangerous demonstrations of his self-sufficiency.

It seemed that he positively liked getting away with his absent-mindedness and irresponsibility. The insouciance with which he failed to cooperate was sometimes infuriating and once or twice caused embarrassment to those who made arrangements for him that, when the time came, he preferred to know nothing about. But this was no new trait of his character; accepting it, his Australian friends were able, in retrospect, to laugh at their own discomforture. His attitude may be illustrated by incidents occurring on his arrival and at his departure in Australia in 1953. It was the same attitude that earned the comment that his blindness was selective: he could always see the man he wanted to talk to. This was not true, and yet there was truth in it.

On arrival at Sydney, having left his raincoat on the aircraft, he made inquiries for it at the QANTAS office. The clerk brought out an enormous volume in which to check the item and asked what color the coat was. Earlier Fisher had mentioned the color but now the formality of the proceeding annoyed him. "Colour?" he replied, "Tell me the colours by which you classify your coats and I will tell you which it is." Two women standing by the counter smiled at this retort, and Fisher turned to engage them in conversation on the silly questions officials asked, while Helen Turner whispered to the clerk that the coat was fawn, and he produced it. Fisher had left his fountain pen in the plane, too, and when they were buying a new one, he drew out an enormous ring of keys from his pocket, tossed them lightly in the air and said, "Ah, I still have these; it will cost me £40 if I lose them." They were the set of keys for Caius College and college property, useless in Australia. A few days later he broke the wing off his reading glasses, and it emerged that he had not brought a spare pair with him. Helen Turner took the broken frame to be repaired and vividly recalls the astonishment and incredulity of the optician that any man who needed so huge a correction should have traveled 10,000 miles from home without a spare pair.

Then he announced that he had lost the return half of his air ticket; while he was getting a reissue, he decided to change the return route booked on the ticket. These "simple" requests somewhat exasperated the clerks; they had to ring to London to obtain the number on the lost ticket, but they managed to deal with the problems in time and issued a new round-the-world ticket, as desired. The sequel followed when Cornish, returning from the railway station after he had seen Fisher off on the first leg of his journey back to England, noticed one of the crumpled scraps of paper in the fireplace of the room Fisher had used and, smoothing it out, saw it was the long-lost ticket, found again and quietly discarded.

The departure had not gone smoothly. They had to make an early start, about 5 a.m. The Wolseley was out of order and Cornish used the 1928

Dodge. The car was loaded and ready to go when Fisher asked, "Where's my bag?" It took Cornish a few minutes to discover it, where Fisher had laid it down outside the front door (he had gone out that way first and the door had shut behind him; leaving the bag, he had then made his way through the undergrowth to the back door by the garage to pick up hat, coat, and stick). Backing out in the darkness, the car ran over a child's scooter in the driveway which jammed up over the fan blade. They extricated the scooter and drove off in a hurry. The fan blade had been bent, however and, by a tin-opener effect, had sliced open the bottom end of the radiator, so that the car broke down halfway to the railway station. Fisher got out of the car, carrying his case. Cornish, now anxious about time, instructed him in firm tones: "Wait here for me. Just *stay here* while I go and get a taxi." When he returned, of course, Fisher was nowhere to be seen. As they transferred the luggage to the taxi, the cabbie observed "an old codger with glasses" coming toward them, who hailed them: "Ah, Alf, you found a taxi." But where was his brief case? It was too late now to ask. They rushed to the station, brought out tickets, paid off the taximan; Cornish found Fisher halted talking with a porter, hauled him off, put him aboard, and told the conductor, with relief, "He's all yours!" Fisher popped his head out of the window: "Alf, if you find it send it on to me in Cambridge, will you?" He sat down, picked up a paper, brought out a pipe, and relaxed.

Cornish returned home. At about 9:30 a.m. the phone rang. The caller had noticed a bag by the curbside as he drove by and had picked it up. Inside he read the notice, written large, which a thoughtful secretary had attached before Fisher had left England: "IF FOUND ANYWHERE IN AUSTRALIA RETURN TO DR. E. A. CORNISH" and giving his address. So the bag was returned to Cornish and Cornish sent it by air freight to Sydney and telegraphed Helen Turner to that effect. He did not mention which airline it was on but she telephoned them both repeatedly, until the case was secured.

Next morning rain was falling heavily and Miss Turner, unable to take Fisher to the airport herself, decided to check with QANTAS to find out if his flight would actually depart and, at the same time, to assure herself that Oliver Lancaster, the professor of statistics who was to see Fisher off, had arrived at the airport and booked Fisher in. The response from the desk was all she could desire: "Relax, relax, Madam, we have lost one passenger but your man is safely aboard." Not long after, she received a call from Lancaster, stuttering in agitation. Fisher had departed all right, but before going aboard, he had asked for Helen Turner's address, so that he might send her flowers. Lancaster had cast about in his memory and turned up not her address but the fact that her telephone number was of a certain form, both halves being divisible by seven. In the telephone book they had found H. Turner listed at 4900 and the flowers had been sent there. Afterwards he recalled that the right number was 4928 and was now ringing to tell her that a bouquet from Fisher was on its way to her namesake.

□ □ □

Not allowing himself to be constrained more than was convenient by the need to keep track of time and place and possessions, Fisher was the more free to engage himself with the activities he did enjoy, in particular, meeting and talking with scientists and sharing their interest and knowledge of the local scene. If one should select a single quality by which to characterize him during this visit to Australia in 1959, it would be the vitality of his interest in everything about him. He not only toured all the capital cities of the continent, meeting new faces, new problems, new experiences, giving lectures, making expeditions to explore the local fauna and flora, and being entertained, but he positively enjoyed it. He was happily involved, if not with arrangements made for him by others then with his own explorations, in the local Botanic Gardens where he would observe the different species of *Araucaria* trees and feed the ducks and swans, or in the local bars where discussion turned to local affairs, Australian beers, licensing hours, and horse racing.

Even at the end of his 6-month visit to Australia, during the 2 weeks he toured New Zealand with his daughter, his enthusiasm did not flag; he was as alive to the interest of the great sows at Ruakura Animal Research Station at the end of the second week as he had been to the splendors of the Southern Alps at the beginning of the first. Yet the days were crammed. On Saturday, August 5 they flew from Sydney to Christchurch, Christchurch to Dunedin, transferred to an Austin four-seater to fly inland, almost scraping the tops of the Remarkable Range on the way to Queenstown and the lakeside home of Dick Purves. On Sunday they saw the intervening country (moors and gorge and reservoir, etc.) as they drove back to Dunedin. On Monday Fisher gave a talk in the Medical School at Dunedin, followed by three conferences with medicos, geneticists and mathematicians, then flew on to Christchurch. On Tuesday he was picked up at 9 a.m. for a conference with Crop Research Officers, followed by a private conversation with a mathematician, before driving out to the Agricultural College for lunch. In the afternoon he lectured in the University, had formal tea with the vice-chancellor, and, with some professors, visited the new engineering college. Then there was a fairly long drive out to Lyttleton where he boarded the night ferry for Wellington. On Wednesday and Thursday he was in Wellington fulfilling the engagements arranged for him by R. M. Williamson of DSIR and the University of Wellington, and he spent one happy evening discussing astronomy and theories of the universe. On Thursday afternoon he was driven to Palmerston North, to stay at the famous hostelry of Massey College. On Friday morning he took a brief opportunity to walk in the lovely gardens there. He had informal conversations that morning, met reporters after lunch, lectured in the afternoon, dined out with a party that evening.

He was tired at the end of the first week and showed it by a slight irritability but the weekend provided relief from social functions. On Saturday they drove up into the mountains and across the volcanic center of North Island to Taupo and, walking out, picked up pumice on the shores of Lake Taupo (25 × 17 miles in extent), apparently the crater of an extinct volcano. On Sunday they visited the Wairakei project in the hot springs area, where Mr. Banwell entertained them in his office, with the help of a map of the geology of the district that covered the whole floor, and models of the bore holes, showing their depth and temperature, projecting in a corresponding position from the ceiling, and with other models and maps. Then they went to see the valley themselves. On Monday came the drive down from the mountains, following the course of the River Waikato, at first in a series of falls and cascades, then spreading itself quietly in the lowlands and resting in the reservoir not far from Hamilton, where Fisher spent Tuesday and Wednesday at Ruakura Animal Research Station before returning to Wellington by train and, the following day, sailing for Sydney.

His imagination too was engaged: he had only to take a walk along the gorge of the River Torrens north of Adelaide, where city waste was dumped, to imagine its transformation by a series of locks above the city, raising the level of the water from a trickle to cover the river bed, and with canoes and punts going up and down from the steps of waterside residences, a scene of beauty and amenity. He had only to visit the Seppelt winery in the Barossa Valley, famous for its wines, to imagine a vast new export business, with the London vintners (and his own college) selecting from Australian vintages no less famous than those of France. The Australian wineries, he felt, should be too proud to allow Australian wines to be dismissed abroad, because of the poor reputation of one or two cheap brands known in England, while they kept such excellent wines for their own consumption.

Coming fresh and enthusiastically to such experiences, he saw possibilities in them that to native Australians were not real possibilities, for he ignored economic facts they accepted as binding. His arguments were provocative. Implicitly, he challenged their assumptions; he implied that men need not accept man-made systems, like economics, as their masters, but have the option of adapting them to meet their real needs.

Soon after his arrival in Australia he made quite a sensation in the local press, with repercussions in the British press, by suggesting that in the case of an infant being diagnosed as imbecile and incurable, no one but the parents had the right to put an end to its existence. He implied that parents do have the moral right and should be given the legal right, with proper legal and medical safeguards, and that if such children died it would be a mercy to the parents and a relief to society. In this case he challenged doctors to reconsider their man-made code and admit they had the option of adapting it to meet the real needs of the people they served in contemporary society. He was reported rather crudely as advocating euthanasia and mercy killing, and his suggestion was met by some rather hysterical correspondence in the papers and by outrage on the part of some of the medical profession.

□ □ □

Fisher had already made himself unpopular with some of the medical profession by challenging the conclusions drawn in 1956 from a study of smoking and lung cancer among British doctors. The study was launched in 1952, and two-thirds of the doctors of Great Britain had sent in returns to the questionnaire; an association was shown to exist between cigarette smoking and lung cancer. But an annotation published in the *British Medical Journal* in 1956 went beyond the evidence and, assuming smoking had been proven to be the cause of lung cancer, in Fisher's words, led up to "the almost shrill conclusion that it was necessary that every device of modern publicity should be employed to bring home to the world at large this terrible danger."

The campaign against smoking, thus announced, violated the whole trend of Fisher's philosophy. First, it assumed something to be known that was not known. As a statistician, he recognized that the information obtained by the British investigation and others of the same sort proved no more than a correlation, in which case there are three logical possibilities: *A* may be the cause of *B*, *B* may be the cause of *A*, or something else may be the cause of both. There was a prima facie case for further investigation of smoking and of other possible causes of lung cancer. But, second, assuming the answer would close the mind to the range of possibilities open to investigation. To a geneticist, the most plausible alternative to the theory as it stood was genetical causation, but he feared even this might not be investigated; if the cause of lung cancer were assumed to be known, all scientific research into the question would be stultified. Finally, but not unimportant, Fisher was an individualist and thought people should be shown the options open to them and be allowed to judge for themselves. Just as he would have trusted parents of a hopelessly handicapped infant to decide freely on the evidence, without moral suasion, about the life of their child, so he did not think it good "to plant fear in the minds of perhaps a hundred million smokers throughout the world—to plant it with the aid of all the means of modern publicity backed by public money, without knowing for certain that they have anything to be afraid of in the particular habit against which the propaganda is to be directed."

From the time Fisher first expressed his views, he was under fire. He was taking an unpopular line in opposing medical authority. If his arguments could not be discredited, it was felt to be all the more morally outrageous to put them forward, to encourage people, as it seemed, to kill themselves by smoking. The emotional tone of the first as of succeeding campaigns against smoking was one of religious fervor; it was mounted on simple absolutism and any agnosticism appeared as moral turpitude. No ambiguity, no doubt was admissible.

Moreover, Fisher had opened himself to attack, for in 1956 he had accepted the invitation of the Tobacco Manufacturers Standing Committee to be their scientific consultant. It was easy to insinuate that he served the tobacco companies and so to discredit what he said. In July 1957 he took part in a BBC program on smoking. The interview was prerecorded; he did not know until the broadcast was put on the air that he was to be introduced with the words: "Now Sir Ronald Fisher is the scientific consultant to the Tobacco Manufacturers' Standing Committee and we put this question to him . . ." He protested strongly to the BBC against the implication that his qualification to speak was his consultant status and that he spoke for the Standing Committee. He also wrote to the Chairman of the Standing Committee, reiterating his position:

I am free under our agreement to say what I like, however much financial damage it may do to the interests of the tobacco companies. In fact I do not receive the small fee I have accepted as payment for saying anything but for giving expert advice to the Committee in question, whenever they are pleased to ask for it. Consequently, I intend to be quite unresponsive if ever it occurs to you that I can usefully say or do anything.

And the chairman concurred: Fisher's statement set out their relation "exactly as we always understood it to be."

Fisher neither stopped serving the Standing Committee, nor ceased to speak out about the current evidence on the subject of smoking and lung cancer. Already he was in communication with F. von Verschuer in Berlin about the possibility of gathering data on the smoking habits of twins, with a view to finding out whether genotypic differences existed. By the autumn of 1957 Cavalli Sforza was proposing to undertake similar research in Italy. In March 1958, von Verschuer's data arrived, and in April Fisher sent a letter to the *British Medical Journal* [C P 275], delayed only in order to check some doubtful twin cases in Berlin. Of 82 recorded pairs of males, 51 were monozygotic and genetically identical, 31 dizygotic and as closely related therefore as normal brothers. Of the identical twins 12 pairs, less than one-quarter, showed distinct differences of smoking habit, whereas more than one-half of the fraternal twins were dissimilar to the same extent. The data could be rearranged in several ways, according to the extent to which attention was given to minor variations, but the general picture remained the same; there could be little doubt that the genotype exercised a considerable influence on smoking and on the particular habit of smoking adopted.

In May 1958 Fisher received "priceless" data on English twins collected by Dr. Eliot Slater of the Maudsley Hospital. His summary of this material appeared in August. In this case the twins were girls, 71 pairs, of which 53 were identical. The fraternal twins were equally divided in smoking habit (9 pairs alike and 9 unlike), whereas the identical twins were more similar (44 pairs alike, 9 unlike), a confirmation of the German data. However, Slater's data referred to twin pairs, half of whom had been separated at birth. The data on identical twins could be subdivided to show what was the effect of mutual influence on the smoking habits of twins, and this appeared to be negligible. Fisher concluded:

There is nothing to stop those who greatly desire it from believing that lung cancer is caused by smoking cigarettes. They should also believe that inhaling cigarette smoke is a protection.* To believe either is, however, to run the risk of failing to recognize, and therefore failing to prevent, other and more geniune causes.

This sentiment was echoed 13 years later by Professor Theodor D. Sterling [99] in a report generated by the Washington University project on the review of crucial data bearing on the smoking and health issue. Having devoted 80 pages to an impartial and critical review of the evidence available by 1971, and having found it still unconvincing and full of ambiguities, Sterling went on to suggest:

It would be very desirable if the antecedent for lung cancer turned out to be or only depended on such a simple event as smoking. The readiness with which the existing evidence has been accepted as demonstrating causality for cigarette smoking perhaps is the best measure of the desire to keep our world simple and orderly. But cancer is a complex disease. . . . There is a real danger that, having cast cigarettes as the prime villain, scientific interest and effort will turn to needs that appear more pressing.

Fisher was doing his best to incite inquiry. His talk entitled Cigarettes, Cancer, and Statistics [C P 274], given in Minneapolis in 1957, was published in the spring number of *The Centennial Review* (1958). He addressed the Cambridge University Medical Society at a meeting on "Combustion and the Products Formed" at which R. Doll also spoke. He spoke on "Smoking and

*The original study of R. Doll and A. Bradford Hill in 1950 showed a negative correlation between inhaling and lung cancer. In the study on doctors the question about inhaling was not asked.

Lung Cancer" at the Annual Meeting of the British Association for the Prevention of Tuberculosis. In October he addressed the Liverpool Medical School on the same topic. In November the supply of offprints of his notes in *Nature* ran out, and he began contemplating the preparation of a pamphlet in which his earlier notes in *Nature* and letters to the *British Medical Journal* might be bound together. They appeared the following May, together with some new material.

On the question of air pollution, he was trying to obtain data on urban and rural incidence of lung cancer; he was checking out expert opinion on whether there was any differentiation in the histological types of cancer occurring in these different environments. He was taking considerable interest in the question of diesel and automobile fumes, suspected of containing possible carcinogens. On the question of inhaling, he was asking Bradford Hill for the actual data collected in the early study, to analyze them himself. He managed to get these data just before leaving England in March 1959. He analyzed them on receipt, ready for his pamphlet.

The actual observations were well worth the trouble of obtaining them. He felt he was making some progress in his main object of getting questions asked, and the figures, showing a clearly significant negative correlation between inhaling and cancer expected from the various classes of cigarette smokers, gave him a provocative question raiser.

In Australia his lectures on the subject were frankly intended to raise questions. He spoke of "teasing" the British Medical Association and of being "deliberately provocative." He obviously enjoyed giving these talks, which so clearly illustrated points he wanted to make about inference: the consequences of not randomizing, the need to check on the assumptions implicit in any hypothesis, and the need to admit all of the data, *particularly* those which seemed anomalous. Audiences were quick to laugh when he pointed out the negative correlation between lung cancer and inhaling; and there were other anomalies: the positive correlation between smoking habit and genotype, the lack of success at that time in isolating any carcinogenic substance from tobacco, the difference in incidence of cancer between smokers of cigarettes and smokers of tobacco in other forms, between people in different environments, between the sexes. It became obvious that more and better data were needed before any hypothesis could be considered to have been convincingly demonstrated. If he seemed to some the devil's advocate, his reasonableness, good-humor and wit were hard to resist. His talks were generally well received.

At the same time he was the target of a good deal of abuse for his stand. He took comfort at the number of correspondents who wrote appreciatively of his pamphlet and at the news gleaned in Australia from an old *Spectator*, "that opinion in England has been somewhat softened up and several of my points seem to be taken seriously."

Genetic studies proliferated. First von Verschuer's data came in, then Slater's. A little later in 1958 G. F. Todd and J. I. Mason presented a summary of current evidence from their inquiry. In January 1959 L. Friberg of the Institute of Hygiene, Stockholm, sent a paper on "Smoking Habits of Twins in Sweden" for Fisher to submit to *Nature,* and Tage Kemp wrote about a similar inquiry he proposed to make in Denmark. Later that year, Cavalli Sforza presented the results of his twin study in Italy to the Italian Association of Genetics.

Apart from twin studies, Fisher thought it worthwhile to look into possible association between smoking and identifiable human genotypes. In September 1958 he was in Brussels for the meetings of the International Statistical Institute and, at the invitation of L. Martin, spoke to the International Pharmacological Society also on the subject of scientific inference, a topic intimately connected with the smoking controversy. With Martin he discussed the serological investigations Martin and Hubinat were planning to start with the cooperation of the Belgian National Transfusion Service, in hopes of identifing any serological differences that might exist in respect of smoking habits. Fisher arranged sponsorship of this investigation by the Tobacco Manufacturers' Standing Committee. At the same time, he was urging on that committee the desirability of making a population survey of smoking habits with the collection of data on a large scale on taste testing, the secretor factor, and blood groups.

In 1958 Fisher was brought into discussion of the evidence in the United States in connection with legal suits expected to be brought to trial against tobacco manufacturers for personal damage caused by their products. Early in 1960 he visited the United States at the invitation of a legal firm representing an American tobacco company, whose case was brought to trial in April that year. Other suits were either not brought to trial or were unsuccessful, and the legal pressure on tobacco companies was relieved for a time, but the question of the causation of lung cancer continued unresolved.

□ □ □

Fisher welcomed the chance to visit America to see some of his friends there. He spent several weeks with George and Joan Box in Madison, and got to know his new son-in-law better. Incidentally, he was able to avoid some of the summer heat of Australia, which was a real consideration with him.

He spent most of 1960 in Adelaide, his new home, working on new editions of his books, consulting on statistical and genetical problems in a wide range of research projects in Australia, and making his own contributions to the literature. He was happy there, unhampered by the responsibilities of administrative duties and sheltered from the bitterness of scientific controversy abroad. But he did not seek it as seclusion: even in 1960 he got away to the meetings of the International Statistical Institute in Tokyo and had a good time traveling in Japan, lecturing and sight-seeing, much of the time in the company of George Box.

In 1961 he spent a great deal of time abroad. In January and February of 1961, and again 1962, he arranged to visit the Mahalanobis home at the Indian Statistical Institute in Calcutta, with an ulterior motive of avoiding again the summer heat of Australia. Whenever he was with the Mahalanobis household he was given the same apartments among the familiar physical surroundings; he enjoyed the freedom of a household run along familiar lines and by people he knew personally, for the chief house servants did not change, and his beloved Ranee and Prasanta Mahalanobis saw that he treated it like home and were glad. In April he passed through the United States on his way to England for a lengthy visit. He was in France for the International Statistical Institute meeting at Paris, and in Italy for the International Congress of Human Genetics at Rome. At the end of September he returned across the United States and thence flew to Australia.

During these travels he visited Madison twice and met his granddaughter Helen Box; in September saw her take her first steps on tiptoe from her mother toward him. (Plate 25). Others of his children visited him in Cambridge, including the Cox family on furlough from Kenya with his grandchildren Stephen and Jenny, and he visited his sister in Argyllshire. He visited Leeds University to receive an honorary doctorate of science and Rome to be inducted into the Pontifical Academy of the Vatican. He saw many friends both at meetings and in their own homes. He stayed a few days with E. B. Ford at All Souls College, Oxford. This was a rather special visit, for Ford was then able to read aloud about half the chapters of his book *Ecological Genetics* (1962) [100]. It was 30 years since he had first discussed with Fisher his plan to do the research work, and to write the book; the realization of this plan was a great achievement. Ford recalls:

Fisher expressed himself as satisfied with the progress of the work, planned, as he well knew, so long before. When I received his approval I put the question I had held in store for many years. I asked if he would allow me to dedicate the book to him. He agreed, and I believe the compliment, the best I had to offer, brought him pleasure. The dedication had in the end to be slightly altered for, by the time the book was published, he was dead.

While at Oxford on this visit, they planned to meet in Italy some weeks later, for both had business in Rome, and Fisher was anxious to show Ford the pavilion in the Vatican Gardens where the Academy is housed, and the Vatican gardens themselves. "So on the last morning we drove to the station happily enough, with our plans to see each other again quite soon. Yet, as I stood by his train to say good-bye, I suddenly knew, quite certainly, that I should never see him again. Our meeting in Italy had to be cancelled for reasons we could not have foreseen. He went back to Australia and died there."

On his return to Adelaide, Fisher took up residence at St. Mark's College. Accommodation there had improved and the Master, R. B. Lewis, was most helpful in arranging a few simple additions, like better lighting, to make Fisher's room more comfortable. When he visited India at the beginning of 1962, it was clear to Mrs. Mahalanobis that he was not well, but he refused to discuss the matter with her. Back in Adelaide, his appearance gave his friends no cause for alarm. He said nothing of ill-health, and the Master recalls the regularity of his habits, his sitting upright as ever at breakfast and setting off punctually and almost jauntily to walk to the office, just before nine every morning.

About the end of the first week in July, however, all that changed. Fisher forgot an appointment he had made for an X-ray examination and was away from the office when the doctor's assistant rang to find out whether he intended to come that day. Alf Cornish took the phone call and, realizing the matter was serious, confronted Fisher with the phone message. Fisher admitted that he was sure the examination would show he had cancer of the bowel; by then he was glad to have Cornish and Bennett in his confidence, and to have their support when X-ray and pathological examination confirmed his opinion.

After very thorough consultation, he decided to have the operation done locally and at once. The operation was successfully performed on July 21 by the hand of R. P. Jepson, F.R.C.S., professor of surgery at Adelaide University. There were some anxious moments during the following week, for the wound became infected by colon bacteria, but this was soon controlled by antibiotics. He recovered from the setback, and by the end of the week seemed well out of the woods. He was in his usual spirits and enjoyed the cossetting. The nurses were absolutely devoted to him and gave him everything they could. They were "his angels." The physician, Dr. David Phear from Caius (son of Caius' former Praelector), was also devoted to him.

For Alf Cornish it was an anxious time. His wife also was in hospital and very ill. Yet he faithfully looked after Fisher's interests and lent him his strength. He visited Fisher on the afternoon of Sunday, July 29 and later recalled: "He had gripped my forearm and held tightly for practically the whole of the time I was with him; his mind was active and I had to assure him and reassure him that his correspondence had first priority on my time."

Henry and Lilian Bennett looked after him with incomparable tenderness and care. Lilian took charge of his laundry and gave him much comfort by her visits. She brought Belinda and Nicki to see him once, just long enough to give him a kiss and a little present. They were planning for him to spend his convalescence in their home. Henry visited often. On the Sunday he went into hospital to see him three times. He found him in good spirits and alert, and he read several chapters of a C. P. Snow novel to him till he fell asleep. Each trip meant a 20-mile drive, and no danger to Fisher was apprehended by the doctors. He went because he wanted to be with him. The third time he arrived too late: he had already left home when a phone call came from the hospital to tell him Fisher had suffered an embolism; he was waiting to go in to see him when the doctor came out to say that Sir Ronald Fisher was dead.

□ □ □

He died as he had lived, prepared to meet the contacts and conflicts of life with commitment and courage, as if he would prevail, making no provision against the possibility of failure. He left no will. He told no member of his family of his illness or of the impending surgery. His son was in England; his daughters were scattered in Asia, Africa and America; none was in Australia who might attend his funeral, none to make the decisions or take charge of his affairs. So, by default, Sir Ronald bequeathed to his friends and scientific colleagues in Adelaide the trust of his body and his earthly estate.

The funeral service was held at St. Peter's Cathedral on August 2, 1962. It was made a college service: the six young pallbearers in their college gowns were all students who had special ties with Sir Ronald and came from the genetics department, the CSIRO and St. Mark's College. They bore in the casket with Sir Ronald's velvet cap and scarlet gown folded above it. The Master of St. Mark's College was lector and read from the forty-fourth chapter of the book of *Ecclesiasticus*. Alf Cornish spoke briefly and with moving sincerity of the great achievements of Fisher's 50 years of professional life, and of the man who had made them. After the service the cortege moved off under police escort to Springbank Crematorium where the last rites were performed. His ashes are interred in St. Peter's Cathedral beside the pew dedicated to his name. Thus scientists of the generations who had been his students, and were the heirs of his thoughts, gave him honor.

His memorial is in the minds and hearts of such men, his legacy intangible, for it has its being in the freedom and clarity of thought that he has enabled and in the faith that he espoused. As he had said in a BBC broadcast on "Science and Christianity" [C P 263, 1955], this faith is not credulity: "it is necessary to be clear about our ignorance. This is the research scientist's first important step. . . . We must *not* fool ourselves into thinking that we know that of which we have no real evidence, and which, therefore, we do not know, but at most accept"; nor is it irreligion, for it may be inextricably a part of traditional religion, as his faith was of the tradition of the Church of England; "but is a quality, very like courage, which makes one hold fast to that which is good, irrespective of all the discouragement, slander, intimidation, etc. which Christians have learnt to expect in their pilgrimage." With that faith, he was supremely an individualist, and, if the life of any man was guided wholly by the truth, as he conceived it, it was Sir Ronald Fisher.

References

1. Bernard Norton and E. S. Pearson (1976) A note on the background to and refereeing of R. A. Fisher's 1918 paper 'On the correlation between relatives on the supposition of Mendelian inheritance'. *Notes and Records of the Royal Society,* **31**: 1. The referees' reports are given in Appendices (i) and (ii). Appendix (iii) is a transcript of Fisher's 1911 talk, as recorded in the minute book of the Cambridge University Eugenics Society. A corrected version of this talk found among Fisher's possessions and deposited with the Fisher files at Adelaide, is to be reproduced with diagrams in a volume entitled *Natural selection, heredity and eugenics,* J. H. Bennett, Ed., Adelaide University Press.

2. K. Pearson (1903) Mathematical contributions to the theory of evolution. XII: On a generalized theory of alternative inheritance, with special reference to Mendel's Laws, *Phil. Trans.,* **203 A**: 53 – 87.

3. G. U. Yule (1907) On the theory of inheritance of quantitative compound characters on the basis of Mendel's laws—a preliminary note. In *Report of the Third International Conference on Genetics* (London: Spottiswoode) pp. 140 – 142.

4. H. Belloc (1912) *The Four Men.*

5. F. Galton (1883). *Inquiries into Human Faculty and its Development.* London: Cassell.

6. R. A. Fisher, review of *The Shadow on the Universe* (I. M. Clayton), *Eug. Rev.* (1914 – 1915) **7**: 210.

7. R. A. Fisher, review of *Military Selection and Race Deterioration* (V. L. Kellogg), *Eug. Rev.* (1915 – 1916) **8**:264 – 265.

8. R. A. Fisher, Racial Repair, *Eug. Rev.* (1914 – 1915) **7**:204 – 207.

9. R. A. Fisher, review of The Racial Value of Death (P. Popenoe), *Eug. Rev.* (1915 – 1916) **8**:291 – 292.

10. R. A. Fisher, Review of *Citizen's Charter, a scheme of National Organisation* (C. E. Innes). *Eug. Rev.* (1915 – 1916) **8**:274 – 275.

11. R. A. Fisher, Review of *Evolution and Man* (M. M. Metcalf). *Eug. Rev.* (1915–1916) 8:288.

12. A. Quetelet (1835) *Sur l'homme et le développement de ses facultés: Physique sociale,* 2 vols. Brussels: Muquardt.

13. Student (1907) On the probable error of the mean. *Biometrika,* **5:**315.

14. E. S. Pearson (1968) Studies in the history of probability and statistics XX. Some early correspondence between W. S. Gosset, R. A. Fisher, and K. Pearson, with notes and commentary. *Biometrika,* **55:**445.

15. H. E. Soper, A. W. Young, B. H. Cave, A. Lee, and K. Pearson (1916) Cooperative Study. On the distribution of the correlation coefficient in small samples. *Biometrika,* **11:**328–413.

16. K. Pearson (1922) Further note on the χ^2 test of goodness of fit. *Biometrika,* **14:**418.

17. Sir E. John Russell (1966) *A History of Agricultural Science in Great Britain.* London: Allen and Unwin.

18. H. Hotelling (1933) Analysis of a complex of statistical variables into principal components. *J. Educ. Psych.,* **24:**417–441 and 498–520.

19. H. Wold (1966) Non-linear estimation by iterative least squares procedures, *Research Papers in Statistics,* festschrift for J. Neyman, F. N. David, Ed., New York: Wiley.

20. Letters from W. S. Gossett to R. A. Fisher, 1915–1936, with summaries by R. A. Fisher. Privately circulated (1970). Originals now deposited at Statistical Department, University College, London.

21. Student (1925) New tables for testing the significance of observations. *Metron,* **5:**25–32.

22. Student (1917) Tables for estimating the probability that the mean of a unique sample of observations lies between $-\infty$ and any given distance of the mean of the population from which the sample is drawn. *Biometrika,* **11:**414.

23. L. I. [Isserlis] (1926) Review of *Statistical Methods for Research Workers* (R. A. Fisher). *J. Roy. Statist. Soc.,* **89,** Part 1:145–146.

24. Review of *Statistical Methods for Research Workers* (R. A. Fisher), *Brit. Med. J.,* **1:**578. See also *Nature* (Dec. 1925) **116:**815, in which the reviewer writes: "in the present work the absence of proof goes rather far and we fear that readers with little knowledge of the most recent statistical work will find the book as a whole difficult to follow while those unfamiliar with the terms used in biological research will have trouble with some examples."

25. E. S. Pearson (1926) Review of *Statistical Methods for Research Workers* (R. A. Fisher), *Sci. Prog.,* **20:**733–734.

26. Student (1926) Review of *Statistical Methods for Research Workers* (R. A. Fisher). *Eug. Rev.,* **18:**148–150.

27. R. A. Fisher (1932) Contribution to discussion of Biological Principles in the Control of Destructive Animals (M. A. C. Hinton). *Proc. Linn. Soc. Lond.,* **144:**124–125.

28. K. Mather and F. Yates (1963) Ronald Aylmer Fisher *Biographical Memoirs of Fellows of the Royal Society of London,* **9:**91–120.

29. L. L. Cavalli Sforza and W. F. Bodmer, (1971) *The Genetics of Human Populations.* San Francisco: W. F. Freeman.

30. R. A. Fisher, Review of *The Disinherited Family* (E. F. Rathbone) *Eug. Rev.* (1924–1925) **16:**150–153. Contribution to discussion of Family Endowment *Eug. Rev.,* **16:**281–283.

31. T. Eden and F. Yates (1933) On the validity of Fisher's z test when applied to an actual example of non-normal data. *J. Agric. Sci.,* **23:**6–17.

32. E. J. Russell (1926) Field experiments: How they are made and what they are. *J. Min. Agric.,* **32:**989 – 1001.

33. R. A. Fisher (1916) Review of "Is selection or mutation the more important agency in evolution?" (W. E. Castle). *Eug. Rev.,* **8:**84 – 85.

34. W. E. Castle (1914) Piebald rats and selection: an experimental test of the effectiveness of selection and of the theory of gametic purity in Mendelian crosses. Carnegie Inst. Publ., No. 195: 56.

35. S. Wright (1929) Fisher's theory of dominance. *Am. Nat.,* **63:**274 – 288.

36. J. B. S. Haldane (1930) A note on Fisher's theory of the origin of dominance and on a correlation between dominance and linkage. *Am. Nat.,* **64:**87 – 90.

37. J. B. Hutchinson (1931) The possible explanation of the apparently irregular inheritance of polydactyly in poultry. *Am. Nat.,* **65:**376 – 379.

38. E. B. Ford (1930) The theory of dominance. *Am. Nat.,* **64:**560 – 566.

39. R. C. Punnett and M. S. Pease (1929) Notes on polydactyly. *J. Genet.,* **21:**341 – 366.

40. Contributions to discussion of *C P* 116 (1934). *Proc. Linn. Soc. Lond.,* **147:**81 ff

41. L. C. Dunn and W. Landauer (1930) Further data of a case of autosomal linkage in the domestic fowl. *J. Genet.,* **22:**95 – 101.

42. C. C. Craig (1929) An application of Thiele's seminvariants to the sampling problem. *Metron,* **7:**3.

43. J. Wishart (1929) A problem in combinatorial analysis giving the distribution of certain moment statistics. *Proc. Lond. Math. Soc. (2)* **29:**309 – 321.

44. W. G. Cochran (1972) Experiments for non-linear functions, Technical Report No. 39, Dept. of Statistics, Harvard University.

45. J. Neyman (1935) Statistical problems in agricultural experimentation. *J. Roy. Statist. Soc., Suppl.,* **2:**154 – 180.

46. F. Yates (1935) Complex experimentation. *J. Roy. Statist. Soc., Suppl.,* **2:**181 – 223. Contributions to discussion, pp. 223 – 247.

47. J. Wishart (1934) The analysis of variance. *J. Roy. Statist. Soc., Suppl.,* **1:**26 – 51. Contributions to discussion, pp. 51 – 61.

48. Student (1936) Cooperation in large-scale experiments. *J. Roy. Statis. Soc., Suppl.,* **3:**115 – 122.

49. Student (1938) Comparison between balanced and random arrangements of field plots. *Biometrika,* **29:**363 – 379.

50. W. G. Cochran (1967) Footnote. *Science,* **156:**1460 – 1462.

51. C. I. Bliss (1969) Communication between biologists and statisticians, a case study. *Am. Statist.,* **23:**15 – 20.

52. R. B. Cattell (1937) Is national intelligence declining? *Eug. Rev.* **28:**181 – 203.
N. Wallace and R. M. W. Travers (1938) A psychometrical study of a group of speciality salesmen. *Ann. Eug.,* **8:**266 – 302. R. M. W. Travers (1938) The elimination of the influence of repetition on the score of a psychological test. *Ann. Eng.,* **8:**303 – 318.

53. Student (1931) The Lanarkshire milk experiment. *Biometrika,* **23:**398 – 406.

54. Hermann J. Muller (1918) Genetic variability, twin hybrids and constant hybrids, in a case of balanced lethal factors. *Genetics,* **3:**122 – 499. Review of above by R. A. Fisher. *Eug. Rev.,* **11:**92 – 94.

55. O. Winge (1931) The inheritance of double flowers and other characters in *Matthiola*. *Zeitschrift fur Zuchtung, Reihe a Pflanzenzuchtung*, **17:**118 – 135.

56. E. M. East (1927) The inheritance of heterostylism in *Lythrum salicaria*. *Genetics*, **12:**393 – 414.

57. J. L. Crosby (1940) High proportions of homostyle plants in populations of *Primula vulgaris*. *Nature*, **145:**672 – 673.

58. W. F. Bodmer (1958) Natural crossing between homostyle plants of *Primula vulgaris*. *Heredity*, **12:**363 – 370 and (1960) The genetics of homostyly in populations of *Primula vulgaris*. *Phil. Trans.*, **B, 242:**517 – 549.

59. J. L. Lush (1972) *Statistical Papers in honor of George W. Snedecor*. Iowa State College Press, Ames.

60. R. S. Koshal (1935) Applications of the method of maximum likelihood to the derivation of efficient statistics for fitting frequency curves. *J. Roy. Stat. Soc.*, **98:**128.

61. K. Pearson (1936) Method of moments and method of maximum likelihood. *Biometrika*, **28:**34 – 59.

62. P. C. Mahalanobis (1933) Revision of Risley's anthropometric data relating to tribes and castes of Bengal. *Sankhyā*, **1:**76 – 105.

63. H. Hotelling (1931) The generalisation of 'Student's' ratio. *Ann. Math. Stat.*, **2:**360 – 378.

64. P. C. Mahalanobis (1930) On tests and measures of group divergence. Part 1, Theoretical formulae. *J. Asiat. Soc. Beng.*, **26:**541 – 588.

65. R. C. Bose (1936) On the exact distribution of the D^2 statistic. *Sankhyā*, **2:**143 – 154.

66. R. C. Bose and S. N. Roy (1938) The exact distribution of the Studentized D^2 statistic, *Sankhyā*, **3:**part 4.

67. R. R. Race, Second R. A. Fisher Memorial Lecture London (1969), unpublished.

68. R. R. Race and Ruth Sanger (1950) *Blood Groups in Man*. Oxford and Edinburgh: Blackwell.

69. A. S. Wiener and H. R. Peters (1940) Haemolytic reactions following transfusions of blood of the homologous group, with three cases in which the same agglutinogen was responsible. *Ann. Int. Med.*, **13:**2306 – 2322.

70. K. Landsteiner and A. S. Wiener (1940) An agglutinable factor in human blood recognized by immune sera for rhesus blood. *Proc. Soc. Exp. Biol. N.Y.*, **43:**223.

71. P. Levine and R. E. Stetson (1939) An unusual case of intragroup agglutination. *J. Am. Med. Ass.*, **113:**126 – 127.

72. P. Levine, P. Vogel, E. M. Katzin and L. Burnham (1941) The role of isoimmunization in the pathogenesis of erythroblastosis fetalis. *Am. J. Obst. Gynec.*, **42:**925 – 937.

73. K. Landsteiner and A. S. Wiener (1941) Studies on an agglutinogen (Rh) in human blood reacting with anti-rhesus sera and with human isoantibodies. *J. Exp. Med.*, **74:**309 – 320.

74. R. R. Race, G. L. Taylor, Elizabeth W. Iken and Aileen M. Prior (1944) The inheritance of the allelomorphs of the Rh gene in 56 families. *Ann. Eugen. Lond.*, **12:**206 – 210.

75. P. Levine (1942) The pathogenesis of fetal erythroblastosis. *N. Y. State J. Med.*, **42:**1928 – 1934.

76. R. R. Race, G. L. Taylor, D. F. Cappell and Marjory N. McFarlane (1944) Recognition of a further Rh genotype in man. *Nature*, **153:**53 – 55.

77. R. R. Race and G. L. Taylor (1943) A serum that discloses the genotype of some Rh positive people. *Nature*, **152:**300.

78. R. R. Race (1944) An 'incomplete' antibody in human serum. *Nature*, **153:**771 – 772.

79. A. E. Mourant (1945) A new rhesus antibody. Nature, **155**:542.
80. Clara van den Bosch (1948) The very rare Rh genotype $R_y r$ (CdE/cde) in a case of erythroblastosis foetalis. Nature, **162**:781.
81. S. B. Holt (1945) A polydactyl gene in mice, capable of nearly regular manifestation. Ann. Eugen., **42**:220 – 249.
82. S. E. Luria and M. Delbruck (1943) Mutations of bacteria from virus sensitivity to virus resistance. Genetics, **28**:491.
83. J. Lederberg and E. L. Tatum (1946) Gene recombination in Escherichia coli. Nature, **158**:558.
84. L. L. Cavalli and G. Magni (1947) Methods of analysing the virulence of bacteria and viruses for genetical purposes. Heredity, **1**:127 – 132.
85. F. Jacob and E. L. Wollman (1961) Sexuality and the genetics of bacteria. New York: Academic Press.
86. A. E. Mourant (1956) Anthropology and natural selection, Proceedings of the First International Congress of Human Genetics (Acta Genetica et Statistica Medica, **6**: No. 4:509 – 514. Discussion 514 – 515.
87. J. A. Fraser Roberts (1956) Associations between blood groups and disease, Proceedings of the First International Congress of Human Genetics (Acta Genetica et Statistica Medica, **6**, No. 4:549 – 560.
88. D. J. Finney (1964) Foreword (p. 237) Biometrics, **20**, No. 2, In Memoriam Ronald Aylmer Fisher 1890 – 1962.
89. L. J. Savage (1970) On rereading R. A. Fisher, Fisher Memorial Lecture published posthumously, J. W. Pratt, Ed., Ann. Math. Stat., (1976) **4**:441 – 500.
90. J. Hospers (1953 – 54) Reversals of the main geomagnetic field, I, II, III, Proc. Kon. Nederl. Akad., Wet. B, **56**:467 – 491 and 477 – 491, **57**:112 – 121.
91. D. T. Griggs (1939) A theory of mountain-building, Am. J. Sci., **237**:611 – 650.
92. E. C. Bullard (1949) The magnetic field within the earth, Proc. Roy. Soc., **A**, **197**:433 – 458.
93. S. K. Runcorn (1962) Palaeomagnetic evidence for continental drift and its geophysical cause. Continental Drift, S. K. Runcorn, Ed., New York: Academic Press.
94. P. V. Sukhatme (1938) On Fisher and Behrens' test of significance for the difference in means of two normal samples. Sankhyā, **4**:39 – 48.
95. Monica A. Creasey (1954) Limits for the ratio of means. J. Roy. Stat. Soc., **B**, **16**:186 – 194.
96. D. A. S. Fraser (1976) Necessary analysis and adaptive inference. J. Am. Stat. Ass., **71**:99 – 110.
D. A. S. Fraser (1976) Confidence, posterior probability and the Buehler example. Ann. Stat., **6**:892 – 898.
97. Graham Wilkinson (1977) On resolving the controversy in statistical Inference, J. Roy. Stat. Soc., **B**, **39**:119 – 171.
98. G. A. Barnard (1977) Pivotal inference and the Bayesian controversy. J. Am. Stat. Soc., in press.
99. D. T. Sterling (1971) A critical reassessment of the evidence bearing on smoking as the cause of cancer. Review of crucial data bearing on the smoking and health issue. Washington University.
100. E. B. Ford (1962) Ecological Genetics. London: Methuen.

Bibliography of R. A. Fisher

1. BOOKS

Statistical Methods for Research Workers. Edinburgh: Oliver and Boyd, 1925, 1928, 1930, 1932, 1934, 1936, 1938, 1941, 1944, 1946, 1950, 1954, 1958, 1970. Also published in French, German, Italian, Japanese, Spanish, and Russian.

The Genetical Theory of Natural Selection. Oxford: University Press, 1930; New York: Dover, 1958.

The Design of Experiments. Edinburgh: Oliver and Boyd, 1935, 1937, 1942, 1947, 1949, 1951, 1960, 1966. Also published in Italian, Japanese, and Spanish.

Statistical Tables for Biological, Agricultural and Medical Research. (with F. Yates) Edinburgh: Oliver and Boyd, 1938, 1943, 1948, 1953, 1957, 1963. Also published in Spanish and Portuguese.

The Theory of Inbreeding. Edinburgh: Oliver and Boyd, 1949, 1965.

Contributions to Mathematical Statistics. New York: Wiley, 1950

Statistical Methods and Scientific Inference. Edinburgh: Oliver and Boyd, 1956, 1959; New York: Hafner, 1973.

2. COLLECTED PAPERS OF R. A. FISHER (Vol. 1-5) J. H. Bennett, Ed. The University of Adelaide, 1971 – 1974.

VOLUME 1 (1912 – 1924)

1912   1.   On an absolute criterion for fitting frequency curves. *Messeng. Math.*, **41**:155 – 160.

1913 2. Applications of vector analysis to geometry. *Messeng. Math.*, **42:** 161–178.
1914 3. Some hopes of a eugenist. *Eugen. Rev.*, **5:** 309 –315.
1915 4. Frequency distribution of the values of the correlation coefficient in samples from an indefinitely large population. *Biometrika*, **10:**507 – 521.
      5. (With C. S. Stock). Cuénot on preadaptation: a criticism. *Eugen Rev.*, **7:**46 – 61.
      6. The evolution of sexual preference. *Eugen Rev.*, **7:**184 – 192.
1916 7. Biometrika. *Eugen. Rev.*, **8:**62 – 64.
1917 8. Positive eugenics. *Eugen. Rev.*, **9:**206 – 212.
1918 9. The correlation between relatives on the supposition of Mendelian inheritance. *Trans. Roy. Soc. Edinb.*, **52:**399 – 433.
      10. The causes of human variability. *Eugen. Rev.*, **10:**213 – 220.
1919 11. The genesis of twins. *Genetics*, **4:**489 – 499.
1920 12. A mathematical examination of the methods of determining the accuracy of an observation by the mean error, and by the mean square error. *Mon. Not. Roy. Ast. Soc.*, **80:**758 – 770.
      13. Review of *Inbreeding and Outbreeding*. (E. M. East and D. F. Jones) *Eugen. Rev.*, **12:**116 –119.
1921 14. On the "probable error" of a coefficient of correlation deduced from a small sample. *Metron*, **1:**3 – 32.
      15. Studies in crop variation. I. An examination of the yield of dressed grain from Broadbalk. *J. Agric. Sci.*, **11:**107 –135.
      16. Some remarks on the methods formulated in a recent article on the quantitative analysis of plant growth. *Ann. Appl. Biol.*, **7:**367 – 372.
      17. Review of The Relative Value of the Processes Causing Evolution. (A. L. and A. C. Hagedoorn) *Eugen. Rev.*, **13:**467 – 470.
1922 18. On the mathematical foundations of theoretical statistics. *Phil. Trans.*, **A, 222:**309 – 368.
      19. On the interpretation of χ^2 from contingency tables, and the calculation of *P*. *J. Roy. Stat. Soc.*, **85:**87 – 94.
      20. The goodness of fit of regression formulae, and the distribution of regression coefficients. *J. Roy. Stat. Soc.*, **85:**597 – 612.
      21. (With W. A. Mackenzie). The correlation of weekly rainfall. *Q. J. Roy. Met. Soc.*, **48:** 234 – 242.
      22. (With H. G. Thornton and W. A. Mackenzie). The accuracy of the plating method of estimating the density of bacterial populations. *Ann. Appl. Biol.*, **9:**325 – 359.
      23. Statistical appendix to a paper by J. Davidson on Biological studies of *Aphis rumicis*. *Ann. Appl. Biol.*, **9:**142 –145.
      24. On the dominance ratio. *Proc. Roy. Soc. Edinb.*, **42:**321 – 341.
      25. The systematic location of genes by means of crossover observations. *Amer. Nat.*, **56:**406 – 411.
      26. Darwinian evolution by mutations. *Eugen. Rev.*, **14:**31 – 34.
      27. New data on the genesis of twins. *Eugen. Rev.*, **14:**115 –117.
      28. The evolution of the conscience in civilised communities. *Eugen. Rev.*, **14:**190 –193.
      29. Contribution to a discussion on the inheritance of mental qualities, good and bad. *Eugen. Rev.*, **14:**210 – 213.
1923 30. Note on Dr. Burnside's recent paper on errors of observation. *Proc. Camb. Phil. Soc.*, **21:**655 – 658.
      31. Statistical tests of agreement between observation and hypothesis. *Economica*, **3:**139 –147.

32. (With W. A. Mackenzie). Studies in crop variation. II. The manurial response of different potato varieties. *J. Agric. Sci.*, **13**:311 – 320.
33. Paradoxical rainfall data. *Nature*, **111**:465.
1924   34. The conditions under which χ^2 measures the discrepancy between observation and hypothesis. *J. Roy. Stat. Soc.*, **87**:442 – 450.
35. The distribution of the partial correlation coefficient. *Metron*, **3**:329 – 332.
36. On a distribution yielding the error functions of several well known statistics. *Proc. In. Cong. Math., Toronto*, **2**:805 – 813.
37. The influence of rainfall on the yield of wheat at Rothamsted. *Phil. Trans.*, **B**, **213**:89 – 142.
38. A method of scoring coincidences in tests with playing cards. *Proc. Soc. Psych. Res.*, **34**:181 – 185.
39. (With S. Odén). The theory of the mechanical analysis of sediments by means of the automatic balance. *Proc. Roy. Soc. Edinb.*, **44**:98 – 115.
40. The elimination of mental defect. *Eugen. Rev.*, **16**:114 – 116.
41. The biometrical study of heredity. *Eugen. Rev.*, **16**:189 – 210.

VOLUME II (1925 – 1931)

1925   42. Theory of statistical estimation. *Proc. Camb. Phil. Soc.*, **22**:700 – 725.
43. Applications of "Student's" distribution. *Metron*, **5**:90 – 104.
44. Expansion of "Student's" integral in powers of n^{-1}. *Metron*, **5**:109 – 120.
45. (With P. R. Ansell). Note on the numerical evaluation of a Bessel function derivative. *Proc. Lond. Math. Soc., Series 2*, **24**:liv-lvi.
46. Sur la solution de l'équation intégrale de M. V. Romanovsky. *C. R. Acad. Sci., Paris*, **181**:88 – 89.
47. The resemblance between twins, a statistical examination of Lauterbach's measurements, *Genetics*, **10**:569 – 579.
1926   48. The arrangement of field experiments. *J. Min. Agric. G. Br.*, **33**:503 – 513.
49. Bayes' theorem and the fourfold table. *Eugen. Rev.*, **18**:32 – 33.
50. On the random sequence. *Q. J. Roy. Met. Soc.*, **52**:250.
51. On the capillary forces in an ideal soil: correction of formulae given by W. B. Haines. *J. Agric. Sci.*, **16**:492 – 503.
52. (With E. B. Ford). Variability of species. *Nature*, **118**:515 – 516.
53. Eugenics: Can it solve the problem of decay of civilisations? *Eugen. Rev.*, **18**:128 – 136.
54. Modern eugenics. *Sci. Prog.*, **21**:130 – 136; *Eugen. Rev.*, **18**:231 – 236.
55. Periodical health surveys. *J. State Med.*, **34**:446 – 449.
1927   56. (With J. Wishart). On the distribution of the error of an interpolated value, and on the construction of tables. *Proc. Camb. Phil. Soc.*, **23**:912 – 921.
57. (With T. Eden). Studies in crop variation. IV. The experimental determination of the value of top dressings with cereals. *J. Agric. Sci.*, **17**:548 – 562.
58. (With H. G. Thornton). On the existence of daily changes in the bacterial numbers in American soil. *Soil Sci.*, **23**:253 – 259.
59. On some objections to mimicry theory—statistical and genetic. *Trans. Roy. Ent. Soc. Lond.*, **75**:269 – 278.
60. The actuarial treatment of official birth records. *Eugen. Rev.*, **19**:103 – 108.
1928   61. The general sampling distribution of the multiple correlation coefficient. *Proc. Roy, Soc. A*, **121**:654 – 673.

62. On a property connecting the χ^2 measure of discrepancy with the method of maximum likelihood. *Atti Cong. Int. Mat., Bologna,* **6**:95 – 100.
63. (With L. H. C. Tippett). Limiting forms of the frequency distribution of the largest or smallest member of a sample. *Proc. Camb. Phil. Soc.,* **24**:180 – 190.
64. Further note on the capillary forces in an ideal soil. *J. Agric. Sci.,* **18**:406 – 410.
65. (With T. N. Hoblyn). Maximum- and minimum-correlation tables in comparative climatology. *Geogr. Ann.,* **10**:267 – 281.
66. Correlation coefficients in meteorology. *Nature,* **121**:712.
67. The effect of psychological card preferences. *Proc. Soc. Psych. Res.,* **38**:269 – 271.
68. The possible modification of the response of the wild type to recurrent mutations. *Am. Nat.,* **62**:115 – 126.
69. Two further notes on the origin of dominance. *Am. Nat.,* **62**:571 – 574.
70. Triplet children in Great Britain and Ireland. *Proc. Roy. Soc.,* **B, 102**:286 – 311.
71. (With B. Balmukand). The estimation of linkage from the offspring of selfed heterozygotes. *J. Genet.,* **20**:79 – 92.
72. (With E. B. Ford). The variability of species in the Lepidoptera, with reference to abundance and sex. *Trans. Roy. Entom. Soc. Lond.,* **76**:367 – 379.
73. The differential birth rate: new light on causes from American figures. *Eugen. Rev.,* **20**:183 – 184.

1929 74. Moments and product moments of sampling distributions. *Proc. Lond. Math. Soc., Series 2,* **30**:199 – 238.
75. Tests of significance in harmonic analysis. *Proc. Roy. Soc.,* **A, 125**:54 – 59.
76. The sieve of Eratosthenes. *Math. Gaz.,* **14**:564 – 566.
77. A preliminary note on the effect of sodium silicate in increasing the yield of barley. *J. Agric. Sci.,* **19**:132 – 139.
78. (With T. Eden). Studies in crop variation. VI. Experiments on the response of the potato to potash and nitrogen. *J. Agric. Sci.,* **19**:201 – 213.
79. The statistical method in psychical research. *Proc. Soc. Psych. Res.,* **39**:189 – 192.
80. Statistics and biological research. *Nature,* **124**:266 – 267.
81. The evolution of dominance: a reply to Professor Sewall Wright. *Am. Nat.,* **63**:553 – 556.
82. The over-production of food. *Realist,* **1**:45 – 60.

1930 83. The moments of the distribution for normal samples of measures of departure from normality. *Proc. Roy. Soc.,* **A, 130**: 16 – 28.
84. Inverse probability. *Proc. Camb. Phil. Soc.,* **26**:528 – 535.
85. (With J. Wishart). The arrangement of field experiments and the statistical reduction of the results. *Imp. Bur. Soil Sci. Tech. Comm.,* **10**:23 pp.
86. The distribution of gene ratios for rare mutations. *Proc. Roy. Soc. Edinb.,* **50**: 205 – 220.
87. The evolution of dominance in certain polymorphic species. *Am. Nat.,* **64**:385 – 406.
88. Mortality amongst plants and its bearing on natural selection. *Nature,* **125**:972 – 973.
89. Note on a tri-colour (mosaic) mouse. *J. Genet.,* **23**:77 – 81.

1931 90. (With J. Wishart). The derivation of the pattern formulae of two-way partitions from those of simpler patterns. *Proc. Lond. Math. Soc., Series 2,* **33**:195 – 208.
91. The sampling error of estimated deviates, together with other illustrations of the properties and applications of the integrals and derivatives of the normal error function. *Brit. Assn. Math. Tab.,* **1**:xxvi-xxxv (3rd ed., xxviii-xxxvii, 1951).

92;   (With S. Bartlett). Pasteurised and raw milk. *Nature*, **127:**591 – 592.
93.   The evolution of dominance. *Biol. Rev.*, **6:**345 – 368.
94.   The biological effects of family allowances. *Family Endowment Chronicle*, 1:21 – 25.

VOLUME III (1932 – 1936)

1932   95.   Inverse probability and the use of likelihood. *Proc. Camb. Phil. Soc.*, **28:**257 – 261.
96.   (With F. R. Immer and O. Tedin) The genetical interpretation of statistics of the third degree in the study of quantitative inheritance. *Genetics*, **17:**107 – 124.
97.   The evolutionary modification of genetic phenomena. *Proc. 6th Int. Congr. Genet.*, **1:**165 – 172.
98.   The bearing of genetics on theories of evolution. *Sci. Prog.*, **27:**273 – 287.
99.   *The social selection of human fertility.* The Herbert Spencer Lecture, 32 pp. Oxford: Clarendon Press.
100.   Family allowances in the contemporary economic situation. *Eugen. Rev.*, **24:**87 – 95.
101.   Inheritance of acquired characters. *Nature*, **130:**579.
1933   102.   The concepts of inverse probability and fiducial probability referring to unknown parameters. *Proc. Roy. Soc.*, **A, 139:**343 – 348.
103.   The contributions of Rothamsted to the development of the science of statistics. *Annual Report Rothamsted Experimental Station*, 43 – 50.
104.   On the evidence against the chemical induction of melanism in Lepidoptera. *Proc. Roy. Soc.*, **B, 112:**407 – 416.
105.   Selection in the production of ever-sporting stocks. *Ann. Bot.*, **47:**727 – 733.
106.   Number of Mendelian factors in quantitative inheritance. *Nature*, **131:**400 – 401.
107.   Contribution to a discussion on mortality among young plants and animals. *Proc. Linn. Soc. Lond.*, **145:** 100 – 101.
1934   108.   Two new properties of mathematical likelihood. *Proc. Roy. Soc.*, **A, 144:**285 – 307.
109.   Probability, likelihood and quantity of information in the logic of uncertain inference. *Proc. Roy. Soc.*, **A, 146:**1 – 8.
110.   (With F. Yates). The 6 × 6 Latin squares. *Proc. Camb. Phil. Soc.*, **30:**492 – 507.
111.   Randomisation, and an old enigma of card play. *Math. Gaz.*, **18:**294 – 297.
112.   Appendix to a paper by H. G. Thornton and P. H. H. Gray on the numbers of bacterial cells in field soils. *Proc. Roy. Soc.*, **B, 115:**540 – 542.
113.   The effect of methods of ascertainment upon the estimation of frequencies. *Ann. Eugen.*, **6:**13 – 25.
114.   The amount of information supplied by records of families as a function of the linkage in the population sampled. *Ann. Eugen.*, **6:**66 – 70.
115.   The use of simultaneous estimation in the evaluation of linkage. *Ann. Eugen.*, **6:**71 – 76.
116.   Some results of an experiment on dominance in poultry, with special reference to polydactyly, *Proc. Linn. Soc. Lond.*, **147:**71 – 81.
117.   Crest and hernia in fowls due to a single gene without dominance. *Science*, **80:**288 – 289.
118.   (With C. Diver). Crossing-over in the land snail *Cepaea nemoralis. Nature*, **133:**834 – 835.
119.   Professor Wright on the theory of dominance. *Am. Nat.*, **68:**370 – 374.

120. The children of mental defectives. In the *Report of Departmental Cttee. on Sterilisation*, 60 – 74, H.M.S.O.
121. Indeterminism and natural selection. *Philos. Sci.*, **1**:99 – 117.
122. Adaptation and mutations. *School Sci. Rev.*, **15**:294 – 301.

1935 123. The mathematical distributions used in the common tests of significance. *Econometrica*, **3**:353 – 365.
124. The logic of inductive inference. *J. Roy. Stat. Soc.*, **98**:39 – 54.
125. The fiducial argument in statistical inference. *Ann. Eugen.*, **6**:391 – 398.
126. The case of zero survivors in probit assays. *Ann. Appl. Biol.*, **22**:164 – 165.
127. Statistical tests. *Nature*, **136**:474.
128. Contribution to a discussion of J. Neyman's paper on statistical problems in agricultural experimentation. *J. Roy. Stat. Soc., Suppl.*, **2**:154 – 157, 173.
129. Contribution to a discussion of F. Yates' paper on complex experiments. *J. Roy. Stat. Soc., Suppl.*, **2**:229 – 231.
130. On the selective consequences of East's theory of heterostylism in *Lythrum. J. Genet.*, **30**:369 – 382.
131. The detection of linkage with "dominant" abnormalities. *Ann. Eugen.*, **6**:187 – 201.
132. The detection of linkage with recessive abnormalities. *Ann. Eugen.*, **6**:339 – 351.
133. The sheltering of lethals. *Am. Nat.*, **69**:446 – 455.
134. The inheritance of fertility: Dr. Wagner-Manslau's tables. *Ann. Eugen.* **6**:225 – 251.
135. Dominance in poultry, *Philos. Trans.*, **B**, **225**:197 – 226.
136. Eugenics, academic and practical. *Eugen. Rev.*, **27**:95 – 100.

1936 137. Uncertain inference. *Proc. Am. Acad. Arts Sci.*, **71**:245 – 258.
138. The use of multiple measurements in taxonomic problems. *Ann. Eugen.*, **7**:179 – 188.
139. (With S. Barbacki). A test of the supposed precision of systematic arrangements. *Ann. Eugen.*, **7**:189 – 193.
140. The half-drill strip system agricultural experiments. *Nature*, **138**:1101.
141. "The coefficient of racial likeness" and the future of craniometry. *J. Roy. Anthropol. Inst.*, **66**:57 – 63.
142. Heterogeneity of linkage data for Friedreich's ataxia and the spontaneous antigens. *Ann. Eugen.*, **7**:17 – 21.
143. Tests of significance applied to Haldane's data on partial sex linkage. *Ann. Eugen.*, **7**:87 – 104.
144. Has Mendel's work been rediscovered? *Ann. Sci.*, **1**:115 – 137.
145. (With K. Mather). A linkage test with mice. *Ann. Eugen.*, **7**:265 – 280.
146. (With K. Mather). Verification in mice of the possibility of more than fifty per cent recombination. *Nature*, **137**:362 – 363.
147. The measurement of selective intensity. *Proc. Roy. Soc.*, **B**, **121**:58 – 62.

VOLUME IV (1937 – 1947)

1937 148. (With E. A. Cornish). Moments and cumulants in the specification of distributions. *Rev. Inst. Int. Stat.*, **5**:307 – 322.
149. Professor Karl Pearson and the method of moments. *Ann. Eugen.*, **7**:308 – 318.
150. (With B. Day). The comparison of variability in populations having unequal means. An example of the analysis of covariance with multiple dependent and independent variates. *Ann. Eugen.*, **7**:333 – 348.

151. On a point raised by M. S. Bartlett on fiducial probability. *Ann. Eugen.*, **7**:370 – 375.
152. The wave of advance of advantageous genes. *Ann. Eugen.*, **7**:355 – 369.
153. The relation between variability and abundance shown by the measurements of the eggs of British nesting birds. *Proc. Roy. Soc.*, **B**, **122**:1 – 26.
154. (With H. Gray). Inheritance in man: Boas's data studied by the method of analysis of variance. *Ann. Eugen.*, **8**:74 – 93.

1938 155. The statistical utilization of multiple measurements. *Ann. Eugen.*, **8**:376 – 386.
156. Quelques remarques sur l'estimation en statistique. *Biotypologie*, **6**:153 – 158.
157. On the statistical treatment of the relation between sea-level characteristics and high-altitude acclimatization. *Proc. Roy. Soc.*, **B**, **126**:25 – 29.
158. The mathematics of experimentation. *Nature*, **142**:442 – 443.
159. Presidential address, Indian statistical conference, *Sankhyā*, **4**:14 – 17.
160. Comment on D. McGregor's note on the distribution of the three forms of *Lythrum salicaria*. *Ann. Eugen.*, **8**:177.
161. Dominance in poultry: feathered feet, rose comb, internal pigment and pile. *Proc. Roy. Soc.*, **B**, **125**:25 – 48.

1939 162. The comparison of samples with possibly unequal variances. *Ann. Eugen.*, **9**:174 – 180.
163. The sampling distribution of some statistics obtained from non-linear equations. *Ann. Eugen.*, **9**:238 – 249.
164. A note on fiducial inference. *Ann. Math. Stat.*, **10**:383 – 388.
165. "Student". *Ann. Eugen.*, **9**:1 – 9.
166. The precision of the product formula for the estimation of linkage. *Ann. Eugen.*, **9**:50 – 54.
167. Selective forces in wild populations of *Paratettix texanus*. *Ann. Eugen.*, **9**:109 – 122.
168. Stage of development as a factor influencing the variance in the number of off-spring, frequency of mutants and related quantities. *Ann Eugen.*, **9**:406 – 408.
169. (With G. L. Taylor). Blood groups in Great Britain. *Brit. Med. J.*, **2**:826.
170. (With E. B. Ford and J. Huxley). Taste-testing the anthropoid apes. *Nature*, **144**:750.
171. (With J. Vaughan). Surnames and blood-groups. *Nature*, **144**:1047.
172. The Galton Laboratory. *Times*, 29 September; *Science*, **90**:436.

1940 173. On the similarity of the distributions found for the test of significance in harmonic analysis, and in Steven's problem in geometrical probability. *Ann. Eugen.*, **10**:14 – 17.
174. An examination of the different possible solutions of a problem in incomplete blocks. *Ann. Eugen.*, **10**:52 – 75.
175. The precision of discriminant functions. *Ann. Eugen.*, **10**:422 – 429.
176. The estimation of the proportion of recessives from tests carried out on a sample not wholly unrelated. *Ann. Eugen.*, **10**:160 – 170.
177. (With W. H. Dowdeswell and E. B. Ford). The quantitative study of populations in the Lepidoptera. I. *Polyommatus icarus*. *Ann. Eugen.*, **10**:123 – 136.
178. (With K. Mather). Non-lethality of the mid factor in *Lythrum salicaria*. *Nature*, **146**:521.
179. (With G. L. Taylor). Scandinavian influence in Scottish ethnology. *Nature*, **145**:590.
180. The Galton Laboratory. *Science*, **91**:44 – 45.

1941 181. The asymptotic approach to Behrens's integral, with further tables for the d test of significance. *Ann. Eugen.*, **11**:141–172.
      182. The negative binomial distribution. *Ann. Eugen.*, **11**:182–187.
      183. The interpretation of experimental four-fold tables. *Science*, **94**:210–211.
      184. The theoretical consequences of polyploid inheritance for the mid style form of *Lythrum salicaria. Ann. Eugen.*, **11**:31–38.
      185. Average excess and average effect of a gene substitution. *Ann. Eugen.*, **11**:53-63.
1942 186. New cyclic solutions to problems in incomplete blocks. *Ann. Eugen.*, **11**:290–299.
      187. Completely orthogonal 9 × 9 squares—a correction. *Ann. Eugen.*, **11**:402–403.
      188. The likelihood solution of a problem in compounded probabilities. *Ann. Eugen.*, **11**:306–307.
      189. The theory of confounding in factorial experiments in relation to the theory of groups. *Ann. Eugen.*, **11**:341–353.
      190. Some combinatorial theorems and enumerations connected with the numbers of diagonal types of a Latin square. *Ann. Eugen.*, **11**:395–401.
      191. The polygene concept. *Nature*, **150**:154.
      192. (With K. Mather). Polyploid inheritance in *Lythrum salicaria. Nature*, **150**:430.
1943 193. A theoretical distribution for the apparent abundance of different species. *J. Anim. Ecol.*, **12**:54–58.
      194. Note on Dr. Berkson's criticism of tests of significance. *J. Am. Stat. Assn.*, **38**:103–104.
      195. (With W. R. G. Atkins). The therapeutic use of vitamin C. *J. Roy. Army Med. Corps*, **83**:251–252.
      196. (With K. Mather). The inheritance of style length in *Lythrum salicaria. Ann. Eugen.*, **12**:1–23.
      197. (With J. A. Fraser Roberts). A sex difference in blood-group frequencies. *Nature*, **151**:640–641.
      198. The birthrate and family allowances. *Agenda*, **2**:124–133.
1944 199. (With S. B. Holt). The experimental modification of dominance in Danforth's short-tailed mutant mice. *Ann. Eugen.*, **12**:102–120.
      200. Allowance for double reduction in the calculation of genotype frequencies with polysomic inheritance. *Ann. Eugen.*, **12**:169–171.
      201. (With R. R. Race and G. L. Taylor). Mutation and the Rhesus reaction. *Nature*, **153**:106.
1945 202. A system of confounding for factors with more than two alternatives, giving completely orthogonal cubes and higher powers. *Ann. Eugen.*, **12**:283–290.
      203. The logical inversion of the notion of the random variable. *Sankhyā*, **7**:129–132.
      204. Recent progress in experimental design. In *L'application du calcul des probabilités*, 19–31. *Proc. Int. Inst. Intell. Coop., Geneva*, (1939).
      205. A new test for 2 × 2 tables. *Nature*, **156**:388.
      206. (With L. Martin). The hereditary and familial aspects of exophthalmic goitre and nodular goitre. *Q. J. Med.*, **14**:207–219.
1946 207. Testing the difference between two means of observations of unequal precision. *Nature*, **158**:713.
      208. A system of scoring linkage data, with special reference to the pied factors in mice. *Am. Nat.*, **80**:568–578.
      209. (With R. R. Race). Rh gene frequencies in Britain. *Nature*, **157**:48–49.

210. The fitting of gene frequencies to data on Rhesus reactions. Ann. Eugen. 13:150–155.

1947 211. The analysis of covariance method for the relation between a part and the whole. Biometrics, 3:65–68.

212. Development of the theory of experimental design. Proc. Int. Statist. Conf., 3:434–439.

213. The theory of linkage in polysomic inheritance. Philos. Trans., B, 233:55–87.

214. The Rhesus factor: a study in scientific method. Am. Sci., 35:95–102, 113.

215. Note on the calculation of the frequencies of Rhesus allelomorphs. Ann. Eugen., 13:223–224.

216. The science of heredity. Listener, 37:662–663.

217. The renaissance of Darwinism. Listener, 37:1001, 1009; Parents' Rev. 58:183–187.

218. (With V. C. Martin). Spontaneous occurrence in Lythrum salicaria of plants duplex for the short-style gene. Nature, 160:541.

219. (With E. B. Ford). The spread of a gene in natural conditions in a colony of the moth Panaxia dominula. Heredity, 1:143–174.

220. Number of self-sterility alleles. Nature, 160:797–798.

221. (With M. F. Lyon and A. R. G. Owen). The sex chromosome in the house mouse. Heredity, 1:355–365.

VOLUME V (1948–1962)

1948 222. Conclusions fiduciaires. Ann. Inst. Henri Poincaré, 10:191–213.

223. (With D. Dugué). Un résultat assez inattendu d'arithmétique des lois de probabilité. C. R. Acad. Sci., Paris, 227:1205–1206.

224. Biometry. Biometrics, 4:217–219.

225. A quantitative theory of genetic recombination and chiasma formation. Biometrics, 4:1–9.

226. (With G. D. Snell). A twelfth linkage group of the house mouse. Heredity, 2:271–273.

227. (With V. C. Martin). Genetics of style-length in Oxalis. Nature, 162:533.

228. Modern genetics. Brit. Sci. News, 1:2–4.

229. What sort of man is Lysenko? Soc. Freed. Sci., Occasl. Pamp., 9:6–9; Listener, 40:874–875.

1949 230. A biological assay of tuberculins. Biometrics, 5:300–316.

231. Note on the test of significance for differential viability in frequency data from a complete three-point test. Heredity, 3:215–219.

232. (With W. H. Dowdeswell and E. B. Ford). The quantitative study of populations in the Lepidoptera. 2. Maniola jurtina. Heredity, 3:67–84.

233. A preliminary linkage test with agouti and undulated mice. Heredity, 3:229–241.

234. A theoretical system of selection for homostyle Primula. Sankhyā, 9:325–342.

235. The linkage problem in a tetrasomic wild plant, Lythrum salicaria. Proc. 8th Int. Congr. Genet. (suppl. to Hereditas) 225–233.

1950 236. The significance of deviations from expectation in a Poisson series. Biometrics, 6:17–24.

237. A class of enumerations of importance in genetics. Proc. Roy. Soc., B, 136:509–520.

238. Gene frequencies in a cline determined by selection and diffusion. Biometrics, 6:353–361.

239. (With E. B. Ford). The "Sewall Wright effect". *Heredity*, **4**:117–119.
240. Polydactyly in mice. *Nature*, **165**:407, 796.
241. Creative aspects of Natural Law. The Eddington Memorial Lecture. 23 pp. Cambridge: Cambridge University Press.
1951 242. Statistics, in *Scientific Thought in the Twentieth Century*, A. E. Heath, Ed. London: Watts, pp. 31–55.
243. Standard calculations for evaluating a blood-group system. *Heredity*, **5**:95–102.
244. A combinatorial formulation of multiple linkage tests. *Nature*, **167**:520.
245. Limits to intensive production in animals. *Brit. Agric. Bull.*, **4**:217–218.
246. (With L. Martin). The hereditary and familial aspects of toxic nodular goitre (secondary thyrotoxicosis). *Q. J. Med.*, **20**:293–297.
1952 247. Sequential experimentation. *Biometrics*, **8**:183–187.
248. Statistical methods in genetics. The Bateson Lecture, 1951, *Heredity*, **6**:1–12.
1953 249. Dispersion on a sphere. *Proc. Roy. Soc.*, A, **217**:295–305.
250. Note on the efficient fitting of the negative binomial. *Biometrics*, **9**:197–199.
251. The expansion of statistics. *J. Roy. Stat. Soc.*, A, **116**:1–6; *Am. Sci.*, **42**:275–282, 293.
252. Population genetics. The Croonian Lecture, 1953, *Proc. Roy. Soc.*, B, **141**:510–523.
253. The variation in strength of the human blood group P. *Heredity*, **7**:81–89.
254. The linkage of polydactyly with *leaden* in the house mouse. *Heredity*, **7**:91–95.
255. (With W. Landauer). Sex differences of crossing-over in close linkage. *Am. Nat.*, **87**:116.
1954 256. The analysis of variance with various binomial transformations. *Biometrics*, **10**:130–139.
257. Contribution to a discussion of a paper on interval estimation by M. A. Creasy. *J. Roy. Statist. Soc.*, B, **16**:212–213.
258. Retrospect of the criticisms of the theory of natural selection In *Evolution as a Process*, J. S. Huxley, A. C. Hardy, and E. B. Ford, Eds. London: Allen and Unwin, pp. 84–98.
259. A fuller theory of "junctions" in inbreeding. *Heredity*, **8**:187–197.
260. The experimental study of multiple crossing over. *Proc. 9th Int. Congr. Genet.*, *Caryologia*, **6**: Suppl., 227–231.
1955 261. Statistical methods and scientific induction. *J. Roy. Stat. Soc.*, B, **17**:69–78.
262. (With V. C. Fyfe). Double reduction at the rosy, or pink, locus in *Lythrum salicaria*. *Nature*, **176**:1176.
263. Science and Christianity. *Friend*, **113**:(42), 2 pp.
1956 264. On a test of significance in Pearson's *Biometrika tables* (no. 11). *J. Roy. Stat. Soc.*, B, **18**:56–60.
265. (With M. J. R. Healy). New tables of Behrens' test of significance. *J. Roy. Stat. Soc.*, B, **18**:212–216.
266. Blood groups and population genetics. *Proc. 1st Int. Congr. Human Genet., Acta Genet.*, **6**:507–509.
1957 267. The underworld of probability. *Sankhyā*, **18**:201–210.
268. Comment on the notes by Neyman, Bartlett and Welch in this Journal. (*18*, 288–302) *J. Roy. Stat. Soc.*, B, **19**:179.
269. Dangers of cigarette-smoking. *Brit. Med. J.*, **2**:43.
270. Dangers of cigarette-smoking. *Brit. Med. J.*, **2**:297–298.
271. Methods in human genetics. *Proc. 1st Int. Congr. Human Genet., Acta Genet.*, **7**:7–10.

1958 272. The nature of probability. *Centennial Rev.*, **2**:261–274.
      273. Mathematical probability in the natural sciences. *Proc. 18th Int. Congr. Pharmaceut. Sci.; Metrika*, **2**:1–10; *Technometrics*, **1**:21–29; *La Scuola in Azione*, **20**:5–19.
      274. Cigarettes, cancer, and statistics. *Centennial Rev.*, **2**:151–166.
      275. Lung cancer and cigarettes? *Nature*, **182**:108.
      276. Cancer and smoking. *Nature*, **182**:596.
      277. Polymorphism and natural selection. *Bull. Inst. Int. Stat.*, **36**:284–289. *J. Ecol.*, **46**:289–293.
      278. The discontinuous inheritance. *Listener*, **60**:85–87.
1959 279. Natural selection from the genetical standpoint. *Aust. J. Sci.*, **22**:16–17.
      280. An algebraically exact examination of junction formation and transmission in parent-offspring inbreeding. *Heredity*, **13**:179–186.
1960 281. (With E. A. Cornish). The percentile points of distributions having known cumulants. *Technometrics*, **2**:209–225.
      282. Scientific thought and the refinement of human reasoning. *J. Oper. Res. Soc. Japan*, **3**:1–10.
      283. On some extensions of Bayesian inference proposed by Mr. Lindley. *J. Roy. Stat. Soc.*, B, **22**:299–301.
1961 284. Sampling the reference set. *Sankhyā*, **23**:3–8.
      285. The weighted mean of two normal samples with unknown variance ratio. *Sankhyā*, **23**:103–114.
      286. Possible differentiation in the wild population of *Oenothera organensis*. *Aust. J. Biol. Sci.*, **14**:76–78.
      287. A model for the generation of self-sterility alleles. *J. Theoret. Biol.*, **1**:411–414.
1962 288. The simultaneous distribution of correlation coefficients. *Sankhyā*, **24**:1–8.
      289. Some examples of Bayes' method of the experimental determination of probabilities a priori. *J. Roy. Stat. Soc.*, B, **24**:118–124.
      290. The place of the design of experiments in the logic of scientific inference. *Colloq. Int. Cent. Nat. Recherche Scientifique, Paris*, **110**:13–19; *La Scuola in Azione*, **9**:33–42 (in Italian).
      291. Confidence limits for a cross-product ratio. *Aust. J. Stat.*, **4**:41.
      292. Enumeration and classification in polysomic inheritance. *J. Theoret. Biol.*, **2**:309–311.
      293. Self-sterility alleles: a reply to Professor D. Lewis. *J. Theoret. Biol.*, **3**:146–147.
      294. The detection of a sex difference in recombination values using double heterozygotes. *J. Theoret. Biol.*, **3**:509–513.

3. PUBLISHED MATERIAL NOT REPRINTED IN THE *COLLECTED PAPERS OF R. A. FISHER*

Letters to Journals

1915 (With C. S. Stock). The eugenic aspect of the employment of married women. *Eugen. Rev.*, **6**:313–315.
1923 The evolution of the conscience in civilised communities—a reply to C. K. Millard. *Eugen. Rev.*, **15**:374.

1924   The Darwinian theory. *Sci. Prog.*, **18**:289 – 290.
       The Darwinian theory. *Sci. Prog.*, **18**:466 – 467.
1927   Statistical methods for research workers. *Sci. Prog.*, **21**:340 – 341.
1928   *Ethics and Economics of Family Endowment.* (E. Rathbone) *Eugen. Rev.*, **20**:72.
1930   The balance of births and deaths. (R. R. Kuczynski) *Eugen. Rev.*, **21**:236.
       Genetics, mathematics and natural selection. *Nature*, **126**:805 – 806.
1931   Signing reviews. *Sci. Prog.*, **25**:689 – 690.
1932   Note on L. C. Dunn's new series of allelomorphs in mice. *Nature*, **129**:130.
1933   Bearing of genetics on theories of evolution—reply to J. T. Cunningham. *Sci. Prog.*, **27**:701 – 702.
1934   Reply to W. L. McAtee on protective adaptations. *Proc. Roy. Ent. Soc. Lond.*, **9**:26 – 27.
1936   Income-tax and birth-rate; family allowances. *Times*, 30 April.
       Curve-fitting. *Nature*, **138**:934.
1938   Local varieties and species—natural selection. *Times*, 3 May.
1939   Citizens of the future; burden of a falling birthrate; the case for family allowances. *Times*, 10 April.
1945   Note on recognition of Rhesus genotypes. *Nature*, **155**:543.

Contributions to Discussions

1921   On the time-correlation problem, with especial reference to the variate-difference correlation method. (G. U. Yule) *J. Roy. Stat. Soc.*, **84**:534 – 536.
1922   An investigation of sickness data of public elementary school teachers in London, 1904 – 1919. (J. Y. Hart) *J. Roy. Stat. Soc.*, **85**:405.
       Birth control. *Eugen. Rev.*, **12**:297.
1925   Family endowment. *Eugen Rev.*, **16**:281 – 283.
1932   Biological principles in the control of destructive animals. (M. A. C. Hinton) *Proc. Linn. Soc. Lond.*, **144**:124 – 125.
       Protective adaptations of animals—especially insects. *Proc. Entom. Soc. Lond.*, **7**:87 – 89.
1934   On the two different aspects of the representative method. (J. Neyman) *J. Roy. Stat. Soc.*, **97**:614 – 619.
       Statistics in agricultural research. (J. Wishart) *J. Roy. Stat. Soc. Suppl.*, **1**:51 – 53.
       Discrimination by specification. (B. H. Wilsdon) *J. Roy Stat. Soc., Suppl.*, **1**:198 – 199.
1936   Cooperation in large scale experiments. (W. S. Gosset) *J. Roy. Stat. Soc., Suppl.*, **3**:122 – 124.
1937   On the value of Royal Commissions in sociological research. (M. Greenwood) *J. Roy. Stat. Soc.*, **100**:403 – 404.
1939   Some aspects of the teaching of statistics. (J. Wishart) *J. Roy. Stat. Soc.*, **102**:554 – 556.
       Statistical treatment of animal experiments. (J. Wishart) *J. Roy. Stat. Soc., Suppl.*, **6**:12 – 13.
       Long term agricultural experiments (W. G. Cochran) *J. Roy. Stat. Soc., Suppl.*, **6**:143 – 144.
1946   A review of recent statistical developments in sampling and sample surveys (F. Yates) *J. Roy. Stat. Soc.*, **109**:31.

498 The Life of a Scientist

1948 The validity of comparative experiments. (F. J. Anscombe) *J. Roy. Stat. Soc.*, **A. 111:**202–203.

1949 Some statistical problems arising in genetics. (J. B. S. Haldane) *J. Roy. Stat. Soc.*, **B. 11:**9–10.

1955 The Macmillan gap and the shortage of risk capital. (Lord Piercy) *J. Roy. Stat. Soc.*, **A. 118:**7–8.

Miscellaneous

1916 Racial repair. *Eugen. Rev.*, **7:**204–207.
Ethnology and the war. *Eugen Rev.*, **7:**207–209.
Note on bibliography of eugenic literature. *Eugen. Rev.*, **8:**158.
After the war problems. *Eugen. Rev.*, **8:**357.

1917 Disabled soldiers and marriage. *Eugen. Rev.*, **9:**55.

1921 Opening of new building for the Department of Applied Statistics at University College, London. *Eugen. Rev.*, **12:**237.

1927 The effect of family allowances on population; some French data on the influence of family allowances on fertility. *Family Endowment Society Pamp.*, 7–11.

1928 Income tax. *Eugen. Rev.*, **19:**231–233.
Income-tax rebates; the birth-rate and our future policy. *Eugen. Rev.*, **20:**79–81.

1933 Foreward to H. J. Buchanan-Wollaston's paper, some modern statistical methods, their application to the solution of herring race problems. *J. Int. Counc. Explor. Sea*, **8:**7–8.
The general formula of heredity. (H. H. Laughlin) *Nature*, **132:**1012.

1934 The "Viceroy" (*Basilarchia archippus* Cram.) mistaken for its model the "Monarch" (*Danaus plexippus* Linn.) *Proc. Roy. Entom. Soc. Lond.*, **9:**97.

1935 Foreword. *Ann. Eugen.*, **6:**1.
Linkage studies and the prognosis of hereditary ailments. *Int. Congr. Life Assur. Med. Lond.*

1936 The significance of regression coefficients; Tests of significance in harmonic analysis; Inverse probability. (Abstracts.) Cowles commission research conference on economics and statistics. *Colorado Coll. Publ. Gen. Ser.*, **208:**63–67.
Statistical Inference and the Testing of Hypotheses. (mimeographed notes of lectures and discussions) *U.S. Dept. Agric. Graduate School*, 51 pp.

1937 The character of W. F. Sheppard's work. *Ann. Eugen.*, **8:**11–12.

1938 *Statistical Theory of Estimation.* (Readership lectures.) Calcutta University Press, 45 pp.

1943 Foreword to *Statistical Analysis in Biology.* (K. Mather) London: Methuen.

1945 G. L. Taylor. (Obituary.) *Brit. Med. J.*, **1:**463–464.
The Indian Statistical Institute. *Nature*, **156:**722.

1948 T. H. Morgan. (Obituary.) *Roy. Soc. Obit. Not. Fellows*, **5:**451–454.

1950 The sub-commission on statistical sampling of the United Nations. *Bull. Inst. Int. Stat.*, **32:**207–209.
Foreword to *Blood Groups in Man.* (R. R. Race and R. Sanger) Oxford: Blackwell.

1957 Space travel and ageing. *Discovery*, **18:**56–57, 174.

1965 Commentary and assessment of Gregor Mendel's paper, in *Experiments in Plant Hybridisation*, J. H. Bennett, Ed. Edinburgh: Oliver and Boyd.

Reviews Published in *Eugenics Review*

VOLUME 6 (1914–1915) *Mechanism, Life, and Personality.* (J. S. Handane) 165; *The Family in its Sociological Aspects.* (J. Q. Dealey) 165–166.
VOLUME 7 (1915–1916) *The Shadow on the Universe.* (I. M. Clayton) 210; *Kinship and Social Organisation.* (W. H. R. Rivers) 210–211; *The Progress of Eugenics.* (C. W. Saleeby) 213–214.
VOLUME 8 (1916–1917) On the birth-rates in various parts of England and Wales. (T. A. Welton) 82: Is selection or mutation the more important agency in evolution? (W. E. Castle, 84–85; L'évolution des êtres organisés, par sauts brusques. (H. de Vries) 182; A few notes on the neolithic Egyptians and the Ethiopians. (V. Giuffrida-Ruggieri) 184; *Losses of Life in Modern Wars.* (G. Bodart); *Military Selection and Race Deterioration.* (V. L. Kellogg) 264–265; Citizens charter. A scheme of national organisation. (C. E. Innes) 274–275; The suppression of characters on crossing. (R. H. Biffen) 280; Inheritance of artistic and musical ability. (H. Drinkwater) 280; Colour and pattern transference in pheasants. (R. H. Thomas) 280; A criticism of the hypothesis of linkage and crossing over. (A. H. Trow) 281: The anthropology of the Jew. (L. D. Covitt) 281; The Society of Friends. 284–285; War, immigration, eugenics, 287; Eugenics and agriculture. (O. F. Cook) 287; Consanguineous marriage. 287–288; Inheritance of baldness. (D. Osborn) 288; Evolution and Man. (M. M. Metcalf) 288; The environment of the ape man. (E. W. Berry) 288–289; The racial value of death. (P. Popenoe) 291–292; *A Manual of Mendelism.* (J. Wilson) 363–364; Marriage and the population question. (B. Russell) 385; Non-Mendelian heredity. Croisements et mutations. (H. de Vries) 385–386.
VOLUME 9 (1917–1918) *Genetics: an Introduction to the Study of Heredity.* (H. E. Walter) 61–62; Constructive aspect of birth-control. (R. J. Sprague) 80; The present status of instruction in genetics. (E. E. Barker) 80–81; The relationship systems of the Nandi, Masai and Thonga. (B. Z. Seligman) 156–157; L'origine d'éspeces nouvelles selon la théorie du croisement. (R. Pinotta) 168–169; Variability under inbreeding and cross-breeding (W. E. Castle) 169; Hybridism and the rate of evolution in Angiosperms. (E. C. Jeffrey) 169–170; On the effect of continued administration of certain poisons to the domestic fowl, with special reference to the progeny. (R. Pearl) 256–257; A scale for grading neighbourhood conditions. (J. H. Williams) 260; The borderline of mental deficiency. (S. C. Kohs) 260; Large families. (P. Popenoe) 263–264; Piebald rats and multiple factors. (E. C. MacDowell) 264; Observations on the inheritance of anthocyan pigment in paddy varieties. (G. P. Hector) 367.
VOLUME 10 (1918–1919) On the factors governing the sex of the eggs of the honey-bee. (O. Morganthaler) 58; Hybridisation tests between spelt and wheat in Holland. (H. M. Gmelin) 58–59; The colour of the seed of a natural hybrid of two varieties of *Phaseolus vulgaris.* (J. Lundberg and A. Akerman) 59; The effects of age on the hybridisation of *Pisum sativum* (E. Zederbaum) 59; Selecting dairy bulls by permormance. (W. E. Carroll) 59; The mathematical theory of population, of its character and fluctuations, and of the factors which influence them. (G. H. Knibbs) 111–112; Linked quantitative characters in wheat crosses. (G. F. Freeman) 117; Inheritance of endosperm colour in maize. (O. E. White) 117; Maternal inheritance in the soybean. (H. Teras) 177; On the sterility of hybrids between the pheasant and the gold campine fowl. (D. W. Cutler) 240; On the future of statistics. (H. Westergaar) 242; Recent Mendelian results in Indian rice. 246.
VOLUME 11 (1919–1920) Genetic variability, twin hybrids and constant hybrids, in a case of balanced lethal factors. (H. Muller) 92–94; Immigration restriction and world eugenics. (P. F. Hall) 97–98; Disease and natural selection. (H. C. and M. A. Soloman) 98; Some present aspects of immigration. 98; Better American families, II. (W. E. Key) 98; Good qualities are correlated. (F. A. Woods) 98; Race mixture in Hawaii. (V. McLangley) 98; The racial limitation of Bolshevism. (F. A. Woods) 160.

VOLUME 12 (1920 – 1921) The check to the fall of the phthisis death-rate since the discovery of the tubercle *Bacillus*. (K. Pearson) 60; Eugenics in relation to economics and statistics. (L. Darwin) 60 – 61; A preliminary study of the effects of administering ethyl alcohol to the Lepidopterous insect, *Selenia bilunaria*, with particular reference to the offspring. (J. W. Harrison) 61 – 62; Inheritance of wing colour in Lepidoptera. II. Melanism in *Tephrosia consonaria*. (H. Onslow) 62; Racial studies in fishes. III. Diallel crossings with trout. (J. Schmidt) 62; The criterion of goodness of fit of psychophysical curves. (G. H. Thomson) 62 – 63; On the degree of perfection of hierarchical order among correlation coefficients. (G. H. Thomson) 63; Constitution and byelaws of the Eugenics Educational and Social Club of St. Louis, Mo. 129 – 130; Registrar General's (U. K.) Report for 1918. 232 – 233; Report upon the physical examination of men of military age by National Service Medical Boards. 233 – 234; Notes on the inheritance of colours and markings in pedigree Hereford cattle. (F. Pitt) 246; The relation of life-tables to the Makeham Law. (H. L. Trachtenberg) 313; Census of the United States, 1920, 313; Multiple allelomorphs and limiting factors in inheritance in the stock (E. R. Sounders) 313; The inheritance of specific iso-agglutinins in human blood (J. H. Learmonth) 313 – 314; Density and death rate: Farr's Law. (J. Brownlee) 314; Notes on a case of linkage in *Paratettix*. (J. B. S. Haldane) 314.

VOLUME 13 (1921 – 1922) Genetics and cytological examination of the phenomena of primary non-disjunction in *Drosophila melanogaster*. (S. R. Safir) 429; Studies in inheritance in the Japanese Convolvulus. (B. Miyazawa) 430; The evolution of climate in north-west Europe. (C. E. P. Brooks) 489.

VOLUME 14 (1922 – 1923) *A Treatise on Probability*. (J. M. Keynes) 46 – 50; Population emigration. (H. Bunle and F. Leurence) 63 – 64; Conception control. (F. E. Barrett) 281 – 282; Genetic analysis, schemes of cooperation and multiple allelomorphs of *Linum usitatissimum*. (T. Tammes) 297; Age distribution of Germans killed in 5 years of war; in Wirtschaft und Statistik, July 1922. 298 – 299.

VOLUME 15 (1923 – 1924) *The Riddle of Unemployment*. (C. Palmer) 364; Alcoholism and the growth of white rats. (E. C. MacDowell) 368; The correlation between the egg production of the various periods of the year in the Rhode Island Red breed of domestic fowl. (J. A. Harris and H. D. Goodale) 368; Studies on inheritance in pigeons. IV. Checks and bars and other modifications of black. (S. v. H. Jones) 368; Selection through the choice of seeds from dominant plants of the allogamous population. (R. B. Robbins) 369; Inheritance of attached ear-lobes. (K. Hilden) 369; The genotypical response of the plant species to the habitat. (G. Turesson) 369 – 370; Postaxial polydactylism in six generations of a Norwegian family. (A. Sverdrup) 370; Diallel crossings with the domestic fowl. (J. Schmidt) 370; the coloration of the testa of the poppy seed. (H. M. Leake and B. R. Pershad) 371; Inheritance in *Ricinus communis* L. (S. C. Harland) 371; Changes in the birth rate and in legitimate fertility in London, 1911 – 1921. (T. T. S. de Jastrzebski) 371; The hereditary factor in the etiology of tuberculosis. (A. Govaerts) 441; *The Biology of Death*. (R. Pearl) 442 – 443; *Biometrika*, **14:** parts 3 and 4. 445 – 446; Recent contributions to the theory of "two factors". (C. Spearman) 446; Infant mortality, results of a field study in Gary, Ind. (E. Hughes) 534; *Primitive Society*. (R. H. Lowie) 535 – 536; *Woman: a Vindication*. (A. M. Ludovici) 610 – 611; *The Physiology of Twinning*. (H. H. Newman) 661 – 612; The social and geographical distribution of intelligence in Northumberland. (J. F. Duff and G. H. Thomson) 632 – 633.

VOLUME 16 (1924 – 1925) *Anomalies and Diseases of the Eye*. (J. Bell) 55; *Studies in Evolution and Genetics*. (S. J. Holmes) 56 – 57; On sex chromosomes, sex determination, and preponderance of females in some dioecious plants. (O. Winge) 81 – 82; The progeny of a cross between a yellow wrinkled and a green rounded seeded pea. 83; La predominance des naissances masculines. (H. W. Methorst) 83 – 84; *The Disinherited Family*. (E. F. Rathbone) 150 – 153; Inheritance of white wing colour, a sex-linked (sex-controlled) variation in yellow pierid butterflies. (J. H. Gerould) 168; The association of size differences with seed coat pattern and pigmentation in *Phaseolus vulgaris.* (K. Sax) 168; Inheritance of white seedlings in maize. (M. Demerec) 168; The inheritance of sickle cell anaemia in man. (W. H. Talliaferro and J. G. Huck) 168; Variations of linkage in rats and mice. (W. E. Castle and W. L. Wachter) 168 – 169; Inheritance of egg size in *Drosophila melanogaster.* (D. C. Warren) 169; Genetics of *Primula sinensis.* (R. P. Gregory, D. de Winton and W. Bateson) 169; *Les Jumeaux.* (E. Apert) 217; *Genetics and Eugenics.* (W. E. Castle) 285; *A Record of Measurements, Weights and Other Facts, Relating to Man.* (J. F. Tocher) 307.

VOLUME 17 (1925 – 1926) *Race and National Solidarity.* (C. C. Josey) 126 – 127; Zur Kentniss der menschlichen Kopfform in genitischer Hinsicht. (K. Hilden) 132; The plant species in relation to habitat and climate. (G. Turesson) 133; *La Methode Statistique.* (A. Niceforo) 202 – 203; Nationalité et acroissement de la population en Finlande. (W. Backman) 216 – 217; *Annals of Eugenics,* **1:** parts I and II, 326.

VOLUME 18 (1926 – 1927) *The Third Chromosome Group of Mutant Characters of Drosophila melanogaster.* (C. B. Bridges and T. H. Morgan) 62; Linkage relations of yellow pigment in maize. (E. W. Lindstrom); Linkage in the Japanese Morning-glory. (Y. Imai) 72; *Genetics,* **10:** no. 4, 72 – 73; Note on Vermilion duplication. (G. Bonnier) 73 – 74; *Journal of Genetics,* **16:** no. 2, 74 – 75; Hyperglycaemia as a Mendelian recessive character in mice. (P. J. Cammage and H. A. H. Howard) 175; Genetic studies in *Brassica oleracea.* (M. S. Pease) 175; On the avuncular relationship. (K. Pearson) 264 – 265; *Hereditas,* **8:** no. 1 – 2, 267.

VOLUME 19 (1927 – 1928) Family allowances and the skilled worker. (O. Vlasto) 86; *Genetics,* July 1926, 87; *Genetics,* September 1926, 87; The health of London in the eighteenth century. (J. Brownlee) 93; *Journal of Genetics,* **18:** no. 1, 167 – 168; *Journal of Genetics,* **18:** no. 2, 168; Inheritance of four o'clocks (*Mirabilis galapa*). (F. P. Kiernan and O. E. White) 169; The effect of chromosome aberrations on development in *Drosophila melanogaster.* (Ju-Ghi Li) 248 – 249; The inheritance of heterostyly in *Lythrum salicaria* (E. M. East); A genetic study of certain chlorophyll deficiencies in maize. (W. A. Carver); Mendelian inheritance of chromosome shape in *Matthiola.* (M. M. Leslie and H. B. Frost) 336 – 337.

VOLUME 20 (1928 – 1929) *Sex in Man and Animals.* (J. R. Baker); *Organic Inheritance in Man.* (F. A. E. Crew); *Animal Biology.* (J. B. S. Haldane and J. Huxley) 38 – 39; *Twins and Orphans: the Inheritance of Intelligence.* (A. H. Wingfield) 57; *Civilisation or Civilisations: an Essay in the Spenglerian Philosophy of History.* (E. H. Goddard and P. A. Gibbons) 120; *The Species Problem.* (G. C. Robson) 129; Linkage studies in rice. (Lien Fang Chao) 137; Hereditary growth anomaly of the thumb. (O. Thomsen) 138; Fresh evidence on the inheritance factor in tuberculosis. (P. Stocks) 210 – 211; On the relative values of the factors which influence infant welfare. (E. M. Elderton) 211; Genetic studies in *Ricinus communis* L. (J. E. Peat) 215 – 216; The inheritance of dwarfing in *Gammarus chevreuxi.* (E. B. Ford) 216; Polymorphism in the moth *Acalla comariana* Zeller. (J. C. F. Fryer) 296; Sexual difference of linkage in *Gammarus chevreuxi.* (J. S. Huxley) 296; Amputated, a recessive lethal in cattle with a discussion of the bearing of lethal factors on the principle of livestock breeding. (C. Wriedt and O. L. Mohr) 296; Criteria for distinguishing identical and fraternal twins. (Taku Komai); The gene. (R. Goldschmidt) 299 – 300.

VOLUME 21 (1929 – 1930) *The Decline of the West,* Vol. 2 (O. Spengler) 64; A consideration of variegation. (Yoshitaku Iwai) 73; A sex difference in chromosome length in the Mammalia. (H. M. Evans and O. Swezy) 73; An eight-factor cross in the guinea-pig. (S. Wright) 73; The effect on crossing-over and non-disjunction of X-raying the anterior and posterior halves of *Drosophila* pupae. (J. W. Mavor) 157; The genetics of *Platypoecilus.* (A. C. Fraser and M. Gordon) 157; A cytological and genetical study of triploid maize. (B. McClintock) 157; Linkage groups of the Japanese Morning Glory. (Y. Imai) 157; Two factors influencing feathering in chickens. (C. H. Danforth) 157 – 158; On the result of simultaneous gametic and environmental correlations in a segregating population. (O. Tedin) 158; On the nature of the sex chromosomes in *Humulus.* (O.

Winge) 158; A genetical investigation in *Scolopendrium vulgare*. I, Andersson-Kottë) 158; The recessive mutant "engrailed" in *Drosophila melanogaster*. (R. Eker) 158; *Race and Population Problems*. (H. G. Duncan) 222; Inheritance of resistance to the Danysz bacillus in the rat. (M. R. Irwin) 231–232; The minute reaction in the development of *Drosophila melanogaster*. (J. Schultz) 232; *Man and Civilisation*. (J. Storck); *Evolution and the Spirit of Man*. (J. P. Milum); *The Renewal of Culture*. (L. Ringbom) 313–314.

VOLUME 22 (1930–1931) On a new theory of progressive evolution. (K. Pearson) 207–208 (also in *Nature*, **126**:246–247); *Danger Spots in World Population*. (W. S. Thompson) 67–68; Further studies on the chromosome mechanism responsible for unisexual progenies in Sciara. (C. W. Metz and M. L. Schmuck) 81; Memorandum on differential fertility, focundity and sterility. (F. A. E. Crew and E. Moore) 150–151; Experiments on the genetics of wild populations. (J. W. Gregor) 152; Further data on a case of autosomal linkage in the domestic fowl. (L. C. Dunn and W. Landauer) 152–153; The inheritance of frizzled plumage. (F. B. Hutt) 153; *Journal of Genetics*, **22**: no. 2., 153; Note on the blood-relationship of double cousins. (K. Pearson) 222–223; Articles in **Genetics, 15**: no. 4., 223; Sex linkage in man. (C. B. Davenport) 223–224; The correlation between the sex of human siblings. (J. A. Harris and B. Gunstadt) 224; the inheritance of dormancy and premature germination in maize. (P. C. Mangelsdorf) 224; Parallelism of chromosome ring formation, sterility and linkage in *Pisum*. (C. Hammarlund and A. Hakansson) 224; The true rate of natural increase of population of the United States. Revision on basis of recent data. (L. I. Dublin and A. J. Lotka) 224–225; Are progressive mutations produced by X-rays? (J. T. Patterson and H. J. Muller) 320–321.

VOLUME 23 (1931–1932) Evolution in Mendelian populations. (S. Wright) 88–89; Studies on the creeper fowl. I. Genetics. (W. Landauer and L. C. Dunn) 89; The finger prints of twins. (H. H. Newman) 89; Genetics and cytology of the tetraploid form of *Primula sinensis*. (A. S. Sömme) 89–90; Linkage in the tetraploid *Primula sinensis*. (D. de Winton and J. B. S. Haldane) 90; Studies on sex determination and the sex chromosome mechanism in Sciara. (C. W. Metz and M. L. Schmuck) 181–182; Step allelomorphism in *Drosophila melanogaster*. (I. J. Agol) 182; The side-chain theory of the structure of the gene. (D. H. Thompson) 182; Behavior of two mutable genes of *Delphinium ajacis*. (M. Demerec); Further notes on the Tortricid moth *Acalla comariana* Zeller. (J. C. F. Fryer) 183; *Biology and Mankind*. (S. A. McDowall) 362.

VOLUME 24 (1923–1933) *The Declining Birth Rate in Rotterdam*. (J. Sanders) 58–59; The machanism of organic evolution. (C. B. Davenport) 62; *The Balance of Births and Deaths*, Vol. 2. (R. R. Kuczynski) 309–310; A theoretical and experimental study on the changes in the crossing-over values. (G. Eloff) 335–336; Further studies on the genetics of abnormalities appearing in the descendants of X-rayed mice. (C. C. Little and B. W. MacPheters) 336.

VOLUME 25 (1933–1934) Genetical evidence for a cytological abnormality in man. (J. B. S. Haldane) 61; A case of conditioned dominance in *Drosophila obscura*. (F. A. E. Crew) 61.

VOLUME 26 (1934–1935) *Biometrika*, **26**: part 1–2. 237–238; The fertility of families on relief. (S. A. Stouffer) 306–307.

VOLUME 43 (1952) *Hereditary Genius* (F. Galton) (2nd ed. repr.) 37.

Other Reviews

1920   *Mathematics for Collegiate Students of Agriculture and General Science*. (A. M. Kenyon and W. V. Lovitt) *Nature*, **105**:131.

1927   *The Theory of the Gene*. (T. H. Morgan) *Sci. Prog.*, **21**:159–160.
        The Principles and Pratice of Yield Trials. (F. L. Engledow and G. U. Yule) *Nature*, **120**:145–147.

1929   *The Balance of Births and Deaths*, v. 1: *Western and Northern Europe*. R. R. Kuczynski; *The Shadow of the World's Future*. (G. H. Knibbs) *Nature*, **123**:357–358.
Frequency Curves and Correlation. (W. P. Elderton) *Sci. Prog.*, **23**:349–350.

1930   *Mathematics Preparatory to Statistics and Finance.* (G. N. Bauer) *Nature*, **125**:379.

1932   *Genetic Principles in Medicine and Social Science.* (L. Hogben) *Health and Empire*, **7**:147–150.

1933   *Tables for Statisticians and Biometricians.* Parts 1–2. K. Pearson, Ed. *Nature*, **131**:893–894.

1935   *Dynamics of Population.* (F. Lorimer and F. Osborn) *Nature*, **135**:46–48.
Biomathematics. (W. M. Feldman) *Ann. Eugen.*, **6**:252.

1938   *Twins—a Study of Heredity and Environment.* (H. H. Newman, F. N. Freeman and K. J. Holzinger) *Ann. Eugen.*, **8**:216–218.
The design and analysis of factorial experiments. (F. Yates) *Nature*, **142**:90–92.

1940   *Statistical Methods for Medical and Biological Students.* (G. Dahlberg); *Tables of Random Sampling Numbers.* (M. G. Kendall and B. B. Smith); *Tests of Significance.* (J. H. Smith); A Bibliography of Statistical and Other Writings of Karl Pearson. (G. M. Morant) *Nature*, **145**:1010.
British Association Mathematical Tables, Vol. 7. *Ann. Eugen.*, **10**:120–121.
Statistical Method from the Viewpoint of Quality Control. (W. A. Shewhart) *Nature*, **146**:150.

1942   *The Fundamental Principles of Mathematical Statistics.* (H. H. Wolfenden); *Sampling Methods in Forestry and Range Management.* (F. X. Schumacher and R. A. Chapman) *Nature*, **150**:196.
Aims and objects of eugenic researchers in Bengal. (S. S. Sarkar) *Ann. Eugen.*, **11**:405.

1943   *The Advanced Theory of Statistics*, 1. (M. G. Kendall) *Nature*, **152**:431.
The Advanced Theory of Statistics, 1. (M. G. Kendall) *Camb. Rev.*, 23rd October.

1948   *Advances in Genetics*, 1. M. Demerec, Ed. *Sci. Prog.*, **36**:176–177.

1950   The report of the Royal Commission on population. *Camb. J.*, **3**:32–39.

1951   *Inheritance in Dogs, with Special Reference to Hunting Breeds.* (O. Winge) *Heredity*, **5**:149–150.

1953   *Quantitative Inheritance.* E. C. R. Reeve and C. H. Waddington Eds., *Heredity* **7**:293.

1954   *The Distribution of the Human Blood Groups.* (A. E. Mourant) *Brit. Med. J.*, **2**:1034.

1956   *Experimental Design and its Statistical Basis.* (D. J. Finney) *Heredity*, **10**:275.

1957   *Foundations of Inductive Logic.* (R. F. Harrod) *Observer*, 27th January.
Scientific Inference, 2nd ed. (H. Jeffreys) *Camb. Rev.* 18th May.

1958   *The Autobiography of Charles Darwin.* N. Barlow, Ed. *Nature*, **182**:71.

Index